METHODS IN MOLECULAR BIOLOGY

Series Editor
John M. Walker
School of Life and Medical Sciences
University of Hertfordshire
Hatfield, Hertfordshire, UK

For further volumes:
http://www.springer.com/series/7651

For over 35 years, biological scientists have come to rely on the research protocols and methodologies in the critically acclaimed *Methods in Molecular Biology* series. The series was the first to introduce the step-by-step protocols approach that has become the standard in all biomedical protocol publishing. Each protocol is provided in readily-reproducible step-by-step fashion, opening with an introductory overview, a list of the materials and reagents needed to complete the experiment, and followed by a detailed procedure that is supported with a helpful notes section offering tips and tricks of the trade as well as troubleshooting advice. These hallmark features were introduced by series editor Dr. John Walker and constitute the key ingredient in each and every volume of the *Methods in Molecular Biology* series. Tested and trusted, comprehensive and reliable, all protocols from the series are indexed in PubMed.

Yeast Functional Genomics

Methods and Protocols

Second Edition

Edited by

Frédéric Devaux

Institut de Biologie Paris Seine, Sorbonne Université, Paris, France

Humana Press

Editor
Frédéric Devaux
Institut de Biologie Paris Seine
Sorbonne Université
Paris, France

ISSN 1064-3745 ISSN 1940-6029 (electronic)
ISBN 978-1-0716-2259-9 ISBN 978-1-0716-2257-5 (eBook)
https://doi.org/10.1007/978-1-0716-2257-5

This Humana imprint is published by the registered company Springer Science+Business Media, LLC part of Springer Nature.
The registered company address is: 1 New York Plaza, New York, NY 10004, U.S.A.

Preface

Five years after the first edition of the *Yeast Functional Genomics: Methods and Protocols* issue of the *Methods in Molecular Biology* book series came out, it seems that the topic is more relevant and livelier than ever. Not only are the basics of yeast functional genomics still extensively used by hundreds of laboratories, but new techniques appeared that scaled up the power of genome-wide approaches of yeast functioning.

Third-generation sequencing devices gave a new start to yeast transcriptomics by giving access to full-length cDNAs, and even allow direct sequencing of RNAs and their chemical modifications, paving the way to yeast epitranscriptomics.

Transposon saturation, RNAi and CRISPR/cas assays empowered the already high potential for global genetic screens in yeast, looking for protein-protein interactions, and essential genes or gene-drug interactions. At the same time, yeast phenomics, i.e., high-throughput yeast phenotypic profiling, which appeared at the end of the first decade of the century, has been automatized and now takes advantage of bar-coded strain collections allowing for accurate lineage tracing.

Alternatives to the classic chromatin immunoprecipitation-sequencing (ChIP-seq) approach, together with single cell transcriptomics and with the development of various techniques to detect RNA polymerase II activity all along the genome, have been changing the way we envision yeast transcriptional networks.

Specific ribosome profiling, tailored and more quantitative proteomics with smarter way of analyzing native protein complexes are pushing our knowledge of translational regulatory networks to the same level as that of their transcriptional counterparts.

The number of genomes available for yeast species jumped from 40 to more than 300, giving access to comprehensive comparative transcriptomics studies.

Finally, the time dedicated to in silico analyses in the daily practice of yeast functional "genomicists" has increased again, and the exponential increase in the amount of available data of various types (transcriptomics, proteomics, interactomics, ...) raises the crucial question of their integration into comprehensive and analytical databases, to prepare for their future exploration using machine learning approaches.

The rationale guiding the elaboration of this second edition was to reflect all these recent evolutions of the field and to make it as complementary to the first edition as possible. Therefore, most of the chapters in the present volume present new techniques and protocols that were not included in the first edition. Compared with the previous version, the part dedicated to in silico analyses increased and most protocols in this volume include a significant bioinformatics part. Some authors even provided their informatics scripts as supplementary material available through the online version of the book.

As with the first edition, the main goal of the second edition was to make it interesting and useful to the community of yeast researchers. I do hope that we reached this goal (again).

Paris, France ***Frédéric Devaux***

Contents

Preface . . . . . . *v*
Contributors . . . . . . *xi*

PART I TRANSCRIPTOMIC ANALYSES

1 Single-Cell RNA Sequencing in Yeast Using the 10× Genomics Chromium Device 3
Lieselotte Vermeersch, Abbas Jariani, Jana Helsen, Benjamin M. Heineike, and Kevin J. Verstrepen

2 Genome-Wide Profiling of Transcription Initiation with STRIPE-seq 21
Robert A. Policastro and Gabriel E. Zentner

3 A Modified Cross-Linking Analysis of cDNAs (CRAC) Protocol for Detecting RNA–Protein Interactions and Transcription at Single-Nucleotide Resolution 35
Drice Challal, Jessie Colin, Tommaso Villa, and Domenico Libri

4 High-Resolution Deep Sequencing of Nascent Transcription in Yeast with BioGRO-seq 57
Antonio Jordán-Pla and José E. Pérez-Ortín

5 Direct Sequencing of RNA and RNA Modification Identification Using Nanopore 71
Thidathip Wongsurawat, Piroon Jenjaroenpun, and Intawat Nookaew

6 UNAGI: Yeast Transcriptome Reconstruction and Gene Discovery Using Nanopore Sequencing 79
Mohamad Al kadi, Nicolas Jung, and Daisuke Okuzaki

7 Identification of Taxonomically Restricted Transcripts from Illumina RNA Sequencing Data 91
William R. Blevins

PART II DNA REPLICATION AND PROTEIN/DNA INTERACTIONS

8 FORK-seq: Single-Molecule Profiling of DNA Replication 107
Magali Hennion, Bertrand Theulot, Jean-Michel Arbona, Benjamin Audit, and Olivier Hyrien

9 CUT&RUN Profiling of the Budding Yeast Epigenome 129
Sandipan Brahma and Steven Henikoff

10 ChIP-SICAP: A New Tool to Explore Gene-Regulatory Networks in *Candida albicans* and Other Yeasts 149
Lasse van Wijlick, Ansh Goyal, Sophie Bachellier-Bassi, and Christophe d'Enfert

Part III Translation Dynamics, Protein Complexes, and Proteomics

11 Purification of Ribosome-Nascent-Chain Complex for Ribosome Profiling and Selective Ribosome Profiling 179
Hagit Bar-Yosef, Johannes Venezian, Kevin Klann, and Ayala Shiber

12 Single-Step Affinity Purification (ssAP) and Mass Spectrometry of Macromolecular Complexes in the Yeast *S. cerevisiae* 195
Christian Trahan and Marlene Oeffinger

13 Composition and Dynamics of Protein Complexes Measured by Quantitative Mass Spectrometry of Affinity-Purified Samples 225
Abdelkader Namane and Cosmin Saveanu

14 Deep Mutational Scanning of Protein–Protein Interactions Between Partners Expressed from Their Endogenous Loci In Vivo 237
Alexandre K. Dubé, Rohan Dandage, Soham Dibyachintan, Ugo Dionne, Philippe C. Després, and Christian R. Landry

15 Proteomic Mapping by APEX2-Catalyzed Proximity Labeling in *Saccharomyces cerevisiae* Semipermeabilized Cells 261
Birgit Singer-Krüger and Ralf-Peter Jansen

16 Quantitative Proteomics in Yeast: From bSLIM and Proteome Discoverer Outputs to Graphical Assessment of the Significance of Protein Quantification Scores 275
Nicolas Sénécaut, Pierre Poulain, Laurent Lignières, Samuel Terrier, Véronique Legros, Guillaume Chevreux, Gaëlle Lelandais, and Jean-Michel Camadro

17 A Strong Cation Exchange Chromatography Protocol for Examining N-Terminal Proteoforms 293
Esperanza Fernández, Annelies Bogaert, Evy Timmerman, An Staes, Francis Impens, and Kris Gevaert

Part IV Genotypic Screens and Phenotypic Profiling

18 RNA Interference (RNAi) as a Tool for High-Resolution Phenotypic Screening of the Pathogenic Yeast *Candida glabrata* 313
Andreas Tsouris, Joseph Schacherer, and Olena P. Ishchuk

19 High-Throughput Gene Mutagenesis Screening Using Base Editing 331
Philippe C. Després, Alexandre K. Dubé, Nozomu Yachie, and Christian R. Landry

20 SAturated Transposon Analysis in Yeast (SATAY) for Deep Functional Mapping of Yeast Genomes 349
Agnès H. Michel and Benoît Kornmann

21 High-Throughput, High-Precision Colony Phenotyping with *Pyphe* 381
Stephan Kamrad, Jürg Bähler, and Markus Ralser

22 Bulk-Fitness Measurements Using Barcode Sequencing Analysis in Yeast 399
Claire A. Chochinov and Alex N. Nguyen Ba

PART V IN SILICO INTEGRATION OF FUNCTIONAL GENOMICS DATA

23 Prediction of Gene and Genomic Regulation in *Candida* Species, Using the PathoYeastract Database: A Comparative Genomics Approach 419
Pedro Pais, Jorge Oliveira, Romeu Viana, Inês V. Costa, Isabel Sá-Correia, Pedro T. Monteiro, and Miguel C. Teixeira

24 Enabling Studies of Genome-Scale Regulatory Network Evolution in Large Phylogenies with MRTLE 439
Shilu Zhang, Sara Knaack, and Sushmita Roy

25 Omics Analyses: How to Navigate Through a Constant Data Deluge 457
Thomas Denecker and Gaëlle Lelandais

Index *473*

Contributors

MOHAMAD AL KADI • *Department of Bacterial Infections, Research Institute for Microbial Diseases, Osaka University, Osaka, Japan*
JEAN-MICHEL ARBONA • *Université de Lyon, CNRS, INSERM, Laboratory of Biology and Modelling of the Cell, Lyon, France*
BENJAMIN AUDIT • *Univ Lyon, ENS de Lyon, CNRS, Laboratoire de Physique, Lyon, France*
SOPHIE BACHELLIER-BASSI • *Unité Biologie et Pathogénicité Fongiques, Département de Mycologie, Institut Pasteur, USC 2019 INRAE, Paris, France*
JÜRG BÄHLER • *Department of Genetics, Evolution and Environment, Institute of Healthy Ageing, University College London, London, UK*
HAGIT BAR-YOSEF • *Faculty of Biology, Technion—Israel Institute of Technology, Haifa, Israel*
WILLIAM R. BLEVINS • *Single Cell Genomics Group, Centro Nacional de Análisis Genómico (CNAG), Barcelona, Spain*
ANNELIES BOGAERT • *VIB Center for Medical Biotechnology, Ghent, Belgium; Department of Biomolecular Medicine, Ghent University, Ghent, Belgium*
SANDIPAN BRAHMA • *Basic Sciences Division, Fred Hutchinson Cancer Research Center, Seattle, WA, USA; Howard Hughes Medical Institute, Seattle, WA, USA*
JEAN-MICHEL CAMADRO • *Mitochondria, Metals, and Oxidative Stress Group, Institut Jacques Monod, Université de Paris—CNRS, Paris, France; ProteoSeine@IJM, Institut Jacques Monod, Université de Paris—CNRS, Paris, France*
DRICE CHALLAL • *Université de Paris, CNRS, Institut Jacques Monod, Paris, France; Institut Curie, PSL Research University, CNRS UMR 3348, Orsay, France*
GUILLAUME CHEVREUX • *ProteoSeine@IJM, Institut Jacques Monod, Université de Paris—CNRS, Paris, France*
CLAIRE A. CHOCHINOV • *Department of Cell and Systems Biology, University of Toronto, Toronto, ON, Canada; Department of Biology, University of Toronto at Mississauga, Mississauga, ON, Canada*
JESSIE COLIN • *Université de Paris, CNRS, Institut Jacques Monod, Paris, France; Ecole Pratique des Hautes Etudes – PSL Research University, Paris, France; Unité Biologie des ARN des Pathogènes Fongiques, Institut Pasteur, Paris, France*
INÊS V. COSTA • *iBB—Institute for Bioengineering and Biosciences, Biological Sciences Research Group, Instituto Superior Técnico, Universidade de Lisboa, Lisbon, Portugal; Department of Bioengineering, Instituto Superior Técnico, Universidade de Lisboa, Lisbon, Portugal*
CHRISTOPHE D'ENFERT • *Unité Biologie et Pathogénicité Fongiques, Département de Mycologie, Institut Pasteur, USC 2019 INRAE, Paris, France*
ROHAN DANDAGE • *Département de Biochimie, Microbiologie et Bio-informatique, Faculté de Sciences et Génie, Université Laval, Québec, QC, Canada; PROTEO, le regroupement québé cois de recherche sur la fonction, l'ingénierie et les applications des protéines, Université Laval, Québec, QC, Canada; Centre de Recherche en Données Massives (CRDM), Université Laval, Québec, QC, Canada; Institut de Biologie Intégrative et des Systèmes, Université Laval, Québec, QC, Canada; Département de Biologie, Faculté de Sciences et Gé nie, Université Laval, Québec, QC, Canada*

Thomas Denecker • *CNRS, Institut Français de Bioinformatique, IFB-core, UMS 3601, Évry, France*

Philippe C. Després • *Département de Biochimie, Microbiologie et Bio-informatique, Faculté de Sciences et Génie, Université Laval, Québec, QC, Canada; PROTEO, le regroupement québécois de recherche sur la fonction, l'ingénierie et les applications des protéines, Université Laval, Québec, QC, Canada; Centre de Recherche en Données Massives (CRDM), Université Laval, Québec, QC, Canada; Institut de Biologie Intégrative et des Systèmes, Université Laval, Québec, QC, Canada*

Soham Dibyachintan • *Département de Biochimie, Microbiologie et Bio-informatique, Faculté de Sciences et Génie, Université Laval, Québec, QC, Canada; PROTEO, le regroupement québécois de recherche sur la fonction, l'ingénierie et les applications des protéines, Université Laval, Québec, QC, Canada; Centre de Recherche en Données Massives (CRDM), Université Laval, Québec, QC, Canada; Département de Biologie, Faculté de Sciences et Génie, Université Laval, Québec, QC, Canada; Department of Chemical Engineering, Indian Institute of Technology Bombay (IIT), Powai, Mumbai, Maharashtra, India*

Ugo Dionne • *PROTEO, le regroupement québécois de recherche sur la fonction, l'ingénierie et les applications des protéines, Université Laval, Québec, QC, Canada; Centre de Recherche en Données Massives (CRDM), Université Laval, Québec, QC, Canada; Institut de Biologie Intégrative et des Systèmes, Université Laval, Québec, QC, Canada; Centre de recherche du Centre Hospitalier Universitaire (CHU) de Québec, Université Laval, Québec, QC, Canada; Centre de recherche sur le cancer de l'Université Laval, Québec, QC, Canada*

Alexandre K. Dubé • *Département de Biochimie, Microbiologie et Bio-informatique, Faculté de Sciences et Génie, Université Laval, Québec, QC, Canada; PROTEO, le regroupement québécois de recherche sur la fonction, l'ingénierie et les applications des protéines, Université Laval, Québec, QC, Canada; Centre de Recherche en Données Massives (CRDM), Université Laval, Québec, QC, Canada; Institut de Biologie Intégrative et des Systèmes, Université Laval, Québec, QC, Canada; Département de Biologie, Faculté de Sciences et Génie, Université Laval, Québec, QC, Canada*

Esperanza Fernández • *VIB Center for Medical Biotechnology, Ghent, Belgium; Department of Biomolecular Medicine, Ghent University, Ghent, Belgium*

Kris Gevaert • *VIB Center for Medical Biotechnology, Ghent, Belgium; Department of Biomolecular Medicine, Ghent University, Ghent, Belgium*

Ansh Goyal • *Unité Biologie et Pathogénicité Fongiques, Département de Mycologie, Institut Pasteur, USC 2019 INRAE, Paris, France; Northwestern University Feinberg School of Medicine, Chicago, IL, USA*

Benjamin M. Heineike • *Molecular Biology of Metabolism Laboratory, Francis Crick Institute, London, UK; Quantitative Gene Expression Research Group, MRC London Institute of Medical Sciences (LMS), London, UK; Quantitative Gene Expression Research Group, Faculty of Medicine, Imperial College London, Institute of Clinical Sciences (ICS), London, UK*

Jana Helsen • *VIB Laboratory for Systems Biology, VIB-KU Leuven Center for Microbiology, Leuven, Belgium; CMPG Laboratory of Genetics and Genomics, Department M2S, KU Leuven, Leuven, Belgium; CMPG Laboratory of Predictive Genetics and Multicellular Systems, Department M2S, KU Leuven, Leuven, Belgium*

Steven Henikoff • *Basic Sciences Division, Fred Hutchinson Cancer Research Center, Seattle, WA, USA; Howard Hughes Medical Institute, Seattle, WA, USA*

MAGALI HENNION • *Epigenetics and Cell Fate, CNRS, Université de Paris, Paris, France*
OLIVIER HYRIEN • *Institut de Biologie de l'Ecole Normale Supérieure (IBENS), Ecole Normale Supérieure, CNRS, INSERM, Université PSL, Paris, France*
FRANCIS IMPENS • *VIB Center for Medical Biotechnology, Ghent, Belgium; Department of Biomolecular Medicine, Ghent University, Ghent, Belgium; VIB Proteomics Core, Ghent, Belgium*
OLENA P. ISHCHUK • *Department of Biology and Biological Engineering, Systems and Synthetic Biology, Chalmers University of Technology, Gothenburg, Sweden*
RALF-PETER JANSEN • *Interfaculty Institute of Biochemistry, University of Tübingen, Tübingen, Germany*
ABBAS JARIANI • *VIB Laboratory for Systems Biology, VIB-KU Leuven Center for Microbiology, Leuven, Belgium; CMPG Laboratory of Genetics and Genomics, Department M2S, KU Leuven, Leuven, Belgium*
PIROON JENJAROENPUN • *Division of Bioinformatics and Data Management for Research, Department of Research and Development, Faculty of Medicine Siriraj Hospital, Mahidol University, Bangkok, Thailand*
ANTONIO JORDÁN-PLA • *Facultad de Biológicas, Departamento de Bioquímica y Biología Molecular, Institut de Biotecnología i Biomedicina (Biotecmed), Universitat de València, Burjassot, Valencia, Spain*
NICOLAS JUNG • *Department of Infection Metagenomics, Research Institute for Microbial Diseases, Osaka University, Osaka, Japan*
STEPHAN KAMRAD • *Department of Genetics, Evolution and Environment, Institute of Healthy Ageing, University College London, London, UK; Molecular Biology of Metabolism Laboratory, The Francis Crick Institute, London, UK; Department of Biochemistry, Charité Universitätsmedizin Berlin, Berlin, Germany*
KEVIN KLANN • *Faculty of Medicine, Institute of Biochemistry II, Goethe University, Frankfurt am Main, Germany*
SARA KNAACK • *Wisconsin Institute for Discovery, University of Wisconsin-Madison, Madison, WI, USA*
BENOÎT KORNMANN • *Department of Biochemistry, University of Oxford, Oxford, UK*
CHRISTIAN R. LANDRY • *Département de Biochimie, Microbiologie et Bio-informatique, Faculté de Sciences et Génie, Université Laval, Québec, QC, Canada; PROTEO, le regroupement québécois de recherche sur la fonction, l'ingénierie et les applications des proté ines, Université Laval, Québec, QC, Canada; Centre de Recherche en Données Massives (CRDM), Université Laval, Québec, QC, Canada; Institut de Biologie Intégrative et des Systèmes, Université Laval, Québec, QC, Canada; Département de Biologie, Faculté de Sciences et Génie, Université Laval, Québec, QC, Canada*
VÉRONIQUE LEGROS • *ProteoSeine@IJM, Institut Jacques Monod, Université de Paris—CNRS, Paris, France*
GAËLLE LELANDAIS • *Institute for Integrative Biology of the Cell (I2BC), CEA, CNRS, Université Paris-Saclay, Gif-sur-Yvette, France*
DOMENICO LIBRI • *Université de Paris, CNRS, Institut Jacques Monod, Paris, France*
LAURENT LIGNIÈRES • *ProteoSeine@IJM, Institut Jacques Monod, Université de Paris—CNRS, Paris, France*
AGNÈS H. MICHEL • *Department of Biochemistry, University of Oxford, Oxford, UK*
PEDRO T. MONTEIRO • *INESC-ID, Knowledge Discovery and Bioinformatics Group, Lisbon, Portugal; Department of Computer Science and Engineering, Instituto Superior Técnico, Universidade de Lisboa, Lisbon, Portugal*

ABDELKADER NAMANE • *Génétique des Interactions Macromoléculaires (UMR3525-CNRS), Institut Pasteur, Paris, France*

ALEX N. NGUYEN BA • *Department of Cell and Systems Biology, University of Toronto, Toronto, ON, Canada; Department of Biology, University of Toronto at Mississauga, Mississauga, ON, Canada*

INTAWAT NOOKAEW • *Department of Biomedical Informatics, University of Arkansas for Medical Sciences, Little Rock, AR, USA*

MARLENE OEFFINGER • *RNP Biochemistry Laboratory, Center for Genetic and Neurological Diseases, Institut de recherches cliniques de Montréal, Montréal, QC, Canada; Département de biochimie et médicine moléculaire, Faculté de médecine, Université de Montréal, Montré al, QC, Canada; Division of Experimental Medicine, Faculty of Medicine, McGill University, Montréal, QC, Canada*

DAISUKE OKUZAKI • *Integrated Frontier Research for Medical Science Division, Institute for Open and Transdisciplinary Research Initiatives, Osaka University, Osaka, Japan; Single Cell Genomics, Human Immunology, WPI Immunology Frontier Research Center, Osaka University, Osaka, Japan; Genome Information Research Center, Research Institute for Microbial Diseases, Osaka University, Osaka, Japan*

JORGE OLIVEIRA • *INESC-ID, Knowledge Discovery and Bioinformatics Group, Lisbon, Portugal; Department of Computer Science and Engineering, Instituto Superior Técnico, Universidade de Lisboa, Lisbon, Portugal*

PEDRO PAIS • *iBB—Institute for Bioengineering and Biosciences, Biological Sciences Research Group, Instituto Superior Técnico, Universidade de Lisboa, Lisbon, Portugal; Department of Bioengineering, Instituto Superior Técnico, Universidade de Lisboa, Lisbon, Portugal*

JOSÉ E. PÉREZ-ORTÍN • *Facultad de Biológicas, Departamento de Bioquímica y Biología Molecular, Institut de Biotecnología i Biomedicina (Biotecmed), Universitat de València, Burjassot, Valencia, Spain*

ROBERT A. POLICASTRO • *Department of Biology, Indiana University, Bloomington, IN, USA*

PIERRE POULAIN • *Mitochondria, Metals, and Oxidative Stress Group, Institut Jacques Monod, Université de Paris—CNRS, Paris, France*

MARKUS RALSER • *Molecular Biology of Metabolism Laboratory, The Francis Crick Institute, London, UK; Department of Biochemistry, Charité Universitätsmedizin Berlin, Berlin, Germany*

SUSHMITA ROY • *Wisconsin Institute for Discovery, University of Wisconsin-Madison, Madison, WI, USA; Department of Biostatistics and Medical Informatics, University of Wisconsin-Madison, Madison, WI, USA*

ISABEL SÁ-CORREIA • *iBB—Institute for Bioengineering and Biosciences, Biological Sciences Research Group, Instituto Superior Técnico, Universidade de Lisboa, Lisbon, Portugal; Department of Bioengineering, Instituto Superior Técnico, Universidade de Lisboa, Lisbon, Portugal*

COSMIN SAVEANU • *Génétique des Interactions Macromoléculaires (UMR3525-CNRS), Institut Pasteur, Paris, France*

JOSEPH SCHACHERER • *Université de Strasbourg, CNRS, GMGM UMR 7156, Strasbourg, France; Institut Universitaire de France (IUF), Paris, France*

NICOLAS SÉNÉCAUT • *Mitochondria, Metals, and Oxidative Stress Group, Institut Jacques Monod, Université de Paris—CNRS, Paris, France*

AYALA SHIBER • *Faculty of Biology, Technion—Israel Institute of Technology, Haifa, Israel*

BIRGIT SINGER-KRÜGER • *Interfaculty Institute of Biochemistry, University of Tübingen, Tübingen, Germany*

An Staes • *VIB Center for Medical Biotechnology, Ghent, Belgium; Department of Biomolecular Medicine, Ghent University, Ghent, Belgium; VIB Proteomics Core, Ghent, Belgium*

Miguel C. Teixeira • *iBB—Institute for Bioengineering and Biosciences, Biological Sciences Research Group, Instituto Superior Técnico, Universidade de Lisboa, Lisbon, Portugal; Department of Bioengineering, Instituto Superior Técnico, Universidade de Lisboa, Lisbon, Portugal*

Samuel Terrier • *ProteoSeine@IJM, Institut Jacques Monod, Université de Paris—CNRS, Paris, France*

Bertrand Theulot • *Institut de Biologie de l'Ecole Normale Supérieure (IBENS), Ecole Normale Supérieure, CNRS, INSERM, Université PSL, Paris, France; Sorbonne Université, Collège Doctoral, Paris, France*

Evy Timmerman • *VIB Center for Medical Biotechnology, Ghent, Belgium; Department of Biomolecular Medicine, Ghent University, Ghent, Belgium; VIB Proteomics Core, Ghent, Belgium*

Christian Trahan • *RNP Biochemistry Laboratory, Center for Genetic and Neurological Diseases, Institut de recherches cliniques de Montréal, Montréal, QC, Canada*

Andreas Tsouris • *Université de Strasbourg, CNRS, GMGM UMR 7156, Strasbourg, France*

Lasse van Wijlick • *Unité Biologie et Pathogénicité Fongiques, Département de Mycologie, Institut Pasteur, USC 2019 INRAE, Paris, France*

Johannes Venezian • *Faculty of Biology, Technion—Israel Institute of Technology, Haifa, Israel*

Lieselotte Vermeersch • *VIB Laboratory for Systems Biology, VIB-KU Leuven Center for Microbiology, Leuven, Belgium; CMPG Laboratory of Genetics and Genomics, Department M2S, KU Leuven, Leuven, Belgium*

Kevin J. Verstrepen • *VIB Laboratory for Systems Biology, VIB-KU Leuven Center for Microbiology, Leuven, Belgium; CMPG Laboratory of Genetics and Genomics, Department M2S, KU Leuven, Leuven, Belgium*

Romeu Viana • *iBB—Institute for Bioengineering and Biosciences, Biological Sciences Research Group, Instituto Superior Técnico, Universidade de Lisboa, Lisbon, Portugal; Department of Bioengineering, Instituto Superior Técnico, Universidade de Lisboa, Lisbon, Portugal*

Tommaso Villa • *Université de Paris, CNRS, Institut Jacques Monod, Paris, France*

Thidathip Wongsurawat • *Division of Bioinformatics and Data Management for Research, Department of Research and Development, Faculty of Medicine Siriraj Hospital, Mahidol University, Bangkok, Thailand*

Nozomu Yachie • *School of Biomedical Engineering, Faculty of Applied Science and Faculty of Medicine, University of British Columbia, Vancouver, BC, Canada; Research Center for Advanced Science and Technology, Synthetic Biology Division, University of Tokyo, Tokyo, Japan; Department of Biological Sciences, Graduate School of Science, University of Tokyo, Tokyo, Japan; Institute for Advanced Biosciences, Keio University, Tsuruoka, Japan*

Gabriel E. Zentner • *Department of Biology, Indiana University, Bloomington, IN, USA; Indiana University Melvin and Bren Simon Comprehensive Cancer Center, Indianapolis, IN, USA; eGenesis, Inc., Cambridge, MA, USA*

Shilu Zhang • *Wisconsin Institute for Discovery, University of Wisconsin-Madison, Madison, WI, USA*

Part I

Transcriptomic Analyses

Chapter 1

Single-Cell RNA Sequencing in Yeast Using the 10× Genomics Chromium Device

Lieselotte Vermeersch, Abbas Jariani, Jana Helsen, Benjamin M. Heineike, and Kevin J. Verstrepen

Abstract

Single-cell RNA sequencing (scRNA-seq) is emerging as an essential technique for studying the physiology of individual cells in populations. Although well-established and optimized for mammalian cells, research of microorganisms has been faced with major technical challenges for using scRNA-seq, because of their rigid cell wall, smaller cell size and overall lower total RNA content per cell. Here, we describe an easy-to-implement adaptation of the protocol for the yeast *Saccharomyces cerevisiae* using the 10× Genomics platform, originally optimized for mammalian cells. Introducing Zymolyase, a cell wall–digesting enzyme, to one of the initial steps of single-cell droplet formation allows efficient in-droplet lysis of yeast cells, without affecting the droplet emulsion and further sample processing. In addition, we also describe the downstream data analysis, which combines established scRNA-seq analysis protocols with specific adaptations for yeast, and R-scripts for further secondary analysis of the data.

Key words Single-cell RNA sequencing, Transcriptomics, Single-cell omics, Yeast, *Saccharomyces cerevisiae*, 10× Genomics

1 Introduction

High-throughput single-cell RNA sequencing (scRNA-seq) has become a well-established technique to study cell-to-cell heterogeneity, rare phenotypes and their effect on how populations function. Various methods have been developed for scRNA-seq, with different single cell isolation and library preparation strategies depending on the specific organisms involved [1–10]. Cell isolation methods range from micromanipulation and index sorting to more recent microfluidics-based methods. While each method has its strengths and weaknesses, droplet-based methods, such as Drop-seq [11, 20] or the 10× Genomics' microfluidic platform, are often considered the gold standard because they generally yield higher

Lieselotte Vermeersch and Abbas Jariani have contributed equally to this work.

Frédéric Devaux (ed.), *Yeast Functional Genomics: Methods and Protocols*, Methods in Molecular Biology, vol. 2477,
https://doi.org/10.1007/978-1-0716-2257-5_1,

throughput while reducing cost and workload. On the flipside, these droplet-based methods do require specialized microfluidics equipment and have been shown to generally detect a lower number of genes per cell [12]. Currently, the droplet-based method from 10× Genomics has become one of the most commonly used commercial systems [4, 13–16]. The platform uses Gel Bead-In-Emulsions (GEMs), trapping single cells in separate emulsion droplets along with reverse-transcription reagents and a uniquely labeled primer hydrogel bead. This allows in-droplet barcoding and reverse transcription into cDNA before pooling and bulk-level sequencing.

The use of this technique to study microbial cells has, however, been limited because of practical hurdles, such as the existence of a rigid cell wall and lower total amount of mRNA per cell. Here, we provide an easy-to-implement adaptation to the existing 10× Genomics' protocol that allows using this technology for scRNA-seq in *Saccharomyces cerevisiae* by including an in-droplet cell wall digesting step. In addition, the bioinformatic pipelines for further analysis of the data are also described in detail.

2 Materials

2.1 Growth Media and Sampling Buffers

1. Desired growth medium: autoclaved and filter-sterilized (0.2 μm filter) (*see* **Notes 1** and **2**).
2. 50% glycerol: 1260 g/L glycerol. To make 500 mL, add 250 mL of dH_2O to 250 mL (equivalent to 315 g) of glycerol and filter-sterilize (0.2 μm filter).

2.2 In-Droplet Spheroplasting and Cell Lysis

1. 100× Zymolyase stock solution: 100 mg/mL Zymolyase 100 T (Amsbio) dissolved into buffer of a reverse transcriptase (*see* **Notes 3** and **4**). Depending on the version of the kit, a different stock solution might be required. Filter-sterilize (low retention syringe filter—0.2 μm pores), keep on ice (alternatively, aliquot the Zymolyase and freeze. Thaw when ready to use and avoid multiple freeze-thaw cycles).
2. PBS: Phosphate buffered saline, filter-sterilized, pH 7.2. Make a 10× PBS solution and dilute to 1× in MilliQ H_2O. To make 1 L 10× PBS, add into 800 mL MilliQ water: 81 g NaCl, 2.0 g KCl, 14.4 g Na_2HPO_4, 2.4 g KH_2PO_4. Add a stirrer bar, adjust to pH 7.2 (with HCl solution) and volume up to 1 L with MilliQ water. Filter-sterilize (*see* **Note 5**).
3. 10× Genomics Chromium Single-Cell Reagent Kit protocol, can be found on the 10× Genomics' support site, under https://support.10xgenomics.com/

2.3 Software Installations

1. R Statistical Programming Language
   https://www.r-project.org
2. R packages.
   (a) Seurat. More information on https://satijalab.org/seurat/.
   (b) ggplot2.
   (c) dplyr.
3. Cell Ranger.
   The software can be found on the 10× Genomics' support site, https://support.10xgenomics.com/.
4. bcl2fastq.
   The software can be found on the Illumina support site, https://support.illumina.com/sequencing/sequencing_software/bcl2fastq-conversion-software/downloads.html.
5. A BASH Shell.
   Should be preinstalled for users on Linux/MacOS.

3 Methods

For sample and single-cell library preparation, follow all standard precautions usually taken to perform RNA- and single-cell work, for example, dedicated space, pipettes and reagents, and use of filter tips and nuclease-free reagents.

3.1 Sample Preparation

1. Grow yeast in media of interest, until density of interest is reached (*see* **Notes 1**, **2**, and **6**).
2. Freeze 0.5 mL of cell culture mixed with 0.5 mL of ice-cold 50% glycerol at −80 °C until droplet generation using the 10× Genomics Chromium device (*see* **Notes** 7 and **8**).

3.2 Single-Cell Library Preparation

To accommodate in-droplet spheroplasting and lysis of the yeast cells, a Zymolyase solution is added to the reverse transcriptase mix before running the Chromium device in the 2.0 or 3.0 version of the 10× kit and protocol. For version 3.1, cells take up to 18 min at room temperature to be encapsulated with beads (rather than 6 to 8.5 min for v2.0–3.0), causing premature lysis of cells and high levels of background mRNA. Adding Zymolyase solution to the GEM beads therefore works better for v3.1. This approach may also work for v2.0 and v3.0, but we have not tested this, as older versions are being phased out by 10× and are not readily available.

1. Thaw cell cultures on ice, measure cell count, and pellet the cells (900 × *g*, 3′, 4 °C) (*see* **Note 9**). Resuspend into ice-cold filter-sterilized PBS to reach the desired cell count, according to the 10× Genomics Chromium Single Cell protocol (*see*

Note 10). In order to keep the proportion of hydrogel beads that contain more than one cell low, it is advisable to aim for 1000 to 2000 recovered cells.

2. From here on, follow the well-documented protocol of the 10× Genomics Chromium device, except for adding the Zymolyase solution to the single-cell master mix (v2.0 and 3.0) or to the beads (v3.1). To add the Zymolyase solution to the single-cell master mix in v2.0 and 3.0 (which is essentially the reverse transcription master mix), replace 1 μL of water in the master mix with Zymolyase stock solution at the appropriate concentration. For v2.0, the total reaction volume will be 100 μL, and the Zymolyase stock solution should be at 100× (100 mg/mL). For v3.0, the total reaction volume is 80 μL, hence the Zymolyase stock solution should only be 80× (80 mg/mL). Then, prepare the microfluidic Chromium Single Cell Chip, add the cell suspension to the single-cell master mix and immediately transfer to the microfluidics chip. Add hydrogel beads and partitioning oil according to the protocol, and commence encapsulating the cells with hydrogel beads using the 10× Genomics Chromium device.

   To add the Zymolyase solution to the beads (v3.1, possibly earlier versions), defrost 70 μL aliquots of hydrogel beads, remove 1 μL of beads and add 1 μL of 70× Zymolyase. Then vortex the beads for 30 s, spin down for 5 s and load 50 μL of beads to the appropriate well according to the Chromium v3.1 protocol. Add cell suspension and reverse transcription master mix to the chip according to the protocol, and commence encapsulating the cells with hydrogel beads.
3. After emulsification, perform reverse transcription, cleanup, cDNA amplification and library construction (*see* **Note 11**) according to the protocol.
4. After library construction and QC, sequence the prepared libraries (*see* **Note 12**).

3.3 Cell Ranger Pipeline

Cell Ranger consists of a set of analysis pipelines developed by 10× Genomics (downloadable from the 10× Genomics website) to process the single-cell RNA-sequencing output (Fig. 1). For some commands, it is useful to set limits on memory and CPU usage. We used a Linux 64 bit machine with 32 GB of RAM and Intel Core i7–4800 CPU to execute the following analysis pipeline.

3.3.1 Demultiplex Raw Base Call (BCL) into FASTQ Files

1. Create a sample sheet, as an input comma-separated (csv) file, containing the following three columns: the sequencer lane number (“Lane”), the sample name (“Sample”) and the index position (used in the library construction) from the GEM well of the 10× 96-well plate (“Index”) (*see* **Note 13**).

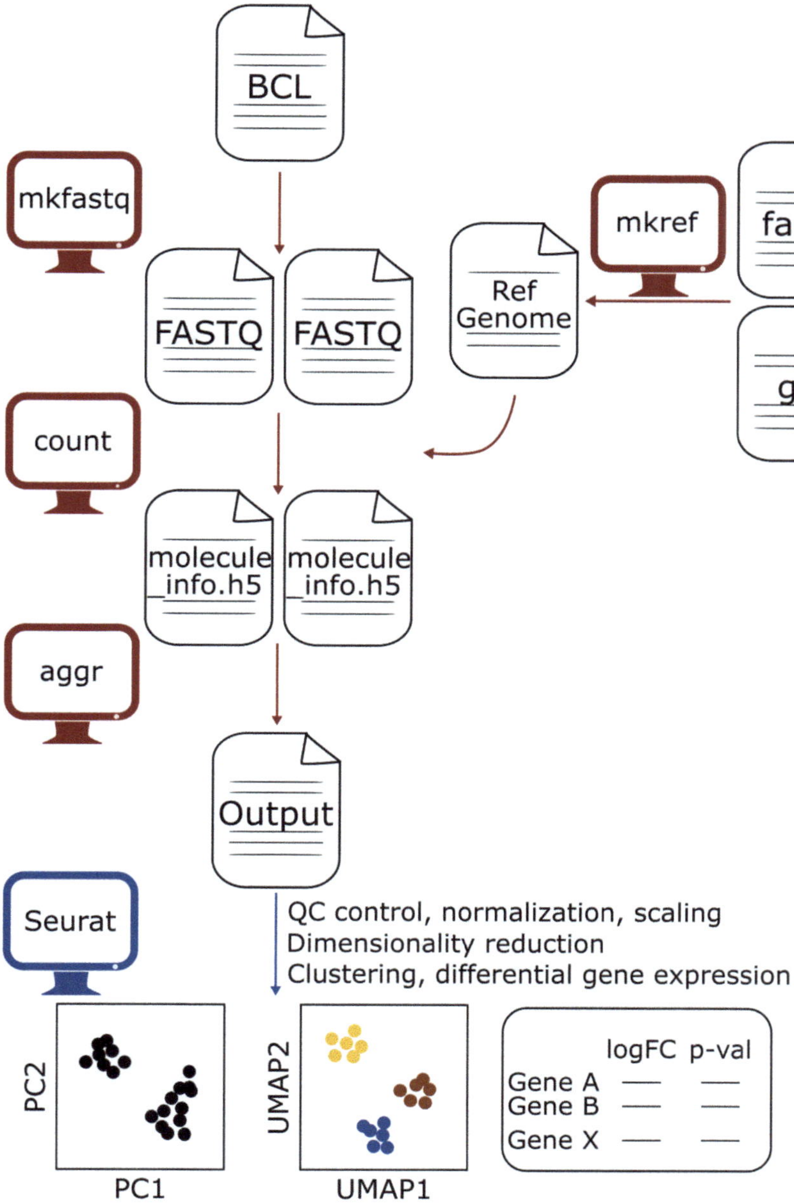

Fig. 1 Overview of the general bioinformatics pipeline to analyze 10× Genomics single-cell RNA-seq data. The first part of the analysis consists of a set of pipelines from the Cell Ranger software (10× Genomics) to process the scRNA-seq raw BCL files into output ready for secondary analysis. Further secondary data analysis mainly makes use of the Seurat R-package for data exploration and visualization

2. Convert raw BCL files from the Illumina sequencer to FASTQ files, separated for each sample and created in the "outs" directory, using "cellranger mkfastq" (*see* **Note 14**). You can set the maximal GB the pipeline may request using the "localmem" option.

```
$ cellranger mkfastq --run=/path/to/folder_with_BCL_files/ \
                     --id=mkfastqOut \
                     --csv=/path/to/sample_sheet_mkfastq.csv \
                     --localmem=3
```

3.3.2 Mapping to Reference Genome and UMI Counting

1. Build a custom yeast reference genome package using "cellranger mkref." Following input files are needed: (1) fasta file containing the genomic DNA sequence for each chromosome, (2) gtf file annotating the feature loci on the reference genome (*see* **Note 15**).

```
 $ cellranger mkref --genome=S288C \
-- fasta=/path/to/fastafile.fa \
--genes=/path/to/gtffile.gtf
```

2. Make a separate directory per sample to generate single cell feature counts. Using 'cellranger count', map the reads to the reference genome, create bam files and carry out UMI (Unique Molecular Identifier, represents absolute number of observed transcripts) counting for each gene (*see* **Notes 16**, **17**, **18**, and **19**).

```
$ cellranger count --id=unique_run_ID_name \
--fastqs=mkfastqOut/outs/fastq_path \
--sample=sample_name \
--transcriptome=S288C \
--expect-cells=2000 \
--localcores=15 \
--localmem=40 \
--chemistry=SC3Pv2
```

3.3.3 Aggregate Data from Multiple Samples

1. Create an input csv file which specifies a list of the "cellranger count" output files; containing 1 row per sample, with a sample name ("library_id") and the path to the output of the "cellranger count" command ("molecule_h5") (*see* **Note 20**).
2. Aggregate the counts from the different samples using "cellranger aggr" (*see* **Note 21**) to create multiple output files, among which the filtered feature-barcode matrices MEX file.

```
$ cellranger aggr --id=unique_run_ID_string \
--csv=aggregation_input.csv
--normalization=none
```

3.4 Secondary Analysis in R

There are multiple tools for secondary data analysis, including Monocle, Seurat, Scanpy, and Cell Ranger R. We use the Seurat R package for secondary analysis (*see* **Note 22**), the input of which is the output from the Cell Ranger pipeline (Fig. 1).

3.4.1 Read-In of the Data, QC and Filtering, Normalization

1. Read the output generated in the last step of the Cell Ranger pipeline (more specifically the filtered feature-barcode matrices MEX) into a matrix, using the 'Read10x' function of the Seurat package.

```
data_merge <- Read10X(data.dir='/path/to/filtered_feature_-
bc_matrix')
```

This creates a data matrix where each column corresponds to a single cell, denoted by a molecular barcode followed by a numerical index (indicating which of the samples this particular cell comes from), for example, "AAACCTGAGTGATCGG-1". Each row in this matrix corresponds to a gene based on the exon annotation of the .gtf file used in mkref, for example, "COS7".

2. Create a Seurat object from this matrix and assign sample names (based on the numerical index) through a metadata column. We assigned sample names based on the time point of sampling. Therefore, we introduced "timepoint" in the metadata. This can be adapted to the specific setup of the experiment. When creating the Seurat object, it can be useful to prefilter out the barcodes (potential cells) with no reads and with data in less than, for example, 5 cells using "min.cells" and "min.features" respectively.

```
min_cells = 1
min_features = 5
data_merge_obj <- CreateSeuratObject(counts=data_merge, min.
cells=min_cells, min.features=min_features)
data_merge_obj@meta.data$timepoint <- 'na'
data_merge_obj@meta.data[grepl('-1$',colnames
(data_merge)),'timepoint'] <- 'sample_name_1'
data_merge_obj@meta.data[grepl('-2$',colnames
(data_merge)),'timepoint'] <- 'sample_name_2'
data_merge_obj@meta.data[grepl('-3$',colnames
(data_merge)),'timepoint'] <- 'sample_name_3'
...
data_merge_obj@meta.data[grepl('-x$',colnames
(data_merge)),'timepoint'] <- 'sample_name_x'
```

3. Plot and inspect major QC metrics (percentage of mitochondrial reads, possible doublets, etc.) through violin plots of distribution values. Seurat automatically calculates QC metrics (in the metadata: nCount_RNA is number of UMIs, nFeature_RNA is features per cell). The proportion of mitochondrial reads could be an important metric for downstream analyses and interpretation, but needs to be annotated and calculated manually. In *S. cerevisiae*, mitochondrial gene names start with 'Q', hence we first define all those genes starting with 'Q' as mitochondrial, and add this to the metadata (*see* **Note 23**).

```
data_merge_obj[['percent.mt']] <- PercentageFeatureSet(da-
ta_merge_obj, pattern = '^Q')
plot1 <- VlnPlot(data_merge_obj, features = c('nFeature_RNA',
'nCount_RNA'), ncol = 2, group.by='timepoint')
```

4. The thresholds for filtering out cells can be determined by looking at the violin plots. We provide an example based on previously published data [17], in which yeast undergoing a shift from glucose to maltose were studied. High numbers of detected genes (nFeature_RNA) could indicate possible doublets. Therefore, the threshold for maximum total number of genes detected was set to 2000 for sample 1 (Fig. 2) (*see* **Note 24**). These thresholds can, within reason, be determined per sample.

```
data_merge_obj <- data_merge_obj[,(grepl('-1$', colnames(da-
ta_merge_obj)) & data_merge_obj$nFeature_RNA < 2000) |
  (grepl('-2$', colnames(data_merge_obj)) & data_merge_obj
$nFeature_RNA < 2000) |
  (grepl('-4$', colnames(data_merge_obj)) & data_merge_obj
$nFeature_RNA < 1500) |
  (grepl('-3$', colnames(data_merge_obj)) & data_merge_obj
$nFeature_RNA < 1500) |
  (grepl('-5$', colnames(data_merge_obj)) & data_merge_obj
$nFeature_RNA < 2000)]
```

5. Normalize the data using the default global-scaling "LogNormalize" option, to correct for differences in capture efficiency between cells and differences in read depth between samples (*see* **Note 25**).

```
data_merge_obj <- NormalizeData(data_merge_obj)
```

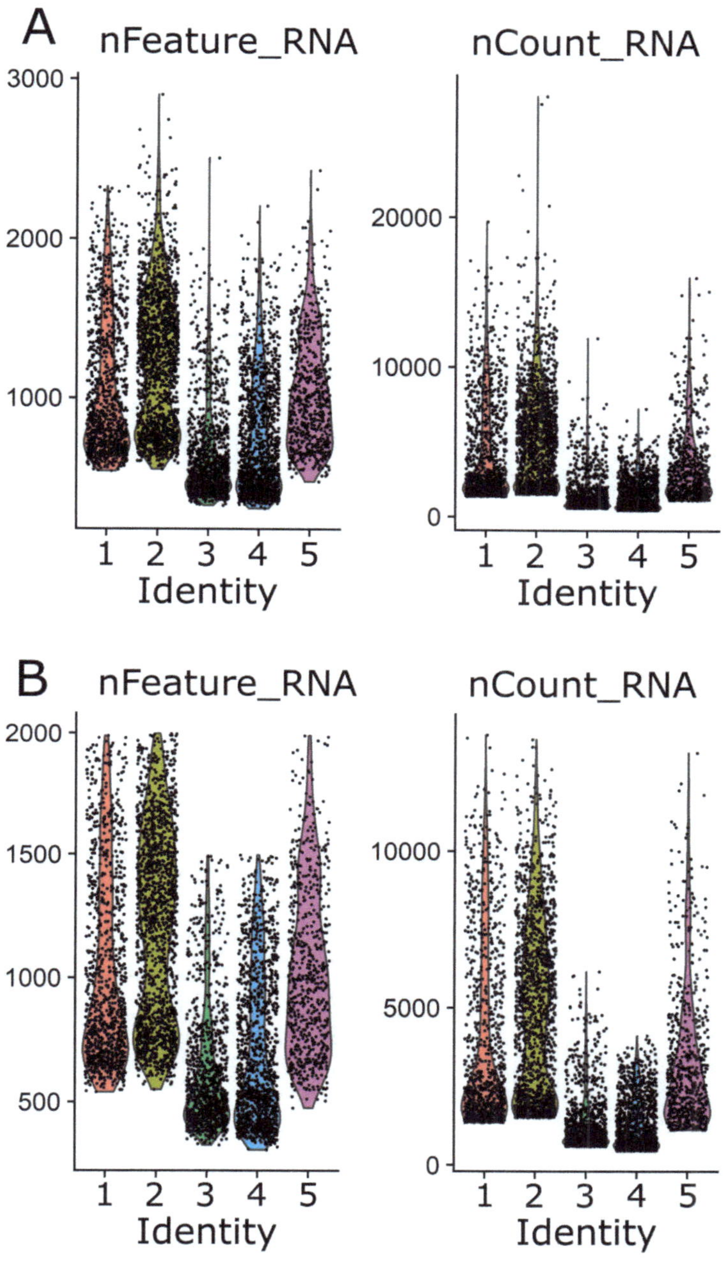

Fig. 2 QC plots before and after quality control and filtering for possible doublets in 5 scRNA-seq samples from [17]. Violin plots for number of detected genes and number of counts (**a**) before and (**b**) after filtering

6. Detect highly variable genes across the single cells using "FindVariableFeatures," using the default selection method option "vst." The selection method defines how to choose the top variable features. You can decide on "nfeatures" depending on how many variable genes you want to use to cluster the cells. The downstream analysis will focus on these genes which will help highlight biological signals in the data.

```
data_merge_obj <- FindVariableFeatures(data_merge_obj, selection.method = 'vst', nfeatures = 300)
```

7. To improve downstream dimensionality reduction and clustering, we need to scale (linearly transform) the data, in order to prevent highly expressed genes from dominating the next steps of the analysis (*see* **Note 26**).

```
data_merge_obj <- ScaleData(data_merge_obj, verbose = FALSE)
```

3.4.2 Dimensional Reduction and Visual Exploration of Data

1. Run linear dimensional reduction using "RunPCA" on the scaled data (*see* **Note 27**).

```
data_merge_obj <- RunPCA(data_merge_obj, npcs = 20, verbose = FALSE)
```

2. Run UMAP on the dimensionally reduced data, and visualize using "DimPlot" (Fig. 3). You can also visualize the data with cells colored by a quantitative feature (such as expression levels of specific genes/growth markers) using "FeaturePlot" (Fig. 4).

```
data_merge_obj <- RunUMAP(data_merge_obj, reduction = 'pca', dims = 1:20 , n.epochs = 1000)
DimPlot(data_merge_obj, reduction = 'umap', group.by = 'timepoint', pt.size = 0.1)
FeaturePlot(data_merge_obj, c('GENE_NAME1','GENE_NAME2','GENE_NAMEX'), pt.size = 0.5, cols = c('gray50','red'), reduction= 'umap')
```

3.4.3 Clustering and Differential Gene Expression

1. Cluster the cells using a graph-based approach. This method first calculates the *k*-nearest neighbors, then constructs the

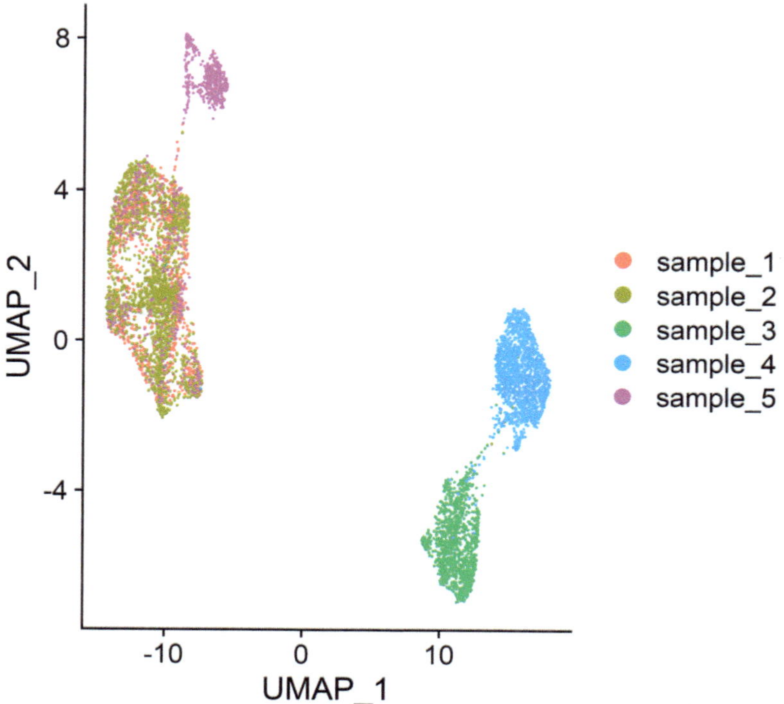

Fig. 3 Visualization of dimensionally reduced data, colored by sample using "DimPlot." UMAP dimensions were created using 20 principal components

shared nearest neighbors' graph. The resolution parameter (*see* **Note 28**) determines the number of clusters (Fig. 5).

```
data_merge_obj <- FindNeighbors(data_merge_obj, dims = 1:20)
data_merge_obj <- FindClusters(data_merge_obj, resolution =
0.1)
DimPlot(data_merge_obj, pt.size = 0.1)
```

2. Find specific biomarkers that define clusters via differential expression. The parameter "logfc.threshold" is used to limit the testing to genes which show a specific fold-difference between the groups of cells. You can test for markers of a specific cluster (specified in ident.1) or you can find markers distinguishing clusters from each other (e.g., ident.1 = 2, ident.2 = 3 in this example will find markers differing between cluster 2 and cluster 3). You can further limit the list of markers by extracting only genes with specific average log-fold changes or adjusted *p*-values you define. Export as a csv-file.

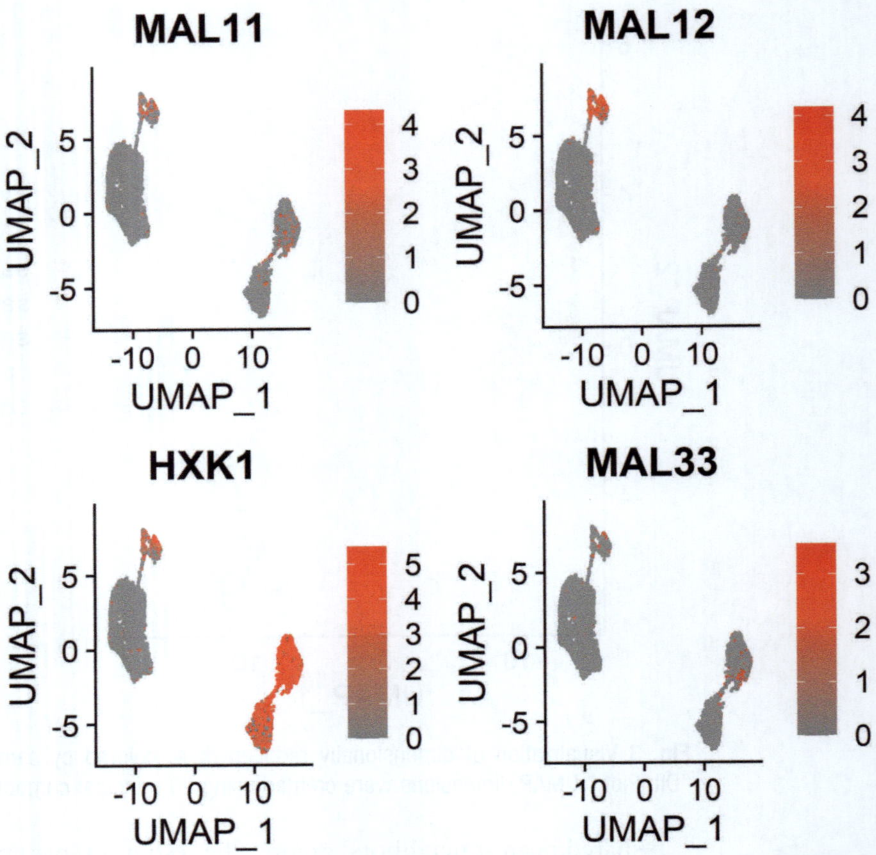

Fig. 4 Visualization of dimensionally reduced data, with cells colored by expression level of the specific marker genes *MAL11*, *MAL12*, *HXK1*, and *MAL33*, using "FeaturePlot"

```
diff_exp_res <- FindMarkers(data_merge_obj, ident.1 =
2, ident.2 = 3, logfc.threshold = 0.01)
diff_exp_res <- diff_exp_res[abs(diff_exp_res$avg_logFC)>
0.25 & diff_exp_res$p_val_adj<0.05, ]
write.csv(diff_exp_res, 'diff_exp_res.csv', row.names = F)
```

3. Upload this csv-file to a Gene Ontology webtool, such as Gorilla (http://cbl-gorilla.cs.technion.ac.il/) to identify and visualize enriched GO terms in your gene list.

4 Notes

1. We never autoclave sugar stock solutions to avoid Maillard reactions and to obtain accurate sugar concentrations.
2. To avoid possible clogging of the 10× Genomics device, it is recommended to filter-sterilize all growth media and buffers.

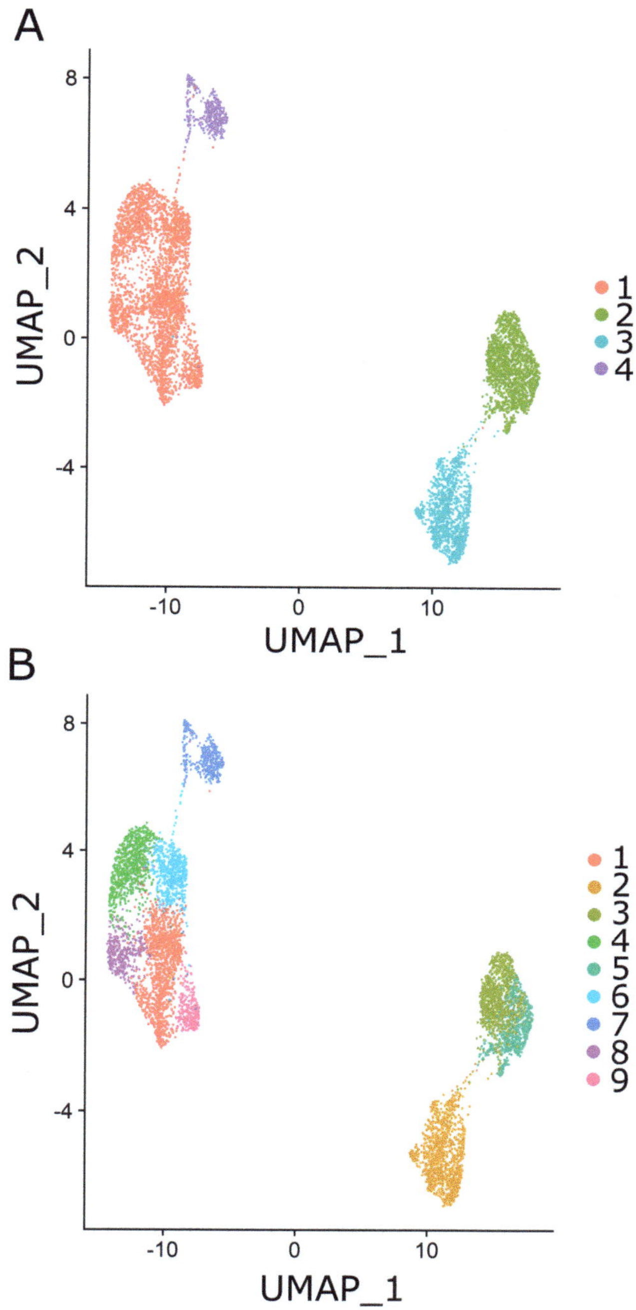

Fig. 5 Clustering of cells based on k-nearest neighbors. The top graph (**a**) shows clustering using a resolution of 0.1, resulting in a lower number of clusters retrieved compared to the bottom graph (**b**) using a resolution of 0.5

3. Zymolyase cannot be dissolved in water at our target concentration. We specifically used the Quantiscript RT Buffer from the QuantiTect reverse transcription kit, and diluted the buffer from a 5× to a 1× concentration before use. Alternatively, dissolving in PBS should also work.
4. After dissolving the Zymolyase, the solution should be dark but completely homogeneous with no visible solid particles. If filter-sterilizing does not work well and clogs the filter, mixing the concentrated Zymolyase well before adding to the master mix without filter-sterilizing seems to work as well.
5. Alternatively, you can use PBS + 400 μg/mL BSA as recommended for sample preparation in the 10× Cell Preparation Guide. Both seem to work equally well.
6. When setting up the experiment, make sure to calculate in advance the minimal cell concentration and volume you would need to sample to recover 1000–2000 cells (according to the Cell Suspension Volume Calculator Table in the 10× protocol).
7. Freezing the cells using liquid nitrogen is possibly even better, but from our experience −80 °C when working with ice-cold buffers should be sufficient.
8. Make sure to mix the sample well and extract an aliquot to obtain an accurate cell count before freezing.
9. Limit time between thawing and the Chromium run as much as possible. If needed, wash once in PBS (+ BSA) buffer prior to counting and diluting. If you encounter clumps of cells during cell counting, dependent on your yeast strains and experimental conditions, sonicating for 40–60 s in a water bath sonicator can help resolve this. If time needed for thawing, resuspension and counting becomes quite long, it might be a good idea to use RNAlater (Qiagen) [4].
10. If you have multiple samples you might want to bring all samples to a similar cell count so you can combine all samples with the reverse transcription master mix at the same time.
11. Make sure you write down the 10× Sample Index name while constructing the library, this is important information for the downstream analysis, especially if you are running multiple samples.
12. Take into consideration that all lanes on the sequencer should be 10× samples, since this sequencing run has specific parameters especially for these types of samples. In terms of read depth, we found that we got about 90% sequence saturation with 20,000 reads/cell.

13. If one 10× library has been sequenced on different flow cells/ sequencers, the "cellranger mkfastq" command will have to be run separately for each sequencing run using separate sample sheets.
14. The "cellranger mkfastq" command is in essence a wrapper around Illumina's "bcl2fastq" software. Therefore, before running this command, you should make sure you have installed the 'bcl2fastq' software (downloadable from the Illumina website).
15. Both the fasta file and the gtf file can be downloaded from UCSC genome browser (fasta file from the 'Sequence and Annotation Downloads' section, the gtf file from the 'Table-Browser' section). Note that Cell Ranger only uses the exon feature in the gtf for its calculations. In *S. cerevisiae*, the exon annotation does not contain the 3'UTR which results in reads, that fall entirely within the 3'UTR, not being assigned. Losing these reads can be avoided by including 3'UTR annotations in the exon feature of the .gtf file, similar to the approach used in [4].
16. You can run the "cellranger count" command in parallel for different samples in separate terminal windows.
17. Set the expected number of cells option to the expected value based on the calculation table in the protocol of the 10× Genomics Chromium device. If you need to combine multiple sequencing runs of the same GEM well, you need to input multiple fastq files, separated by a comma here.
18. You can put limits on memory and cpu usage of this command. This is the most computationally intensive step of the analysis. On a laptop with a Core i7–4800 CPU, and 32GB of RAM, these commands of "cellranger count" took about 30 h to run. Of course, this will be dependent on your read depth, and possibility of using parallel processing.
19. The "chemistry" option allows you to specify which assay configuration and reagent kit you used. In principle, it should be detected automatically.
20. This path would be something like: unique_run_ID_name/ outs/molecule_info.h5.
21. In principle, you can normalize across samples to equalize the average read depth per cell between groups before merging. We did not use aggregation normalization between samples since we knew that the RNA content of cells drops during the time course of our specific sampling scheme, and were therefore able to verify that the total UMI count generally decreased during the time course as expected. We later normalize each cell by the total UMI count prior to clustering and calculating differential expression.

22. ggplot2 and dplyr are two additional R packages that we use here. Most of the analysis using Seurat is based on the, well-explained, Seurat Guided Clustering Tutorial (https://satijalab.org/seurat/articles/pbmc3k_tutorial.html).
23. High levels of mitochondrial reads are generally flagged in mammalian cells as a sign of apoptosis, and therefore a threshold for proportion of mitochondrial reads is usually set as an extra filter to remove dead cells [18]. However, in yeast, large variation in mitochondrial reads more often indicates biological variability, which is why we here do not filter but rather include the proportion of mitochondrial reads for further downstream analysis.
24. Another, perhaps more intuitive, option would be filtering on the total number of counts instead. Having doublets would result in more mRNA in a droplet, higher counts and thus detection of more genes, making nCount_RNA and nFeature_RNA highly correlated. Here again, using a per sample filter, based on the distribution of nCount, would be best.
25. If the total mRNA per sample changes in your experiment, then this normalization step will cause relative differences between samples for specific genes that may not have absolute changes. If needed, you could further remove technical noise with imputation [19].
26. Here, the default method has been used. There is an option "vars.to.regress" to regress out specific features to remove unwanted sources of variation (such as mitochondrial gene expression, number of detected molecules), which might be valuable to use.
27. We decided on the number of PCs (here 20) by visual inspection of the UMAP clustering generated from a few different options. We chose the minimal number that maintains the expected UMAP projection shape. The section on determining dimensionality of the dataset from the Seurat Guided Clustering Tutorial provides a discussion on other approaches including the use of Jack Straw plots (JackStrawPlot), Elbow plots (ElbowPlot), and analyzing dimensionality heatmaps (DimHeatmap).
28. Based on results from previous experiments, we tweaked this parameter to have the 4 expected clusters. According to the Seurat Guided Clustering Tutorial, a range between 0.4 and 1.2 usually gives good clustering.

Acknowledgments

LV, AJ, and KJV would like to thank Thomas Van Brussel, Bernard Thienpont, and Diether Lambrechts at the VIB Center for Cancer Biology for their support in setting up the 10× sequencing protocol. LV, AJ, and KJV experiments were partly funded by the European Research Council (ERC CoG682009), FWO, KU Leuven, and VIB. LV acknowledges FWO SB Grant SB/1S07117N. JH acknowledges FWO SB Grant SB/1S17117N. BH would like to thank Markus Ralser at the Francis Crick Institute for helpful advice and for funding support from the Francis Crick Institute which receives its core funding from Cancer Research UK (FC001134), the UK Medical Research Council (FC001134), and the Wellcome Trust (FC001134), and specific funding from Wellcome Trust (IA 200829/Z/16/Z), to M. Ralser. BH would like to thank Ivan Andrew, Laurence Game, and Samuel Marguerat at the MRC London Institute of Medical Sciences for their support in setting up his 10× sequencing protocol. BH experiments were partly funded by the UK Medical Research Council (MC-A658-5TY20).

References

1. Gasch AP, Yu FB, Hose J et al (2017) Single-cell RNA sequencing reveals intrinsic and extrinsic regulatory heterogeneity in yeast responding to stress. PLoS Biol 15:e2004050. https://doi.org/10.1371/journal.pbio.2004050
2. Nadal-Ribelles M, Islam S, Wei W et al (2019) Sensitive high-throughput single-cell RNA-seq reveals within-clonal transcript correlations in yeast populations. Nat Microbiol 4:683–692. https://doi.org/10.1038/s41564-018-0346-9
3. Saint M, Bertaux F, Tang W et al (2019) Single-cell imaging and RNA sequencing reveal patterns of gene expression heterogeneity during fission yeast growth and adaptation. Nat Microbiol 4:480–491. https://doi.org/10.1038/s41564-018-0330-4
4. Jackson CA, Castro DM, Saldi G-A et al (2020) Gene regulatory network reconstruction using single-cell RNA sequencing of barcoded genotypes in diverse environments. Elife 9:e51254. https://doi.org/10.7554/eLife.51254
5. Islam S, Zeisel A, Joost S et al (2014) Quantitative single-cell RNA-seq with unique molecular identifiers. Nat Methods 11:163–166. https://doi.org/10.1038/nmeth.2772
6. Hwang B, Lee JH, Bang D (2018) Single-cell RNA sequencing technologies and bioinformatics pipelines. Exp Mol Med 50:96. https://doi.org/10.1038/s12276-018-0071-8
7. Yin Y, Jiang Y, Lam K-WG et al (2019) High-throughput single-cell sequencing with linear amplification. Mol Cell 76:676–690.e10. https://doi.org/10.1016/j.molcel.2019.08.002
8. Cao J, Packer JS, Ramani V et al (2017) Comprehensive single-cell transcriptional profiling of a multicellular organism. Science 357:661–667. https://doi.org/10.1126/science.aam8940
9. Imdahl F, Vafadarnejad E, Homberger C et al (2020) Single-cell RNA-sequencing reports growth-condition-specific global transcriptomes of individual bacteria. Nat Microbiol 5:1202–1206. https://doi.org/10.1038/s41564-020-0774-1
10. Bettauer V, Massahi S, Khurdia S et al (2020) *Candida albicans* exhibits distinct cytoprotective responses to anti-fungal drugs that facilitate the evolution of drug resistance. bioRxiv. https://doi.org/10.1101/2020.01.21.914549
11. Macosko EZ, Basu A, Satija R et al (2015) Highly parallel genome-wide expression profiling of individual cells using Nanoliter

droplets. Cell 161:1202–1214. https://doi.org/10.1016/j.cell.2015.05.002

12. Skinnider MA, Squair JW, Foster LJ (2019) Evaluating measures of association for single-cell transcriptomics. Nat Methods 16: 381–386. https://doi.org/10.1038/s41592-019-0372-4
13. Adamson B, Norman TM, Jost M et al (2016) A multiplexed single-cell CRISPR screening platform enables systematic dissection of the unfolded protein response. Cell 167: 1867–1882.e21. https://doi.org/10.1016/j.cell.2016.11.048
14. Kaufmann E, Sanz J, Dunn JL et al (2018) BCG educates hematopoietic stem cells to generate protective innate immunity against tuberculosis. Cell 172:176–190.e19. https://doi.org/10.1016/j.cell.2017.12.031
15. Yan KS, Janda CY, Chang J et al (2017) Non-equivalence of Wnt and R-spondin ligands during Lgr5+ intestinal stem-cell self-renewal. Nature 545:238–242. https://doi.org/10.1038/nature22313
16. Zheng GXY, Terry JM, Belgrader P et al (2017) Massively parallel digital transcriptional profiling of single cells. Nat Commun 8: 14049. https://doi.org/10.1038/ncomms14049
17. Jariani A, Vermeersch L, Cerulus B et al (2020) A new protocol for single-cell RNA-seq reveals stochastic gene expression during lag phase in budding yeast. Elife 9:e55320. https://doi.org/10.7554/eLife.55320
18. Osorio D, Cai JJ (2020) Systematic determination of the mitochondrial proportion in human and mice tissues for single-cell RNA-sequencing data quality control. Bioinformatics 37(7):963–967. https://doi.org/10.1093/bioinformatics/btaa751
19. Tang W, Bertaux F, Thomas P et al (2019) bayNorm: Bayesian gene expression recovery, imputation and normalization for single-cell RNA-sequencing data. Bioinformatics 36(4): 1174–1181. https://doi.org/10.1093/bioinformatics/btz726
20. Urbonaite G, Lee JTH, Liu P et al (2021) A yeast-optimized single-cell transcriptomics platform elucidates how mycophenolic acid and guanine alter global mRNA levels. Communications Biology 4(1) https://doi.org/10.1038/s42003-021-02320-w

Chapter 2

Genome-Wide Profiling of Transcription Initiation with STRIPE-seq

Robert A. Policastro and Gabriel E. Zentner

Abstract

Transcription start site (TSS) usage is a critical factor in the regulation of gene expression. A number of methods for global TSS mapping have been developed, but barriers of expense, technical difficulty, time, and/or cost have limited their broader adoption. To address these issues, we developed Survey of TRanscription Initiation at Promoter Elements with high-throughput sequencing (STRIPE-seq). Requiring only three enzymatic steps with intervening bead cleanups, a STRIPE-seq library can be prepared from as little as 50 ng total RNA in ~5 h at a cost of ~$12 (US). In addition to profiling TSS usage, STRIPE-seq provides information on transcript levels that can be used for differential expression analysis. Thanks to its simplicity and low cost, we envision that STRIPE-seq could be employed by any molecular biology laboratory interested in profiling transcription initiation.

Key words Transcription start site, TSS, Transcription, Transcription initiation, STRIPE-seq, CAGE

1 Introduction

A transcription start site (TSS) is the first base of a given transcript produced by an RNA polymerase. Most genes do not have a single TSS, but rather clusters of TSSs known as transcription start regions (TSRs). Alternative TSR usage can alter expression or function of a transcript in various ways, such as the differential inclusion of sequence elements that impact translation and stability, particularly upstream open reading frames (uORFs) [1–3]. It can also impact the protein isoform produced from a given transcript by including or excluding exonic sequences [4, 5]. Indeed, alternative TSRs are proposed to be a major contributor to human transcript isoform diversity [6]. Alternative usage of TSRs has also been reported in development and disease [7–10].

Given the impact of TSR selection on gene expression, many techniques for global TSS profiling have been developed. Such methods must enrich for 5′-complete cDNA molecules, which

Frédéric Devaux (ed.), *Yeast Functional Genomics: Methods and Protocols*,
Methods in Molecular Biology, vol. 2477, https://doi.org/10.1007/978-1-0716-2257-5_2,

may be accomplished through a number of means. Cap-trapping uses the 5′ 7-methylguanosine cap of mRNAs and various other RNA species (e.g., lncRNAs) as a target for pulldown of 5-′-complete cDNA–RNA hybrids (as in Cap Analysis of Gene Expression (CAGE)) [11] or pre–reverse transcription pulldown of capped RNA (as in MAPCap) [12]. Oligo capping, used in a wide variety of TSS mapping methods [13–19], involves enzymatic removal of the cap and ligation of a synthetic oligo to facilitate isolation of 5′-complete cDNA molecules [20]. A third approach, template-switching reverse transcription (TSRT), upon which STRIPE-seq is based, leverages the unusual property of MMLV-derived reverse transcriptases to switch RNA templates during reverse transcription. TSRT is a well-established mechanism of retroviral recombination [21] and has been used to synthesize full-length cDNAs [22]. Mechanistically, it is widely assumed that, upon encountering the 5′ end of a transcript, particularly one bearing a cap, the reverse transcriptase acts as a terminal transferase that adds 3–4 nontemplated residues to the cDNA, most of which are Cs. This single-stranded C overhang on the cDNA then allows annealing of a template-switching oligo (TSO) bearing a three-riboguanosine (rGrGrG) overhang. After TSO annealing, reverse transcriptase switches templates from the RNA to the TSO, incorporating the TSO sequence into the cDNA molecule. Despite this generally accepted mechanism, recent work has indicated that the mechanism of TSRT is likely more complex than is widely appreciated. On synthetic templates, three Cs did not represent the majority of overhangs, and modification of TSO ribonucleotide overhangs to be complementary to the most frequently occurring sequences had no substantial impact on TSRT efficiency versus a TSO with an rGrGrG overhang [23]. Furthermore, this work suggests that nontemplated nucleotide addition and template switching are concurrent and competing processes. While the generally accepted mechanism of TSRT does not appear to be fully correct and there is thus more work to be done in this area, the fact remains that it is a simple and effective tool for capturing the 5′ ends of capped RNA species. Indeed, TSRT has been incorporated into several TSS mapping methods [24–29] as well as single-cell RNA-seq approaches [30–35].

Here, we present a detailed protocol for STRIPE-seq, our TSRT-based protocol for TSS mapping [36]. The protocol is simple, requiring only three enzymatic steps (digestion of uncapped RNA, TSRT, and library PCR) with bead-based reaction cleanups, and can be performed with modest amounts of total RNA (50–200 ng) from any eukaryotic source (*see* Fig. 1). Thanks to its simplicity, STRIPE-seq library preparation is inexpensive and can be performed in approximately 5 h, and we anticipate that it can be easily adopted by any researchers looking to globally profile TSS usage.

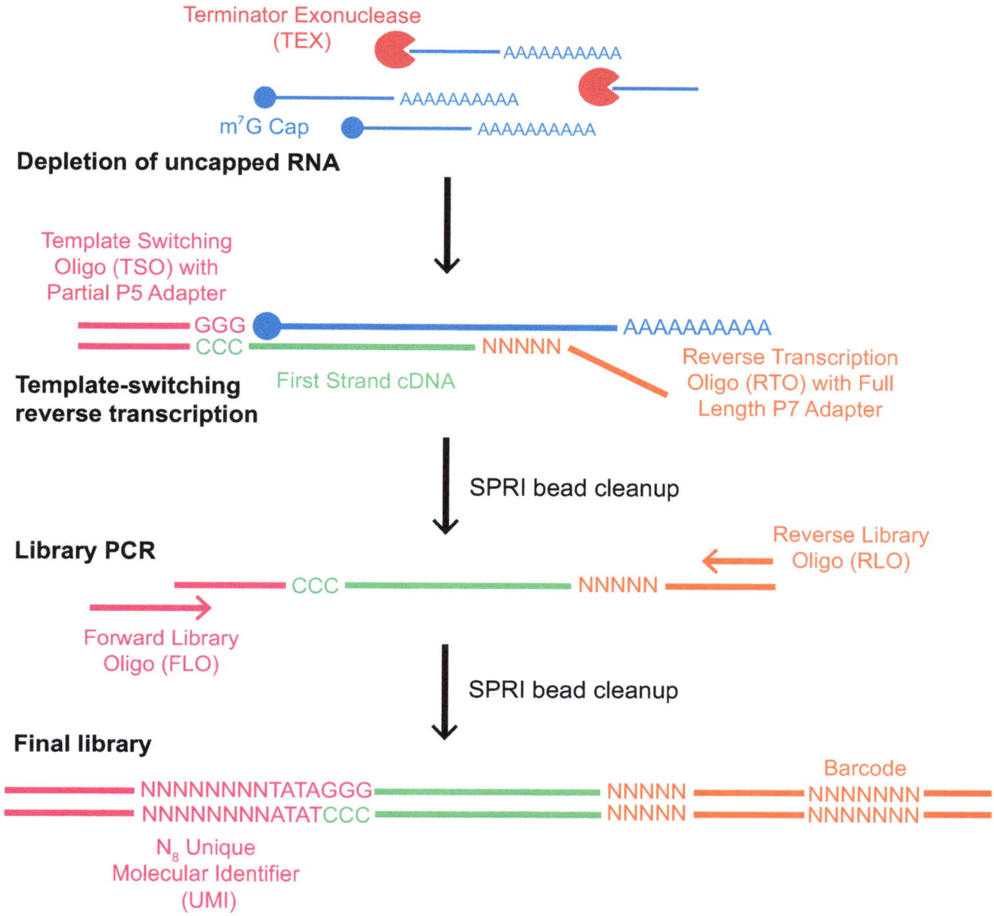

Fig. 1 Overview of the STRIPE-seq library construction workflow. Total RNA is treated with TEX to deplete uncapped RNA (predominantly rRNA). TEX-treated RNA is then used as input for TSRT, which adds a partial Illumina P5 adapter and the complete Illumina P7 adapter to the 5′ and 3′ ends of the first-strand cDNA, respectively. SPRI bead cleanup is then performed to remove excess TSO and RTO. First-strand cDNA is then used as input for library PCR, which completes the 5′ P5 adapter. SPRI bead cleanup is then used to remove excess FLO and RLO, and refine the library size distribution for sequencing

2 Materials

2.1 Reagents and Buffers

1. Nuclease-free water.
2. Terminator 5′-phosphate-dependent exonuclease (TEX).
3. 10× Terminator Reaction Buffer A.
4. TEX master mix: 0.2 μL TEX and 0.2 μL TEX Reaction Buffer A (per sample).
5. 3.3 M sorbitol/0.66 M trehalose solution. Add 2 mL nuclease-free water and 8.02 g trehalose to a 50 mL tube, then add 3 mL nuclease-free water and mix by pipetting up and down (*see*

Note 1). Add 17.8 g sorbitol and 5.5 mL nuclease-free water and mix by pipetting up and down. Bring the total volume of the solution to 30 mL with nuclease-free water. Transfer solution to an RNase-free glass bottle and autoclave at 121 °C for 30 min. Store 1.5 mL aliquots protected from light at room temperature.

6. dNTPs (10 mM each).
7. 5 M betaine.
8. SuperScript II Reverse Transcriptase.
9. 5× SuperScript II First Strand Buffer.
10. 100 mM DTT.
11. RNAClean XP beads (*see* **Note 2**).
12. 70% ethanol.
13. Nuclease-free water.
14. 2× KAPA HiFi HotStart ReadyMix.

2.2 Oligonucleotides

1. Reverse transcription oligo (RTO): 5′-CAAGCAGAAGACGG CATACGAGAT [barcode reverse complement] GTGACTG GAGTTCAGACGTGTGCTCTTCCGATCTNNNNN-3′.

   Insert the reverse complement of the desired barcode sequence where indicated. N_5 represents a random pentamer used to prime reverse transcription. Sequences of 72 uniquely barcoded RTOs are provided in Supplemental Table S5 of [36]. Software to generate a desired number of uniquely barcoded RTOs is available at https://github.com/rpolicastro/barcode-generator.

2. Template switching oligo (TSO): 5′-biotin- CCTACAC GACGCTCTTCCGATCTNNNNNNNNTATArGrGrG-3′.

   N_8 represents an 8 bp unique molecular identifier (UMI). rG indicates riboguanosine.

3. Forward library oligo (FLO): 5′-AATGATACGGCGACCACC GAGATCTACACTCTTTCCCTACACGACGCTCTTCCG - 3′.
4. Reverse library oligo (RLO): 5′- CAAGCAGAAGACGGCA TACG-3′.

2.3 Equipment

1. Magnetic rack.
2. Agilent TapeStation 2200/4200.

3 Methods

3.1 Depletion of Uncapped RNA

1. Prepare TEX master mix (per reaction: 0.2 μL TEX, 0.2 μL 10× Terminator Reaction Buffer A) with sufficient volume for the number of reactions to be performed +1 to account for volume loss during pipetting.
2. In a 0.2 mL PCR tube, combine 0.4 μL TEX master mix and 50–200 ng total RNA in a maximum volume of 1.6 μL. Bring the volume to 2 μL with nuclease-free water, if necessary. Mix by gently pipetting up and down.
3. Incubate TEX reactions at 30 °C for 1 h in a thermal cycler. Hold at 4 °C until ready to proceed.

3.2 Template--Switching Reverse Transcription

1. Prepare RTO annealing mixes (per reaction: 1.5 μL Sorbitol/trehalose stock, 1 μL 10 μM RTO, 0.5 μL dNTPs mix (10 mM each) in 0.2 mL PCR tubes. Each reaction should have a uniquely barcoded RTO.
2. Add 2 μL TEX reaction to 3 μL RTO annealing mix. Vortex to mix and flash spin in a minifuge.
3. Incubate the RTO annealing mixes at 65 °C for 5 min and 4 °C for 2 min. Hold at 4 °C until ready to proceed.
4. Prepare TSRT master mix (per reaction: 2 μL 5 M betaine, 2 μL 5× SuperScript II First Strand Buffer, 0.5 μL 100 mM DTT, 0.5 μL SuperScript II reverse transcriptase) with sufficient volume for the number of reactions to be performed +1 to account for volume loss during pipetting.
5. Add 5 μL TSRT master mix to each 5 μL RTO annealing mix. Vortex to mix and flash spin in a minifuge.
6. Incubate TSRT reactions at 25 °C for 10 min and 42 °C for 5 min in a thermal cycler.
7. Add TSO. Open thermal cycler and, to each sample, add 0.25 μL 400 μM TSO, vortex quickly to mix, and spin down in a minifuge.
8. Return samples to thermal cycler and incubate at 42 °C for 25 min and 70 °C for 10 min. Hold at 4 °C until ready to proceed.
9. Transfer each TSRT reaction to a 0.5 mL tube.
10. Add 8 μL RNAClean XP beads to each TSRT reaction and pipet up and down 10× to mix. Incubate at room temperature for 5 min.
11. Place tubes in magnetic rack and incubate at room temperature for 5 min.

12. Carefully aspirate supernatant, leaving ~2 μL to avoid taking up beads.
13. While tubes are still in the rack, add 175 μL 70% ethanol and immediately discard without incubation.
14. Air-dry beads at room temperature for 5 min.
15. Remove tubes from rack and resuspend beads in 12 μL nuclease-free water.
16. Place tubes in magnetic rack and incubate at room temperature for 1 min.
17. Transfer 11 μL supernatant to new 0.2 mL PCR tubes.

3.3 Library PCR

1. Prepare library PCR master mix (per reaction: 12.5 μL 2× KAPA HiFi HotStart ReadyMix, 0.75 μL 10 μM FLO, 0.75 μL 10 μM RLO) with sufficient volume for the number of reactions to be performed +0.5 to account for volume loss during pipetting. Vortex to mix and spin down in a minifuge.
2. Add 14 μL library PCR master mix to the 11 μL of cleaned TSRT product from **step 17** in Subheading 3.2. Vortex to mix and spin down in a minifuge.
3. Run library PCR reaction (*see* Table 1).

3.4 Small Fragment Removal

1. Transfer library PCR products into 0.5 mL tubes.
2. Add 16.3 μL RNAClean XP beads to each library PCR reaction and pipet up and down 10× to mix. Incubate at room temperature for 5 min.
3. Place tubes in magnetic rack and incubate at room temperature for 5 min.
4. Carefully aspirate supernatant, leaving ~2 μL to avoid taking up beads.
5. While tubes are still in the rack, add 175 μL 70% ethanol and immediately discard without incubation.
6. Air-dry beads at room temperature for 5 min.
7. Remove tubes from rack and resuspend beads in 12 μL nuclease-free water.
8. Place tubes in magnetic rack and incubate at room temperature for 1 min.
9. Transfer 15 μL supernatant to new 0.5 mL tubes.
10. If analysis of library size distribution after small fragment removal is desired, reserve 1 μL of the remaining supernatant.

3.5 Large Fragment Removal

1. Add 8.3 μL RNAClean XP beads to the 15 μL of cleaned PCR product from **step 9** in Subheading 3.4 and pipet up and down 10× to mix. Incubate at room temperature for 10 min.

Table 1
Library PCR thermal cycling parameters

Step	Temperature (°C)	Time
1	95	3 min
2	98	20 s
3	63	15 s
4	72	45 s
5	Repeat steps 2–4	16–20 cycles
6	72	2 min
7	4	Hold

2. Place tubes in magnetic rack and incubate at room temperature for 10 min.
3. Transfer 22 μL supernatant to new 0.5 mL tubes.
4. Add 22 μL RNAClean XP beads to each supernatant and pipet up and down 10× to mix. Incubate at room temperature for 5 min.
5. Carefully aspirate supernatant, leaving ~2 μL to avoid taking up beads.
6. While tubes are still in the rack, add 175 μL 70% ethanol and immediately discard without incubation.
7. Air-dry beads at room temperature for 5 min.
8. Remove tubes from rack and resuspend beads in 16 μL nuclease-free water.
9. Transfer 15 μL supernatant to new 0.5 mL tubes.
10. Check final library size distribution on an Agilent TapeStation instrument using a High Sensitivity D5000 ScreenTape (*see* Fig. 2).
11. Proceed to sequencing (*see* **Note 3**) and data processing and analysis (*see* **Note 4**).

3.6 Computational Analysis

Initial STRIPE-seq data processing can be broken down into three main steps: adapter trimming, genome alignment, and alignment quality processing (*see* Fig. 3). We have developed a custom workflow, GoSTRIPEs, to automate this process, but will describe the main steps below (*see* **Note 5**).

Adapter trimming. The first 15 bases of the R1 read for STRIPE-seq libraries have the structure 'N(8)TATAGGG', with N_8 being the UMI, TATA the spacer, and GGG the bases derived from the three rGs of the TSO. The first consideration for the analysis is whether the UMI should be used for PCR duplicate

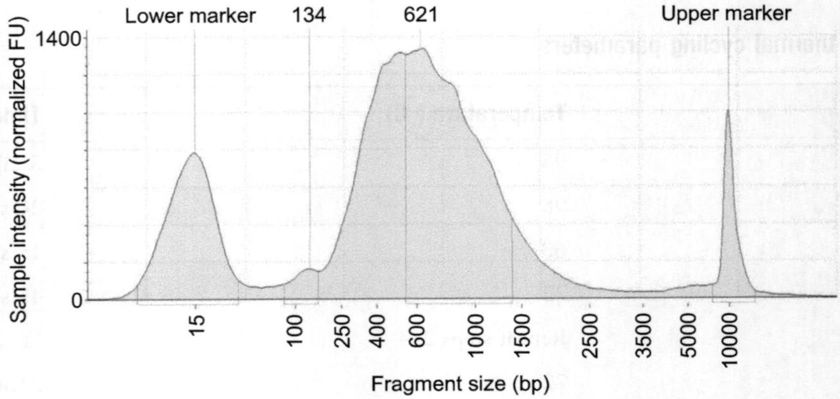

Fig. 2 TapeStation analysis of a STRIPE-seq library. A TapeStation trace of a STRIPE-seq library is shown, illustrating a typical size distribution

removal, as this is only necessary for single-end libraries. For paired-end libraries, Samtools [37] is used to remove duplicates after alignment.

1. Initial FASTQ quality control. FASTQ read quality should first be assessed using FastQC [38] (≥ v0.11.9). Expect the report to indicate a sequence composition bias in bases 9 through 15 of the R1 read due to the fixed base structure of "TATAGGG" at these positions. Furthermore, GC content skews are usually seen if rRNA contamination is present.
2. (Single-end data only) Add the UMI to the FASTQ read header and trim the UMI from the R1 read. We recommend *extract* from UMI-tools (≥ v1.1.1) [39] for this task due to the ability of its network based approach to correct for base errors in the UMI introduced through PCR or sequencing. The following additional settings are recommended: -p NNNNNNNN.
3. Ensure proper R1 read structure. R1 reads should always contain 'N(8)TATAGGG' as the first 15 bases. Cutadapt [40] (≥ v3.2) can both enforce this requirement for a read to be retained and trim this sequence from reads that do contain it. The following settings are recommended if UMI-tools *extract* was run: -g "^TATAG{3}" -O 6 -e 1 --discard-untrimmed. If UMI-tools *extract* was not run, these additional settings are preferred: -g "^N{8}TATAG{3}" -O 14 -e 1 --discard-untrimmed.
4. (Optional) Correct for template switching chaining. Rarely, template switching will occur again after the initial template switching and reverse transcription of the TSO, resulting functionally in two back-to-back adapters in the R1 read with structure "N(8)TATAGGG[adapter]N(8)TATAGGG". If the

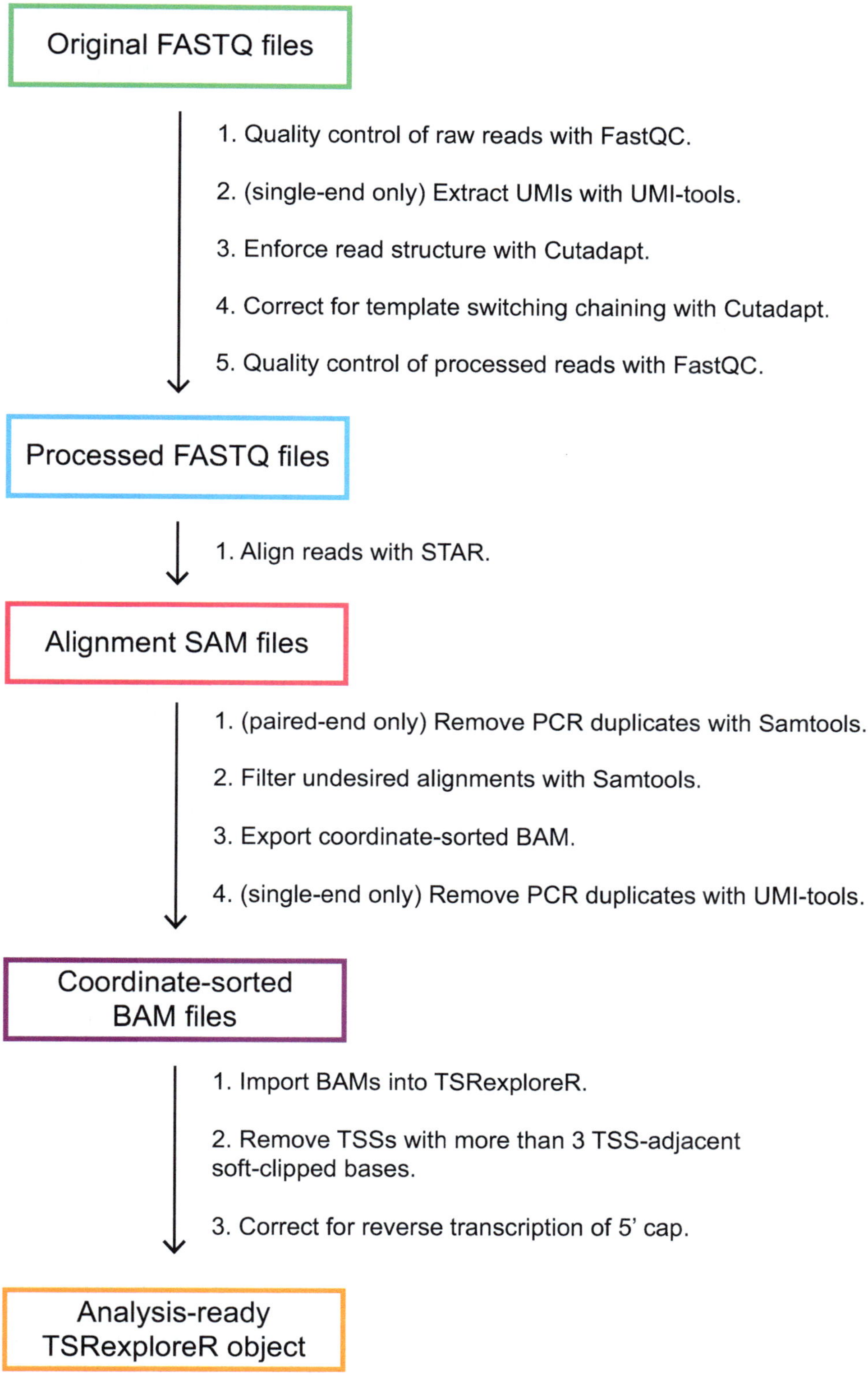

Fig. 3 Overview of STRIPE-seq data processing. This flowchart summarizes the computational steps performed by GoSTRIPEs for preprocessing, alignment, and postalignment quality control of STRIPE-seq data

R1 read length is long enough, these artifacts can be detected and removed with a second pass of Cutadapt using settings '-g "^CCTACACGACGCTCTTCCGATCTN{8}TATAG{3}" -O 21 -e 2 -m 15'.

5. Quality control of processed FASTQ files. Another round of FastQC is recommended before alignment to assess library quality after trimming and filtering of the original FastQC reads.

6. Read alignment. After processing of FASTQ files, a standard genome alignment using STAR [41] (≥ v2.7.7a) is recommended. STAR requires both a genome assembly in FASTA format and a genome annotation in GTF format. These files can be obtained through databases such as ENSEMBL (https://useast.ensembl.org/index.html) or GenBank (https://www.ncbi.nlm.nih.gov/genbank/).

7. Genome index generation. Before alignment, STAR creates a suffix array index when --runMode genomeGenerate is set, using the genome annotation and assembly. Default settings are recommended, but note that for smaller genomes --genomeSAindexNbases may need to be lowered from its default value of 14. For instance, a value of 10 is recommended for *S. cerevisiae*.

8. Read alignment. Once the STAR genome index is generated, the samples can then be aligned to the genome. Default settings are recommended.

9. Alignment processing. After alignment, PCR duplicates should be removed, as well as uninformative alignments such as secondary alignments or paired-end reads without a mate. Samtools [37] (≥ v1.11) is used to filter and coordinate sort the BAM files. For paired-end data, Samtools will additionally be used to remove PCR duplicates, whereas for single-end data UMI-tools is used. Following BAM processing, cleaned BAM files should be indexed using Samtools *index*.

   Recommended paired-end workflow: samtools sort -n < sam file> | samtools fixmate -m - - | samtools sort - | samtools markdup - - | samtools view -F 3852 -f 3 -O BAM -o < output bam file>

   Recommended single-end workflow: samtools sort <sam file> | samtools view -F 2820 -O BAM -o < out bam>. UMI-tools *dedup* using default settings can then deduplicate the processed BAM file using the UMIs stored previously in the read headers.

   After the preliminary processing of the BAM files, the aligned reads can be imported into our R package TSRexploreR [42] for final processing and downstream analysis (*see* **Note 4**). Remaining processing steps include removing reads with too many 5′ soft-clipped bases and correcting for spurious 5′ Gs caused by reverse transcription of the cap.

10. Soft-clipped base filtering. Some soft-clipped bases adjacent to the TSS should be tolerated, due to the 5′ cap frequently being reversed transcribed to add an extra C to the first-strand cDNA [26, 43] and because TSRT can sometimes add more than three nontemplated bases [44]. The *import_bams* function of TSRexploreR has an argument, *soft_remove*, that by default will only keep reads with 3 or fewer soft-clipped 5′ adjacent bases. We recommend leaving this setting as the default.
11. G correction. As previously stated, it is common for the 5′ cap to be reverse transcribed in TSS mapping technologies such as CAGE and STRIPE-seq. By chance, a fraction of these Gs will match the genome, and a correction must therefore be performed to account for this. The TSRexploreR function *G_correction* will stochastically remove templated 5′ Gs based on the observed frequency of soft-clipped (that is, nontemplated) 5′ Gs.
12. Downstream analysis with TSRexploreR [45]. An in-depth description of STRIPE-seq analysis using TSRexploreR is beyond the scope of this protocol. We direct interested readers to the documentation section of the TSRexploreR GitHub repository (https://github.com/zentnerlab/TSRexploreR/tree/main/documentation), which contains detailed vignettes describing import and processing of BAM files, a standard STRIPE-seq analysis workflow, analysis of differential TSS and TSR usage, detection of TSS cluster shifts, and data conditioning.

4 Notes

1. The sorbitol/trehalose solution is quite viscous and thus should be mixed slowly.
2. Homemade SPRI beads [46] may be used for cleanup steps, provided they are nuclease-free.
3. The low sequence diversity of STRIPE-seq libraries at the TATAGGG positions of the R1 read can cause problems with cluster detection, particularly on Illumina platforms using two-color chemistry (e.g., NextSeq and NovaSeq) [47, 48]. To increase sequence diversity, STRIPE-seq runs should be spiked with 20%–25% PhiX or more diverse library types (e.g., RNA-seq or ChIP-seq).
4. We developed two pieces of software to facilitate STRIPE-seq data analysis: GoSTRIPEs (https://github.com/rpolicastro/gostripes) and TSRexploreR (https://github.com/

zentnerlab/TSRexploreR) [42]. GoSTRIPEs performs read quality control, alignment, and postalignment processing, generating cleaned BAM files ready for further analysis as described above. TSRexploreR is a highly flexible R package with functions for TSS processing, correlation, visualization, differential feature analysis, TSS shifting detection, and more.

5. The automated GoSTRIPEs workflow is based on the exact analysis steps performed in the original STRIPE-seq paper [36]. The workflow detailed in this chapter represents a streamlined and updated workflow.

References

1. Calvo SE, Pagliarini DJ, Mootha VK (2009) Upstream open reading frames cause widespread reduction of protein expression and are polymorphic among humans. Proc Natl Acad Sci 106:7507–7512. https://doi.org/10.1073/pnas.0810916106
2. Malabat C, Feuerbach F, Ma L, Saveanu C, Jacquier A (2015) Quality control of transcription start site selection by nonsense-mediated-mRNA decay. eLife 4:e06722. https://doi.org/10.7554/eLife.06722
3. Kurihara Y, Makita Y, Kawashima M, Fujita T, Iwasaki S, Matsui M (2018) Transcripts from downstream alternative transcription start sites evade uORF-mediated inhibition of gene expression in Arabidopsis. Proc Natl Acad Sci 115:7831–7836. https://doi.org/10.1073/pnas.1804971115
4. Mejía-Guerra MK, Li W, Galeano NF, Vidal M, Gray J, Doseff AI, Grotewold E (2015) Core promoter plasticity between maize tissues and genotypes contrasts with predominance of sharp transcription initiation sites. Plant Cell 27:3309–3320. https://doi.org/10.1105/tpc.15.00630
5. Ushijima T, Hanada K, Gotoh E, Yamori W, Kodama Y, Tanaka H, Kusano M, Fukushima A, Tokizawa M, Yamamoto YY, Tada Y, Suzuki Y, Matsushita T (2017) Light controls protein localization through Phytochrome-mediated alternative promoter selection. Cell 171:1316–1325.e12. https://doi.org/10.1016/j.cell.2017.10.018
6. Reyes A, Huber W (2018) Alternative start and termination sites of transcription drive most transcript isoform differences across human tissues. Nucleic Acids Res 46:582–592. https://doi.org/10.1093/nar/gkx1165
7. Pal S, Gupta R, Kim H, Wickramasinghe P, Baubet V, Showe LC, Dahmane N, Davuluri RV (2011) Alternative transcription exceeds alternative splicing in generating the transcriptome diversity of cerebellar development. Genome Res 21:1260–1272. https://doi.org/10.1101/gr.120535.111
8. Haberle V, Li N, Hadzhiev Y, Plessy C, Previti C, Nepal C, Gehrig J, Dong X, Akalin A, Suzuki AM, van Ijcken WFJ, Armant O, Ferg M, Strahle U, Carninci P, Muller F, Lenhard B (2014) Two independent transcription initiation codes overlap on vertebrate core promoters. Nature 507:381–385. https://doi.org/10.1038/nature12974
9. Boyd M, Thodberg M, Vitezic M, Bornholdt J, Vitting-Seerup K, Chen Y, Coskun M, Li Y, Lo BZS, Klausen P, Jan Schweiger P, Pedersen AG, Rapin N, Skovgaard K, Dahlgaard K, Andersson R, Terkelsen TB, Lilje B, Troelsen JT, Petersen AM, Jensen KB, Gögenur I, Thielsen P, Seidelin JB, Nielsen OH, Bjerrum JT, Sandelin A (2018) Characterization of the enhancer and promoter landscape of inflammatory bowel disease from human colon biopsies. Nat Commun 9:1661. https://doi.org/10.1038/s41467-018-03766-z
10. Demircioğlu D, Cukuroglu E, Kindermans M, Nandi T, Calabrese C, Fonseca NA, Kahles A, Lehmann K-V, Stegle O, Brazma A, Brooks AN, Rätsch G, Tan P, Göke J (2019) A pan-cancer transcriptome analysis reveals pervasive regulation through alternative promoters. Cell 178:1465–1477.e17. https://doi.org/10.1016/j.cell.2019.08.018
11. Murata M, Nishiyori-Sueki H, Kojima-Ishiyama M, Carninci P, Hayashizaki Y, Itoh M (2014) Detecting expressed genes using CAGE. In: Miyamoto-Sato E, Ohashi H, Sasaki H, Nishikawa J, Yanagawa H (eds) Transcription factor regulatory networks: methods and protocols. Springer, New York, NY, pp 67–85
12. Bhardwaj V, Semplicio G, Erdogdu NU, Manke T, Akhtar A (2019) MAPCap allows

high-resolution detection and differential expression analysis of transcription start sites. Nat Commun 10:3219. https://doi.org/10.1038/s41467-019-11115-x

13. Wakaguri H, Yamashita R, Suzuki Y, Sugano S, Nakai K (2007) DBTSS: database of transcription start sites, progress report 2008. Nucleic Acids Res 36:D97–D101. https://doi.org/10.1093/nar/gkm901
14. Ni T, Corcoran DL, Rach EA, Song S, Spana EP, Gao Y, Ohler U, Zhu J (2010) A paired-end sequencing strategy to map the complex landscape of transcription initiation. Nat Methods 7:521–527. https://doi.org/10.1038/nmeth.1464
15. Arribere JA, Gilbert WV (2013) Roles for transcript leaders in translation and mRNA decay revealed by transcript leader sequencing. Genome Res 23:977–987. https://doi.org/10.1101/gr.150342.112
16. Pelechano V, Wei W, Steinmetz LM (2013) Extensive transcriptional heterogeneity revealed by isoform profiling. Nature 497: 127. https://doi.org/10.1038/nature12121
17. Gu W, Lee H-C, Chaves D, Youngman EM, Pazour GJ, Conte D, Mello CC (2012) Cap-Seq and CIP-TAP identify pol II start sites and reveal capped small RNAs as C. elegans piRNA precursors. Cell 151:1488–1500. https://doi.org/10.1016/j.cell.2012.11.023
18. Park D, Morris AR, Battenhouse A, Iyer VR (2014) Simultaneous mapping of transcript ends at single-nucleotide resolution and identification of widespread promoter-associated non-coding RNA governed by TATA elements. Nucleic Acids Res 42:3736–3749. https://doi.org/10.1093/nar/gkt1366
19. Core LJ, Martins AL, Danko CG, Waters C, Siepel A, Lis JT (2014) Analysis of nascent RNA identifies a unified architecture of initiation regions at mammalian promoters and enhancers. Nat Genet 46:1311–1320. https://doi.org/10.1038/ng.3142
20. Kazuo M, Sumio S (1994) Oligo-capping: a simple method to replace the cap structure of eukaryotic mRNAs with oligoribonucleotides. Gene 138:171–174. https://doi.org/10.1016/0378-1119(94)90802-8
21. Negroni M, Buc H (2001) Retroviral recombination: what drives the switch? Nat Rev Mol Cell Biol 2:151–155. https://doi.org/10.1038/35052098
22. Zhu Y, Machleder E, Chenchik A, Li R, Siebert P (2001) Reverse transcriptase template switching: a SMART approach for full-length cDNA library construction. BioTechniques 30: 892–897
23. Wulf MG, Maguire S, Humbert P, Dai N, Bei Y, Nichols NM, Corrêa IR, Guan S (2019) Non-templated addition and template switching by Moloney murine leukemia virus (MMLV)-based reverse transcriptases co-occur and compete with each other. J Biol Chem 294:18220–18231. https://doi.org/10.1074/jbc.RA119.010676
24. Zhang Z, Dietrich FS (2005) Mapping of transcription start sites in Saccharomyces cerevisiae using 5′ SAGE. Nucleic Acids Res 33:2838–2851. https://doi.org/10.1093/nar/gki583
25. Batut P, Dobin A, Plessy C, Carninci P, Gingeras TR (2013) High-fidelity promoter profiling reveals widespread alternative promoter usage and transposon-driven developmental gene expression. Genome Res 23:169–180. https://doi.org/10.1101/gr.139618.112
26. Cumbie JS, Ivanchenko MG, Megraw M (2015) NanoCAGE-XL and CapFilter: an approach to genome wide identification of high confidence transcription start sites. BMC Genomics 16:597. https://doi.org/10.1186/s12864-015-1670-6
27. Poulain S, Kato S, Arnaud O, Morlighem J-É, Suzuki M, Plessy C, Harbers M (2017) NanoCAGE: a method for the analysis of coding and noncoding 5′-capped transcriptomes. In: Napoli S (ed) Promoter associated RNA: methods and protocols. Springer, New York, NY, pp 57–109
28. Cole C, Byrne A, Beaudin AE, Forsberg EC, Vollmers C (2018) Tn5Prime, a Tn5 based 5′ capture method for single cell RNA-seq. Nucleic Acids Res 46:e62–e62. https://doi.org/10.1093/nar/gky182
29. Schon MA, Kellner MJ, Plotnikova A, Hofmann F, Nodine MD (2018) NanoPARE: parallel analysis of RNA 5′ ends from low-input RNA. Genome Res. https://doi.org/10.1101/gr.239202.118
30. Islam S, Kjällquist U, Moliner A, Zajac P, Fan J-B, Lönnerberg P, Linnarsson S (2011) Characterization of the single-cell transcriptional landscape by highly multiplex RNA-seq. Genome Res 21:1160–1167. https://doi.org/10.1101/gr.110882.110
31. Picelli S, Björklund ÅK, Faridani OR, Sagasser S, Winberg G, Sandberg R (2013) Smart-seq2 for sensitive full-length transcriptome profiling in single cells. Nat Methods 10: 1096. https://doi.org/10.1038/nmeth.2639
32. Hagemann-Jensen M, Ziegenhain C, Chen P, Ramsköld D, Hendriks G-J, Larsson AJM, Faridani OR, Sandberg R (2020) Single-cell RNA counting at allele and isoform resolution using smart-seq3. Nat Biotechnol 38:708–714.

https://doi.org/10.1038/s41587-020-0497-0

33. Bagnoli JW, Ziegenhain C, Janjic A, Wange LE, Vieth B, Parekh S, Geuder J, Hellmann I, Enard W (2018) Sensitive and powerful single-cell RNA sequencing using mcSCRB-seq. Nat Commun 9:2937. https://doi.org/10.1038/s41467-018-05347-6
34. Macosko EZ, Basu A, Satija R, Nemesh J, Shekhar K, Goldman M, Tirosh I, Bialas AR, Kamitaki N, Martersteck EM, Trombetta JJ, Weitz DA, Sanes JR, Shalek AK, Regev A, McCarroll SA (2015) Highly parallel genome-wide expression profiling of individual cells using nanoliter droplets. Cell 161(5):1202–1214. https://doi.org/10.1016/j.cell.2015.05.002
35. Zheng GXY, Terry JM, Belgrader P, Ryvkin P, Bent ZW, Wilson R, Ziraldo SB, Wheeler TD, McDermott GP, Zhu J, Gregory MT, Shuga J, Montesclaros L, Underwood JG, Masquelier DA, Nishimura SY, Schnall-Levin M, Wyatt PW, Hindson CM, Bharadwaj R, Wong A, Ness KD, Beppu LW, Deeg HJ, McFarland C, Loeb KR, Valente WJ, Ericson NG, Stevens EA, Radich JP, Mikkelsen TS, Hindson BJ, Bielas JH (2017) Massively parallel digital transcriptional profiling of single cells. Nature Communications 8(1) https://doi.org/10.1038/ncomms14049
36. Policastro RA, Raborn RT, Brendel VP, Zentner GE (2020) Simple and efficient profiling of transcription initiation and transcript levels with STRIPE-seq. Genome Res 30:910–923. https://doi.org/10.1101/gr.261545.120
37. Li H, Handsaker B, Wysoker A, Fennell T, Ruan J, Homer N, Marth G, Abecasis G, Durbin R, Genome Project Data Processing S (2009) The sequence alignment/map format and SAMtools. Bioinformatics 25:2078–2079. https://doi.org/10.1093/bioinformatics/btp352
38. Andrews S (2010) FastQC: a quality control tool for high throughput sequence data. https://www.bioinformatics.babraham.ac.uk/projects/fastqc/
39. Smith TS, Heger A, Sudbery I (2017) UMI-tools: modelling sequencing errors in unique molecular identifiers to improve quantification accuracy. Genome Res 27(3):491–499. https://doi.org/10.1101/gr.209601.116
40. Martin M (2011) Cutadapt removes adapter sequences from high-throughput sequencing reads. EMBnet J 17(3). https://doi.org/10.14806/ej.17.1.200
41. Dobin A, Davis CA, Schlesinger F, Drenkow J, Zaleski C, Jha S, Batut P, Chaisson M, Gingeras TR (2013) STAR: ultrafast universal RNA-seq aligner. Bioinformatics 29:15–21. https://doi.org/10.1093/bioinformatics/bts635
42. Policastro RA, McDonald DJ, Brendel VP, Zentner GE (2021) Flexible analysis of TSS mapping data and detection of TSS shifts with TSRexploreR. *NAR Genom and Bioinforma* 3:lqab051. https://doi.org/10.1093/nargab/lqab051
43. Carninci P, Sandelin A, Lenhard B, Katayama S, Shimokawa K, Ponjavic J, Semple CA, Taylor MS, Engstrom PG, Frith MC (2006) Genome-wide analysis of mammalian promoter architecture and evolution. Nat Genet 38:626–635. https://doi.org/10.1038/ng1789
44. Zajac P, Islam S, Hochgerner H, Lönnerberg P, Linnarsson S (2014) Base preferences in non-templated nucleotide incorporation by MMLV-derived reverse Transcriptases. PLoS One 8:e85270. https://doi.org/10.1371/journal.pone.0085270
45. Policastro RA, McDonald DJ, Brendel VP, Zentner GE, (2021) Flexible analysis of TSS mapping data and detection of TSS shifts with TSRexploreR. NAR Genomics and Bioinformatics 3(2) https://doi.org/10.1093/nargab/lqab051
46. Rohland N, Reich D (2012) Cost-effective, high-throughput DNA sequencing libraries for multiplexed target capture. Genome Res 22:939–946. https://doi.org/10.1101/gr.128124.111
47. Krueger F, Andrews SR, Osborne CS (2011) Large scale loss of data in low-diversity Illumina sequencing libraries can be recovered by deferred cluster calling. PLoS One 6:e16607. https://doi.org/10.1371/journal.pone.0016607
48. Mitra A, Skrzypczak M, Ginalski K, Rowicka M (2015) Strategies for achieving high sequencing accuracy for low diversity samples and avoiding sample bleeding using illumina platform. PLoS One 10:e0120520–e0120520. https://doi.org/10.1371/journal.pone.0120520

Chapter 3

A Modified Cross-Linking Analysis of cDNAs (CRAC) Protocol for Detecting RNA–Protein Interactions and Transcription at Single-Nucleotide Resolution

Drice Challal, Jessie Colin, Tommaso Villa, and Domenico Libri

Abstract

Detecting protein–RNA interactions in vivo is essential for deciphering many important cellular pathways. Several methods have been described for this purpose, among which cross-linking analysis of cDNA, CRAC. This method relies on a first step of UV cross-linking of living yeast cells and several subsequent steps of purification of the protein–RNA complexes, some of which under denaturing condition. Without altering the general principle of the method, we have modified and improved the protocol, with the specific aim of sequencing the nascent RNA isolated from transcription complexes and generate high-resolution and directional transcription maps.

Key words RNA–protein interaction in vivo, Genome-wide RNAPII mapping, Transcription, CRAC (cross-linking analysis of cDNAs)

1 Introduction

Because of differences in the stability of RNA molecules, the steady state abundance of transcripts is often a poor indicator of transcriptional activity. Many RNAs that are not, or barely, detected in wild type cells derive from genes that are transcribed with comparable intensity to genes producing abundant steady state transcripts. More accurate measures of transcription levels are generally based on the detection of RNA polymerase associated to its template, presumably during RNA production. Detection of RNA polymerase by chromatin immunoprecipitation (ChIP) is a method that does not provide a directional output and lacks resolution. Several approaches have been developed based on the sequencing of the nascent RNA that provide higher resolution mapping data. One

Drice Challal, Jessie Colin and Tommaso Villa contributed equally.

Frédéric Devaux (ed.), *Yeast Functional Genomics: Methods and Protocols*,
Methods in Molecular Biology, vol. 2477, https://doi.org/10.1007/978-1-0716-2257-5_3,

such method, nascent transcript sequencing (NETseq) [1] relies on the purification of the RNA polymerase II (RNAPII) transcription elongation complex under native condition and the sequencing of the associated RNA. Mapping of the 3′-OH of recovered nonpolyadenylated reads is a generally reliable indicator of RNAPII position. However, native purification conditions are not very efficient in eliminating nonnascent or contaminating RNAs of various sources. Indeed, cotranscriptional processing RNA intermediates associated with the elongation complex are also detected by this method and can be misinterpreted as indicators of RNAPII genomic position. For instance, the presence of reads derived from the 5′-exon splicing intermediate generates a strong signal immediately before the 5′-splice site [1] (see also below), which can be mis-interpreted as a sign of RNAPII pausing.

A similar method, RNAPII PAR-CLIP (photoactivable ribonucleoside-enhanced cross-linking and precipitation), requires the previous incorporation of a photoactivable ribonucleoside analogue (4-thio-uracil, 4tU) into the nascent RNA that is thereby efficiently cross-linked to the Rpb2 subunit of RNAPII. Thanks to the formation of a covalent RNA–protein adduct, purification steps can be conducted under denaturing conditions, which limits copurification of associated, nonnascent RNA molecules. The extent to which the addition of the toxic 4tU [2] can alter the physiology of the cell and the transcription process has not been fully investigated. Other methods also rely on the incorporation of 4tU in RNA and the purification of the newly transcribed molecule [3, 4]; however, the extent to which 4tU-containing RNAs are nascent (and therefore signal the position of RNAPII) as opposed to post-transcriptional is debatable, particularly so in the compact and largely intron-less genome of *S. cerevisiae*. Indeed, the length of many genes is below the number of nucleotides that RNAPII can transcribe in the timeframe of the labeling pulse. CRAC [5] is analogous to PAR-CLIP in that the protein of interest is cross-linked to the bound RNA by exposure of the cells to short wavelength UV light for a short time lapse in the absence of 4tU. Cross-linking is likely less efficient in this condition than when photoactivable nucleotide analogs are used, but there is less concern about the impact of the latter on transcription. As with PAR-CLIP, the subsequent purification includes denaturing steps, which decreases the amount of associated but not cross-linked RNAs. The position of the cross-link can also be identified by this method because transcribing through the cross-link site during the reverse transcription step induces microdeletions or misincorporation in the cDNA that can be easily identified and mapped.

Here, we describe a modified protocol (Fig. 1) that we have extensively used to generate high resolution and directional transcription maps [6, 7] but also to analyze genome-wide the association of RNA binding proteins with their RNA targets [8–10]. To

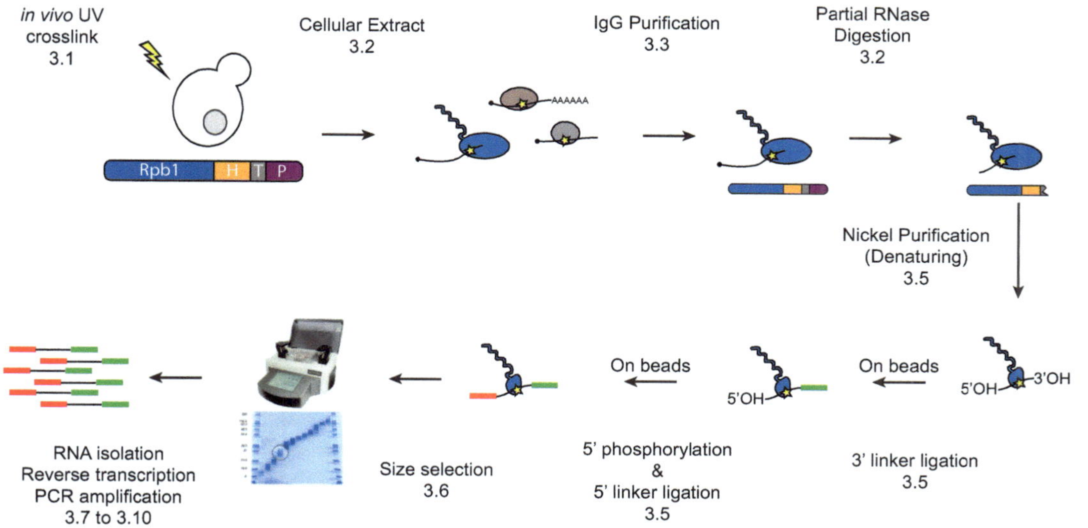

Fig. 1 Overview of the CRAC procedure. Living cells are UV cross-linked and the protein of interest (Rpb1 for transcription analysis) is purified under native conditions. A limited RNase digestion allows reduction of the size of the cross-linked RNA fragment. A second purification under denaturing conditions is performed and linkers are added to the cross-linked RNA while the complex is still attached to the beads, which allows extensive washing between each step. After elution, the RNA–protein complex is further subjected to a denaturing, size-selection purification using the GELFREE 8100 system. The RNA from the complex is finally isolated, reverse-transcribed, and the library prepared by PCR amplification is sequenced using Illumina technology. Each step of the procedure is numbered (Subheadings 3.1–3.10) and detailed in the main text

improve detection of the 3′-end of each read (defining the last position of RNAPII) we modified the original adapters so that sequencing occurs from the 3′ end. Because some proteins bind polyadenylated reads, we added an RNase H/oligo-dT degradation step to limit the size of poly(A) tails that might be present on a subset of reads. Polyadenylated reads can still be detected by sequencing as oligo-dT directed RNase H cleavage cannot fully degrade the poly(A) tail. We also modified the purification protocol to improve yields, and eliminated the original radioactive labeling of the protein-bound RNA. Our protocol provides a very robust and high-resolution detection of RNAPII distribution (Fig. 2a) which is further improved by mapping only the 3′-end of each read, eventually with a smoothing treatment of the signal (Fig. 2a). Metagene analyses obtained using RNAPII CRAC are similar to the ones obtained using NET-seq (Fig. 2b). Differences are observed in the detection of the 5′-exon splicing intermediate (data not shown) and also at the very position of poly(A) tail addition (Fig. 2b) most likely due to the detection in NET-seq of a 3′-end cleavage intermediate that has not yet been polyadenylated and remains associated to the transcription elongation complex. Differences with RNAPII PAR-CLIP data are more marked as the

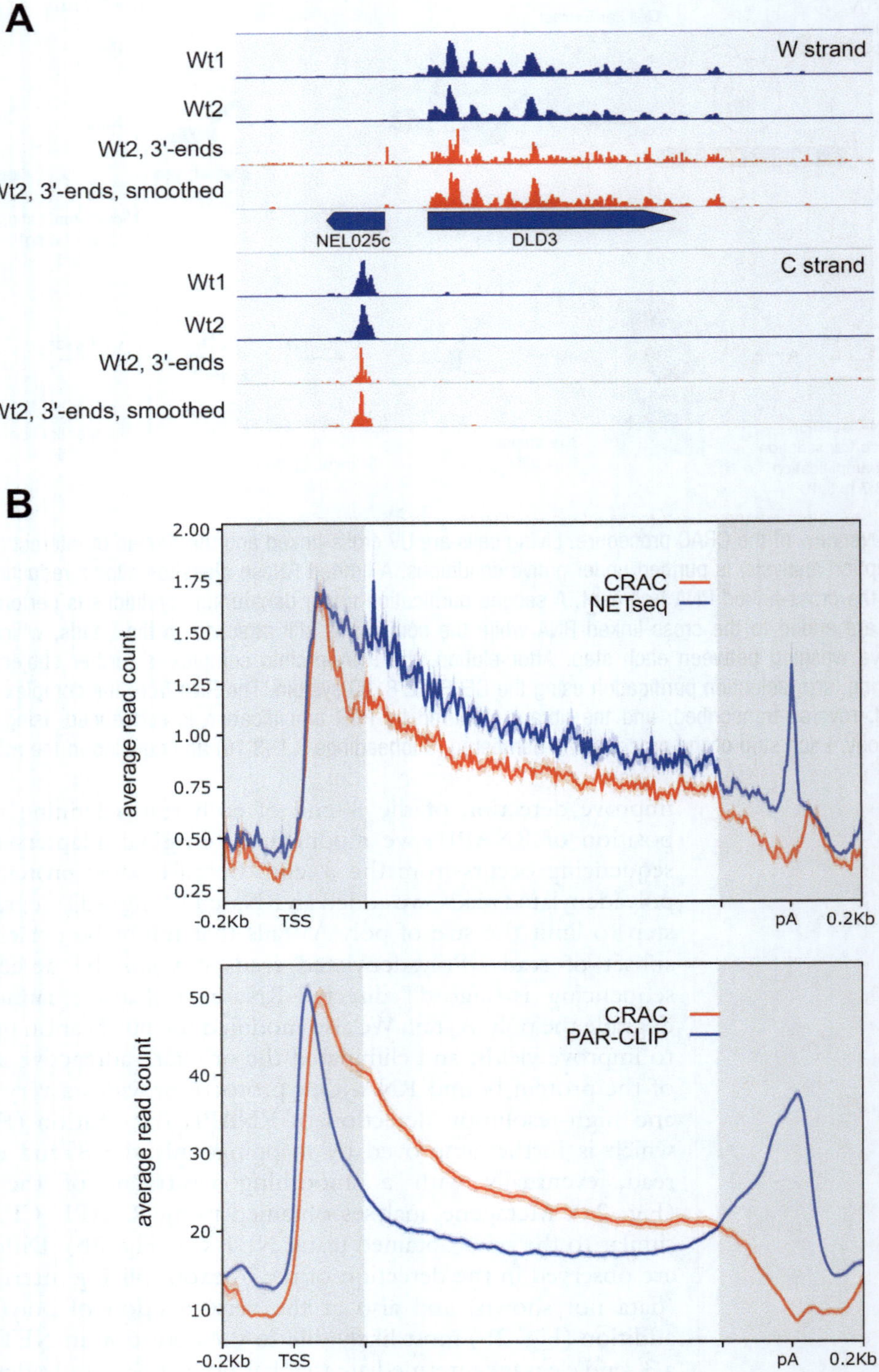

Fig. 2 (**a**) Snapshots showing the distribution of the RNAPII CRAC signal at the *DLD3-NEL025C* (SUT503-CUT542) locus. Wt1 and Wt2 tracks (blue) represent two independent experiments with wild type cells performed by different operators. In red, only the 3′-end of reads from Wt2 are shown, as raw and smoothed signals as indicated. (**b**) Metagene analysis to compare RNAPII signals on all yeast genes for CRAC and

latter method reveals high signals in the 3′-end of many genes (Fig. 2b). The reasons for this difference between NET-seq and CRAC on one side and PAR-CLIP on the other side is unclear. The similarity of NET-seq and CRAC signals tends to exclude the occurrence of strong RNAPII accumulation in the 3′-end of genes, but some peculiarities of PAR-CLIP RNAPII might reveal events that escape detection by the other methods.

For RNAPII CRAC, we mainly used cells expressing a C-terminally tagged Rpb1, but very similar results have been obtained with C-terminally tagged Rpb2. Interestingly, tagged Rpb3 did not cross-link efficiently to the nascent RNA in our conditions, which testifies to the specificity of the method.

2 Materials

Solutions are prepared with autoclaved deionized water and can be stored at 4 °C. Protease inhibitors and beta-mercaptoethanol are freshly added to solutions before usage. RNase-free plastic equipment and reagents are used.

2.1 Yeast Strain and Media

1. Yeast strain expressing HTP (6×His, TEV cleavage site, Protein A)-tagged protein of interest (Rpb1 or Rpb2 for RNAPII CRAC).
2. CSM-TRP media.

2.2 Reagents and Enzymes

1. cOmplete EDTA-free protease inhibitor cocktail tablets.
2. Pefabloc SC-Protease-Inhibitor [AEBSF].
3. DNase I recombinant, RNase-free.
4. Dynabeads M-280 Tosylactivated.
5. IgG from rabbit serum.
6. RNase H.
7. Recombinant GST-TEV protease.
8. RNace-IT Ribonuclease Cocktail.

Fig. 2 (continued) NET-seq (top) or CRAC and PAR-CLIP (bottom). In the top panel only the 3′-ends of the CRAC reads have been plotted for homogeneity with NET-seq data. Note that the peak observed at the 3′ of genes in the NET-seq profile is constituted by reads ending precisely at the polyadenylation site, and presumably represent molecules that have been cleaved but not yet polyadenylated and are associated at the RNAPII elongation complex. This is distinct from the diffuse peak observed in the PAR-CLIP profile in a similar position. The gray shadowed areas represent nonscaled regions of 200 nt around the anchoring point. NET-seq data from ref. [1]. PAR-CLIP data from ref. [15]

9. Guanidine hydrochloride.
10. Imidazole.
11. 0.3 M NaCl.
12. Ni-NTA Agarose.
13. Spin columns—Snap Cap.
14. RNaseOUT Recombinant Ribonuclease Inhibitor.
15. Quick CIP.
16. T4 RNA Ligase 2, truncated KQ.
17. T4 polynucleotide kinase.
18. T4 RNA Ligase 1 (ssRNA ligase).
19. Vivacon 500.
20. SilverQuest™ Silver Staining Kit.
21. Amersham™ Quick Stain Protein Labeling Kit.
22. Rabbit Peroxidase Anti-Peroxidase Soluble Complex.
23. Mouse monoclonal anti-polyHistidine (HIS-1).
24. Proteinase K, recombinant, PCR grade.
25. Ethanol.
26. 3 M NaOAc pH 5.5.
27. Phenol–chloroform–isoamyl alcohol 25:24:1, Saturated with 10 mM Tris–HCl, pH 8.0, 1 mM EDTA.
28. Chloroform.
29. 5PRIME Phase Lock Gel Heavy, 2 mL tubes.
30. GlycoBlue Coprecipitant.
31. SuperScript IV Reverse Transcriptase.
32. RNase H.
33. LightCycler FastStart DNA Master SYBR Green I.
34. LightCycler capillaries.
35. LA Taq.
36. Low Molecular Weight DNA Ladder.
37. Exonuclease I.
38. Gelfree 8100 5% Tris Acetate Cartridge Kit.
39. NucleoSpin Gel and PCR cleanup.
40. Qubit dsDNA HS Assay Kit.

2.3 Buffers

1. TN150 buffer: 50 mM Tris–HCl pH 7.8, 150 mM NaCl, 0.1% NP-40, 5 mM beta-mercaptoethanol.
2. TN1000 buffer: 50 mM Tris–HCl pH 7.8, 1 M NaCl, 0.1% NP-40, 5 mM beta-mercaptoethanol.

3. Wash buffer I: 6.0 M guanidine hydrochloride, 50 mM Tris–HCl pH 7.8, 1 M NaCl, 0.1% NP-40, 10 mM imidazole, 5 mM beta-mercaptoethanol.
4. Wash buffer I-1 M: 6.0 M guanidine hydrochloride, 50 mM Tris–HCl pH 7.8, 1 M NaCl, 0.1% NP-40, 10 mM imidazole, 5 mM beta-mercaptoethanol.
5. 1× PNK buffer: 50 mM Tris–HCl pH 7.8, 10 mM $MgCl_2$, 0.1% NP-40, 5 mM beta-mercaptoethanol.
6. 5× PNK buffer: 250 mM Tris–HCl pH 7.8, 50 mM $MgCl_2$, 50 mM beta-mercaptoethanol.
7. Wash buffer II: 50 mM Tris–HCl pH 7.8, 50 mM NaCl, 0.1% NP-40, 10 mM imidazole, 5 mM beta-mercaptoethanol.
8. Elution buffer: 50 mM Tris–HCl pH 7.8, 50 mM NaCl, 0.1% NP-40, 150 mM imidazole, 5 mM beta-mercaptoethanol.
9. Commercial ultrapure PCR-grade H_2O (Roche).

2.4 Equipment

1. Megatron W5 UV cross-linking unit (UVO3 Ltd).
2. Mixer Mill MM 400 (Retsch).
3. Bioruptor Standard (Diagenode).
4. GELFREE 8100 Fractionation Station (Expedeon).
5. Qubit Fluorometer (Thermo Fisher Scientific, Invitrogen).
6. LightCycler 2.0 (Roche).

2.5 Oligonucleotides

Oligonucleotides were designed to sequence from the 3′-end of RNA. 3′ linkers are 5′-riboadenylated (rApp), allowing for efficient and specific ligation to RNA substrates by T4 RNA ligase 2 in the absence of ATP, and are 3′-blocked (ddC). A three nucleotides barcode (underlined, lowercase) is included for multiplexing and six random nucleotides are used for eliminating PCR duplicates. The 5′ linker contains an inverted dideoxythymidine at its 5′-end to prevent self-ligation and protect from degradation. For the reverse transcriptase step, the original primer described in Bohnsack et al. [5] was modified as we noted a possible recognition of endogenous rRNA.

2.5.1 3′ Linkers

L3-6N-gag /5rApp/GCT<u>ctc</u> NNNNNNAGATCGGAAGAGCGTCGTGTAGGGAAAGAGTGT/3ddC/

L3-6N-acu /5rApp/GCT<u>agt</u> NNNNNNAGATCGGAAGAGCGTCGTGTAGGGAAAGAGTGT/3ddC

L3-6N-ugc /5rApp/GCT<u>gca</u> NNNNNNAGATCGGAAGAGCGTCGTGTAGGGAAAGAGTGT/3ddC/

L3-6N-cua /5rApp/GCT<u>tag</u> NNNNNNAGATCGGAAGAGCGTCGTGTAGGGAAAGAGTGT/3ddC/

L3-6N-cgu /5rApp/GCT<u>acg</u> NNNNNNAGATCGGAAGAGCGTCGTGTAGGGAAAGAGTGT/3ddC/

L3-6N-guc /5rApp/GCTgacNNNNNNAGATCGGAAGAG CGTCGTGTAGGGAAAGAGTGT/3ddC/

L3-6N-uca /5rApp/GCTtgaNNNNNNAGATCGGAAGAG CGTCGTGTAGGGAAAGAGTGT/3ddC/

L3-6N-agu /5rApp/GCTact NNNNNNAGATCGGAA GAGCGTCGTGTAGGGAAAGAGTGT/3ddC/

2.5.2 5′ Linker

L5miRCat /5InvddT/CTTGrGrCrArCrCrCrGrArGrArArUrUrCr CrA-3.

2.5.3 RT and PCR Oligonucleotides

RT L3–2 5-ACACTCTTTCCCTACACGACGCTCTTCCG-3

P5_3′5-AATGATACGGCGACCACCGAGATCTACACTCT TTCCCTACACGACGCTCTTCCGATCT-3

miRCat_PCR2 5-CAAGCAGAAGACGGCATACGAgatcCTT GGCACCCGAGAAT-3

2.6 Software and Algorithms

The cracalyze pipeline is used for demultiplexing, mapping and normalizing reads using *S. pombe* spike-in. The PeakCcall suite [6] is used for extracting 5′ or 3′ read positions, mismatches and deletion, for smoothing the data and for peak calling.

3 Methods

3.1 Yeast Growth and UV Cross-Linking

1. Grow HTP-tagged strains in 2 L of CSM-TRP medium (*see* **Notes 1** and **2**) to $OD_{600} = 0.6$ at 30 °C.
2. UV-irradiate cells at 254 nm for 50 s (*see* **Note 3**) using the Megatron W5 UV cross-linking unit.
3. Harvest cells by centrifugation in 1 L bottles, 5 min at ~9000 × *g* at 4 °C.
4. Wash 2 L pellets in 40 mL cold PBS and transfer in 50 mL conical tube, spin down cells in tabletop centrifuge for 5 min at 3200 × *g* at 4 °C, and weigh pellets.

3.2 Preparation of Cell Lysate

1. Resuspend pellets in 2.4 mL/g of cells of TN150 buffer supplemented with cOmplete EDTA-free Protease Inhibitor Cocktail and 2 mM AEBSF (*see* **Note 4**).
2. Flash-freeze the suspension in droplets. Cell "pop corn" can be indefinitely stored at −80 °C.
3. Mechanically break cells through 5 cycles of 3 min at 20 Hz in a Mixer Mill MM400 and store resulting cell powders at −80 °C until use (*see* **Note 5**).
4. Thaw cell powders by immersing tubes in cold water and measure lysate volume (*see* **Note 6**).
5. Solubilize chromatin of lysates with DNase I (165 U/g of cells) in the presence of 10 mM $MnCl_2$ for 1 h in Infors shaker at

16 °C and 180 rpm. This step can be replaced by sonication for 15 min on a water/ice mixture (45 s on–45 s off) and is required for releasing proteins associated with chromatin.

6. Preclear solubilized lysate by centrifuging for 20 min at 4600 × *g* at 4 °C in tabletop centrifuge (*see* **Note 7**).
7. Clarify cell lysate by centrifuging for 20 min at 20,000 × *g* at 4 °C in SS-34 rotor.
8. Transfer cleared lysates to 15 mL tubes on ice by decanting/pipetting avoiding carryover of lipidic layer and measure volume (*see* **Note 8**). Save a small amount (~50 μL) for western blot analysis of Input fraction. Use same amount of lysate for all samples.

3.3 Affinity Purification of the Complex Using IgG Beads

1. Wash necessary (*see* **Note 9**) amount of 50 mg/mL rabbit IgG-coupled M-280 tosylactivated Dynabeads per sample twice with 1 mL–2 mL TN150 buffer and once more with 1–2 mL of TN150 buffer supplemented with cOmplete EDTA-free Protease Inhibitor Cocktail, using a magnetic device.
2. Resuspend beads in 100 μL/sample of TN150 buffer supplemented with protease inhibitors.
3. Add extracts to beads and rotate in 15 mL tubes at 4 °C for 2 h.
4. Collect beads using the magnet and save a small amount of the supernatant for western blot analysis of flow-through fraction.
5. Wash beads three times with 10 mL TN1000 supplemented with protease inhibitor for 2–5 min, and once with 10 mL TN150 (*see* **Note 10**).
6. Transfer beads to 1.5 mL tubes using TN150, remove the buffer and resuspend in 600 μL TN150.
7. Elute His-tagged protein–RNA complexes from IgG beads by cleaving the TEV site with the addition of 5 μL (20–25 U) of homemade GST-TEV protease to the 600 μL of TN150 and incubation for 2 h at 18 °C with 1000 rpm shaking, 1 min on–20 s off in a Thermomixer. Here, an optional step of RNase H/oligo-dT cleavage can be performed to eliminate poly(A) tails from polyadenylated reads. The reaction is carried out in the same buffer and at the same time as TEV elution (*see* **Note 11**).
8. Collect beads and transfer eluates to fresh tubes (*see* **Note 12**).

3.4 Partial RNase Digestion

1. Treat eluate with 0.05–0.2 U (1 μL of a 1/200 to 1/50 dilution; *see* **Note 13**) RNace-IT RNase cocktail for 5 min at 37 °C to shorten the protein-bound RNA. Vortex briefly after addition of the enzyme, move to Thermomixer and start counting time. Stagger when handling multiple samples.

2. Quench the RNase reaction by quickly transferring 550 μL to tubes containing 400 mg guanidine hydrochloride and vortex to dissolve. Save leftover RNase reaction (~30–60 μL) for western blot analysis of TEV Eluate fraction.
3. Equilibrate the solution (typically ~830 μL) for nickel affinity purification by adjusting to 0.3 M NaCl and 10–15 mM imidazole pH 7.9 (typically, 45 μL of 3 M NaCl and 10 μL of 1 M imidazole are added) (*see* **Note 14**).

3.5 Affinity Purification of the Complex Using Nickel Beads and on-Beads Linker Ligation

3.5.1 Binding of Protein–RNA Complexes

1. Transfer 100 μL (for each sample) of Ni-NTA agarose nickel beads slurry into Snap Cap spin columns lodged in their collection tubes and pack beads by centrifuging for 1 min at 1000 × *g*.
2. Wash beads three times with 700 μL Wash buffer I supplemented with protease inhibitors (*see* **Note 15**).
3. Apply equilibrated samples to Ni-NTA agarose nickel beads in bottom-capped Snap Cap spin columns lodged in their collection tubes, close top of column, Parafilm around, and rotate at 4 °C overnight.
4. Uncap top and bottom of Snap Cap spin columns and discard flowthrough (*see* **Note 16**) by centrifuging for 1 min at 1000 × *g* max.
5. Wash beads twice with 700 μL Wash buffer I-1 M and then three times with 700 μL 1× PNK buffer, each time by inverting top-capped Snap Cap spin columns lodged in their collection tubes and centrifuging for 1 min at 1000 × *g*.

3.5.2 On-Beads 3′-End Dephosphorylation (Optional)

The partial RNase digestion leads to the formation of phosphorylated RNA 3′-ends. A dephosphorylation step is generally required for the subsequent ligation of the 3′ linker. However, when sequencing the nascent RNAs, the 3′-OH end in the catalytic center of the RNAPII enzyme is protected from the action of RNases. We have indeed observed identical results with and without phosphatase treatment for RNAPII CRAC.

1. Add 80 μL per sample of CIP mix (16 μL 5× PNK buffer, 5 μL (25 U) Quick CIP, 2 μL (80 U) RNaseOUT, 57 μL H_2O) on bottom-capped spin column, Parafilm top of column (*see* **Note 17**) and incubate at 37 °C for 30 min at 900 rpm in Thermomixer.
2. Wash beads once with 400 μL Wash buffer I-1 M and then three times with 400 μL 1× PNK buffer, each time by inverting top-capped Snap Cap spin columns lodged in their collection tubes and centrifuging for 1 min at 1000 × *g*.

3.5.3 On-Beads 3′ Linker Ligation

1. Add 72 μL per sample of T4 RNA ligase 2 mix (16 μL 5× PNK buffer, 26.7 μL 50% PEG 8000, 4 μL (800 U) T4 RNA ligase 2 truncated KQ, 2 μL (80 U) RNaseOUT, 23.3 μL H_2O) on bottom-capped spin column.
2. Add a different adaptor to each sample by adding 8 μL of 10 μM solution of 3′ linker (1 μM final concentration), Parafilm top of column and incubate at 25 °C for 5–6 h at 900 rpm, 30 s on–1.5 min off in Thermomixer.
3. Wash beads once with 400 μL Wash buffer I-1 M and then three times with 400 μL 1× PNK buffer, each time by inverting top-capped Snap Cap spin columns lodged in their collection tubes and centrifuging for 1 min at 1000 × *g*.

3.5.4 On-Beads 5′-End Phosphorylation

RNase treatment leaves 5′-OH ends on RNA fragments. These must be phosphorylated to allow the addition of the 5′ linker by T4 RNA ligase (*see* Subheading 3.5.5).

1. Add 80 μL per sample of T4 PNK mix (16 μL 5× PNK buffer, 2 μL (20 U) T4 PNK, 1 μL 100 μM ATP (1.25 μM final concentration), 2 μL (80 U) RNaseOUT, 59 μL H_2O) on bottom-capped spin column, Parafilm top of column and incubate at 37 °C for 45 min at 900 rpm in a Thermomixer (without collector tubes).
2. Wash beads twice with 400 μL Wash buffer I-1 M and then three times with 400 μL 1× PNK buffer, each time by inverting top-capped Snap Cap spin columns lodged in their collection tubes and centrifuging for 1 min at 1000 × *g*.

3.5.5 On-Beads 5′ Linker Ligation

1. Add 80 μL per sample of T4 RNA ligase 1 mix (16 μL 5× PNK buffer, 8 μL 10 mM ATP (1 mM final concentration), 4 μL (40 U) T4 RNA ligase 1, 1 μL 100 μM L5 miRCat 5′ linker (1.25 μM final concentration), 2 μL (80 U) RNaseOUT, 49 μL H_2O) on bottom-capped spin column, Parafilm top of column and incubate at 16 °C overnight at 700 rpm, 30 s on–1.5 min off in Thermomixer (without collector tubes).
2. Wash beads three times with 400 μL Wash buffer II, each time by inverting top-capped Snap Cap spin columns lodged in their collection tubes and centrifuging for 1 min at 1000 × *g*.

3.5.6 Purification of Protein–RNA Complexes

1. Move Snap Cap spin columns to clean 1.5 mL Eppendorf tubes and add 200 μL Elution buffer on beads. Incubate on bench 5 min with occasional gentle hand-shaking, then centrifuge for 1 min at 1000 × *g*.
2. Repeat this step by keeping the Snap Cap spin columns in the same 1.5 mL tube and adding further 200 μL Elution buffer on beads. Centrifuge for 1 min at 1000 × *g*. Save a small amount (about 10 μL) for silver or Cy5 stain analysis of Ni-NTA Eluate

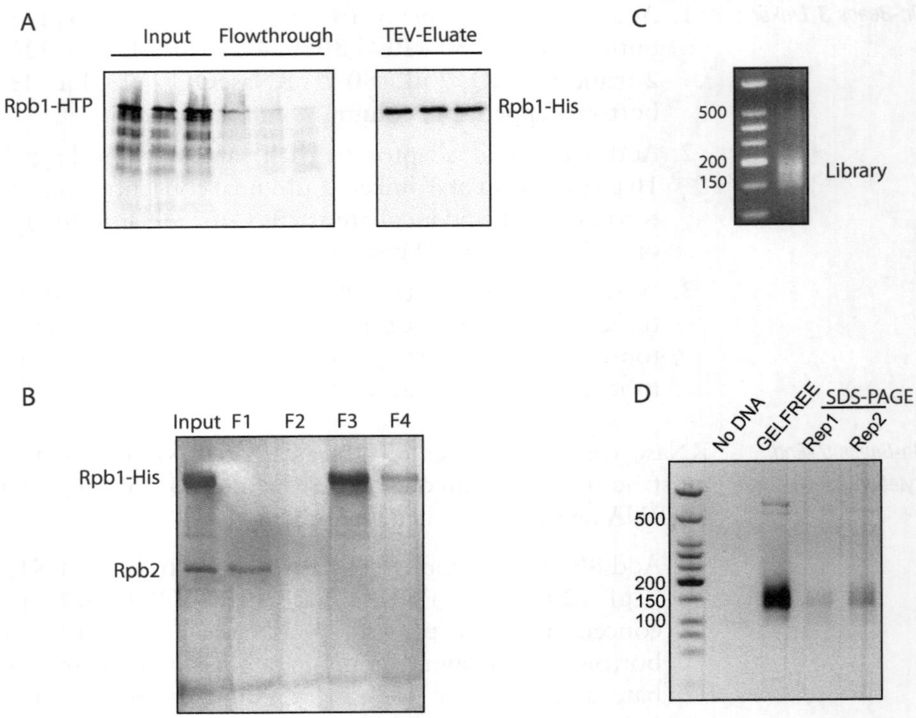

Fig. 3 Representative examples of typical controls to verify the efficiency of the protein–RNA complex purification and library construction. (**a**) Western blot to verify the efficiency of the first purification step on the IgG column. Shown are proteins from three independent preparations probed with an anti-His antibody. Input (total extract), flowthrough and TEV-eluate fractions are analyzed as indicated. (**b**) Four purification fractions from GELFREE 8100 (F1 to F4) are analyzed by PAGE and silver staining together with a fraction of the input (eluate from Subheading 3.5.6, see main text). Using this fractionation procedure Rpb1-6×His is typically found in fractions 3 and 4. The faster migrating band present in the input and in fraction F1 is most likely a small amount of Rpb2 that survived the first denaturing purification step. (**c**) Analysis of the cDNA library average size after PCR amplification using LA Taq polymerase (25 cycles, analyzed on 2% agarose gel stained with Ethidium bromide, *see* Subheading 3.9, **step 2**). The typical profile is a smear of fragments in the 150–200 bp size range. (**d**) Comparison of typical yields of cDNA library using either the original procedure (lanes SDS-PAGE, two independent replicates, Rep1 and Rep2) or the GELFREE 8100 system (lane GELFREE). Equivalent amount of TEV eluate were used for the two purification methods. Libraries were analyCTTGrGrCrArCrCrCrGrArGrArArUrUrCzed as in C

fraction (*see* **Note 18**) or western blot. Keep eluates on ice during protein analysis before next step.

3. The original CRAC procedure includes a radioactive labeling step of the RNA and a size selection step based on SDS-PAGE and transfer on nitrocellulose membrane [5]. We found this step to be time-consuming and the recovery of the RNA from the membrane to be a major source of material loss. This step was replaced by a different method to recover the protein–RNA complex that does not use radioactivity and allows combining the samples for a common elution (Figs. 1 and 3). We presently use an apparatus called GELFREE 8100 that allows

fractionation of complex protein samples. The eluted samples from step 2 are combined and concentrated using a Vivacon 500 ultrafiltration spin column. The concentrated, mixed sample is applied to several GELFREE cassette lanes (typically 2–3 lanes for 6 samples total) using manufacturer conditions for the approximate size of the protein–RNA complex to be purified. 5% gel cassettes have been used for Rpb1 isolation. The eluted fractions are analyzed by PAGE as described in **Note 18** for identifying the fraction containing the protein of interest. Nevertheless, this purification step has been skipped without a major increase in the level of background for some proteins (*see* **Note 19**).

3.6 RNA Isolation

1. Process fractions containing protein–RNA complexes. Add SDS to 0.5% final concentration, and treat with 100 μg of proteinase K at 50 °C for 1 to 2 h at 800 rpm, 1 min on–1 min off in Thermomixer.
2. Extract proteinase K-digested fractions twice with phenol–chloroform–isoamyl alcohol (25:24:1) and once with chloroform by vortexing (or mixing, *see* **Note 20**) and centrifuging for 5–10 min at 16,000 × *g* at 10 °C, each time being careful to avoid the interphase.
3. Precipitate RNA by adding NaOAc pH 5.5 to 0.3 M final, 1 μL GlycoBlue carrier, 2.5 volumes ethanol (or 1 volume Isopropanol) to the aqueous phase 2 h to overnight at −20 °C.
4. Pellet RNA by centrifuging for 30 min at 16,000 × *g* at 4 °C, carefully remove supernatant, wash once with 0.5 mL–1 mL 70% ethanol, centrifuge for 10 min at 16,000 × *g* at 4 °C, carefully remove supernatant, speed-vac dry pellet for 5 min without heating.

3.7 Synthesis of the cDNA Library

1. Resuspend the RNA pellet in 5 μL water at room temperature for 5 min at 900 rpm in Thermomixer and wash tube walls by pipetting (*see* **Note 21**).
2. Add 8 μL RT Mix I (2.6 μL 5× SSIV buffer, 0.65 μL 100 mM DTT, 1 μL 10 μM RT L3–2 primer, 3.75 μL H_2O) and incubate sample at 70 °C for 10 min in Thermomixer to anneal RT primer.
3. Set temperature to 50 °C and leave sample to continue annealing in Thermomixer until it reaches temperature.
4. Briefly spin sample, then add 7 μL RT Mix II (1.4 μL 5× SSIV buffer, 0.35 μL 100 mM DTT, 2 μL 5 mM dNTP mix, 1 μL RNaseOUT, 1 μL SuperScript IV Reverse Transcriptase, 1.25 μL H_2O) and incubate sample at 50 °C for 1 h in Thermomixer.
5. Inactivate reaction by incubating sample at 80 °C for 10 min.

6. Add 1 μL (5 U) RNase H and incubate cDNA sample at 37 °C for 30 min in Thermomixer in order to digest residual RNA.
7. Mix cDNA from the different GELFREE lanes (or different purification fractions) and dilute the sample two-fold (*see* **Note 22**).

3.8 Amplification of the cDNA Library

The cDNA library is amplified by PCR for sequencing. We calculate the minimum number of PCR cycles required to obtain sufficient material for Illumina NextSeq500 sequencing (currently 10 μL of a 0.2 ng/μL solution, roughly 2 nM, assuming an average size of 150 bp for the library fragments). This is done in two steps: (1) estimation of the absolute concentration of the starting cDNA library and (2) total amplification of the library and purification of the fragments. A step for analyzing the size distribution of the library is also included.

3.8.1 Estimate of the Minimal Number of Cycles for Amplifying the cDNA Library

1. Mix 2 μL of cDNA sample or of appropriate dilutions of the reference library with 8 μL qPCR mix (0.8 μL 25 mM $MgCl_2$, 0.134 μL 1a, 0.857 μL 1b, 1 μL 10 μM miRCat_PCR#2 primer (1 μM final concentration), 1 μL 10 μM P5–3′ primer (1 μM final concentration), 4.2 μL H_2O) in capillaries. Spin capped capillaries 1 min at 400 × *g*. Run qPCR reactions with Library Amplification program: 5 min 95 °C–(10 s 95 °C–40 s 60 °C–30 s 72 °C) 30×.
2. Estimate the number of cycles necessary for La Taq amplification of cDNA sample to obtain appropriately concentrated library for sequencing. Defining the minimal amount of PCR cycles is important for limiting the number of PCR duplicates and avoiding amplification biases. We first estimate the absolute concentration of the cDNA library before amplification by qPCR, comparing it to a standard of similar size and known concentration that can be amplified with the same primers. We find that using a capillary LightCycler qPCR apparatus gives the most reliable result, but other methods can be used, for instance Droplet Digital PCR.

We amplify 2 μL of the cDNA library in duplicate in a same reaction as a serial dilution of the standard. This gives a Ct that corresponds to the starting concentration of the library (CtStart). Using the standard for measuring PCR efficiency we also estimate the Ct number that is expected for a cDNA library that would be at the expected final concentration (CtFinal). CtFinal—CtStart is the number of PCR cycles that would be required to bring the starting cDNA library to the expected concentration. This is clarified in the example shown in Fig. 4.

From these data the PCR efficiency for this couple of oligos can be calculated (in this case $10^{-0.2658} = 1.84$) and, by extrapolation,

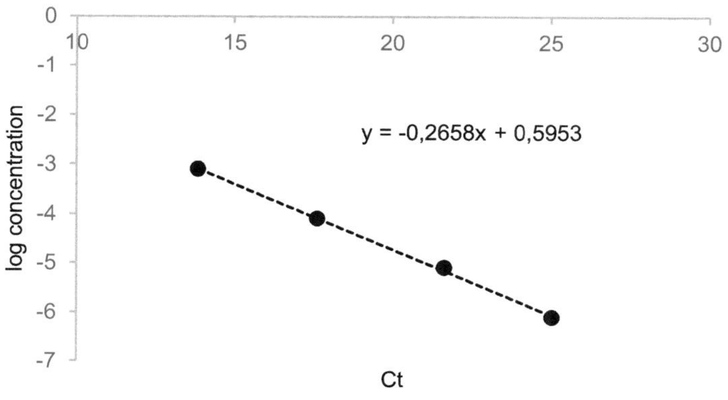

Fig. 4 Example of the estimation of the Ct number that is expected for a cDNA library that would be at the expected final concentration

Table 1
Correspondence between known standard concentration and qPCR cycles

Standard concentration (ng/μL)	Log (concentration)	Mean qPCR values
0.0008	-3.10	13.85
0.00008	-4.10	17.62
0.000008	-5.10	21.64
0.0000008	-6.10	25.04

the Ct expected for a solution at 0.2 ng/μL (CtFinal, here 4.87). Our test sample has given, in the same reaction, an average Ct (CtStart) of 15.20. 10.33 cycles (i.e., 15.20–4.87) would be required for the final PCR reaction (Table 1). However, because the latter produces large volumes that are concentrated in the final step, a concentration factor should be considered for determining the number of cycles required. The final PCR is performed using a tabletop apparatus in 25 μL reactions. In our hands, PCR amplification using La Taq polymerase (TaKaRa) in 25 μL reactions has very comparable efficiency as qPCR using capillary LightCycler, but this should be carefully verified and considered for the final calculation.

The total volume of the final PCR reaction depends on that of the starting cDNA library. For a 40 μL RT reaction (*see* Subheading 3.7) we use 8 × 25 μL reactions (5 μL of cDNA mix per PCR reaction). We generally set up a separate RT reaction per each protein–RNA eluate and using GELFREE we typically obtain between 1 and 3 independent size fractionation eluates (*see* **Note 19**). Therefore, the total volume of the starting cDNA library is between 40 and 120 μL, and the total volume of the final PCR reaction will be between 200 and 600 μL. In the final step (*see* **step**

6 of Subheading 3.9) samples are concentrated to a final volume of 40 μL. If we consider a volume of 400 μL for the final PCR reaction of our example, this corresponds to a ten-fold concentration. Because of this final concentration factor, less PCR cycles are required to obtain the required concentration for sequencing. This "gain" in PCR cycles depends on the PCR efficiency that in our hands is very similar to the one measured by qPCR for this couple of oligos. Therefore:

- Vol. PCR: 400 μL.
- Vol. elution: 40 μL.
- Enrichment: 10.
- "gain" in PCR cycles (PCR efficiency 1.84): 3.76.

In our example, the minimal number of PCR cycles required is therefore 10.33–3.76 = 6.57. A couple of additional amplification cycles are typically added to this number to account for possible errors in the estimated PCR efficiency and to obtain excess material for quality control of the library by the sequencing facility when needed.

The library concentration in the final PCR reaction is verified by qPCR (LightCycler) before purification. The Ct of this test reaction should be inferior to the expected Ct for a 0.2 ng/μL reaction (4.87 in our example) plus the cycles corresponding to the concentration factor (3.76), that is, 8.63 for our example.

3.8.2 Analysis of the Size Profile of cDNA Library

1. Mix 2 μL of cDNA sample or H_2O control with 23 μL PCR mix (2.5 μL 10× Mg^{2+} plus LA PCR buffer, 2 μL 2.5 mM dNTP mix (0.2 mM final concentration), 1 μL 10 μM miR-Cat_PCR#2 primer (0.4 μM final concentration), 1 μL 10 μM P5–3′ primer (0.4 μM final concentration), 0.5 μL (2.5 U) LA Taq, 16 μL H_2O). Amplify for 25 cycles with the following program: 2 min 95 °C - (30 s 95 °C - 45 s 58 °C - 1 min 72 °C) 25× - 5 min 72 °C.
2. Control LA Taq amplification on 2% agarose by running 5 μL of sample and control reactions with a low molecular weight DNA ladder. An optimal profile appears as a smeary population spreading above 150 bp and up to 200 bp without defined bands. A band of roughly 100 nt would indicate the presence of primer dimers or very small inserts.

3.9 Final PCR and Library Purification

1. Based on total cDNA sample volume, set up N separate 25 μL LA Taq PCR reactions and amplify for calculated number of cycles (typically 8–11). Mix 5 μL of cDNA sample (*see* **Note 23**) with 20 μL PCR mix (2.5 μL 10× Mg^{2+} plus LA PCR buffer, 2 μL 2.5 mM dNTP mix (0.2 mM final concentration), 1 μL 10 μM miRCat_PCR#2 primer (0.4 μM final

concentration), 1 μL 10 μM P5–3′ primer (0.4 μM final concentration), 0.5 μL (2.5 U) LA Taq, 13 μL H_2O). Amplify with the following program: 2 min 95 °C - (30 s 95 °C - 45 s 58 °C - 1 min 72 °C) $_{Nx}$ - 5 min 72 °C.

2. Pool LA Taq PCR reactions in one tube and keep on ice during the subsequent qPCR test.
3. Test CRAC library amplification by capillary qPCR. Mix 2 μL of CRAC library or H_2O control with 8 μL qPCR mix as in **step 1** of Subheading 3.8.1. This should give a number of cycles at least equal to calculated theoretical number (*see* Subheading 3.9, **step 1** and **Note 24**).
4. Digest primers in CRAC library sample with Exonuclease I. Add 2.5 μL (50 U) per 200 μL PCR reaction and incubate at 37 °C for at least 1 h in Thermomixer.
5. Inactivate reaction by incubating sample at 80 °C for 20 min.
6. Purify CRAC library with NucleoSpin Gel and PCR cleanup. Load sample several times on the same unique column if needed. Dry columns with two 1 min and 30 s centrifugations, and a 3 min incubation of the open column at 45 °C. Recover CRAC library with two successive 20 μL elutions with water or elution buffer, each with incubation at room temperature for 5 min, followed by centrifugation for 1–1.5 min.
7. Quantify CRAC library with a Qubit Fluorometer and Qubit dsDNA HS Assay Kit by reading 4 μL of sample. Typical final concentration ranges from 0.5 to 2 ng/μL.
8. Store CRAC library at-20 °C until ready for sequencing.

3.10 Data Analysis

Analysis of the data obtained is based on the pyCRACutility suite [11] modified as described [6, 7]. Briefly, CRAC samples are demultiplexed using the pyBarcodeFilter script and subsequently, the 5′ linker (Subheading 2.5.2) is clipped with Cutadapt [10, 12]. The resulting insert is quality trimmed from the 3′ end using Trimmomatic rolling mean clipping [13] (window size = 5, minimum quality = 25). At this stage we use the pyCRAC script pyFastqDuplicateRemover to collapse PCR duplicates using a 6 nucleotides random tag that is part of the 3′ linker (Subheading 2.5.1). During demultiplexing, pyBarcodeFilter retains this information in the header of each sequence. Reads containing more than 70% of Ts are removed. These are common source of artefactual signals due to the unspecific mapping of stretches of As (sequencing is from the 3′-end of cDNAs). The resulting sequences are reverse complemented with Fastx reverse complement (part of the fastx toolkit, http://hannonlab.cshl.edu/fastx_toolkit/) and mapped to the required genome with bowtie2 (using "-N 1" option) [14]. Unmapped reads are searched for the presence of poly (A) tails and polyadenylated reads are mapped independently.

When the *S. pombe* spike-in is present in the sample, reads that do not map on *S. cerevisiae* are mapped on the *S. pombe* genome after filtering for reads shorter than 40 nt to avoid cross mapping of *S. cerevisiae* reads on *S. pombe*. The amount of mapped spike-in reads is used for normalization purposes. The output (directional wig and bigwig files) provides the raw mapped read counts, the read counts normalized to the total number of mapped reads per each sample and the read counts normalized to the *S. pombe* spike-in mapped reads.

Because the whole read is considered for the read counts, the resolution of the signal depends on the average size of the reads. To increase signal resolution, we also generated tools to only map the read 3′-end (which defines the last position of RNAPII during transcription or backtracking after RNA cleavage) or its 5′-end. Cross-linking induces deletions or misincorporations during reverse transcription of the RNA, which can be used as proxy for the position of protein–RNA contact. This information can also be used to distinguish bona fide signals from background due to highly abundant RNA molecules (generally tRNAs and rRNAs). We have generated tools to extract reads containing mismatches or short deletions (the latter appear to be more frequent), as well as smoothing procedures. These scripts are available upon request.

4 Notes

1. Other appropriate drop-outs can be used to grow cells. However, tryptophan should be excluded from growth medium as it quenches UV light, thus reducing cross-linking efficiency [5].
2. *S. pombe* cells used to spike-in experiments can be grown in lower volumes of EMM + xAA (2–5 amino acids are added) minimal medium, added in the desired proportion to *S. cerevisiae* cultures if cross-linked together (usually 0.5%–1% of *S. cerevisiae* cells) or cross-linked separately for lysate mixing.
3. Optimal UV cross-linking time should be determined empirically: increased cross-linking will reduce the efficiency of the reverse transcription step and might induce DNA damage response. In our hands 50 s is best for RNAPII.
4. For unstable proteins, lysis buffer is supplemented with benzamidine at 2.5 mM final concentration.
5. Frequency can be as low as 10 Hz and be increased up to 20 Hz. Higher frequency will give better lysis but might be detrimental to some proteins. Preparation of extracts can be alternatively done with a Carver Press, in which case cell pellets resuspension volumes need to be adjusted to Carver Press cylinders' capacity and extracts need to be processed

immediately. Initial TN150 lysis buffer can be supplemented with 5%–10% glycerol (or even substituted with classic IP lysis buffers). Glycerol must be included if extracts are prepared using Carver Press.

6. Thawing of lysates can be done in Infors shaker set at 16 °C and 180 rpm or under lukewarm water (risky for unstable proteins).
7. Preclearing of lysates after chromatin solubilization can be omitted and lysates subjected only to high-speed centrifugation. However, preclearing helps avoiding excessive lipidic carryover.
8. A Bradford assay can be carried out on cleared lysates before the first immunoprecipitation to control for protein concentration. This can be tricky due to the residual presence of lipids and pigments.
9. 100 μL (for average abundance proteins) to 250 μL (for high abundance proteins such as Rpb1) M-280 tosylactivated Dynabeads coupled with rabbit IgG (5 mg–15 mg of beads per samples).
10. We have observed that cOmplete protease inhibitor inhibits to some extent TEV activity. For this reason, the final wash of Dynabeads as well as the elution step are performed in TN150 without protease inhibitor.
11. A fraction of the reads associated to some proteins are polyadenylated, which, if the poly(A) tail is too long, might prevent reading the sequence of the fragment (sequencing is done from the 3′-end). For this reason, an optional RNase H/oligo-dT degradation step can be performed in order to partially digest poly(A) tails. The same 600 μL of TN150 in which TEV digestion is performed are in this case supplemented with 0.4 μM poly-dT oligonucleotide, 3 nM $MgCl_2$, and 2 μL (10 U) RNase H.
12. If needed, dried beads can be stored at −20 °C for western blot analysis.
13. Conditions for partial RNase digestion should be determined empirically and both enzyme concentration and digestion time can vary. For RNAPII CRAC, 0.2 U of RNace-IT cocktail are used.
14. Imidazole must be present at roughly 10–20 mM concentration for specific binding to Ni-NTA agarose beads.
15. When working with Snap Cap spin columns, leaving the top cap open when centrifuging will prevent foam formation.
16. The flowthrough of Ni-NTA columns cannot be used for western blot analysis as it contains guanidine hydrochloride, which affect PAGE running.

17. For each enzymatic step, for which the reaction volume is low, be sure to cap only the bottom of the tube and close the top with Parafilm in order not to squeeze out the reaction mix.
18. Side-by-side analysis of TEV and Ni-NTA Eluates (or at least of Ni-NTA Eluates if analysis of the first immunoprecipitation has been carried out separately) should be performed to ensure successful elution and estimate eventual loss of material along the procedure, as well as general profile (degradation fragments, contaminants, etc.). This is particularly important if samples are subsequently mixed, as this is the last point in the procedure to check samples individually. Fractions obtained after GELFREE purification are also checked in the same way. Analysis can be done by Silver Staining, or by Quick Stain Cy5 (staining efficiency may vary according to proteins, protein solution composition, and relative amounts), and/or by western blot. Samples for Quick Stain Cy5 analysis are treated according to manufacturer's protocol and can be stored at −20 °C. SDS to 0.1% should be added to samples from GST-TEV and Ni-NTA elution steps before Cy5 staining.
19. Unfortunately, GELFREE is being discontinued by Abcam. We are studying alternative solutions based on similar principles (e.g., elution from PAGE or other fractionation methods).
20. Phase Lock Gel Heavy tubes can be used for second phenol–chloroform–isoamyl alcohol (25:24:1) and subsequent chloroform extractions (in the same tube if volumes are compatible). In this case do not vortex samples (see also manufacturer's instructions).
21. Resuspension of RNA pellets in H_2O rather than in RT Mix I prevents foaming up of the sample due to detergents in 5× SSIV buffer. Alternatively, this can be done by directly adding 13 μL of RT Mix I (in this case prepared with 8.75 μL H_2O).
22. Sample two-fold dilution is expected to reduce the impact of the RT buffer on the subsequent amplification steps and improve accuracy due to the pipetting of larger volumes.
23. Use of 5 μL of cDNA sample in 25 μL represents a scale up of the qPCR reaction for determining the concentration of the library. This is important to approach similar conditions of amplification. We use 25 μL reaction volumes to minimize thermal inertia and independent reactions to minimize PCR artefacts that might arise from the stochastic early amplification of fragments.
24. If qPCR test of CRAC library results in significantly different number of cycles than expected, additional LA Taq PCR amplification cycles can be performed, provided the sample has not been treated with Exonuclease I.

References

1. Churchman LS, Weissman JS (2011) Nascent transcript sequencing visualizes transcription at nucleotide resolution. Nature 469:368–373
2. Burger K, Mühl B, Kellner M et al (2013) 4-thiouridine inhibits rRNA synthesis and causes a nucleolar stress response. RNA Biol 10:1623–1630
3. Baejen C, Andreani J, Torkler P et al (2017) Genome-wide analysis of RNA polymerase II termination at protein-coding genes. Mol Cell 66:38–49.e6
4. Schwalb B, Michel M, Zacher B et al (2016) TT-seq maps the human transient transcriptome. Science 352:1225–1228
5. Bohnsack MT, Tollervey D, Granneman S (2012) Identification of RNA helicase target sites by UV cross-linking and analysis of cDNA. Meth Enzymol 511:275–288
6. Challal D, Barucco M, Kubik S et al (2018) General regulatory factors control the fidelity of transcription by restricting non-coding and ectopic initiation. Mol Cell 72:955–969.e7
7. Candelli T, Challal D, Briand J-B et al (2018) High-resolution transcription maps reveal the widespread impact of roadblock termination in yeast. EMBO J 37:e97490
8. Luciano P, Jeon J, El-Kaoutari A et al (2017) Binding to RNA regulates set1 function. Cell Discov 3:17040
9. Villa T, Barucco M, Martin-Niclos M-J et al (2020) Degradation of non-coding RNAs promotes recycling of termination factors at sites of transcription. Cell Rep 32:107942
10. Han Z, Jasnovidova O, Haidara N et al (2020) Termination of non-coding transcription in yeast relies on both an RNA pol II CTD interaction domain and a CTD-mimicking region in Sen1. EMBO J 39:e101548
11. Webb S, Hector RD, Kudla G et al (2014) PAR-CLIP data indicate that Nrd1-Nab3-dependent transcription termination regulates expression of hundreds of protein coding genes in yeast. Genome Biol 15:R8
12. Martin M (2011) Cutadapt removes adapter sequences from high-throughput sequencing reads. EMBnet J 17(1). https://doi.org/10.14806/ej.17.1.200
13. Bolger AM, Lohse M, Usadel B (2014) Trimmomatic: a flexible trimmer for illumina sequence data. Bioinformatics 30:2114–2120
14. Langmead B, Salzberg SL (2012) Fast gapped-read alignment with bowtie 2. Nat Methods 9:357–359
15. Schaughency P, Merran J, Corden JL (2014) Genome-wide mapping of yeast RNA polymerase II termination. PLoS Genet 10:e1004632

Chapter 4

High-Resolution Deep Sequencing of Nascent Transcription in Yeast with BioGRO-seq

Antonio Jordán-Pla and José E. Pérez-Ortín

Abstract

RNA biogenesis in eukaryotic cells is a tightly regulated multilayered process in which a diverse set of players act in an orchestrated manner via complex molecular interactions to secure the initial flow of gene expression. Transcription from DNA to RNA is the essential first step in RNA biogenesis, and consists of three main phases: initiation, elongation, and termination. In each phase, transcription factors act on RNA polymerases to modulate their passage along the DNA template in a very precise manner, governed by molecular mechanisms, some of which are not yet fully understood. Genome-scale run-on-based methodologies have been developed with the aim of mapping the position of transcriptionally engaged RNA polymerases. Among them, the BioGRO methodology has been instrumental in advancing our understanding of the transcriptional dynamics in yeast. Here we take the previously known BioGRO method further by coupling it with deep sequencing. BioGRO-seq maps elongating RNA polymerases along the genome with strand specificity and single-nucleotide resolution. BioGRO-seq profiling provides insights into the biogenesis and regulation of not just the canonical protein-coding transcriptome, but also into the often more challenging to study noncoding and unstable transcriptome.

Key words Nascent transcription, RNA polymerase II, RNA polymerase III, Yeast, *Saccharomyces cerevisiae*, Deep sequencing, RNA-seq, Nascent RNA

1 Introduction

Eukaryotic gene expression starts with DNA to RNA transcription and its regulation remains a central challenge in molecular biology [1]. Most RNA synthesis in eukaryotes takes place in the nucleus at the chromatin level conducted by three RNA polymerases (RNA pols). It is considered one of the most tightly regulated existing biological processes, with hundreds of proteins involved [2]. Transcription comprises three main steps, namely initiation, elongation, and termination. Newly synthesized RNA molecules are subjected to a series of maturation processes, both during transcription and posttranscriptionally, and can either remain in the nucleus or be exported to the cytoplasm to form part of the mature messenger

Frédéric Devaux (ed.), *Yeast Functional Genomics: Methods and Protocols*,
Methods in Molecular Biology, vol. 2477, https://doi.org/10.1007/978-1-0716-2257-5_4,

RNA (mRNA) pool that will be used for protein synthesis. Experimental approaches that dissect the different phases of gene expression provide valuable insights into their regulatory mechanisms, regardless of the influence of other processing and degradation steps. Traditional transcriptomic experiments, aimed at studying gene expression by quantifying the total cellular RNA products of a gene's transcription, do not address the act of chromatin transcription per se. To investigate the mechanisms of RNA pol transcription more directly, many in vivo and in vitro procedures have been developed over the years [3, 4]. Chromatin immunoprecipitation (ChIP) for instance detects all RNA pol, regardless of their transcriptional status (active, paused or stalled). In addition, by using specific antibodies [5], it has the potential of distinguishing between the three different RNA pols, or between phosphorylated forms of the same RNA pol. By studying the fraction of RNAs being synthesized in chromatin, known as nascent RNA (nRNA), several high-resolution techniques have been established. Among them, the ones that isolate RNA pol-associated nRNA detect elongating RNA pol forms [4, 6, 7] but are unable to distinguish between active RNA pol II molecules and those backtracked but still retaining an RNA molecule. However, run-on-based approaches [4, 8–10] only detect actively elongating RNA pol.

Run-on based methodologies are one of the most useful experimental approaches to study transcription due to their ability to precisely map the position of elongating RNA pol in chromatin, to estimate nRNA transcription rates, and to capture stable (tRNA, snRNA) and unstable (CUTs, SUTs, antisense) noncoding transcripts [10]. Introduced in the 1970s [11], Nuclear Run-On (NRO), used cells in culture as starting material to isolate nuclei. Isolated nuclei lose their nuclear rNTPs, which stalls transcription. Transcription is then restarted for a few minutes by adding back a mixture of natural rNTPs, spiked with a certain amount of a radiolabeled analogue. The detergent sarkosyl is added to prevent new initiations at the promoter level, and to disrupt the chromatin so that RNA pols are free to elongate through nucleosomal barriers [12]. In those in vitro conditions, transcriptionally active RNA pol extend the nascent transcripts, thus labeling the products and allowing for subsequent autoradiographic detection. Many variants of the original NRO have been developed over the years to study nascent transcription in a wide variety of model organisms [10]. Most methods use next generation sequencing of nRNAs that have been labeled with different nucleotide analogues, such as BrUTP [13] or Biotin-UTP [14], and subsequently purified to remove nonlabeled RNAs. The BioGRO technique [15, 16] was originally designed and optimized for tiling microarray technologies, and had two main advantages over the previous NRO methods: (1) RNase A treatment of permeabilized cells prior to the in vitro transcription step reduced preexisting cellular mRNAs contamination. Moreover, by trimming elongating transcripts and leaving only ~30 base long RNA pol footprints, it also increased the mapping resolution of the active RNA pol (Fig. 1) and (2) cDNA and PCR steps were omitted to avoid potential technical biases and amplification of carryover mature mRNA contamination. BioGRO has been since modified to be coupled to RNA deep sequencing compatible with standard short-read sequencers. Here, we describe in detail the evolved protocol, BioGRO-seq, in which an affinity purification step has been introduced to circumvent the issue of nonlabeled RNA contamination.

2 Materials

Precautions should be taken to minimize DNase and RNase contamination throughout all the protocol steps (*see* **Notes 1** and **2**).

2.1 Equipment

1. Low-speed tabletop centrifuge.
2. Refrigerated microcentrifuge.
3. Temperature-controlled orbital shaker.
4. DNA LoBind 1.5 mL Tubes.
5. Eppendorf Thermomixer® Comfort Heating and Cooling Shaker.
6. Nanodrop ND1000 Spectrophotometer.
7. Thermoblock heater.
8. Savant SPD111V SpeedVac Concentrator.
9. (Optional) GS Gene Linker UV Chamber.
10. Qubit Fluorometer.
11. Bioanalyzer Electrophoresis System (Agilent).
12. Magnet for Dynabeads.
13. Precellys homogenizer (Bertin Instruments).
14. Coulter cell counter (e.g., Beckman Coulter).

2.2 Nascent RNA Biotinylation by Run-on

1. YPD medium: 1% w/v, yeast extract, 2% w/v, peptone, 2% glucose. Store at room temperature (*see* **Note 3**).
2. 0.5% w/v, L-lauroylsarcosine (sarkosyl) in nuclease-free water. Store at room temperature.
3. 2.5× Transcription buffer: 50 mM Tris–HCl, pH 7.7, 50 mM KCl, 80 mM $MgCl_2$. Store at room temperature.
4. ACG mix (ATP, CTP, GTP, 10 mM each). Store frozen.
5. 0.1 M DTT. Store frozen.
6. Biotin-11-UTP (10 mM). Store frozen.

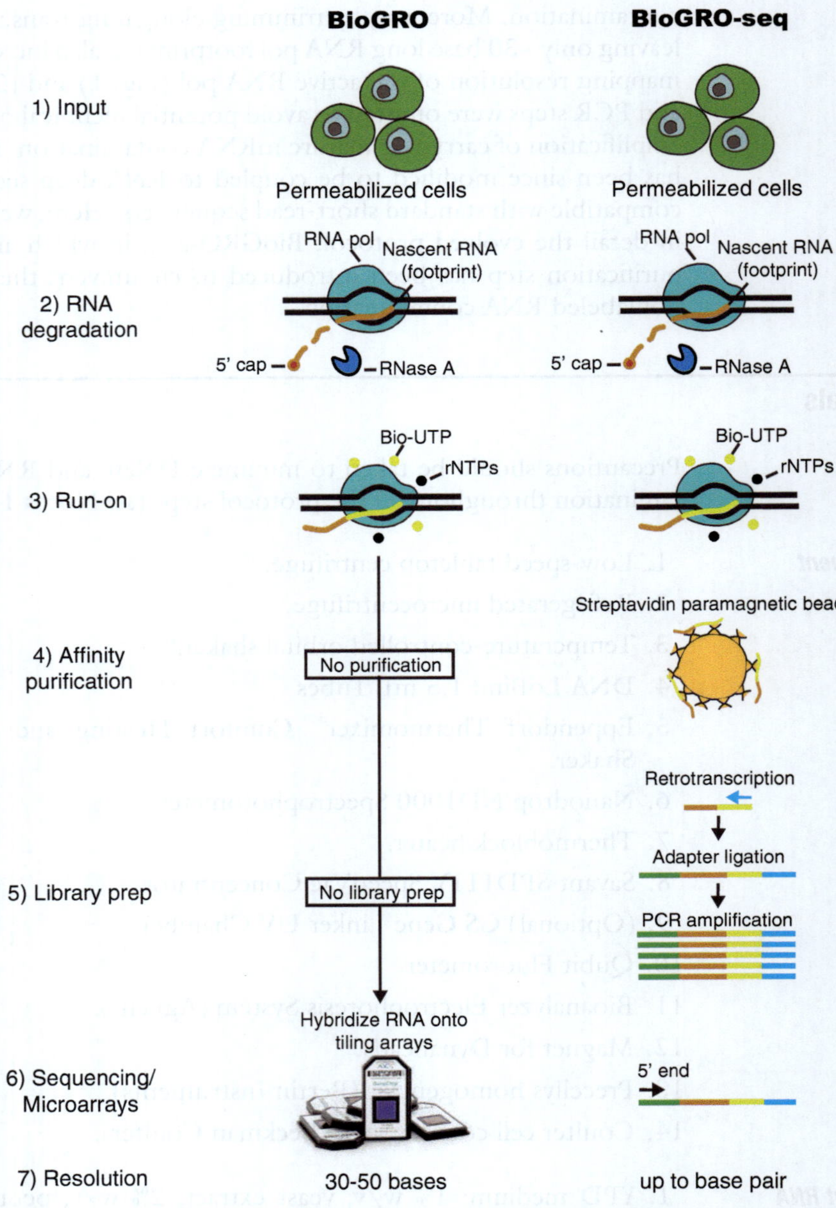

Fig. 1 Comparative workflow of BioGRO and BioGRO-seq. (1) Both methods start by treating exponentially growing yeast cells with sarkosyl. The detergent permeabilises cellular membranes, which depletes NTPs from the nucleus resulting in transcription stalling. (2) RNase A digestion of the 5′ end of nascent RNAs degrades most cellular transcripts and trims the 5′ portions of RNA pol-bound molecules. The only RNAs that survive degradation are short footprints (~ 30 bases) protected by the RNA pol enzyme. (3) A combination of modified/unmodified nucleotides is used to extend the footprints and generate labeled run-on products. (4) Streptavidin-coated magnetic beads are used for affinity purification of Bio-11-UTP labeled RNAs only in BioGRO-seq, as a way of enriching the sample in run-on products. (5) BioGRO-seq includes retrotranscription and PCR amplification as part of the library preparation steps, whereas BioGRO used the run-on products directly for tiling array hybridization. (6) cDNAs are sequenced starting from their 5′ ends. (7) By collapsing the sequences to the first 5′ nucleotide, one can infer the positions of the RNA pol over the DNA template at single nucleotide resolution

7. RNase OUT. Store at −20 °C.
8. RNase A. Store at 4 °C.
9. 11. 1 M Tris–HCl (pH 7.4). Store at room temperature.
10. 13. GlycoBlue, for molecular biology. Store frozen.
11. DNase I, RNase-free. Store frozen.
12. Liquid nitrogen.
13. Ethanol 96%.
14. Nuclease-free water, molecular biology grade.
15. High salt buffer: 2 M NaCl, 50 mM Tris–HCl pH 7, 0.5% Triton X-100.
16. Medium salt buffer: 300 mM NaCl, 10 mM Tris–HCl pH 7.4, 0.1% Triton X-100.
17. Low salt buffer: 5 mM Tris–HCl pH 7.4, 0.1% Triton X-100.
18. Dynabeads M-280 Streptavidin.
19. Dynabeads Binding and washing 2×: 10 mM Tris–HCl pH 7.5, 1 mM EDTA, 2 M NaCl.
20. Dynabeads solution A: 0.1 M NaOH, 0.05 M NaCl.
21. Dynabeads solution B: 0.1 M NaCl.
22. LETS buffer: 0.1 M LiCl, 10 mM EDTA, 0.2% SDS, 10 mM Tris–HCl pH 7.4.
23. Acid PCI: phenol–chloroform–isoamyl alcohol (125:24:1, v/v), pH 4.5.
24. CIAA: chloroform–isoamyl alcohol (24:1, v/v).
25. TRIzol Reagent.
26. T4 Polynucleotide Kinase PNK.
27. T4 DNA ligase buffer 10×.

3 Methods

This protocol describes the BioGRO-seq methodology applied to *S. cerevisiae* cultures, although the procedure is easily applicable to other organisms in which the starting material is similar (e.g., *Drosophila melanogaster* S2 cells grown in flasks or plates). The major steps of this method are outlined in Fig. 1, and general expected results are shown in Figs. 2 and 3.

3.1 Preparation of DNase-Free RNase A

1. Dissolve RNase A at a concentration of 10 mg/mL in 0.01 M sodium acetate (pH 5.2).
2. Heat to 100 °C in a thermoblock for 15 min.
3. Allow it to cool down slowly to room temperature.

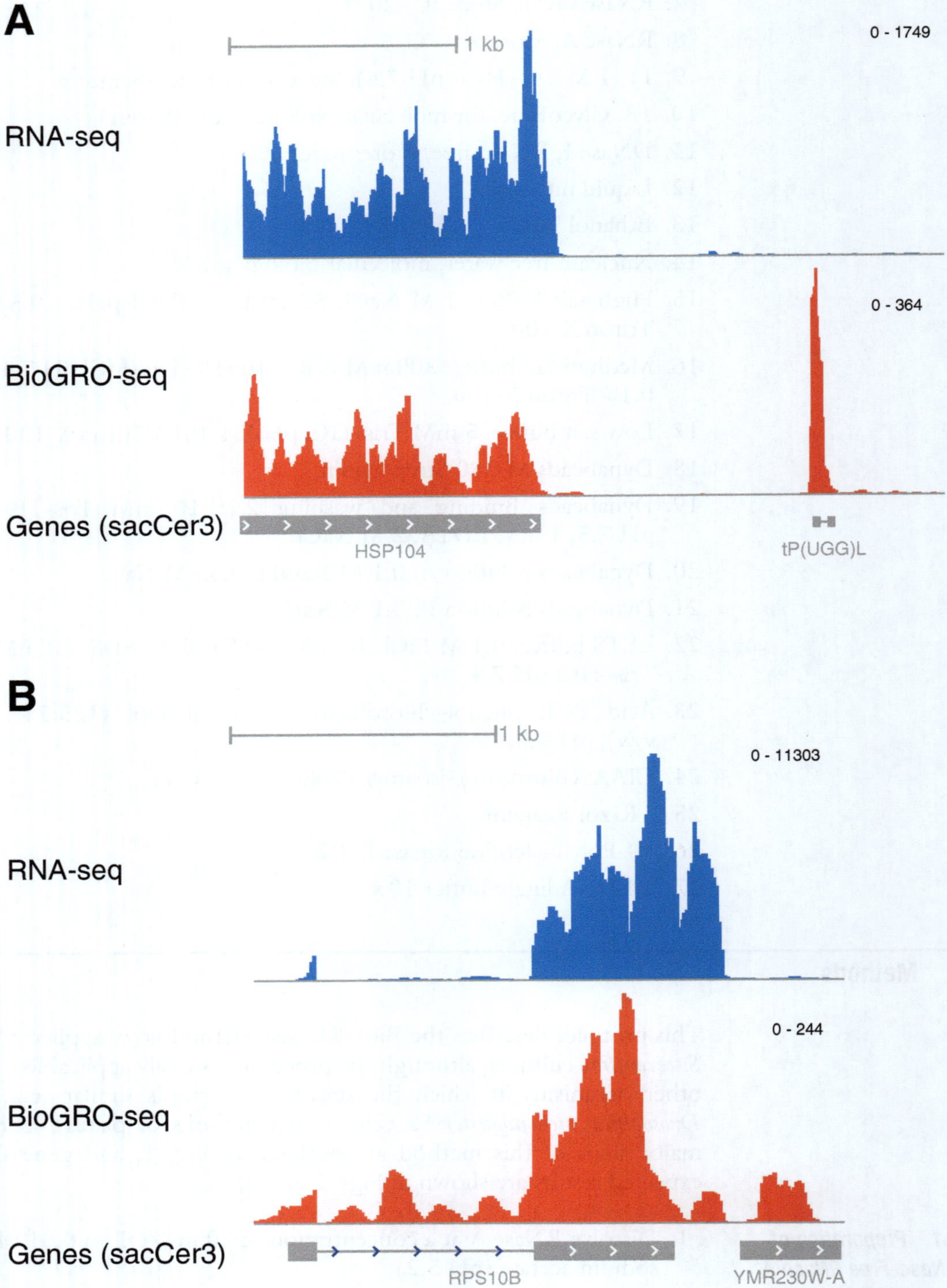

Fig. 2 Example of BioGRO signal along individual genome regions. (**a**) In contrast with traditional RNA-seq experiments, BioGRO-seq enables the study of tRNAs. (**b**) BioGRO-seq samples are enriched in actively transcribing unprocessed nRNAs whereas RNA-seq captures mostly fully processed cytoplasmic mRNAs, as exemplified by the presence of reads in the intronic region of the *RPS10B* gene

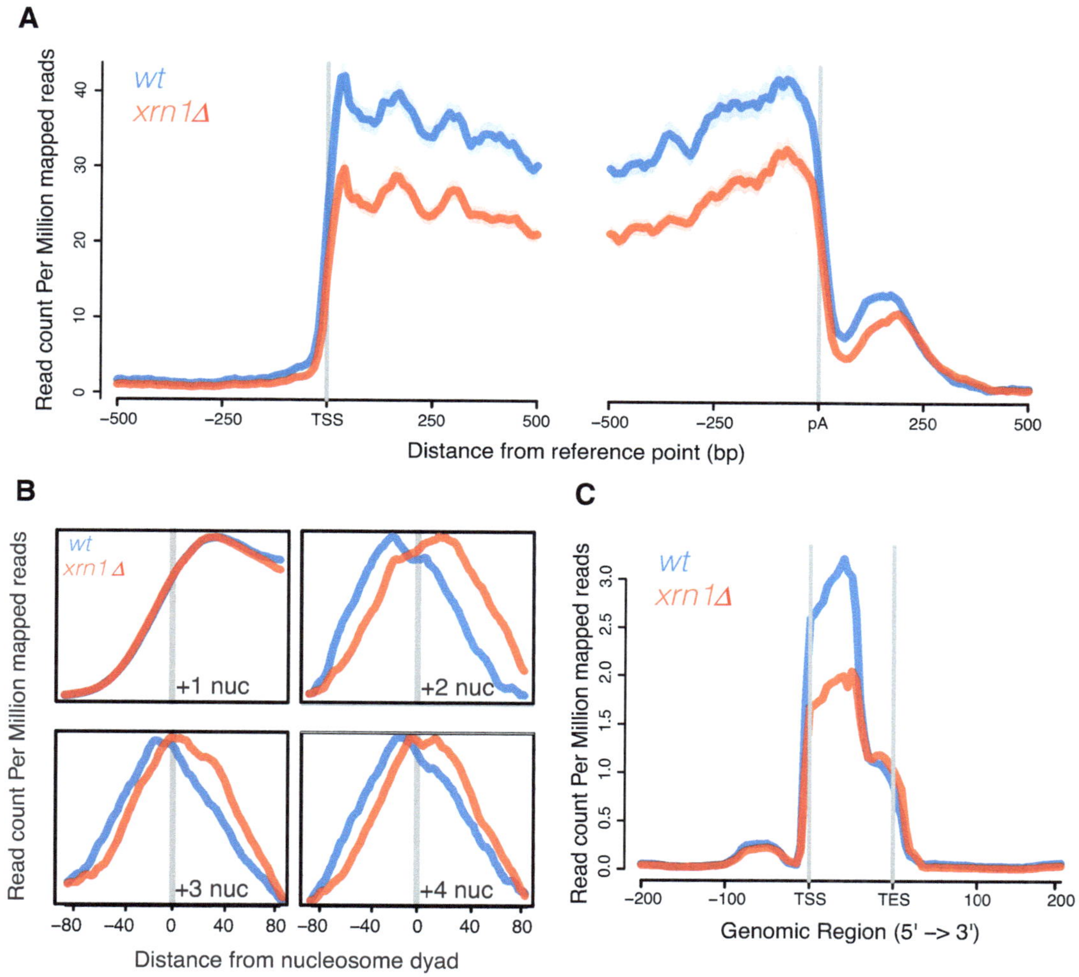

Fig. 3 Average metagene profiles of active RNA pol over specific genomic regions. (**a**) Average metagene profile of the normalized BioGRO-seq read coverage around the transcription start site (TSS) and polyadenylation site (pA) of all annotated protein coding yeast genes (~ 5100 genes). The profile of a wild type strain is shown in blue, and the one of a yeast strain lacking the major cytoplasmic exonuclease *XRN1* (*xrn1Δ*) is shown in red. The average density of active RNA pol II in genes is lower in *xrn1Δ*, which is in line with the reported decrease in total RNA pol occupancy found by other single-gene and genomic approaches [19]. (**b**) The density of active RNA pol around nucleosome dyads is altered in a *xrn1Δ* mutant, which suggest transcription elongation defects. (**c**) BioGRO-seq is able to map RNA pol III at high resolution over tRNA genes. As with protein-coding genes, transcription of tRNAs is also affected in the absence of *XRN1*

4. Adjust the pH by adding 0.1 volume of 1 M Tris–HCl (pH 7.4).
5. Dispense in aliquots and store at −20 °C.

3.2 Preparing Dynabeads for RNA Binding

1. Resuspend the paramagnetic beads by gentle agitation in a rotator or orbital shaker at 4 °C for 15 min.
2. Transfer 100 μL of beads per sample to an Eppendorf tube and place them in the magnet for 1 min.
3. Extract and discard the supernatant by pipetting without disrupting the bead aggregates.
4. Wash twice with 500 μL of Dynabeads solution A and one time with Dynabeads solution B.
5. Resuspend in 500 μL of Binding and washing solution 2×.

3.3 BioGRO-seq Method

1. Allow cells to grow in YPD to the desired OD_{600} (typically 0.5–0.6).
2. For each sample, an aliquot of 250 mL is needed (corresponding to approximately 30×10^8 cells, *see* **Note 4**).
3. Collect cells by centrifugation in five 50 mL falcon tubes at 4400 × *g* for 2.5 min. From now onward, all tubes are processed the same, in parallel, and the extracted RNA is pooled together at the end.
4. Decant the supernatant and submerge the pellet-containing tube in liquid nitrogen for flash freezing (*see* **Note 5**).
5. Transfer the frozen pellet to −20 °C. Keep the tube in the freezer for at least 3 h. This is a safe stopping point, as cells can be stored for longer periods (*see* **Note 6**).
6. Slowly thaw cells on ice and add 10 mL of a 0.5% sarkosyl solution. Mix by inversion.
7. Pellet cells by centrifugation as in **step 3**, and discard the supernatant.
8. Resuspend cells in 3.2 mL of 0.5% sarkosyl and add 32 μL of 10 mg/mL DNase-free RNase A. Mix by gently pipetting up and down several times.
9. Incubate cells with RNase A for 10 min at 30 °C in an orbital shaker to avoid deposition of cells at the bottom of the tube (*see* **Note 7**).
10. Bring the volume up to 45 mL with sarkosyl 0.5%. Mix vigorously by inversion to wash the cells and eliminate RNase A.
11. Recover cells by centrifugation as in **step 3**. Discard the supernatant.
12. Wash cells with 45 mL of 0.5% sarkosyl twice more.
13. Resuspend the pellet in 1 mL of 0.5% sarkosyl and transfer cells to an Eppendorf tube (1.5 mL).
14. Recover cells by centrifugation at 5400 × *g* for 1 min in a microcentrifuge. Carefully remove the supernatant by

pipetting and centrifuge again, if necessary, to eliminate any remaining sarkosyl.

15. Resuspend cells in 113.5 μL of nuclease-free water.
16. Add 5 μL of RNase inhibitor (RNase OUT) to protect the integrity of the labeled nascent RNAs from any residual RNases that might be present after the washes. Mix by pipetting up and down several times. Keep cells on ice until needed.
17. Prepare the transcription reaction mix: 120 μL of 2.5× transcription buffer, 6 μL 0.1 M DTT, 16 μL of AGC mix, and 20.25 μL of 10 mM Biotin-11-UTP (*see* **Notes 8** and **9**).
18. Prewarm both cells and transcription mix at 30 °C for 5 min in the thermomix.
19. Add the transcription mix (162.25 μL) to the cell suspension and mix by gentle pipetting.
20. Perform the run-on reaction by incubating the mix for 5 min at 30 °C in the thermomix, with 550 rpm agitation (*see* **Note 10**).
21. Stop the reaction by adding 1 mL of ice-cold nuclease-free water to the tube. Snap cool and maintain on ice for 5 min.
22. Harvest cells by centrifugation for 1 min at 11,000 × *g* at room temperature. Remove the supernatant containing unincorporated nucleotides.

3.4 RNA Extraction

1. Resuspend each cell pellet in 500 μL LETS buffer and transfer the mix to a new tube containing 500 μL acid PCI + 500 μL glass beads.
2. Disrupt cell walls and membranes to release RNAs with Precellys 24 with 2 cycles of 6500 rpm for 20 s, with a 30 s rest in between.
3. Centrifuge at 12,000 × *g* for 5 min at 4 °C. Transfer the aqueous phase to a new tube.
4. Add an equal volume of CIAA, mix and centrifuge as in **step 3**. Transfer the aqueous phase to a new tube.
5. Precipitate by adding 0.1 vol LiCl 5 M + 2.5 vol of cold 96% EtOH. Keep overnight at −20 °C.
6. Centrifuge 13,400 × *g* for 15 min at 4 °C. Discard supernatant.
7. Wash with 500 μL cold 70% EtOH. Centrifuge 13,400 × *g* for 5 min at 4 °C. Discard supernatant.
8. Dry in the thermomix for 10 min at 45 °C, with the lids open but covered with tin foil. Use SpeedVac for 3 min as an alternative.
9. Dissolve the pellet in each tube in 100 μL RNase-free water by incubating in the thermomix for 10 min at 45 °C, with 500 rpm agitation.

10. Quantify with Nanodrop.
11. Merge the five tubes of each sample into a single Eppendorf LoBind tube (*see* **Note 11**), to a final volume 500 of μL.

3.5 Streptavidin Paramagnetic Bead Pull-Down of Biotinylated RNA

1. Take 100 μL of RNA-ready Dynabeads resuspended in 500 μL of Binding and washing buffer 2× and add 500 μL of biotinylated RNAs from the previous step. Mix them carefully by pipetting and incubate for 15 min on rotation at room temperature.
2. Place tubes in the magnet, wait for 1 min until beads are attached to the walls and remove supernatant with a pipette without touching the beads. Follow the same procedure every time you want to change a solution from the beads.
3. Wash by resuspending the beads in 1 mL High salt buffer with gentle pipetting and remove the supernatant. Wash again with 1 mL Medium salt buffer and one final time with 1 mL Low salt buffer.
4. Resuspend beads in 360 μL RNase-free water and add 300 μL of a mixture of TRIzol reagent +60 μL CIAA. Shake by inversion for 15 s and incubate at room temperature for 3 min.
5. Centrifuge 10,000 × *g* at 4 °C for 10 min. Transfer the aqueous phase to a new LoBind tube.
6. Add a volume of CIAA, mix by inversion, centrifuge 5 more min (as in **step 5**) and transfer the supernatant to a clean LoBind tube.
7. Precipitate RNA by adding 1 μL of GlycoBlue carrier (*see* **Note 12**), 0.1 volume of LiCl 5 M and 2.5 volume of cold 96% EtOH. Keep at −20 °C overnight.
8. Centrifuge 13,400 × *g* for 20 min at 4 °C. Remove supernatant. Wash with cold 70% EtOH.
9. Dry in the thermomix for 10 min at 45 °C, with the lids open and covered with tin foil. Use SpeedVac for 3 min as an alternative.
10. Dissolve the pellet in each tube in 20 μL RNase-free water by incubating in the thermomix for 10 min at 45 °C, with 500 rpm agitation.
11. Use 1 μL for Qubit fluorimetric quantitation (*see* **Notes 13** and **14**).

3.6 End Repair and Library Prep for Deep Sequencing

RNase A trimming products contain 5′-OH ends, which are not compatible with regular index ligation reactions typical of the RNA-seq library prep kits. To phosphorylate 5′ ends we do the following.

1. Mix 18 mL of Qubit quantified BioGRO RNAs from the previous step with 1 μL of T4 PNK (10 U/uL), 1 μL of T4 DNA ligase buffer (10×, contains the required ATP) and 0.5 μL of RNase inhibitor.
2. Incubate at 37 °C for 30 min, then incubate at 65 °C for 10 min to inactivate the T4 PNK.
3. Proceed to library prep with the kit NEBNext® Small RNA Library Prep Set for Illumina® (*see* **Notes 15** and **16**).

3.7 Total RNA Extraction for Conventional Single-Gene or Transcriptomic Analysis (Optional, See Note 17)

The sample collection for total RNA extraction can be performed in parallel with the sample collection for the BioGRO.

1. Allow cells to grow to the desired OD_{600} (typically 0.5–0.6).
2. For each sample, an aliquot of 50 mL cells is needed (corresponding to 6×10^8 cells).
3. Collect cells by centrifugation in a falcon tube at 4400 × *g* for 2.5 min.
4. Discard the supernatant and submerge the pellet-containing tube in liquid nitrogen for flash freezing.
5. Transfer the frozen pellet to −20 °C. Keep the tube in the freezer until needed.
6. Slowly thaw cells on ice and proceed with the same RNA extraction method described in Subheading 3.4.
7. Due to the higher amount of RNA extracted, compared to the BioGRO-seq method, dissolve RNA in 200 μL of nuclease-free water.

4 Notes

1. All homemade buffers and most solutions are routinely autoclaved at 2 kg/cm^2 for 1 h to inactivate RNases.
2. General precautions of working with RNA should be taken. Always use RNase-free water, and prepare all reagents with it. Whenever possible, work with nuclease-free disposable plasticware, such as filter tips. Do not touch anything that is going to be in contact with RNA without gloves, and keep your workbench always clean. You will also minimize potential RNA degradation if you store and work with nuclease-free materials and reagents in a separate area designated for RNA work. Add RNase inhibitors to samples which might contain endogenous RNases.
3. Although YPD is the most common culture medium, other complete or synthetic media may also be used.

4. The cell number in each BioGRO-seq experiment should be very similar between samples to avoid differences in labeling during the run-on. Spectrophotometric measurements (optical density, OD_{600}) can only be used as an estimation of cell density when cell volumes of each yeast strain have been previously taken into account. Cell numbers and volumes can be determined with a Coulter counter.
5. We have observed that the slow freezing of sarkosyl-treated cells causes some RNA degradation. It is recommended to freeze cell pellets immediately in liquid nitrogen or dry ice before storing them at −20 °C.
6. Cell pellets can be stored for hours or days at −20 °C without any negative impact on the run-on reaction performance. We recommend storing at −80 °C for long-term periods (up to several months).
7. The RNA extraction yield of samples treated with RNase A prior to the run-on reaction varies widely. We calculated that using 0.1 μg/μL RNase A was the appropriate concentration to obtain the correct final RNA yield. However, even when using 0.1 μg/μL of RNase A, sometimes the yield is either lower or higher than expected. This variability might be due to two main things: incubation temperature, and RNase A activity. To try to optimize the first aspect we always do the incubation at 30 °C in agitation instead of room temperature. For the second aspect, we recommend testing different RNase A concentrations (for example 1, 5, and 10 μg/μL) and then stick with the one that yields the expected final RNA amount (around 30 μg).
8. For multiple run-on labeling reactions, prepare a 1/tenth excess of the transcription run-on mix.
9. Due to its photoreactive potential, it is better to avoid extended direct light exposure to Biotin-11-UTP. This can be accomplished by keeping it in opaque Eppendorf tubes or by wrapping the tubes with aluminum foil.
10. We have checked that longer incubation times do not result in more transcript labeling.
11. In order to minimize loss of RNA material in the final steps of the BioGRO-seq protocol, we use low nucleic acid retention plastic tubes (such as DNA LoBind Tubes, Eppendorf).
12. Glycogen is an inert polysaccharide that we used as carrier for the precipitation of RNA. Adding a few μg of glycogen significantly increased the recovery of the RNA in alcoholic precipitations. During centrifugation, it forms a visible pellet, which greatly facilitated handling of the precipitated RNA. The existence of colored glycogen variants, such as the GlycoBlue (Ambion) makes it even easier to spot the pellet.

13. To check/optimize the labeling efficiency and RNA trimming parameters, you can run an electrophoresis gel (native, 2% agarose) and then transfer the RNA to a nylon membrane using a standard Northern protocol. Once transferred, cross-link the RNA to the membrane with 50 mJ of UV radiation with a GS Gene Linker (BioRad), or similar, and detect it by incubating with 1:4000 streptavidin–horseradish peroxidase (Streptavidin-HRP, Pierce). Use RNase-free materials and solutions until the RNA is transferred and cross-linked to the membrane.
14. Purity of your biotinylated RNA (proportion of labeled vs. nonlabeled) can be estimated with a dot-blot hybridization assay, comparing your final extracted RNA against a synthetic biotinylated RNA. Briefly: equal starting amounts of each RNA (we typically start with 20 ng) are placed on a nylon membrane as 1 μL dots, followed by a number of serial ½ dilutions in SSC buffer (typically 5–8 dilutions). Once dots are deposited, the membrane is left to dry for 20 min, then cross-linked with a GS Gene Linker (Bio-Rad), or similar, and then the biotin signal is detected as in **Note 13**.
15. Raw sequencing reads should be first inspected by FastQC (https://www.bioinformatics.babraham.ac.uk/projects/fastqc/) and then processed by Cutadapt [17] and Trimmomatic [18] to clip sequencing adapters, remove low quality regions and filter reads shorter than 20 bp which align poorly to the reference genome (sacCer3, R64).
16. RNA transcribed from the ribosomal DNA (rDNA) locus is by far the most abundant type of RNA in a yeast cell. Unwanted carryover of nonlabeled contaminating rDNA transcripts will typically make up for most of the reads of the sequencing run. One can accelerate the bioinformatic processing of the BioGRO-seq datasets by mapping all reads to a genome composed of only the yeast rDNA sequences and keeping the unaligned reads for mapping to the full yeast genome.
17. We include a typical RNA extraction as an optional step aimed at comparing mRNA with nRNA profiles derived from the same sample.

Acknowledgments

The authors are grateful to Ana Miguel Blanco, Manoli Barneo, and María Encarna Pérez for their help with setting up the protocol, and to José García Martínez for reviewing the manuscript. The work in the Valencia laboratory is supported by grants from the Spanish MINECO and the European Union funds (FEDER) (BFU2016-77728-C3-3-P to J.E. P-O).

References

1. Coulon A, Chow CC, Singer RH, Larson DR (2013) Eukaryotic transcriptional dynamics: from single molecules to cell populations. Nat Rev Genet 14(8):572–584
2. Rossi MJ, Kuntala PK, Lai WKM, Yamada N, Badjatia N, Mittal C, Kuzu G, Bocklund K, Farrell NP, Blanda TR, Mairose JD, Basting AV, Mistretta KS, Rocco DJ, Perkinson ES, Kellogg GD, Mahony S, Pugh BF (2021) A high-resolution protein architecture of the budding yeast genome. Nature 592(7853): 309–314
3. Dangkulwanich M, Ishibashi T, Bintu L, Bustamante CJ (2014) Molecular mechanisms of transcription through single-molecule experiments. Chem Rev 114(6):3203–3223
4. Pérez-Ortín JE, de Miguel-Jiménez L, Chávez S (2012) Genome-wide studies of mRNA synthesis and degradation in eukaryotes. Biochim Biophys Acta 1819(6):604–615
5. Visa N, Jordán-Pla A (2018) ChIP and ChIP-related techniques: expanding the fields of application and improving ChIP performance. Methods Mol Biol 1689:1–7
6. Churchman LS, Weissman JS (2011) Nascent transcript sequencing visualizes transcription at nucleotide resolution. Nature 469(7330): 368–373
7. Carrillo-Oesterreich F, Preibisch S, Neugebauer KM (2010) Global analysis of nascent RNA reveals transcriptional pausing in terminal exons. Mol Cell 40(4):571–581
8. García-Martínez J, Aranda A, Pérez-Ortín JE (2004) Genomic run-on evaluates transcription rates for all yeast genes and identifies gene regulatory mechanisms. Mol Cell 15(2): 303–313
9. García-Martínez J, Pelechano V, Pérez-Ortín JE (2011) Genomic-wide methods to evaluate transcription rates in yeast. Methods Mol Biol 734:25–44
10. Jordán-Pla A, Pérez-Martínez ME, Pérez-Ortín JE (2019) Measuring RNA polymerase activity genome-wide with high-resolution run-on-based methods. Methods 159–160: 177–182
11. Marzluff WF (1978) Transcription of RNA in isolated nuclei. Methods Cell Biol 19:317–332
12. Gariglio P, Buss J, Green MH (1974) Sarkosyl activation of RNA polymerase activity in mitotic mouse cells. FEBS Lett 44:330–333
13. Core LJ, Waterfall JJ, Lis JT (2008) Nascent RNA sequencing reveals widespread pausing and divergent initiation at human promoters. Science 322(5909):1845–1848
14. Mahat DB, Kwak H, Booth GT, Jonkers IH, Danko CG, Patel RK, Waters CT, Munson K, Core LJ, Lis JT (2016) Base-pair-resolution genome-wide mapping of active RNA polymerases using precision nuclear run-on (PRO--seq). Nat Protoc 11(8):1455–1476
15. Jordán-Pla A, Gupta I, de Miguel-Jiménez L, Steinmetz LM, Chávez S, Pelechano V, Pérez-Ortín JE (2015) Chromatin-dependent regulation of RNA polymerases II and III activity throughout the transcription cycle. Nucleic Acids Res 43(2):787–802
16. Jordán-Pla A, Miguel A, Serna E, Pelechano V, Pérez-Ortín JE (2016) Biotin-genomic run-on (bio-GRO): a high-resolution method for the analysis of nascent transcription in yeast. Methods Mol Biol 1361:125–139
17. Martin M (2011) Cutadapt removes adapter sequences from high-throughput sequencing reads. EMBnet J 17:10–12
18. Bolger AM, Lohse M, Usadel B (2014) Trimmomatic: a flexible trimmer for Illumina sequence data. Bioinformatics 30:2114–2120
19. Begley V, Jordán-Pla A, Peñate X, Garrido-Godino AI, Challal D, Cuevas-Bermúdez A, Mitjavila A, Barucco M, Gutiérrez G, Singh A, Alepuz P, Navarro F, Libri D, Pérez-Ortín JE, Chávez S (2021) Xrn1 influence on gene transcription results from the combination of general effects on elongating RNA pol II and gene-specific chromatin configuration. RNA Biol 18:1310–1323

Chapter 5

Direct Sequencing of RNA and RNA Modification Identification Using Nanopore

Thidathip Wongsurawat, Piroon Jenjaroenpun, and Intawat Nookaew

Abstract

Direct RNA sequencing (dRNA-seq) simultaneously enables the detection of RNA modifications and characterization of full-length transcripts. In principle, full-length native RNA molecule is translocated through the nanopore by a motor protein while a sensor measures ionic current shifts. Then, the current shifts are interpreted by an algorithm that turn out to RNA sequence. Currently, the standard protocol of dRNA-seq provided by Oxford Nanopore Technologies (ONT) allows to directly ligate and sequence only polyadenylated RNA (poly(A) RNA). Here, we describe a method of dRNA-seq that can be applied for both poly(A) RNA and non-poly(A) tailed-RNA.

Key words RNA, Long-read, Nanopore, Sequencing Technology, Native, Modification, Methylation, Transcriptome, Epitranscriptome, Direct RNA sequencing

1 Introduction

Direct RNA sequencing (dRNA-seq) by Oxford Nanopore Technologies (ONT) offers the unique possibility to sequence single-stranded RNA (ssRNA) in their native form [1–5]. As the native RNA molecule will be translocated through a nanopore, dRNA-seq enables detection of any modifications on the RNA strand that can be inferred from the ionic current signal [1, 2, 5–9]. In addition, without conversion RNA to cDNA and PCR amplification, dRNA-seq can also generate long reads, covering the full-length of a transcript without amplification biases [1–6, 9–13]. The standard protocol provided by ONT, allows only poly(A) RNA such as eukaryotic mRNA, some species of long noncoding RNA, and poly(A) RNA virus to be sequenced. To study RNA modifications using dRNA-seq of other RNA species or non-poly(A) viral RNA, additional steps of design custom adapter specific to the 3'end of the RNA molecules are needed [13]. These steps could be cumbersome.

Frédéric Devaux (ed.), *Yeast Functional Genomics: Methods and Protocols*,
Methods in Molecular Biology, vol. 2477, https://doi.org/10.1007/978-1-0716-2257-5_5,

To overcome the limitation of the standard methodology of dRNA-seq, we introduced a simple step of enzymatic poly-A tailing reaction to polyadenylated RNA. This protocol can be applied for sequencing any ssRNA (at least 200 base pairs in length). Therefore, in this protocol, there are two main steps, including poly-A tailing reaction and dRNA-seq (Fig. 1). This protocol can serve for detecting RNA modification of rRNA, tRNA, mRNA, viral RNA, and in vitro transcripts [1, 3, 10, 14].

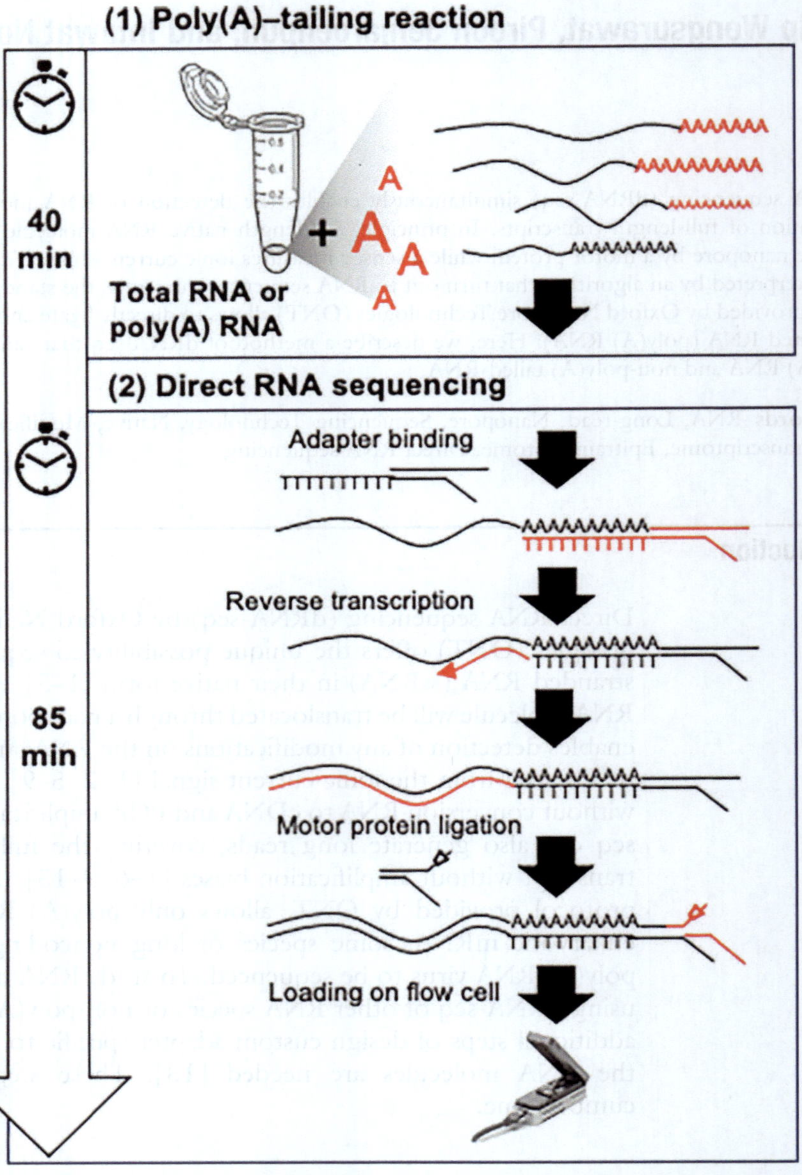

Fig. 1 Illustration of the two main steps and estimated time spent. The red color shows the progression of each step

2 Materials

2.1 Oxford Nanopore Technologies' Kits

1. Direct RNA Sequencing Kit. Thaw all reagents of the kit on ice.
2. Flow Cell Priming Kit.

2.2 Enzymes, Buffers, Water

1. *E. coli* poly(A) polymerase.
2. 10× E. coli Poly(A) Polymerase Reaction Buffer and ATP.
3. SuperScript IV Reverse Transcriptase.
4. NEBNext® Quick Ligation Reaction Buffer.
5. 2000 U/ml T4 DNA Ligase.
6. Nuclease-free water.
7. RNase Inhibitor.
8. Freshly prepared 80% ethanol: 2 ml 100% ethanol (molecular grade), 0.5 ml nuclease-free water. Store at 4 °C.

2.3 Other Reagents and Consumables

1. Agencourt RNAClean XP beads.
2. Qubit RNA HS Assay Kit.
3. 10 mM dNTP solution.

2.4 Equipment

1. Magnetic stand, suitable for 0.2 ml PCR tubes.
2. Microfuge.
3. Vortex mixer.
4. Ice bucket with ice.
5. Thermal cycler.
6. Qubit fluorometer (or equivalent for QC check).
7. Pipettes and pipette tips P2, P10, P20, P200, P1000.
8. 0.2 ml PCR tube.

3 Methods

3.1 Poly(A) Tailing Reaction

1. 17 μl RNA in PCR tube is incubated at 65 °C for 2 min and immediately placed on ice to avoid secondary structure formation (*see* **Note 1**).
2. Add 3.5 μl 10× *E. coli* Poly(A) Polymerase Reaction Buffer into 17 μl RNA in PCR tube, bring to room temperature.
3. Add 4.5 μl ATP into the PCR tube, followed by 4 μl *E. coli* poly (A) polymerase and 1 μl RNase inhibitor into the PCR tube.
4. Mix gently with P200 pipette and incubate reaction at 37 °C for 10 min.

5. Stop the reaction by directly proceed to the bead cleanup step. Mix the Agencourt RNAClean XP bead well so that the reagent appears homogeneous and consistent in color. Add 60 μl homogeneous-Agencourt RNAClean XP bead into PCR tube and mix well using P200 pipette. Incubate for 5 min.
6. Put the PCR tube into the magnetic stand for 2 min.
7. Gently remove supernatant.
8. Wash RNA twice with 80% ethanol by adding 180 μl of 80% ethanol and incubate for 1 min. Then remove the ethanol without disturbing the pellet.
9. Air-dry at room temperature for 2 min (until dry but not crack).
10. Elute by 12 μl nuclease-free water and flick the tube. Incubate for 2 min at room temperature.
11. Put PCR tube into the magnetic stand for 2 min.
12. Pipet 11 μl poly(A) tailed-RNA into a new PCR tube and put the tube on ice.
13. Measure the final RNA concentration (optional, *see* **Note 2**).

3.2 dRNA-seq

Continue the "Direct RNA sequencing" protocol according to the ONT's instructions, except that (i) replace SuperScript III reverse transcriptase by SuperScript IV reverse transcriptase, (ii) replace 70% ethanol with 80% ethanol, and (iii) reduce the reverse transcription time from 50 to 10 min. Carry out all procedures at room temperature, unless otherwise specified.

1. Adapter ligation: Mix 3.0 μl NEBNext Quick Ligation Reaction Buffer into poly(A) tailed-RNA followed by 1.0 μl RT Adapter (RTA) and 1.5 μl T4 DNA Ligase. Mix by pipetting and spin down. Incubate the reaction for 10 min at room temperature.
2. Reverse transcription (*see* **Note 3**): Prepare the reverse transcription master mix by mixing 9.0 μl nuclease-free water, 2.0 μl 10 mM dNTPs, 8.0 μl 5x first-strand buffer, 4.0 μl 0.1 M DTT. Add the master mix to the PCR tube containing the RT adapter-ligated RNA. Mix by 200 μl pipette. Add 2 μl of SuperScript IV reverse transcriptase to the reaction. Place the tube in a thermal cycler and incubate at 50 °C for 10 min, then 70 °C for 10 min, and bring the reaction to 4 °C.
3. RNA cleanup using 80% ethanol: Vortex the Agencourt RNAClean XP beads to resuspend. Add 72 μl of resuspended Agencourt RNAClean XP beads to the reverse transcription reaction and mix by pipetting. Incubate for 5 min at RT. Spin down the sample and pellet on a magnetic stand for 3 min. Keep the tube on the magnet and pipet off the supernatant. Wash RNA twice

with 180 μl of 80% ethanol. Remove the 80% ethanol using a pipette and discard. Spin down and place the tube back on the magnet. Pipet off any residual 70% ethanol. Air-dry at room temperature for 1 min. Remove the tube from the magnetic stand and resuspend the pellet in 20 μl nuclease-free water and flick the tube. Incubate for 5 min at room temperature. Place the tube back on the magnet for 1 min and pipet 20 μl of eluate into a new 0.2 PCR tube.

4. Motor protein ligation: Add 8.0 μl NEBNext Quick Ligation Reaction Buffer into the 20 μl eluted reverse-transcribed RNA. Then add 4.5 μl - 6.0 μl RNA Adapter (RMX) (*see* **Note 4**) followed by 3.0 μl nuclease-free water and 3.0 μl T4 DNA Ligase. Mix by 200 μl pipette. Incubate the reaction for 10 min at room temperature.
5. RNA cleanup using Wash Buffer (WSB): Vortex the Agencourt RNAClean XP beads to resuspend. Add 40 μl of resuspended Agencourt RNAClean XP beads to the reaction tube and mix by pipetting. Incubate for 5 min at RT. Spin down the sample and pellet on a magnetic stand for 3 min. Keep the tube on the magnet and pipet off the supernatant. Add 150 μl of the WSB to the beads. Close the tube lid and gently flick the tube. Spin down and place the tube back on the magnet. Remove the WSB using a pipette, and discard. Repeat adding and removing WSB again. Pipet off any residual WSB. Remove the tube from the magnetic stand and resuspend pellet in 20 μl Elution Buffer and gently flick the tube. Incubate for 10 min at room temperature. Place the tube back on the magnet for 2 min or until the eluate is clear. Pipet 20 μl of eluate into a new 0.2 PCR tube. Add 17.5 μl of nuclease-free water to the prepared RNA library and mix with 200 μl pipette. The total volume is 37.5 μl.
6. Priming and loading the SpotON flow cell: Mix the RNA Running Buffer (RRB), Flush Buffer (FB), and Flush Tether (FLT) tubes thoroughly by vortexing and spin down and keep all solutions at room temperature. To prepare the flow cell priming mix, add 30 μl FLT directly to the tube of FB, and vortex. Slide the priming port cover clockwise to open the priming port. Set a P1000 pipette to 200 μl and insert the tip into the priming port. Turn the wheel until you can see a small volume (20 μl of yellow color) of buffer entering the pipette tip. Load 800 μl of the priming mix into the flow cell via the priming port, avoiding the introduction of air bubbles. Wait for 5 min. During this time, prepare the library for loading.

(a) In a new PCR tube, mix 37.5 μl RRB and 37.5 μl RNA library in nuclease-free water. Gently open the SpotON port cover. Load 200 μl of the priming mix into the priming port (<u>not the SpotON sample port</u>). Set 200 μl pipette to 75 μl. Mix the prepared library by pipetting up and down prior to loading to the sample port. Load all 75 μl of sample to SpotON sample port in a dropwise fashion. Close the SpotON port, close the priming port and close the MinION Mk1B lid.

7. Sequencing: Double-click the MinKNOW icon located on the desktop to open the MinKNOW GUI. Click the "New Experiment" button and fill in the information based on your experiment, such as sample ID/name. Choose the RNA002 kit. Click "Start run." In case of real-time base calling option, ensure the computer has enough resource to run a MinION without deterioration of performance (e.g., memory: 16 GB RAM; CPU: i7 or Xeon with 4+ cores; storage: 1 TB internal SSD or using MinION Mk1C version).

4 Notes

1. Total RNA/poly(A) tail RNA/in vitro transcripts can be an input for this approach. The starting material is up to 6 μg of RNA in 17 μl nuclease-free water. It is very important to use the high-quality RNA, that is, the OD 260/280 ratio should be >1.9 and OD 260/230 should be 2.0–2.2 and intact RNA.
2. Nanodrop or Qubit fluorometer can be used alternatively.
3. Although the reverse transcription step is optional, performing this step is recommended. The complementary DNA synthesized from the reverse transcription step can stabilize the RNA strand increasing overall yield and throughput.
4. RNA input can be viral RNA (intact) which might be very low concentration (<10 ng/ μl). In this case, RNA Adapter (RMX) can be adjusted to 4.5 μl to avoid adapter overload.

Acknowledgments

This research project is supported by Mahidol University (Basic Research Fund: fiscal year 2021), grant ID R016441006, the National Institute of General Medical Sciences of the National Institutes of Health award no. P20GM125503.

References

1. Jenjaroenpun P, Wongsurawat T, Wadley TD et al (2021) Decoding the epitranscriptional landscape from native RNA sequences. Nucleic Acids Res 49:e7
2. Workman RE, Tang AD, Tang PS et al (2019) Nanopore native RNA sequencing of a human poly(a) transcriptome. Nat Methods 16:1297–1305
3. Wongsurawat T, Jenjaroenpun P, Taylor MK et al (2019) Rapid sequencing of multiple RNA viruses in their native form. Front Microbiol 10:260

4. Jenjaroenpun P, Wongsurawat T, Pereira R et al (2018) Complete genomic and transcriptional landscape analysis using third-generation sequencing: a case study of saccharomyces cerevisiae CEN.PK113-7D. Nucleic Acids Res 46: e38
5. Garalde DR, Snell EA, Jachimowicz D et al (2018) Highly parallel direct RNA sequencing on an array of nanopores. Nat Methods 15: 201–206
6. Xu L, Seki M (2020) Recent advances in the detection of base modifications using the nanopore sequencer. J Hum Genet 65:25–33
7. Wang Y, Wang H, Xi F et al (2020) Profiling of circular RNA N(6) -methyladenosine in moso bamboo (*Phyllostachys edulis*) using nanopore-based direct RNA sequencing. J Integr Plant Biol 62:1823–1838
8. Parker MT, Knop K, Sherwood AV et al (2020) Nanopore direct RNA sequencing maps the complexity of arabidopsis mRNA processing and m(6)a modification. elife 9:e49658
9. Zhao L, Zhang H, Kohnen MV et al (2019) Analysis of transcriptome and epitranscriptome in plants using PacBio Iso-Seq and nanopore-based direct RNA sequencing. Front Genet 10: 253
10. Taylor MK, Williams EP, Wongsurawat T et al (2020) Amplicon-based, next-generation sequencing approaches to characterize single nucleotide polymorphisms of orthohantavirus species. Front Cell Infect Microbiol 10:565591
11. Oikonomopoulos S, Bayega A, Fahiminiya S et al (2020) Methodologies for transcript profiling using long-read technologies. Front Genet 11:606
12. Viehweger A, Krautwurst S, Lamkiewicz K et al (2019) Direct RNA nanopore sequencing of full-length coronavirus genomes provides novel insights into structural variants and enables modification analysis. Genome Res 29:1545–1554
13. Keller MW, Rambo-Martin BL, Wilson MM et al (2018) Direct RNA sequencing of the coding complete influenza a virus genome. Sci Rep 8:14408
14. Tan CCS, Maurer-Stroh S, Wan Y et al (2019) A novel method for the capture-based purification of whole viral native RNA genomes. AMB Express 9:45

Chapter 6

UNAGI: Yeast Transcriptome Reconstruction and Gene Discovery Using Nanopore Sequencing

Mohamad Al kadi, Nicolas Jung, and Daisuke Okuzaki

Abstract

Computational approaches are the main approaches used in genome annotation. However, accuracy is low. Untranslated regions are not identified, complex isoforms are not predicted correctly and discovery rate of noncoding RNA is low. RNA-seq has revolutionized transcriptome reconstruction over the last decade. However, fragmentation included in cDNA sequencing leads to information loss, requiring transcripts to be assembled and reconstructed, thus affecting the accuracy of reconstructed transcriptome. Recently, long-read sequencing has been introduced with technologies such as Oxford Nanopore sequencing. cDNA is sequenced directly without fragmentation producing long reads that don't need to be assembled keeping the transcript structure intact and increasing the accuracy of transcriptome reconstruction.

Here we present a protocol and a pipeline to reconstruct the transcriptome of compact genomes including yeasts. It involves generating full-length cDNA and using Oxford Nanopore ligation-based sequencing kit to sequence multiple samples in the same run. The pipeline (1) strands the generated long reads, (2) corrects the reads by mapping them to the reference genome, (3) identifies transcripts including 5′UTR and 3′UTR, (4) profiles the isoforms, filtering out artifacts resulting from low accuracy in sequencing, and (5) improves accuracy of provided annotations. Using long reads improves the accuracy of transcriptome reconstruction and helps in discovering a significant number of novel RNAs.

Key words Nanopore, Long-read sequencing, Transcriptome reconstruction, Full-length cDNA, Pipeline

1 Introduction

Computational approaches are the main approaches used in gene discovery. Genes are predicted based on their structure such as coding regions and splicing sites or based on their homology with close well-annotated species. However, accuracy is low in eukaryotic genomes [1] and there are many challenges such as complex isoform structure, and genes in less-studied species [2]. Moreover, computational approaches performs poorly in terms of identifying untranslated regions and alternative splicing [3] and noncoding RNA gene discovery [4].

Frédéric Devaux (ed.), *Yeast Functional Genomics: Methods and Protocols*,
Methods in Molecular Biology, vol. 2477, https://doi.org/10.1007/978-1-0716-2257-5_6,

RNA-seq is a revolutionary tool that can help in overcoming these challenges. It was shown to increase gene prediction accuracy [5] and to significantly improve well-known genome annotations [6]. RNA-seq involves sequencing the RNAs in a sample providing a picture for the whole transcriptome in that sample. RNA is usually converted into cDNA, fragmented and sequenced to yield short reads that are assembled in silico with or without the guidance of a reference genome. Such sequencing methods are called short-read sequencing. Fragmenting the transcripts leads to information loss making it difficult to identify the transcripts start and end sites. It also yields poor isoform discovery [7] and requires complex algorithm to assemble the short reads [4]. Recently, long-read sequencing methods have been introduced with technologies such as Oxford Nanopore sequencing. cDNA is sequenced without fragmentation keeping the transcript structure intact. Since long reads represent the transcripts themselves, transcriptome reconstruction tools simply collapse long reads with identical splicing sites focusing on producing isoforms [8]. It may be an effective strategy for higher eukaryotes genomes but not for lower eukaryotes such as yeast. Yeast genomes are compact and most of their genes are intron less or just have simple exon-intron structures [9] which are a challenge for current long-read transcriptome reconstruction tools since they are designed mainly for spacious genomes with complex isoforms.

Here we present a protocol and a pipeline, UNAGI, that is mainly designed for transcriptome reconstruction of compact genomes like yeast using Oxford Nanopore cDNA sequencing kit [10] (Fig. 1). This protocol uses SMARTer (Switching Mechanism at 5-End of RNA Template) technology to generate full-length cDNA. Full-length cDNA is recommended to reduce bias introduced by truncated cDNAs [11]. Generated long reads are stranded according to the primer sequences and corrected by mapping to the reference genomes. De novo transcriptome reconstruction and isoforms detection are performed separately. In addition, annotation-aware transcriptome reconstruction can be done in cases where annotations are provided.

2 Materials

2.1 Generating Full-Length cDNA and Library Preparation

1. SMARTer PCR cDNA synthesis kits (*see* **Note 1**).
2. LongAmp® Taq DNA Polymerase.
3. Agencourt AMPure XP Kit.
4. TE buffer: 10 mM Tris–HCl pH 8.0, 0.1 mM EDTA.
5. Ligation Sequencing Kit SQK-LSK109.

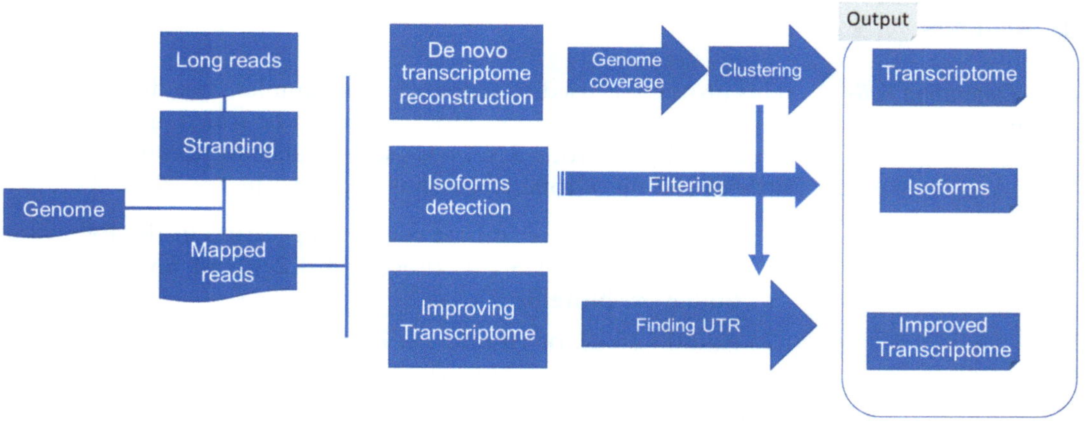

Fig. 1 Schematic overview of the UNAGI pipeline. Reads are stranded then mapped to the genome. The transcriptome is reconstructed using the genome coverage and clustering 3′ and 5′ ends. Isoforms are detected and subjected to many filtering processes. If annotations were given, the annotations will be improved by identifying untranslated regions (UTRs) and combining the newly discovered genes

6. Native Barcoding Expansion 1–12.
7. NEBNext Ultra II End Repair/dA-Tailing Module.
8. Blunt/TA ligase master mix.
9. NEBNext Quick Ligation Module.
10. Flow Cell Priming Kit.
11. Ethanol.
12. Nuclease-free water.
13. 0.5 M EDTA.

2.2 Software

1. MinKNOW.
2. Guppy.
3. Python 3.7.
4. Biopython module.
5. Minimap2.
6. Samtools.
7. Bedtools.

2.3 System Requirement

Although the main code is written in python, most tools used are designed to be used in an UNIX operation system. For Windows 10 users, Ubuntu shell can be installed. Bioinformatic work was done on Ubuntu 18.04.5 LTS (GNU/Linux 4.4.0–18362-Microsoft x86_64).

3 Methods

3.1 Generating Full-Length cDNA

This protocol is used to sequence four samples: two duplicates for two biological conditions. However, it can be scaled up or down as needed. 10 ng of total RNA was used but starting amount can be as little as 1–2 ng of total RNA or as large as 1 μg (*see* **Note 2**).

1. Prepare the master mix at room temperature in the following order. First, in 1.5 mL tube, add 4.2 μL of 5X first-strand buffer (*see* **Note 3**).
2. Add 0.525 μL of 10 mM DTT.
3. Add 2.1 μL of 10 mM dNTP mix.
4. Add 2.1 μL 12 μM SMARter II A oligonucleotide.
5. Add 0.525 μL of RNase inhibitor.
6. In 0.2 mL tubes, add the RNA sample and free-nuclease water up to 1.75 μL so that the total RNA amount is 10 ng.
7. Add 0.5 μL of 12 μM 3′ SMART-Seq CDS primer II A.
8. Incubate in thermal cycler at 72 °C for 3 min then at 42 °C for 2 min.
9. Meanwhile add 2.1 μL of the SMARTScribe Reverse Transcriptase (100 U) to the master mix and mix.
10. Dispense 2.75 μL of master mix to each of the 0.2 tubes and mix gently (*see* **Note 4**).
11. Incubate at 42 °C for 90 min then at 70 °C for 10 min.
12. Add 45 μL of TE buffer to the RT product.
13. Prepare the master mix for cDNA amplification. First, add 46.2 free-nuclease water.
14. Add 16.8 μL of 5′ PCR primer II A.
15. Add 105 μL of LongAmp Taq 2× master mix and mix well.
16. Add 40 μL of the master mix to 10 μL of cDNA-mRNA hybrid.
17. Incubate in the thermal cycler as follows.

95 °C	1 min	1 cycle
95 °C	15 s	27 cycles
65 °C	30 s	
65 °C	3 min	

Pause at 15 cycles and divide each tube into two: optimization (20 μL) and experimental (30 μL). Keep the experimental tube at

4 °C. Continue the thermal cycler program and stop every 3 cycles to take 3 μL for analysis in Qubit and LabChip (*see* **Note 5**).

18. Based on the gel electrophoresis results, choose the best cycle number and subject the experimental tube to this number as above.
19. Add 2 μL of 0.5 M EDTA to each tube to terminate the reaction.
20. Clean the generated cDNA with AMPure ×1.0. Add 30 μL beads to the 30 μL sample and mix.
21. Incubate for 5 min at RT.
22. Spin down and pellet the tube on magnet. Remove the supernatant and wash two times with 200 μL freshly prepared 80% ethanol.
23. Spin down and pellet on the magnet slowly and remove the residual ethanol and let the tubes on the magnet for 2 min to dry.
24. Resuspend pellet with Buffer EB and incubate for 2 min at room temperature. Pellet on the magnet and transfer the eluate to 0.2 mL PCR tubes.

3.2 Library Preparation

3.2.1 End-Prep

1. Analyze the length of cDNA by LabChip, determine the median length and calculate the molarity by using NEBioCalculator https://nebiocalculator.neb.com/ .
2. To 0.2 mL tubes, transfer volume that contains 0.36 pmol and add free-nuclease water up to 50 μL (*see* **Note 6**).
3. End-prep. Prepare a master mix by adding ultra II End-prep reaction buffer 29.4 μL and 12.6 μL ultra II End-prep enzyme mix. Dispense 10 μL in each tube.
4. Mix gently by flicking the tube.
5. Incubate at 20 °C for 5 min then at 65 °C for 5 min.
6. Transfer to LoBind 1.5 mL tube and clean with AMPure XP beads ×0.8 (48 μL of beads to the 60 μL product) and elute with 23.5 μL of free-nuclease water.
7. Dilute 1 μL of End-prepped DNA with 9 μL of EB and measure by Qubit fluorometer.

3.2.2 Barcoding

1. Thaw the native barcodes at room temperature (four barcodes) and place them on ice.
2. Add 2.5 μL of the Native Barcode to the end-prepped DNA (22.5 μL) and mix by inversion.
3. Add 25 μL of Blunt/TA Ligase Master Mix and mix by inversion.
4. incubate for 10 min at room temperature.

5. Clean with AMPure XP beads ×0.8 and elute with 20 μL free-nuclease water.
6. Dilute 1 μL of the barcoded cDNA with 9 μL EB and measure by Qubit fluorometer.

3.2.3 Pooling and Adapter Ligation

1. Pool equimolar amounts of each barcoded sample into a 1.5 mL DNA LoBind tube. Convert from ng/μL (as was measured by Qubit fluorometer) to nM using the median length of the cDNA. Pool the samples to have 0.51 pmol in 65 μL of free-nuclease water.
2. Add 5 μL of Adapter Mix II (AMII) and mix by inversion.
3. Add 20 μL of NEBNext Quick Ligation Reaction Buffer (5×) and mix by inversion.
4. Add 10 μL of Quick T4 DNA Ligase and mix by inversion.
5. Incubate for 10 min at room temperature.
6. Clean up by AMPure XP beads (×0.8) by adding 80 μL of beads, incubating for 5 min on hula mixer, pelleting on magnet and washing two times with 250 μL of Short Fragment Buffer (SFB). Be careful not to use ethanol to avoid adapter damage. Resuspend the pellet with 15 μL of ELB and incubate for 10 min at room temperature then pellet on magnet and transfer the elute to a LoBind tube.
7. Measure the concentration with Qubit fluorometer. Aim for final amount of 0.34 pmol library.

3.2.4 Priming and Loading on the Flow Cell

1. Prepare flow cell priming mix by adding 30 μL of Fluch Tether (FLT) directly to the tube of Flush Buffer (FB), and mix by vortexing at room temperature.
2. Prepare the library by adding 37.5 μL of Sequencing buffer, 25.5 μL of Loading Beads and 12 μL of DNA library.
3. Prime the flow cell with the priming mix.
4. Load 75 μL of the sample to the flow cell.

3.3 UNAGI Pipeline

3.3.1 Setting Up Software

1. UNAGI is deposited at github (https://github.com/iMetOsaka/UNAGI). It can be downloaded using the following command in the shell:

   $ git clone https://github.com/iMetOsaka/UNAGI

   Precompiled binaries of these utilities are already provided with this pipeline but they can be replaced by other versions by changing the path in the file UNAGI/app/conf.ini. These utilities are as follows.
2. Minimap2 is the splice-aware aligner that is optimized to mapping long reads.

3. Samtools will be used for viewing, indexing and sorting the alignment formats sam/bam.
4. Bedtools is a toolset for genome arithmetic such as calculating the genome coverage form an alignment file.
5. UNAGI requires python and Biopython to run. It was tested on python3. To test if python is present on the operating system type the following command:

   $ python3 –version

   If you get a version number you can skip to installing Biopython. Otherwise you have to install it on your operating system. In our example (Ubuntu):

   $ sudo apt-get update

   $ sudo apt-get install python3
6. Biopython Module: The Biopython module will be used to handle the fastq file (reads sequence data). It can be easily installed using the following command.

   $ pip install biopython
7. Porechop can be downloaded and installed from Github:

   $ git clone https://github.com/rrwick/Porechop

   $ cd Porechop

   $ python3 setup.py install

3.3.2 Trimming Adapter

If adapters are not trimmed during basecalling, It's preferred to trim them before the stranding process. Porechop is used by typing the following command.

$ porechop -i input_reads.fastq.gz -o output_reads.fastq.gz

Our test reads are already trimmed.

3.3.3 Running UNAGI

1. A fastq or fastq.gz file that contains the long-read sequence data should be provided. A test file (test.fq) is included with the pipeline.
2. A genome sequence that will be used to reads correction and transcript reconstruction. We used in this example, assembly GCF_000146045.2_R64 of *Saccharomyces cerevisiae*. Genome sequence is included with the pipeline (genome_test.fasta).
3. Annotations should be provided in gff format.

   UNAGI can be started by going to its directory and typing the following.

   $. /unagi -i [path_to_fastq_file] -o [path_to_desired_output] -g [path_to_genome_file]

   You can use the test file first:

$. /unagi -i. /test/test.fq -o. /test/output -g. /test/genome_test.fasta

4. The strand information is lost after generating and amplifying cDNA (Fig. 2) as cDNA now contain both sense and antisense

Fig. 2 Double strand full-length cDNA. The poly(A) tail and the specific sequence added by the enzyme can indicate which strand is the forward and which is the reverse

strands. Taking advantage of keeping the full-length sequence, the strand can be retrieved by simply looking for the poly (A) tail and SMART oligonucleotides flanking the read. Default setting can be kept or changes as needed in the con.ini file.

strand_forwards_identifier is the sequence that identify forward reads.

strand_reverse_identifier is the sequence that identify the reversed reads which will be reversed-complement to the original strand.

strand_start_index and strand_end_index is the region of the read that the software will look for those identifiers to strand the read.

If the reads are already stranded such as in case of direct RNA sequencing reads, "-s True" should be added to the UNAGI command line.

5. De novo transcript reconstruction means here the lack of annotation information but not the genome reference. The genome reference is necessary for correcting the long reads. First, stranded reads will be mapped to the genome using minimap2 with the following parameters: -ax splice -G 5 k --secondary = no. They can be changed in the con.ini file. For example, we chose the parameter -G (Maximum gap on the reference) to be 5 k as longer gaps produced very long artificial introns. The generated alignments file will be converted to bam, sorted and converted to the concise format bed. All intermediary files are saved to a "Transitional" directory in the output directory. From this step forward, reads on sense and antisense strands will be processed separately.
6. Genome coverage will be calculated on each strand separately and transcriptional "islands" will be identified based on genomic coverage. In these "islands," long reads are overlapped making a continuous high genomic coverage over these regions (Fig. 3). In large genomes like human genome, these islands represent separate genes. But in compact genomes like yeast, genes are close to each other. Still, there is usually drop in the coverage between genes. The threshold can be selected in con.ini file. Default is set to be 0 since the transcriptional units will be further separated in the next step. Generated file can be

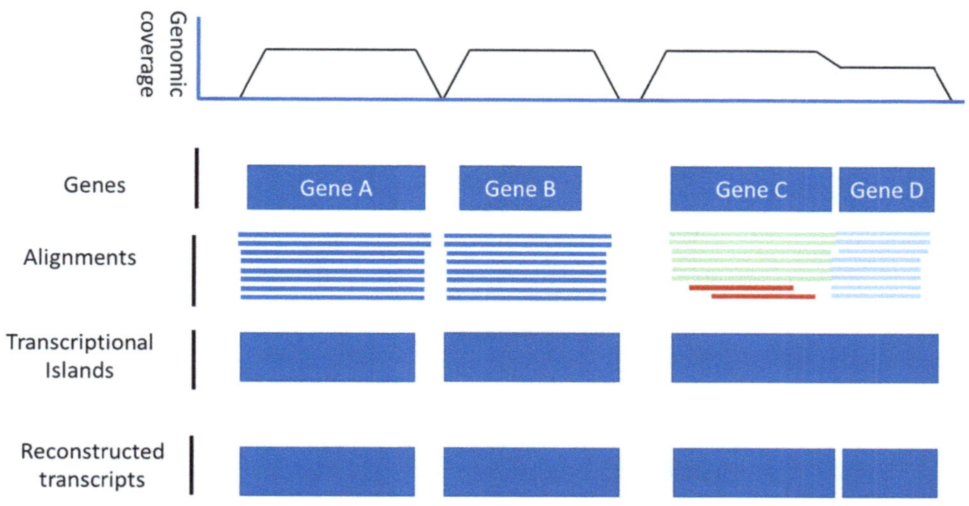

Fig. 3 De novo transcript reconstruction. First reads mapped to the genome and genomic coverage is calculated. Based on genomic coverage transcriptional "islands" are identified. Those can be genes or overlapping genes. In the figure, overlapping reads were colored differently. Clustering 3′ and 5′ ends generates reconstructed transcripts. Irregular reads are filtered out (red reads)

found in the translational directory under the name positive_-genelist_incomplete.bed and negative_genelist_incomplete.bed.

In each transcriptional, reads with 3′ ends close to each other will be clustered together and most frequent genomic position is set to be the gene end. Then the 5′ ends of these reads will be clustered and the gene start will be determined. Overlapped genes would be collapsed and final gene list will be generated.

7. UNAGI cluster reads with identical splice sites together to produce the isoforms. Then, they are subjected to several filtering processes. The isoforms before each filtering process can be found in the transitional directory. First, due to the low accuracy of Nanopore reads, some artificial isoforms are generated. These are filtered out by comparing the splicing site frequency with the genomic coverage at the site. This threshold "max_-coverage_to_splice_ratio" can be changed in the con.ini file. Minimap2 generates false alternative sites because of the poly (A) tail. If a nearby region has poly(A) or poly(T), it generates a short exon that contains this region although it is the poly (A) tail (Fig. 4). UNAGI looks for short exons with high frequency of (A) or (T) and eliminate them. Finally, it will eliminate alterative splicing sites and keep only the highest frequency. In our dataset some of them were really alternative sites and some were artifacts introduced by low accuracy of Nanopore sequencing. Either way, both files are available.

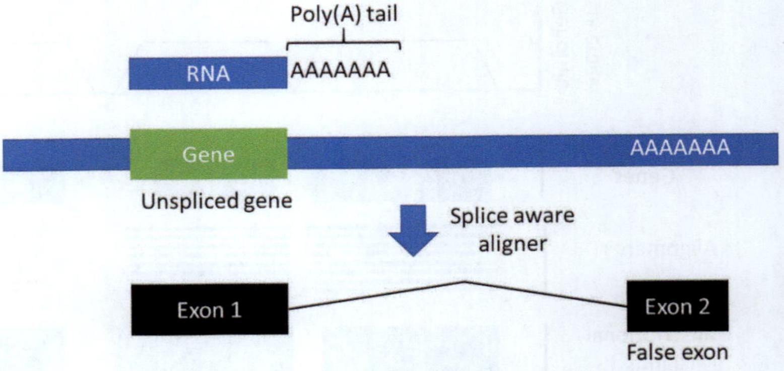

Fig. 4 Artificial exons generated by aligners. The poly(A) tail is not part of the genome. However, the aligner sometimes assigns a close region of poly(A) as an exon

8. If the annotation file was provided, annotation can detect untranslated regions. UNAGI will consider reads overlapping the whole gene body and report the start and end of each gene. In case there are alternative start or end sites, secondary sites will be reported as well. Finally, UNAGI will compare de novo reconstructed transcriptome with the provided annotation and report the genes overlapped with the newly reconstructed transcripts.

4 Notes

1. Some SMARTer kit involves chemical blocking of the primers used in amplification which prevents Nanopore adapters ligation and results in unsuccessful sequencing.
2. SMARTer cDNA synthesis kit has two types standard and pico. If starting amount of RNA is too low, SMARTer Pico PCR cDNA synthesis should be chosen.
3. Through this protocol we made ×0.2 extra of master mixes.
4. RNA and cDNA should handle gently through the protocol to have large proportion of full-length cDNA.
5. We used LabChip but normal gel electrophoresis can be used too.
6. For SQK-LSK kit, Nanopore recommends 100–200 fmol or 150–300 fmol if sequencing is on R10.3 flow cells. However, we increased the amount since our protocol is for cDNA, not genomic DNA.

References

1. Salzberg SL (2019) Next-generation genome annotation: we still struggle to get it right. Genome Biol 20:92
2. Scalzitti N, Jeannin-Girardon A, Collet P et al (2020) A benchmark study of ab initio gene prediction methods in diverse eukaryotic organisms. BMC Genomics 21:293. https://doi.org/10.1186/s12864-020-6707-9
3. Li Z, Zhang Z, Yan P et al (2011) RNA-Seq improves annotation of protein-coding genes in the cucumber genome. BMC Genomics 12: 540. https://doi.org/10.1186/1471-2164-12-540
4. Pertea M, Pertea GM, Antonescu CM et al (2015) StringTie enables improved reconstruction of a transcriptome from RNA-seq reads. Nat Biotechnol 33:290–295. https://doi.org/10.1038/nbt.3122
5. Hoff KJ, Lange S, Lomsadze A et al (2016) BRAKER1: unsupervised RNA-Seq-based genome annotation with GeneMark-ET and AUGUSTUS: table 1. Bioinformatics 32: 767–769. https://doi.org/10.1093/bioinformatics/btv661
6. Beiki H, Liu H, Huang J et al (2019) Improved annotation of the domestic pig genome through integration of Iso-Seq and RNA-seq data. BMC Genomics 20:344. https://doi.org/10.1186/s12864-019-5709-y
7. Puglia GD, Prjibelski AD, Vitale D et al (2020) Hybrid transcriptome sequencing approach improved assembly and gene annotation in cynara cardunculus (L.). BMC Genomics 21: 317. https://doi.org/10.1186/s12864-020-6670-5
8. Tang AD, Soulette CM, van Baren MJ et al (2020) Full-length transcript characterization of SF3B1 mutation in chronic lymphocytic leukemia reveals downregulation of retained introns. Nat Commun 11:1–12. https://doi.org/10.1038/s41467-020-15171-6
9. Proux-Wéra E, Armisén D, Byrne KP, Wolfe KH (2012) A pipeline for automated annotation of yeast genome sequences by a conserved-synteny approach. BMC Bioinformatics 13:237. https://doi.org/10.1186/1471-2105-13-237
10. Al kadi M, Jung N, Ito S et al (2020) UNAGI: an automated pipeline for nanopore full-length cDNA sequencing uncovers novel transcripts and isoforms in yeast. Funct Integr Genomics 20:523–536. https://doi.org/10.1007/s10142-020-00732-1
11. Wang X, You X, Langer JD et al (2019) Full-length transcriptome reconstruction reveals a large diversity of RNA and protein isoforms in rat hippocampus. Nat Commun 10:1–15. https://doi.org/10.1038/s41467-019-13037-0

Chapter 7

Identification of Taxonomically Restricted Transcripts from Illumina RNA Sequencing Data

William R. Blevins

Abstract

In order to perform a well-balanced comparative transcriptomic analysis, the reference genome and annotations for all species included in the comparison must be of a similar quality and completeness. Frequently, comparative transcriptomic analyses include non-model organisms whose annotations are not as well curated; this inequality can lead to biases.

To avoid potential biases stemming from incomplete annotations, a comparative transcriptomic analysis can incorporate de novo transcriptome assemblies for each species, which reduces this disparity. This chapter covers all of the steps which are necessary to run a comparative transcriptomic analysis with de novo transcriptome assemblies, from the first step of the experimental design to the sequencing, and ultimately the bioinformatic analysis.

Key words Comparative transcriptomics, de novo assembly, Taxonomically restricted, Phylogenetic conservation

1 Introduction

Comparative transcriptomic analyses are fundamental to understanding evolution. These analyses can illuminate the mechanisms which drive the conservation of gene regulation and divergence of condition-specific gene programs, as well as identify recently-emerged regulatory elements and transcripts [1, 2]. While whole-genome assemblies are available for hundreds of yeast species, the reference annotations for many of the non-model species of yeast are often of lower-quality and completeness, relative to species like *S. cerevisiae*. The varying quality of reference annotations can have a drastic impact on the accuracy of reference-based comparative analyses which seek to identify taxonomically restricted or

Supplementary Information The online version contains supplementary material available at [https://doi.org/10.1007/978-1-0716-2257-5_7].

Frédéric Devaux (ed.), *Yeast Functional Genomics: Methods and Protocols*,
Methods in Molecular Biology, vol. 2477, https://doi.org/10.1007/978-1-0716-2257-5_7,

species-specific features, such as de novo genes [3]. Therefore, to better estimate the relative conservation of transcripts, one must ensure that the reference annotations used for each species are of comparable completeness and quality. In many cases, this entails creating de novo assemblies to generate transcriptomic profiles for many species in order to bolster any lower-quality reference annotations.

The protocol described here covers several facets of designing a comparative transcriptomic analysis, including the experimental conditions, RNA purification, library preparation, and bioinformatic pipeline to process the raw sequencing data. One main challenge in designing a sequencing experiment involving multiple species of yeast is that yeast are wonderfully diverse; they can occupy vastly different habitats, have distinct nutritional preferences, and are resistant to different stressors [4]. If a few of the species in the study are subjected to inhospitable living conditions, there could be significant changes in gene regulation which would introduce bias and confound the results of any interspecies transcriptomic comparison. One strategy to minimize the impact of this limitation is to select a set of yeast species which can all grow reasonably well in a common rich medium [5] (*see* **Note 1**).

2 Materials

2.1 Growth Media

1. Rich medium suitable for multiple species of yeast: 1.5% yeast extract, 1% peptone, 2% dextrose, 2 g/L SC amino acid mix, 100 mg/L adenine, 100 mg/L tryptophan, 100 mg/L uracil; this formula was taken directly from Tsankov et al. [5] (*see* **Notes 2** and **3**).

2.2 Culture Preparation

1. Belly-dancer
2. Spectrophotometer
3. Ice tray large enough to hold rack for microcentrifuge tubes
4. Microcentrifuge

2.3 RNA Purification

1. Epicentre MasterPure RNA Extraction Kit
2. Vortex shaker
3. Two heat blocks
4. Microcentrifuge
5. Ice tray large enough to hold rack for microcentrifuge tubes
6. Nanodrop
7. Gel electrophoresis machine
8. Bioanalyzer

2.4 Library Preparation

1. TruSeq Stranded mRNA Sample Prep Kit v2
2. 1 μg of total RNA per sample
3. Vortex shaker
4. Microcentrifuge
5. Thermal cycler
6. Streptavidin-coated magnetic beads
7. AMPure XP beads
8. Magnetic stand
9. ERCC Spike-in kit
10. Bioanalyzer

2.5 Bioinformatic Pipeline

1. A reference genome assembly (referred to in the code as $reference_genome.fasta) and gene annotations (referred to in the code as $reference_annotations.gff) for each species included in the experiment, as well as the ERCC spike-in transcripts (*see* **Notes 4** and **5**).
2. Pair-ended and strand-specific RNA sequencing data in the fastq format (referred to in the code as $read_1.fastq and $read_2.fastq, respectively).
3. The following software packages and their dependencies must be installed on a machine with a compatible Linux-based OS (e.g. Debian, Ubuntu, or CentOS): Trimmomatic [6], FastQC [7], Bowtie2 [8], FastQ Screen [9], SAMtools suite [10], Trinity [11], Transrate [12], GMAP [13], Cufflinks suite [14], Salmon [15], BLAST+ suite [16], M-GCAT [17].

3 Methods

3.1 Culture Preparation

1. Streak yeast from frozen glycerol stock onto rich medium Petri dishes (20% agar), seal with Parafilm, then incubate at room temperature.
2. Once individual colonies are large enough to be visible on the plates (24 h–48 h), isolate an isogenic colony and transfer to 5 mL liquid rich medium in culture tubes. Shake at 30 °C. Allow the yeast to grow in the culture tubes until saturation (approximately 48 h).
3. Measure the OD of the saturated culture in the spectrometer (*see* **Note 6**).
4. Inoculate the appropriate amount of diluted saturated culture in 20 mL of sterile medium in a 125 mL flask to achieve a concentration of OD 0.01–0.02 (approximately 1–2 × 10^5 cells of *S. cerevisiae*) (*see* **Note 7**).

5. Measure the initial OD, then continue to measure the OD every hour. Use these measurements to calculate the doubling time during log growth phase (*see* **Notes 8** and **9**).
6. When OD is approaching the desired final OD (0.25 in our experiments [18]), prepare several sterile labeled 1.5 mL microcentrifuge tubes and put them in ice. Immediately after taking the final OD measurement, transfer 1 mL of culture to each microcentrifuge tube on ice, then centrifuge at 4 °C 10,000 × *g* for 1 min.
7. Aspirate the supernatant, leaving the cell pellet at the bottom of the tube. Freeze the pellet to −80 °C.

3.2 RNA Purification

1. Using one frozen cell pellet, purify the RNA using the Epicentre MasterPure RNA Extraction Kit, following the manufacturer's protocol (*see* **Notes 10** and **11**).
2. Measure RNA quality and quantity using Nanodrop, EtBr gel electrophoresis, and/or Bioanalyzer (*see* **Notes 12** and **13**).

3.3 Library Preparation

1. Select for polyA+ mRNAs with streptavidin-coated magnetic beads using 1 μg of total RNA.
2. Add 2 μL diluted ERCC spike-in mix (1:1000 dilution) to the polyA+ mRNA (*see* **Note 14**).
3. Follow the TruSeq Stranded mRNA Sample Prep Kit v2 according to the manufacturer's protocol.
4. Validate the quality of the library via Bioanalyzer.

3.4 Sequencing

Libraries should be sequenced to the appropriate depth on a next-generation sequencing platform such as the Illumina HiSeq 2500 (*see* **Note 15**).

3.5 Bioinformatic Pipeline (Fig. 1)

The commands provided in this section are intended to be run in a Unix shell terminal; variables beginning with "$" represent pre-defined environmental variables, which act as placeholders for the full path of the appropriate files.

3.5.1 Quality Control of Raw Sequencing Data

1. Use FastQC [7] to assess the quality of the raw sequencing data, identify potential problems such as untrimmed adapters or technical artifacts, and take the appropriate steps to address them. The program can be run on the fastq file of each read pair individually as follows:

```
fastqc $read_1.fastq
```

2. Run Trimmomatic [6] to trim any remaining adapter sequences, separate paired reads from unpaired reads, remove

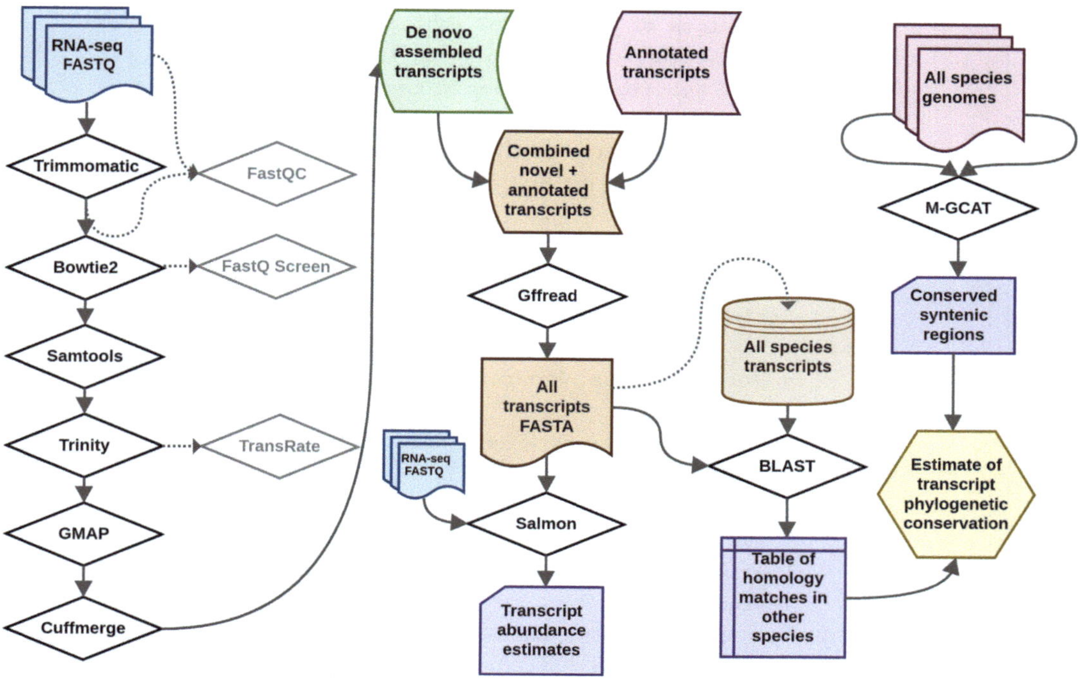

Fig. 1 An overview of the bioinformatic pipeline

low quality reads, and trim low quality bases from the ends of otherwise high-quality reads.

```
java -jar trimmomatic.jar PE -phred33 $read_1.fastq \$read_2.
fastq $trimmed_read_1_paired.fastq \
$trimmed_read_2_paired.fastq $trimmed_read_1_unpaired.fastq \
$trimmed_read_2_unpaired.fastq ILLUMINACLIP: \
$illumina_adapters.txt:2:33:20:2:true LEADING:36 \
TRAILING:32 SLIDINGWINDOW:4:30 MINLEN:35
```

After running Trimmomatic, it is recommended to re-run FastQC to ensure that all adapters and low-quality bases were removed.

3. Screen for potential contamination in the samples using FastQ Screen [9] (*see* **Note 16**). This step can also be used to approximate the fraction of total reads which correspond to transcripts from the ERCC spike-in mix.

```
# Build a bowtie2 index for each species in the experiment, as
well
# as the ERCC spike-in transcripts and potential contaminants
bowtie2-build $reference_genome.fasta $species_name
# Create a configuration file for FastQ Screen which points to
the
```

```
# location of the bowtie2 indices, then run the program
fastq_screen --aligner bowtie2 --subset 100,000 \
--conf $config_file $trimmed_read_1_paired.fastq
```

4. Map the reads to the reference genome and ERCC spike-in transcripts with Bowtie2 [8]:

```
# Combine the ERCC spike-in transcripts and reference genome
cat $ERCC_spike-in.fasta $reference_genome.fasta > \
$reference_genome_with_ERCC.fasta

# Rebuild index for reference genomes including ERCC tran-
scripts
bowtie2-build $reference_genome_with_ERCC.fasta \
$reference_with_ERCC

# Run Bowtie2
bowtie2 -x $reference_with_ERCC -1 $trimmed_read_1_paired.
fastq \
-2 $trimmed_read_2_paired.fastq \
-S $reference_with_ERCC.sam
```

5. Use Perl and Samtools [10] to remove unwanted reads from the mapped reads in the SAM file (*see* **Note 17**):

```
# Remove reads with over 2 mismatches and unpaired reads
perl -ne "print if((/XM:i:[0-2][^0-9]/) || (/^@/));" $refer-
ence_with_ERCC.sam > \
$reference_with_ERCC_filtered.sam.
# Remove unmapped reads and convert to BAM format
samtools view -F 0x04 -Sb $reference_with_ERCC_filtered.sam \
> $reference_with_ERCC_filtered.bam.
# Sort the BAM file
samtools sort $reference_with_ERCC_filtered.bam \
$reference_with_ERCC_filtered_and_sorted.bam.
# Create a BAM index file
samtools index $reference_with_ERCC_filtered_and_sorted.bam
```

3.5.2 Construct Robust Annotations

1. Use Trinity [11] to assemble a transcriptome de novo i.e. without using the reference as a guide. Running Trinity in genome-guided mode, as shown below, can speed up the de novo assembly (*see* **Note 18**).

```
Trinity --seqType fq --left $read_1.fastq \
--right $read_2.fastq --SS_lib_type FR \
--genome_guided_bam $reference_with_ERCC_filtered_and_sorted.
```

```
bam \
--verbose --normalize_max_read_cov 200 --jaccard_clip \
--genome_guided_max_intron 1002 --min_kmer_cov 2 \
--max_reads_per_graph 300,000 --min_glue 5 \
--group_pairs_distance 300 --min_contig_length 200
```

2. Use TransRate [12] to evaluate the quality of the de novo assembly (*see* **Note 19**).

```
# Create a reference transcriptome fasta file from the annota-
tions
gffread -w $reference_annotations.fasta -g $reference_genome.
fasta \
$reference_annotations.gff

# run Transrate
transrate --assembly $trinity_contigs.fasta --left $read_1.
fastq \
--right $read_2.fastq --reference $reference_annotations.fas-
ta
```

3. Align the contigs from the Trinity de novo assembly to the reference genome using GMAP [13] to generate a GFF file (*see* **Notes 20** and **21**):

```
# First, build an index for GMAP using the reference genome
gmap_build -d $species_name_with_ERCC -k 13 $reference_genome_
with_ERCC.fasta.

# Second, align the Trinity contigs to the genome
gmap -t 3 -n 0 -E genomic -d $species_name_with_ERCC \
$trinity_contigs.fasta -f 2 --min-identity 0.9 \
--invertmode 2  >  $trinity_contigs.gff
```

4. Compare the de novo assembly GFF file to the reference annotations GFF file using Cuffmerge [14] to identify which of the de novo assembled transcripts correspond to annotated features, and which are novel. Next, combine the transcripts from the reference annotations with the novel transcripts from the de novo assembly (*see* **Notes 22–24**)

```
# Compare the de novo assembly to the reference annotations
cuffcompare -r $reference_annotations.gff -G $trinity_contigs.
gff \ -o $cuffcompare_output

# Select the novel transcripts, excluding spike-in transcripts
grep "class_code .[xu]" | grep -v "ERCC" \
$cuffcompare_output.combined.gtf > $novel_transcripts.gtf
```

```
# Convert novel gtf file to gff format
gffread -E $novel_transcripts.gtf -o- > $novel_transcripts.gff

# Combine reference annotations and novel transcripts
cat $reference_annotations.gff $novel_transcripts.gff > \
$combined_transcriptome.gff
```

5. Quantify expression with Salmon [15] (*see* **Note 25**)

```
# Create a combined transcriptome fasta file from the
combined gff.
gffread -w $combined_transcriptome.fasta -g $reference_genome.
fasta \
$combined_transcriptome.gff

# Build a Salmon index based on the combined transcriptome
salmon index -t $combined_transcriptome.fasta -i $combined_-
transcriptome_index -k 31

# Convert pre-processed BAM file to fastq files for quantifi-
cation
samtools fastq $reference_with_ERCC_filtered_and_sorted.bam \
-1 $filtered_reads_R1.fastq.gz -2 $filtered_reads_R1.fastq.gz \
-0 /dev/null -s /dev/null -n

# Run salmon in mapping-based mode
salmon quant -i $combined_transcriptome_index -l ISR \
-1 $filtered_reads_R1.fastq.gz -2 $filtered_reads_R2.fastq.gz
\ --validateMappings -o $salmon_output
```

3.5.3 Determine Taxonomic Conservation of Transcripts

1. Search for homologs across the transcriptomes of the other species included in our comparison using BLASTN and TBLASTX [16] (*see* **Note 26**).

```
# Create a multifasta from each transcriptome.
gffread -W -w $combined_transcriptome_multifasta.fa \
-g $reference_genome_with_ERCC.fasta $combined_transcriptome.
gff

# Make a BLAST database for each transcriptome
makeblastdb -in $combined_transcriptome_multifasta.fa dbtype
nucl \
-out $combined_transcriptome_nucl.db -hash_index \ -title
$combined_transcriptome_database
```

```
# Run search for homologs of each transcript in the other
species using BLASTN; the file name placeholders are modified
to highlight that this is comparison is across two different
species
blastn -query $species_A_combined_transcriptome_multifasta.fa
\
-db $species_B_combined_transcriptome_database -out $specie-
s_A_vs_species_B_BLASTN_output.tab \
-dust yes -evalue 0.001 -strand plus -num_threads 1 \
-max_target_seqs 10,000 \
-outfmt "6 qseqid sseqid pident length mismatch gapopen qstart
qend sstart send evalue bitscore sallseqid score nident
positive gaps ppos qframe sframe qlen slen"

# Run search for homologs of each transcript in the other
species using TBLASTX
tblastx $species_A_combined_transcriptome_multifasta.fa \
-db $species_B_combined_transcriptome_database \
-out $species_A_vs_species_B_TBLASTX_output.tab \
-evalue 0.001 -num_threads 1 -max_target_seqs 10,000 -strand
plus \
-seg yes -outfmt "6 qseqid sseqid pident length mismatch
gapopen qstart qend sstart send evalue bitscore sallseqid
score nident positive gaps ppos qframe sframe qlen slen"
```

Using the set of files generated by the pairwise BLASTN comparisons ($species_A_vs_species_B_BLASTN_output.tab) combined with the files from the pairwise TBLASTX comparisons ($species_A_vs_species_B_TBLASTX_output.tab), one can begin to estimate the conservation of each transcript across the phylogeny (*see* **Note 27**).

2. Search for potential orthologs in syntenic blocks between closely-related species using M-GCAT [17]. To ensure that homology detection failure does not bias the classification of taxonomic conservation of transcripts [19, 20], compliment the homology searches with a synteny-based analysis. M-GCAT has an easy-to-use graphical user interface, and can output the genomic coordinates of the syntenic regions conserved between species. Transcripts which appear in the same syntenic region across different species (with the same genomic context) could be considered as orthologous.
3. The results of the pairwise BLAST searches can be combined with the results of the M-GCAT syntenic comparisons to form a more robust estimate of transcript conservation across a phylogeny. The criteria used to define what constitutes a "conserved transcript" depends on what the aims of the study are, but in general, an approach which combines both synteny and homology is more powerful (*see* **Note 28**).

4 Notes

1. Microwell plates can be used to collect data for multiple species at the same time (e.g., viability in a medium, growth rate, and stress tolerance), but if the final experiment is scaled up to larger flasks, these measurements will need to be repeated.
2. Take care to prepare, sterilize, and store the media the same way every time the experiments are repeated.
3. If the lab or common culture room is shared with researchers growing bacteria, after sterilizing the media, 1 mL Ampicillin (stock concentration 100 mg/mL) per 1 L can be added to avoid contamination.
4. It can be difficult to find a genome assembly / reference annotations which has chromosome-level contigs, which is as close to the target species strain as possible, and has well curated annotations. Fortunately, the quality and completeness of reference annotations continue to improve.
5. A more conservative estimation of taxonomic conservation can be achieved by using the NR database or a set of outgroup species to identify transcripts with distant homologs not present in the other species which were sequenced which could have arisen from horizontal gene transfer events.
6. Accurately measuring the OD of saturated culture will likely require a dilution of at least 1:10, taking care to use the same sterile medium as before, as the OD will often exceed the upper range of the spectrophotometer.
7. This concentration is low enough to allow the yeast to bud several times in log growth phase, but not so low that they would fail to start growing.
8. Some species may have doubling times that are several times longer than *S. cerevisiae*; plan the start times of future experiments accordingly.
9. Some species are filamentous or can flocculate at a lower OD, which can lead to inaccurate measurements.
10. We obtained similar results using a hot phenol RNA purification protocol.
11. This stage was performed in batches of 2–4 samples; processing too many samples at a time can lead to delays during key steps, and timing during some steps of RNA purification is crucial.
12. On average, we extracted 2.5 μg of pure total RNA from each cell pellet, and observed Bioanalyzer RIN scores between 6 and 9.

13. The extraction efficiency may vary between species due to phenotypic differences such as the thickness of the cell wall.
14. The amount of ERCC spike-in to add to the libraries will depend on the experimental design. Typically, the goal is to produce enough spike-in transcript reads in order to assess the lower limit of detection and detect potential biases, but without sacrificing the sequencing depth of the sample library too much; we aimed for 2%–4%.
15. In our case (paired-end, stranded, 50 nt sequencing on Illumina HiSeq 2500, v4 chemistry), a dozen libraries were pooled in one lane which yielded approximately 20 million high-quality read pairs per sample.
16. Frequent contaminants of yeast samples can include human, other fungal, and bacterial nucleic acids. This step can also serve to identify potentially mislabeled samples if you include all of the genomes of the species in your study in this screen.
17. One may want to allow for more than two mismatches if, for example, the reference genome was not the same strain as the sample, or if the read length is over 50 nt.
18. Computationally demanding steps like de novo assembly should be run on a machine with at least 24 cores and 120 GB of RAM.
19. It may require several iterations of running Trinity and then TransRate to optimize the values of different parameters (e.g. genome_guided_max_intron, min_contig_length) in the assembly step. The optimal values of these parameters depends on several criteria e.g. the properties of the reference genome of the target species, the fragment size, and the depth of the sequencing data.
20. If this step results in many unmapped transcripts, consider lowering the value of the --min-identity parameter. Unmapped transcripts could also be included in the databases of the homology searches, depending on your approach.
21. The ERCC spike-in transcripts can be used to calculate the lower limit of detection of your pipeline, as well as finding minimum level of expression for which the de novo assembly was able to reliably recover the full-length transcripts.
22. If the quality of the de novo assembly has been optimized e.g. using TransRate, it will be easier to differentiate between transcripts which correspond to novel features and the ones which correspond to annotated features.
23. Inspect the mapped reads, the annotations, and the de novo assembly together in a genome browser to ensure that the novel transcripts were identified correctly.

24. We encountered that Cuffcompare occasionally classified some novel transcripts as class_code "."; thus, manual inspection of class_code "." transcripts may recover additional novel transcripts.
25. There are many programs which use different approaches to quantify transcript expression; it may be worthwhile to compare the output from several programs to see how the computed expression values compare.
26. Selecting the appropriate values for different BLAST parameters, such as the e-value, is important to ensure that your homology detection will return the desired results; notwithstanding, the presence or absence of a homology match is not always biologically relevant [20].
27. The output files from BLAST can be used to generate a table of homology detection for each transcript across the included databases; a pseudo-code example which transforms these files into tables of homology matches is available as supplementary material online. Additional filters can be applied to define what is a reliable homology match (e.g., length, reciprocal matches, identity, e-value, bitscore). The estimate of the conservation of a transcript can also be extended to include homologs of paralogs (homology matches between transcripts in the same species).
28. The criteria used to define which transcripts will be considered as orthologous depends on the aims of the study; for example, some transcript pairs may only partially overlap the syntenic blocks defined by M-GCAT, so one must decide whether to consider these transcripts as orthologous or not.

Acknowledgments

This work was made possible by several grants including PGC2018-094091-B-I00, BFU2015-65235-P, BFU2015-68351-P, BFU2016-80039-R, TIN2015-69175-C4-3-R, and RTI2018-094403-B-C33 from the Spanish Government FEDER (EU), as well as PT17/0009/0014 from the Instituto de Salud Carlos III. We also received support from grants 014SGR1121, 2014SGR0974, 2017SGR1054, and 2017SGR01020 from the Agència de Gestió d'Ajuts Universitaris i de Recerca Generalitat de Catalunya (AGAUR), and MDM-2014-0370 from the "Maria de Maeztu" Program for Units of Excellence in R&D.
We would like to thank Dr. Ksenia Pugach and the Verstrepen lab for donating cultures of several species of yeast. We also thank the sequencing facilities of the Centre for Genomic Regulation (CRG) and Universitat Pompeu Fabra (UPF) for their contributions.

References

1. Wohlbach DJ, Thompson DA, Gasch AP, Regev A (2009) From elements to modules: regulatory evolution in Ascomycota fungi. Curr Opin Genet Dev 19:571–578. https://doi.org/10.1016/j.gde.2009.09.007
2. Brion C, Lutz SM, Albert FW (2020) Simultaneous quantification of mrna and protein in single cells reveals post-transcriptional effects of genetic variation. Elife 9:1–34. https://doi.org/10.7554/eLife.60645
3. Blevins WR, Ruiz-Orera J, Messeguer X et al (2021) Uncovering de novo gene birth in yeast using deep transcriptomics. Nat Commun 12: 604. https://doi.org/10.1038/s41467-021-20911-3
4. Gasch AP (2007) Comparative genomics of the environmental stress response in ascomycete fungi. Yeast. 24:961–976. https://doi.org/10.1002/yea.1512
5. Tsankov AM, Thompson DA, Socha A et al (2010) The role of nucleosome positioning in the evolution of gene regulation. PLoS Biol 8: e1000414. https://doi.org/10.1371/journal.pbio.1000414
6. Bolger AM, Lohse M, Usadel B (2014) Trimmomatic: a flexible trimmer for Illumina sequence data. Bioinformatics 30:2114–2120. https://doi.org/10.1093/bioinformatics/btu170
7. Andrews S, Krueger F, Segonds-Pichon A et al (2010) FASTQC: a quality control tool for high throughput sequence data. Babraham Bioinformatics, pp 5–9
8. Langmead B, Salzberg SL (2012) Fast gapped-read alignment with Bowtie 2. Nat Methods 9: 357–359. https://doi.org/10.1038/nmeth.1923
9. Wingett SW, Andrews S (2018) Fastq screen: A tool for multi-genome mapping and quality control [version 1; referees: 3 approved, 1 approved with reservations]. F1000Res 7:1–13. https://doi.org/10.12688/f1000research.15931.1
10. Li H, Handsaker B, Wysoker A et al (2009) The sequence alignment/map format and SAMtools. Bioinformatics 25:2078–2079. https://doi.org/10.1093/bioinformatics/btp352
11. Grabherr MG, Brian JH, Moran YJ, Levin Z, Thompson DA, Amit I, Adiconis X, Fan L, Raychowdhury R, Zeng Q, Chen Z, Mauceli E, Hacohen N, Gnirke A, Rhind N, Palma F, Birren BW, Friedman N, Regev A (2013) Trinity: reconstructing a full-length transcriptome without a genome from RNA-Seq data. Nat Biotechnol 29:644–652. https://doi.org/10.1038/nbt.1883.Trinity
12. Smith-Unna R, Boursnell C, Patro R et al (2016) TransRate: reference-free quality assessment of de novo transcriptome assemblies. Genome Res 26:1134–1144. https://doi.org/10.1101/gr.196469.115
13. Wu TD, Watanabe CK (2005) GMAP: a genomic mapping and alignment program for mRNA and EST sequences. Bioinformatics 21:1859–1875. https://doi.org/10.1093/bioinformatics/bti310
14. Trapnell C, Roberts A, Goff L et al (2012) Differential gene and transcript expression analysis of RNA-seq experiments with TopHat and cufflinks. Nat Protoc 7:562–578. https://doi.org/10.1038/nprot.2012.016
15. Patro R, Duggal G, Love MI et al (2017) Salmon provides fast and bias-aware quantification of transcript expression. Nat Methods 14: 417–419. https://doi.org/10.1038/nmeth.4197
16. Camacho C, Coulouris G, Avagyan V et al (2009) BLAST+: architecture and applications. BMC Bioinformatics 10:1–9. https://doi.org/10.1186/1471-2105-10-421
17. Treangen TJ, Messeguer X (2006) M-GCAT: interactively and efficiently constructing large-scale multiple genome comparison frameworks in closely related species. BMC Bioinformatics 7:433. https://doi.org/10.1186/1471-2105-7-433
18. Blevins WR, Carey LB, Mar AM (2019) Transcriptomics data of 11 species of yeast identically grown in rich media and oxidative stress conditions. BMC Res Notes 12:250. https://doi.org/10.1186/s13104-019-4286-0
19. Moyers BA, Zhang J (2018) Toward reducing phylostratigraphic errors and biases. Genome Biol Evol 10:2037–2048. https://doi.org/10.1093/gbe/evy161
20. Weisman CM, Murray AW, Eddy SR (2020) Many, but not all, lineage-specific genes can be explained by homology detection failure. PLoS Biol 18:1–24. https://doi.org/10.1371/journal.pbio.3000862

Part II

DNA Replication and Protein/DNA Interactions

Chapter 8

FORK-seq: Single-Molecule Profiling of DNA Replication

Magali Hennion, Bertrand Theulot, Jean-Michel Arbona, Benjamin Audit, and Olivier Hyrien

Abstract

Most genome replication mapping methods profile cell populations, masking cell-to-cell heterogeneity. Here, we describe FORK-seq, a nanopore sequencing method to map replication of single DNA molecules at 200 nucleotide resolution using a nanopore current interpretation tool allowing the quantification of BrdU incorporation. Along pulse-chased replication intermediates from *Saccharomyces cerevisiae*, we can orient replication tracks and reproduce population-based replication directionality profiles. Additionally, we can map individual initiation and termination events. Thus, FORK-seq reveals the full extent of cell-to-cell heterogeneity in DNA replication.

Key words DNA replication, Whole-genome, Single-molecule analysis, Nanopore sequencing, Convolutional neural network, Replication fork direction, Replication origins, Termination sites

1 Introduction

Eukaryotes replicate DNA by stochastically activating numerous replication origins [1]. Cell-population methods provide only an average picture where intercellular variability and rare events are masked. Solving the nature of origins and analyzing fork progression thus requires high-throughput analyses at the single molecule level. Current single molecule methods such as DNA combing can reveal fork progression and spacing by monitoring nonstandard nucleotide incorporation with antibodies [2]. However, they provide no sequence information unless they are combined with DNA probes [3]. This is laborious, low resolution, and low throughput (typically 6 months of work to study one locus of 1 Mb). Hence, high-throughput single molecule genomic profiling of DNA replication is transformative to replication studies [4, 5].

Here, we describe FORK-seq, one such high-throughput genomic protocol to characterize DNA replication on single DNA molecules. As in many previous replication studies, the progression

Frédéric Devaux (ed.), *Yeast Functional Genomics: Methods and Protocols*,
Methods in Molecular Biology, vol. 2477, https://doi.org/10.1007/978-1-0716-2257-5_8,

of replication forks is revealed by monitoring the incorporation of BrdU, a thymidine analog. Originally, we use the nanopore sequencing technology from Oxford Nanopore Technologies (ONT) to obtain the position of each analyzed DNA molecule along the reference genome as well as its BrdU incorporation profile. The former relies on sequence alignment thanks to the robustness of ONT basecalling tools with respect to thymidine substitution by BrdU, while the latter can be achieve using one of two new nanopore current processing pipelines we integrated in the RepNano software suite. Both pipelines use the series of current shifts aligned to the DNA sequence to either (1) assess which thymidine sites are BrdU substituted based on tabulated current shifts matrices (TM) or (2) directly estimate the proportion of substituted sites over 96 bp windows using a trained neural network (CNN) [4].

FORK-seq was successfully applied to a *Saccharomyces cerevisiae* strain whose growth fully depends on the external supply of thymidine or BrdU. The protocol consists in (1) growing yeasts exposing them to a short (contributing to the high resolution of the protocol) BrdU pulse followed by a thymidine chase, (2) extracting genomic DNA and preparing sequencing libraries, (3) sequencing using the MinION from ONT and basecalling the reads, and (4) aligning the reads to the reference genome and quantifying their BrdU incorporation profiles. BrdU incorporation profiles present localized asymmetric motifs where a sharp transition from low to high values is followed by a shallower decrease as the signature of replication forks progression at the time of the BrdU pulse and during the chase, respectively (*see* Fig. 1).

Dedicated post-processing tools from the RepNano software suite can then be used to automatically determine replication fork locations and orientations. When two diverging (resp. converging) forks are detected on the same nanopore read, the location of an initiation (resp. termination) event can also be deduced [4]. All the steps from yeast culture to final lists of forks as well as initiation and termination events on their genomic location are described in this protocol that can be fully completed within 1 week.

Thus, we reported over 1.4 million nanopore reads the identification of 58,651 BrdU content motifs associated to oriented-forks and the location of 4964 and 4485 individual initiation and termination events, respectively [4] (*see* Fig. 1). Bioinformatic analyses showed that replication directionality profile derived from single forks is fully compatible with population-based profiles, and that most events cluster at known origins and fork merging zones, but that 9% and 18% of initiation and termination events, respectively, occur at previously unreported locations. This illustrates that FORK-seq protocol can quantify DNA replication program stochasticity over a complete genome.

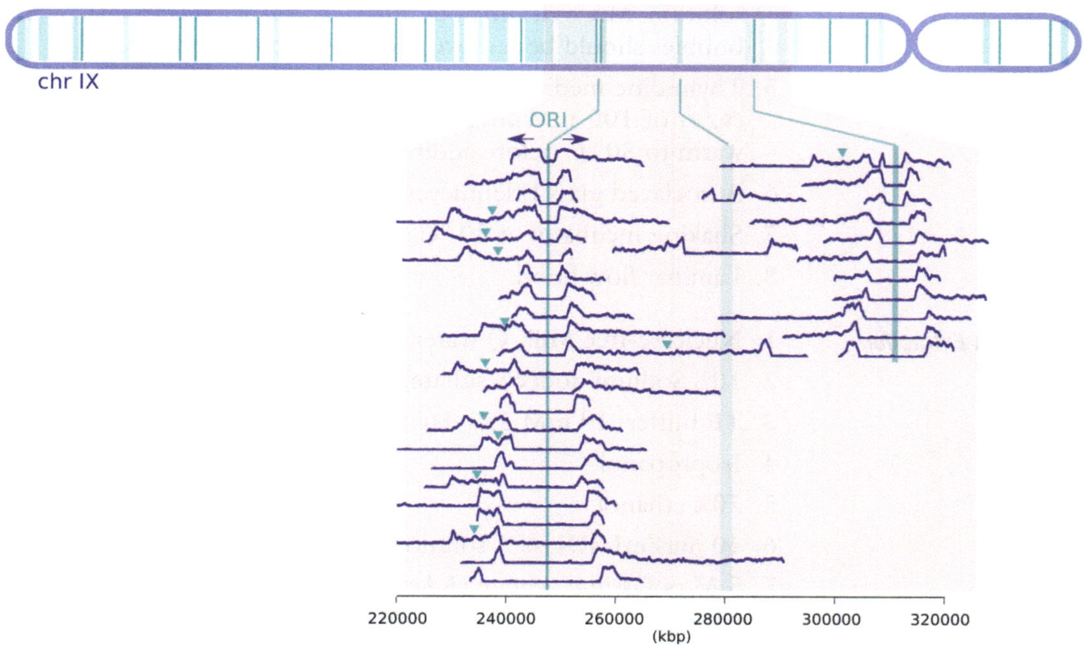

Fig. 1 Illustration of FORK-seq. In purple are shown the BrdU incorporation signals from reads containing an initiation event mapped within a 100 kb window on yeast chromosome IX. Green bars represent known origins from OriDB [11]. Purple arrows show fork orientation and green triangles highlight termination events

2 Materials

2.1 Yeast Strain

MCM869 genotype is *MATa ade2–1 trp1–1 can1–100 leu2–3 his3–11,15 ura3–1::URA3-GPD-TK7x aur1::AUR1-C-ADH-hENT1 bar1::LEU2 cdc21::kanMX* [6] (*see* **Note 1**).

2.2 Yeast Culture and Labelling

1. Liquid YPD medium: Suspend 100 g of YPD powder in 1.8 L of distilled water (*see* **Note 2**). Make up to 2 L with water. Fill five 500 mL bottles with 400 mL of medium. Autoclave for 15 min at 110 °C. Store at RT protected from light.
2. 100 mM thymidine solution: Dissolve 1 g of thymidine (MW 242.23 g.mol^{-1}) into 41.3 mL of ultrapure water. Filter-sterilize with 0.22 μm filter. Store in 1 mL aliquots at −20 °C.
3. 100 mM BrdU solution: Dissolve 1 g of BrdU (MW 307.1 g. mol^{-1}) into 32.6 mL of DMSO. Store in 1 mL aliquots at −20 °C.
4. YPD thymidine plates: Suspend 67 g of YPD agar in 1 L of distilled water. Heat to boiling while stirring to dissolve all ingredients. Autoclave for 15 min at 110 °C. Let the mixture cool down to ~55 °C. Add 1 mL of 100 mM thymidine

solution. Mix well and pour into sterile, labeled petri dishes. Air bubbles should be removed by slow orbital sloshing.

5. Thymidine medium: The day of the experiment, add 1/1000 (v/v) of 100 mM thymidine solution to YPD medium. Pre-warm to 30 °C before adding it to the cells.
6. Autoclaved glass Erlenmeyer flasks (50 and 500 mL).
7. Shaking incubator at 30 °C.
8. Laminar flow hood.

2.3 DNA Extraction

1. Nuclease-free Milli-Q water.
2. 10% Sodium dodecyl sulfate (SDS).
3. TE buffer: 10 mM Tris–HCl pH 8.0, 1 mM EDTA, filtered.
4. Isopropanol.
5. 70% ethanol.
6. 40 mg/mL RNase A solution.
7. 5 M potassium acetate (KAc), pH 7.5.
8. Y1 buffer: 1 M sorbitol, 100 mM EDTA pH 8.0, 14 mM β-mercaptoethanol. For a 1-L solution, dissolve 182.2 g sorbitol in 600 mL distilled water. Add 200 mL of 0.5 M EDTA. Under a chemical hood, add 1 mL of β-mercaptoethanol (14.3 M). Adjust the volume to 1 L with distilled water. Store at 4 °C.
9. 1 U/μL Zymolyase. Prepare a master solution for all your samples just before use by diluting in Y1 buffer to 1 U/μL. Using Zymolyase 20 T (20 U/mg), this corresponds to a 50 mg/mL solution.
10. Lysis buffer: 114 mM Tris–HCl pH 8.0, 115 mM EDTA, 571 mM NaCl, 1.14% PVP40. Prepare a fresh master solution for all your samples. Per sample, mix 500 μL of 1 M Tris–HCl pH 8.0, 500 μL of 0.5 M EDTA and 500 μL of 5 M NaCl. Add 50 mg of PVP 40, and complete with 2.875 mL of water. Warm up 30 min at 65 °C and filter sterilize with 0.45 μm filter.
11. Qubit fluorometer.
12. Qubit dsDNA BR Assay Kit.
13. Pulsed-field gel electrophoresis system or Agilent TapeStation.
14. 50 °C water bath.
15. NanoDrop (Thermofisher).

2.4 Sequencing Libraries

1. Ligation Sequencing Kit (ONT).
2. NEBNext FFPE Repair Mix.
3. NEBNext Ultra II End repair/dA-tailing Module.

4. If multiplexing, Native Barcoding Expansion 1–12 or/and 13–24 (ONT).
5. If multiplexing, NEB Blunt/TA Ligase Master Mix.
6. NEBNext Quick Ligation Module.
7. Agencourt AMPure XP beads.
8. Tris buffer: 10 mM Tris–HCl pH 8.0, filtered.
9. Freshly prepared 70% ethanol in Tris buffer.
10. 1.5 mL Eppendorf DNA LoBind tubes.
11. 0.2 mL PCR tubes.
12. Magnetic rack.
13. Benchtop centrifuge.
14. Thermal cycler or thermoblock.
15. 37 °C water bath.
16. Hula mixer.

2.5 Sequencing

1. R9 flow cell (ONT).
2. Flow Cell Priming Kit (ONT).
3. MinION Mk1B or MinION Mk1C or GridION (ONT) with adapted computational resources.

3 Methods

All yeast manipulations are done under sterile conditions.

3.1 Yeast Culture

1. Streak frozen MCM869 cells with a sterile tip onto a YPD thymidine plate. Let grow in the dark at 30 °C for 48 h to 72 h until you see small colonies. The plate can be stored at 4 °C for a maximum of 2 weeks.
2. The day before the labelling, pick one colony from the plate using a sterile tip and transfer it into 10 mL thymidine medium.
3. Measure OD_{600} and dilute the cells into 100 mL fresh thymidine medium adjusting the volume of cells so that the OD_{600} would reach 1 the next morning (beginning of the exponential phase, *see* **Note 3**). Grow cells in a 500 mL Erlenmeyer flask at 30 °C in the dark under constant agitation (typically 200 rpm).

3.2 Replication Labelling

1. The day of the experiment, check that OD_{600} is around 1. Dilute the cells to 0.1 OD_{600} in 100 mL of fresh thymidine medium. Let grow the cells for 8 h (OD_{600} ~ 0.8).
2. Transfer the cells to two 50 mL conical tubes and centrifuge in a swinging bucket at 30 °C for 2 min at 3000 × *g*. Remove

carefully the supernatant by pouring or by pipetting, taking care of not removing part of the pellet.

3. Wash with 100 mL of YPD medium (without thymidine), repeating the centrifugation and the supernatant removal.
4. Resuspend into 100 mL YPD medium (without thymidine) and put the cells back into the incubator for 30 min at 30 °C (*see* **Note 4**).
5. Add 100 μL of 100 mM BrdU solution with constant shaking, processing quickly to avoid medium cooling. After 4 min at 30 °C (*see* **Note 5**), add 1 mL of 100 mM thymidine and incubate for 45 min at 30 °C.
6. Centrifuge the cells for 5 min at 5000 × *g* (room temperature). Remove the supernatant, wash the pellet with water and put the tubes with cell pellets at −20 °C. They can be stored for several months.

3.3 DNA Extraction

This protocol is derived from [7] with minor changes. It is optimized for a yeast pellet corresponding to 100 mL of OD_{600} ~ 1 suspension, for an expected yield of 10–20 μg of DNA. In the original paper [4], we used a different protocol described in **Note 6**.

3.3.1 Cell Wall Lysis

1. Resuspend the pellet in 4 mL of Y1 buffer, and add 250 μL of Zymolyase solution in a 50 mL tube. Make sure the pellet is well resuspended.
2. Incubate 30 min at 37 °C with gentle shaking (300 rpm). Centrifuge for 3 min at 3000 × *g*. Discard the supernatant.

3.3.2 Spheroplasts Lysis and Protein Precipitation

1. Resuspend the pellet in 3.5 mL of Lysis buffer using a wide bore tip. It is crucial that spheroplasts are well resuspended for proper lysis.
2. Add 500 μL of SDS 10% and 15 μL of RNase A solution (*see* **Note 7**). Flip gently 10 times to homogenize. Incubate 30 min at 50 °C with moderate shaking (300 rpm).
3. Put the lysate tube on ice. Add 10 mL of TE buffer and 5 mL of 5 M KAc, pH 7.5. Invert with precaution 20 times. A white precipitate should appear at this stage. Incubate on ice for at least 5 min.
4. Centrifuge for 10 min at 5000 × *g* (4 °C). Transfer the supernatant to a new 50 mL tube. If some white precipitate remains, repeat the previous centrifugation.

3.3.3 DNA Precipitation

1. Add 1 volume of isopropanol to the supernatant. Invert gently 30 times so that the solution is well homogenized. Bad homogenization will result in low yield. It is possible to see some DNA filaments at this stage.

2. Centrifuge for 20 min at 9500 × *g* at 4 °C. Discard the supernatant.
3. Wash the DNA pellet with 5 mL of 70% ethanol. Centrifuge 5 min at 1000 × *g*. Remove as much ethanol as possible. Repeat the previous centrifugation to remove completely the ethanol. Do not allow the pellet to dry as it will make it harder to resuspend.

3.3.4 Elution, Quantification and Quality Control

1. Add 100 μL of TE. Incubate the pellet for 30 min at 50 °C in a water bath to resuspend it.
2. Transfer to a 1.5 mL tube using a wide bore tip.
3. Quantify the amount of DNA with Qubit fluorometer (dsDNA BR Assay Kit, *see* **Note 8**). The final concentration should not be lower than 40 ng/μL to prepare sequencing libraries.
4. Measure A260/280 and A260/230 purity ratios using NanoDrop. Both ratios should be higher than 1.8 (*see* **Note 9**).
5. The size of the DNA is determined using pulsed-field gel electrophoresis or automatized electrophoresis system such as Agilent TapeStation (*see* **Note 10**). The majority of the DNA should be larger than 50–100 kb (*see* **Note 11**).

3.4 Nanopore Sequencing Libraries

This part follows ONT protocols with minor changes in incubation times, amount of beads and of DNA used at the different steps. Please refer to official documentation at https://nanoporetech.com/resource-centre/protocols for extra details and tips.

3.4.1 DNA Repair and Purification

1. Dilute 2 μg of DNA (*see* **Note 12**) with Tris buffer for a final volume of 48 μL in a PCR tube.
2. Add 3.5 μL of NEBNext FFPE Repair Buffer and 3.5 μL of NEBNext Ultra II End-prep Buffer. Complete with 3 μL of Ultra II End-prep enzyme mix and 2 μL of NEBNext FFPE DNA Repair enzyme mix. Flick gently the tube and incubate 20 min at 20 °C followed by 20 min at 65 °C using a thermal cycler (*see* **Note 13**).
3. Transfer the mix into a 1.5 mL DNA LoBind tube. Add 0.5 volume (30 μL, *see* **Note 14**) of AMPure XP beads previously intensively resuspended by vortexing. Mix by gently flicking the tube.
4. Incubate on a Hula mixer for 10 min at room temperature. Spin down the sample using a mini benchtop centrifuge (full speed for 3 s). Place the tube on a magnetic rack and wait until the eluate is clear.
5. Discard the supernatant and wash the beads twice with 300 μL of 70% ethanol. Spin down and put back on the magnet to remove the ethanol. Repeat this step to remove all residual

ethanol. Never let the beads' pellet dry as it can perturb the elution.

6. Remove the tube from the magnetic rack and add 25 μL of Tris buffer. Incubate 10 min in a 37 °C water bath. Put the tube back on the magnet and wait until the eluate is perfectly clear.
7. Transfer all the eluate in a new 1.5 mL DNA LoBind tube.
8. Quantify by Qubit the amount of DNA recovered using 1 μL of the eluate. The recovery aim is around 1.3 μg of DNA.

3.4.2 Native Barcode Ligation—Facultative

If you want to sequence several samples in one flow cell you have to add a specific barcode to each sample. If not, skip this step and proceed directly with adapter ligation.

1. In a new 1.5 mL DNA LoBind tube, dilute 750 ng of previous eluate in Tris buffer to a final volume of 22.5 μL. Add 2.5 μL of the selected Native Barcode and 25 μL of Blunt/TA Ligase Master Mix. Mix by gently flicking the tube and incubate for 1 h at room temperature.
2. Add 0.5 volume (25 μL) of AMPure XP beads previously intensively resuspended by vortexing. Mix by gently flicking the tube. Incubate on a Hula mixer for 10 min at room temperature.
3. Spin down the sample with a benchtop centrifuge. Place on a magnet and wait until eluate is clear. Discard the supernatant. Wash the beads twice with 300 μL of 70% ethanol. Spin down and put back on the magnet to remove the ethanol. Repeat this step to remove all residual ethanol.
4. Remove the tube from the magnetic rack. Add 25 μL of Tris buffer. Incubate 10 min at 37 °C in a water bath. Put the tube back on the magnet and wait until the eluate is perfectly clear. Transfer the eluate into a new 1.5 mL DNA LoBind tube.
5. Quantify by Qubit the amount of DNA recovered using 1 μL of the eluate. The recovery aim is 500 ng of DNA.
6. In a new DNA LoBind tube, pool the samples you want to sequence in the same flowcell so that you end up with 800 ng of DNA in total. To do so, adjust the volumes to put the same amount of DNA for each sample (or adjust accordingly if you want more reads from specific samples).

3.4.3 Adapter Ligation

This step is identical whether you have multiplexed samples or not, except that the Adapter Mix is different. Use Adapter Mix I if you process a single sample without barcode and Adapter Mix II for multiplexed samples.

1. Dilute 800 ng of previously prepared sample into 65 μL of Tris buffer. Add 5 μL of Adapter Mix (I or II, see above), 20 μL of

NEBNext Quick Ligation Reaction Buffer, and 10 μL of Quick T4 DNA Ligase. Mix by gently flicking the tube. Incubate at least 1 h at room temperature. We did notice significant increase in read length with longer incubation time.

2. Add 0.5 volume (50 μL) of AMPure XP beads previously intensively resuspended by vortexing. Mix by gently flicking the tube. Incubate on a Hula mixer for 10 min at room temperature. Spin down the sample with a benchtop centrifuge. Place on a magnet and wait until eluate is clear. Discard the supernatant.
3. Wash the beads twice with 250 μL Long Fragment Buffer (*see* **Note 15**). Spin down and put back on the magnet to remove the buffer. Repeat this step to remove all residual buffer.
4. Remove the tube from the magnetic rack and add 15 μL of Elution Buffer. Incubate 10 min at 37 °C in a water bath. Put the tube back on the magnet and wait until the eluate is perfectly clear. Transfer the eluate in a new 1.5 mL DNA LoBind tube.
5. Quantify by Qubit the amount of DNA recovered using 1 μL of the eluate. The recovery aim is 400 ng of DNA. Libraries can be stored at 4 °C for several days.

3.4.4 Priming the Flow Cell and Loading the Library

1. Equilibrate the MinION R9.4.1 flow cell at room temperature.
2. Open MinKNOW, insert the flow cell into the MinION device and perform the flow cell check. To do so, choose the flow cell type from the selector box. Then mark the flow cell as "Selected". Click "Check flow cells" at the bottom of the screen and select "FLO-MIN106." Click "Start test." Be sure that the number of pores detected is above the warranty before proceeding any further.
3. Prepare the priming mix by adding 30 μL of mixed Flush Tether into one tube of Flush Buffer.
4. Open the priming port and draw back a small volume to remove any bubble (a few μL, *see* **Note 16**). Load 800 μL of the priming mix through the priming port and wait for 5 min. In the meantime, prepare the library.
5. In a new 1.5 mL DNA LoBind tube, mix 12 μL of DNA (ideally 200 ng, *see* **Note 17**) with 37.5 μL of Sequencing Buffer and 25.5 μL of Loading Beads previously intensively resuspended by vortexing for a final volume of 75 μL.
6. Open the SpotON port, and add 200 μL of the priming mix into the priming port (not into the SpotON port).
7. Using a wide bore tip, resuspend the prepared library and inject it into the SpotON port by droplets. Close the SpotON port

and the priming port and proceed to data acquisition (*see* **Note 18**).

3.5 Nanopore Sequencing and Base Calling

This part is highly dependent on the material available. Using the MinION Mk1C, the base calling and the demultiplexing can be done directly on the device during and after the run (*see* **Note 19**). Using the MinION Mk1B, it is also possible to base call during the run if the computer has enough computational power (please refer to ONT recommendations). If not, the raw FAST5 files can be processed after the run using Guppy (*see* **Note 20**). The analysis is quite resource-consuming and the use of HPC cluster or adapted workstation (*see* **Note 21**) is necessary.

Moreover, one MinION run will typically generate 100–300 Go of raw data. Therefore, appropriate storage capacity must be available.

3.5.1 Concomitant Acquisition and Base Calling

1. For kit selection, select "SQK-LSL109" and if samples are multiplexed, select "EXP-NBD104" or "EXP-NBD114." Configure acquisition keeping default parameters.
2. For the base calling, choose "High-accuracy basecalling," and enable "barcoding" if you have multiplexed samples. This will split both FAST5 and FASTQ files according to their barcode. In "output," keep both FAST5 and FASTQ in the output directory. Keep the filtering value of the quality score (qscore) at 7 (*see* **Note 22**).
3. Start the sequencing and let it run for 72 h.

During the run, MinKNOW displays the activity of the pores and plots a histogram of read lengths. N50 should typically be in the 30–60 kb range. If multiplexing, two barplots showing the number of bases and of reads attributed to each barcode are displayed in real time. FAST5 and FASTQ files are generated. In addition, a `sequencing_summary.txt` file containing useful details about every single read is made (in the FASTQ folder). Using the Mk1C, the base calling will go on for several days after the end of the run.

3.5.2 Acquisition with No Base Calling

If your computer has limited resources do not forget to disable base calling as it could impact the proper acquisition of the raw signal. Once the sequencing is finish you will base call the data with Guppy.

1. Configure acquisition keeping default parameters. For kit selection, select "SQK-LSL109" and if samples are multiplexed, select "EXP-NBD104" or "EXP-NBD114."
2. Disable base calling.
3. Start the sequencing and let it run for 72 h.

During the run, MinKNOW displays the activity of the pores and plots a histogram of read lengths. N50 should typically be in the 30–60 kb range.

3.5.3 Base Calling with Guppy Independently of Data Acquisition

1. Depending on your computational resources, download and install the CPU or GPU version of Guppy (full instructions at https://community.nanoporetech.com/downloads, *see* **Note 23**.)
2. Proceed to base calling with the command line below (if no multiplexing). `GUPPYDIR` refers to Guppy install directory and `WORKDIR` to the project working directory.

```
Base calling with Guppy, no multiplexing|
$ GUPPYDIR/bin/guppy_basecaller -i WORKDIR/RawData/ -s WORK-
DIR/Fastq/ -q 4000 \
--flowcell FLO-MIN106 --kit SQK-LSK109 --cpu_threads_per_cal-
ler 8 --num_callers 1
```

With

- `-i`: location of the raw FAST5 files; all FAST5 files found in this directory will be processed.
- `-s`: location of the output FASTQ files
- `--cpu_threads_per_caller`: number of threads per caller
- `--num_caller`: number of callers.

The last two parameters have to be adapted to your computational resource. Multiple FASTQ files as well as a `sequencing_summary.txt` file will be generated in the chosen directory.

3. If samples are multiplexed, use the same command line simply adding the `barcode_kits` option.

```
Base calling with Guppy, multiplexing
$ GUPPYDIR/bin/guppy_basecaller -i WORKDIR/RawData/ -s WORK-
DIR/Fastq/ \
--flowcell FLO-MIN106 --kit SQK-LSK109 -q 4000 --barcode_kits
EXP-NBD104 \
--cpu_threads_per_caller 8 --num_callers 1
```

The FASTQ files corresponding to each barcode will be saved in the directory `Fastq/barcodeX/`, where `X` is the barcode number.

3.5.4 Demultiplexing of the FAST5 Files

If samples are multiplexed you then have to split the FAST5 files according to the samples. This is performed using the tool ONT FAST5 API (*see* **Note 24**).

1. Make a subset list from the sequencing summary (generated during base calling) with the reads containing a specific barcode. This have to be done for each barcode (replacing `barecodeX` in the commands).

```
Demultiplexing 1/2
$ cd WORKDIR
$ head -n 1 Fastq/sequencing_summary.txt > barcodeX_summary.
txt
$ grep barcodeX Fastq/sequencing_summary.txt >> barcodeX_sum-
mary.txt
```

2. Using this list, you can now split your FAST5 files using the following command (for each barcode). `APIDIR` refers to the installation directory of ONT FAST5 API.

```
Demultiplexing 2/2
$ APIDIR/bin/fast5_subset -i WORKDIR/RawData -s WORKDIR/Raw-
Data/barcodeX \
-l WORKDIR/barcodeX_summary.txt --batch_size 4000
```

3.5.5 Concatenate FASTQ Files

Basecalling during or after the run generates multiple FASTQ files for each sample. To allow subsequent RepNano analysis, it is necessary to generate one FASTQ per sample.

1. Without multiplexing: In the following command, `Sample.fastq` is the name you choose to give to the FASTQ file containing all the reads.

```
Concatenate FASTQ files, no multiplexing
$ cat WORKDIR/Fastq/*.fastq > WORKDIR/Fastq/Sample.fastq
```

2. With multiplexing: `Sample_barcodeX.fastq` is the name you choose to give to the FASTQ file containing all the reads coming from the sample barcoded with barcodeX.

```
Concatenate FASTQ files, if multiplexing
$ cat WORKDIR/Fastq/barcodeX/*.fastq > WORKDIR/Fastq/Sample_-
barcodeX.fastq
```

This has to be done for each barcode (replacing `barecodeX` and `Sample` in the command).

Nanopore reads are now base called and ready for downstream analysis.

3.6 Data Analysis with RepNano

Please refer to RepNano repository for manual and updates (https://github.com/organic-chemistry/RepNano).

3.6.1 Installation

The installation requires Git (*see* **Note 25**). We recommend using a dedicated Conda environment to assure all software dependencies are met prior to running the analysis (*see* **Note 26**). In particular, the preprocessing step uses Tombo (ONT, *see* **Note 27**) package.

```
Installation of the computational environment and of Tombo.
$ conda create --name RepNanoEnv --override-channels -c bio-
conda -c defaults \
 python=3.6 keras pandas numba tqdm joblib ont-tombo matplo-
tlib
```

All further steps assume the `RepNanoEnv` conda environment is activated using:

```
$ conda activate RepNanoEnv
```

RepNano installation `SOFTDIR` refers to the directory where you wish to install RepNano.

```
$ cd SOFTDIR
$ conda activate RepNanoEnv
$ git clone https://github.com/organic-chemistry/RepNano.git
$ cd RepNano
$ python setup.py develop
```

These operations result in a working RepNano installation in directory `SOFTDIR/RepNano`, which will be referred to as `REPNANODIR`. The python module *simplification* is required and can be installed using pip.

```
$ [conda activate RepNanoEnv]
$ pip install simplification
```

3.6.2 BrdU Detection

The pipeline consists in the alignment of the basecalled Oxford Nanopore reads on a reference genome followed by prediction of BrdU content by the neural network (CNN) and the transition matrix (TM) approaches [4]. Sequencing and basecalling results in FAST5 and FASTQ files (*see* Subheading 3.5). For each sample, you have:

- several FAST5 files (raw current values) containing 4000 reads each (*see* **Note 28**),
- one FASTQ file with all the sequences.

A test dataset is available at https://www.opendata.bio.ens.psl.eu/FORK-seq. The analysis procedure is described using this dataset as an example. In order to use it you have to download *TestDataSet.tgz* (1,1G) and decompress it. TESTDIR refers to the directory where the procedure will be run.

```
Download test dataset
$ cd TESTDIR
$ wget https://www.opendata.bio.ens.psl.eu/FORK-seq/TestData-
Set.tgz
$ tar -zxvf TestDataSet.tgz
```

1. **Genome indexing [Facultative]**

   The preprocessing starts with the alignment of FASTQ sequences on a reference genome using Minimap2 [8]. In the `--ref` option, you can either put the FASTA file of your genome or the corresponding Minimap2 index (recommended as it is faster). To generate Minimap2 index you have to run the following command (*see* **Note 29**).

```
Minimap2 index
$ [conda activate RepNanoEnv]
$ cd TESTDIR
$ python REPNANODIR/src/repnano/data/index.py TestDataSet/
chr4.fa TestDataSet/chr4.mmi
```

   This step has to be done only once.

2. **Preprocessing**

   The preprocessing includes the mapping on the reference genome and a realignment of the current values to fit the reference sequence (Tombo resquiggle function). Every FAST5 is processed separately (parallelizable). This will create a `preprocessed_file.fast5` file that will contain the current values aligned on the reference genome (*see* **Note 30**).

```
Preprocessing
$ [conda activate RepNanoEnv]
$ [cd TESTDIR]
$ python REPNANODIR/src/repnano/data/preprocess.py \
--hdf5 TestDataSet/FORKSeq_combined.fast5 --fastq TestData-
Set/FORKSeq_combined.fastq \
--ref TestDataSet/chr4.mmi --output_name TestDataSet/prepro-
cessed_file.fast5 --njobs 6
```

   With

- `-hdf5`: input FAST5 file
- `--fastq`: input FASTQ file
- `--ref`: reference genome (.fa or .mmi)
- `--output_name`: name of the output FAST5 file (*see* **Note 31**)
- `--njobs`: number of jobs run in parallel; to be adapted to the hardware.

  The script outputs a description of the reads that failed to be processed. It usually corresponds to 10%–25% of the reads. The errors are explained below.

- `seq_not_found`: The read ID from the FAST5 file is not found in the FASTQ file (*see* **Note 32**),
- `Alignment not produced`: The read couldn't be mapped on the reference.
- `Read event to sequence alignment extends beyond bandwidth`: The read could be mapped but the alignment of the current values on the reference sequence failed.
- `Found non canonical bases ([N])`: The read mapped on a region of the reference that contains non canonical bases (*see* **Note 33**).

3. **BrdU detection**

   RepNano BrdU detection is then called on the `preprocessed_file.fast5` file.

```
$ [conda activate RepNanoEnv]
$ [cd TESTDIR]
$ python REPNANODIR/src/repnano/models/predict_simple.py \
TestDataSet/preprocessed_file.fast5 --output TestDataSet/
BrdU_calls/output_file.fa \
--overlap 10 --bigf
```

With

- `--output`: output file name.
- `--overlap`: [INT] RepNano BrdU detection with the trained neural network model calculates BrdU ratio on 96 bp windows. This parameter defines the number of overlapping windows used to calculate the final average ratio. This will be done by shifting the signal times with a step.
- `--bigf`: Before 2019, the output of ONT basecallers was different and one FAST5 was generated for each read. Remove this option if you work with this old format.

RepNano prediction step generates two files:

- output_file.fa: a FASTA file of the read sequences where T have been replaced by T, X or B according to the transition matrix (TM) approach [4].
- output_file.fa_ratio_B: a FASTA-like file with the BrdU ratio for each base of the sequence as computed by the neural network (CNN; see overlap argument above).

Mapping information is stored in the definition line (starting with '>') of each read in *output_file.fa* (*see* **Note 34**). Below is the description of the different fields.

- *mapped_chrom*: chromosome.
- *mapped_start*: start coordinate.
- *mapped_end*: end coordinate.
- *mapped_strand*: strand on the reference (+/−).
- *clipped_bases_start*: number of bases clipped at the beginning of the read.
- *clipped_bases_end*: number of bases clipped at the end of the read.
- *num_matches*: number of exact matches between the read and the reference sequence.
- *num_deletions*: number of deletions in the read compared to the reference.
- *num_insertions*: number of insertions in the read compared to the reference.
- *num_mismatches*: number of mismatches between the read and the reference sequence.

3.6.3 Fork Detection

To detect replication forks and their orientations, as well as replication initiation and termination events, you have to run the following command, where BrdU_calls is the folder containing the outputs of RepNano BrdU detection (from previous step, *see* **Note 35**), DetectionFOLDER is the location for the output files and prefix is a prefix you choose for the output file names (it can be a sample ID for instance) (*see* **Note 36**).

```
Fork detection
$ [conda activate RepNanoEnv]
$ python REPNANODIR/src/repnano/detection/ForkPrediction-
CNN-TM.py BrdU_calls DetectionFOLDER prefix
```

This will generate 6 files (forks, initiation and termination events for CNN and TM methods) with the events extracted from all the files contained in the "BrdU_calls" folder. Below is the description of the fields.

- forks: `prefix_CNN.forks` and `prefix_TM.forks`,
  - *chrom*: chromosome.
  - *start*: first coordinate on the chromosome (independent of fork direction).
  - *end*: second coordinate on the chromosome.
  - *direction*: fork direction (R, L for right, left, respectively) relative to the reference sequence.
  - *FASTA*: file containing the read.
  - *read*: read identifier.
  - *strand*: strand on the reference.
  - *JumpScore*: score reflecting the height of the signal, the higher the more confident one can be that this is a true fork and not noise.
  - *AsymScore*: score reflecting the asymmetry of the signal, the higher the more confident one can be about the direction of the fork.
- initiation events: `prefix_CNN.init` and `prefix_TM.init`,
  - *chrom*: chromosome.
  - *middle*: estimated position of the initiation event. This is the midpoint between the starts of the two diverging forks.
  - *end1*: first coordinate of the first fork, for an initiation event this is the end of the leftward-going fork.
  - *start1*: second coordinate of the first fork, for an initiation event this is the start of the leftward-going fork.
  - *start2*: first coordinate of the second fork, for an initiation event this is the start of the rightward-going fork.
  - *end2*: second coordinate of the second fork, for an initiation event this is the end of the rightward-going fork.
  - *FASTA*: file containing the read.
  - *read*: read identifier.
  - *strand*: strand on the reference.
- termination events: `prefix_CNN.term` and `prefix_TM.term`.
  - *chrom*: chromosome.
  - *middle*: estimated position of the termination event. This is the midpoint between the starts of the two converging forks.
  - *start1*: first coordinate of the first fork, for a termination event this is the start of the rightward-going fork.
  - *end1*: second coordinate of the first fork, for a termination event this is the end of the rightward-going fork.

- *end2*: first coordinate of the second fork, for a termination event this is the end of the leftward-going fork.
- *start2*: second coordinate of the second fork, for a termination event this is the start of the leftward-going fork.
- *FASTA*: file containing the read.
- *read*: read identifier.
- *strand*: strand on the reference.

4 Notes

1. FORK-seq relies on the incorporation of BrdU from the external medium into yeast genomic DNA. Wild-type *Saccharomyces cerevisiae* cannot incorporate external nucleosides and therefore cannot be used for FORK-seq experiments. The minimal changes are the addition on a nucleoside transporter to allow the BrdU to enter the cells, and a thymidine kinase (TK) to phosphorylate it and make it a suitable substrate for DNA polymerases. We use the MCM869 yeast strain, which in addition to expressing human ENT1 and *Herpes simplex* TK [9, 10], is deficient for the thymidylate synthase CDC21 [6, 10]. This makes it unable to produce its own thymidine and therefore fully dependent on external supply. In principle, any other strain able to incorporate BrdU at a high rate should be suitable to FORK-seq experiments.
2. The pH may vary depending of the YPD batch. If necessary, adjust it to 6.5–7.
3. MCM869 doubling time is about 150 min.
4. We have noticed that the 30 min thymidine starvation induces stress, as shown by Rad53 phosphorylation, which might be problematic depending on the biological question. Experiments without starvation nevertheless gave very similar BrdU profiles, albeit with very little labeling in early S-phase. Improved strains are under development.
5. In the original paper, we used two steps with different BrdU/thymidine ratios in order to orient the forks, but found out that this was dispensable as the changes in incorporation are not abrupt enough to resolve the steps; in fact, the signal shape is almost identical with a single 4 min BrdU pulse; the orientation of the forks instead relies on the asymmetrical rates of increase and decrease of BrdU incorporation during the pulse and the chase, respectively.
6. DNA extraction using QIAGEN Genomic-tip 20/G and QIAGEN Genomic DNA Buffer Set that contains buffers TE, Y1,

G2, QBT, QC, and QF. Resuspend the frozen pellet into 1 mL Y1 buffer (Please refer to QIAGEN Genomic DNA Handbook for more details; do not use more than the equivalent of 100 mL of $OD_{600} = 1$ cell suspension per column). Transfer to 2 mL-tubes. Add 100 μL of Zymolyase (100 U) and incubate at 37 °C for 1–3 h, inverting the tube from time to time. A white jelly fish should form. Centrifuge at 5000 × *g* for 10 min at 4 °C. Remove and discard the supernatant. Resuspend the pellet by slow pipetting in 2 mL G2 buffer with 10 μL RNase A. Incubate for 15 min at 37 °C. Add 20 μL Proteinase K solution (20 mg/mL stock) and incubate at 50 °C for 1 h. Invert the tube ~10 times to get a clear solution. Centrifuge at 5000 × *g* for 10 min at 4 °C and transfer the supernatant to a new tube. Equilibrate the QIAGEN Genomic-tip 20/G with 2 mL of QBT buffer. Vortex the sample for 10s full speed and load on the column. Wash the column 3 times with 1 mL QC buffer. Put the column on a clean 5 mL tube and elute gDNA with 1 mL QF prewarmed to 50 °C. Repeat this elution step. Precipitate the DNA adding 1,4 mL isopropanol and inverting the tube several times. Centrifuge at 5000 × *g* for 30 min at 10 °C. Remove the supernatant and wash the DNA pellet with 1 mL 70% EtOH. Let the pellet dry and resuspend it in 100 μL TE.

7. Ensure that the RNase used is completely free of DNase activity.
8. Using NanoDrop quantification is not recommended as it often overestimates the concentration.
9. Low A260/280 ratio indicates protein contamination and low A260/230 ratio indicates RNA contamination.
10. You may also use a low concentration agarose gel electrophoresis but the resolution of large DNA fragments is not as good. It still provides information about accidental fragmentation and RNase treatment efficiency.
11. Possible causes for small DNA fragments: Every step has to be performed keeping in mind that high molecular weight DNA is fragile. Be gentle, do not vortex and avoid repeated pipetting. DNase contamination might also occur, be sure to use DNase-free reagents and water.
12. It is recommended to start from 2 μg of DNA. When multiplexing the samples, it is possible to start from less DNA (down to 500 ng) but the number of sequencing reads might be reduced.
13. If no thermal cycler is available, a thermoblock can be used.

14. A smaller volume of beads can be used in order to remove more small DNA fragments, yet we noticed a significant loss of long DNA fragments when doing so.
15. Do not wash with 70% ethanol as it can damage molecular motors and result in low pore occupancy.
16. Set a P1000 pipette to 200 μL, insert the tip into the priming port and turn the wheel until the dial shows 220–230 μL or until you can see a small volume of buffer entering the pipette tip. Care must be taken when drawing back buffer from the flow cell. The array of pores must be covered by buffer at all times. Removing more than 20–30 μL risks a loss of sequencing channels. Be careful not to introduce any air bubble.
17. ONT recommends loading 5–50 fmol of a prepared library. Yet, this quantification depends on the size of the DNA molecules. Using 200 ng, we typically obtain 10–15 Gbp for non-barcoded libraries and 5–7 Gbp for barcoded ones. Barcoding reduces significantly the amount of resulting data and we recommend a maximum of 4 barcodes per run.
18. Pore activity is monitored in real time by MinKNOW. If pore activity becomes too low during the run (i.e., below 30%), it is possible to reload the flow cell with the leftover library. To do so, repeat steps 3–7.
19. It takes typically 6 days to base call a 72 h run on the Mk1C (with Guppy 4.3, please be aware that ONT software evolve very quickly). For high-frequency usage of the sequencer, an independent workstation dedicated to base calling is recommended.
20. If a sequencing core facility is running your samples, inform them that you will need the raw FAST5 files, as they may do immediate archiving (or even deletion) of those big files.
21. Guppy base caller is available for Windows, Mac OS, and Linux and both CPU and GPU versions are possible to get. We have tested only the Linux version using a dedicated workstation with 2× Intel Xeon Silver 4214 (12 cores), a NVIDIA GeForce RTX2080 TI graphic card (providing 4352 CUDA cores), 128 GB of RAM, and two 1 To SSD disks in RAID 0 for the data. Note that Guppy only runs on a selected set of GPUs (see Guppy documentation). With this configuration, the GPU version of Guppy base call a 4000-reads FAST5 file in a few minutes, while it takes more than 10 h using the CPU version and 20 threads on an HPC cluster (depending on read length and CPU speed, typically 1–5 min per read on a single CPU). RepNano BrdU detection is routinely done using an HPC cluster, but we also have successfully tested the workflow on personal computers with Linux and Mac OS. The example provided correspond to using these Unix family operating systems.

22. We haven't tested the filtering based on read length although it seems quite reasonable to discard reads below 5 kb for replication fork detection with the present protocol.
23. You need a full member ONT account to access some of their software, including the Guppy base caller. You will get such account if you buy a sequencer. You can also join an existing account of a collaborator, see https://community.nanoporetech.com/.
24. ONT FAST5 API can be downloaded at https://github.com/nanoporetech/ont_fast5_api
25. To install Git please refer to https://git-scm.com/downloads.
26. To install Conda please refer to https://docs.conda.io/en/latest/miniconda.html.
27. Complete description of Tombo at https://nanoporetech.github.io/tombo/tombo.html.
28. Before 2019, the output of ONT base callers was different and one FAST5 was generated for each read. A procedure to work with the old files is available on our GitHub repository https://github.com/organic-chemistry/RepNano.
29. We use only one chromosome (chr4) for the reference in the test dataset, use a FASTA file containing the sequences of all the chromosomes for real data.
30. On Mac OS, if the command predict_simple.py fails to run and returns the error: *Initializing libiomp5.dylib, but found libomp.dylib already initialized.*, please add `nomkl` to the software environment using

```
$ conda install nomkl
```

(in RepNanoEnv).

31. If the directory where you want to save the output FAST5 does not exist, it will be automatically created.
32. This should not happen, please check that you give the correct path to the FASTQ file corresponding to the FAST5 file.
33. We recommend to replace 'N' bases in the reference FASTA by any canonical base, for instance using

```
$ sed 's/N/A/g' ref_with_N.fa > ref_corrected.fa.
```

This will replace all 'N' by 'A' bases.

34. Both outputs follow the orientation of the read and not the one of the reference.
35. All the files contained in the directory will be processed and a single list will be output.

36. A number of parameters are set up at the beginning of Fork-Prediction-CNN-TM.py and can be modified to make the detection more or less stringent.

Acknowledgments

This work was supported by the Ligue Nationale Contre le Cancer, the Association pour la Recherche sur le Cancer, the Agence Nationale de la Recherche [ANR-15-CE12-0011-01, ANR-18-CE45-0002 and ANR-19-CE12-0028], the Fondation pour la Recherche Médicale [FRM DEI201512344404], and the Cancéropole Ile-de-France [PLBIO16-302]. B.T. was supported by fellowships from the Ministère de l'Enseignement Supérieur et de la Recherche and the Fondation pour la Recherche Médicale [FRM FDT202106013030].

The authors thank the sequencing facility (especially Corinne Cruaud and Stéfan Engelen) of François Jacob Institute of Biology (Genoscope, CEA, Evry) for their great help in setting up ONT nanopore sequencing in OH lab. We also thank IBENS IT support and Genomic Paris Centre platform for technical advices, as well as the computing cluster BioClust (Labex Memolife). Finally, we thank all members of the OH lab, especially Benoît Le Tallec (BT PhD supervisor), Laurent Lacroix, and Etienne Jean for lively and supportive discussions.

References

1. Hyrien O (2015) Peaks cloaked in the mist: the landscape of mammalian replication origins. J Cell Biol 208(2):147–160
2. Técher H, Koundrioukoff S, Azar D et al (2013) Replication dynamics: biases and robustness of DNA fiber analysis. J Mol Biol 425(23):4845–4855
3. Guilbaud G, Rappailles A, Baker A et al (2011) Evidence for sequential and increasing activation of replication origins along replication timing gradients in the human genome. PLoS Comput Biol 7(12):e1002322
4. Hennion M, Arbona JM, Lacroix L et al (2020) FORK-seq: replication landscape of the Saccharomyces cerevisiae genome by nanopore sequencing. Genome Biol 21(1):125
5. Muller CA, Boemo MA, Spingardi P et al (2019) Capturing the dynamics of genome replication on individual ultra-long nanopore sequence reads. Nat Methods 16(5):429–436
6. Ma E, Hyrien O, Goldar A (2012) Do replication forks control late origin firing in Saccharomyces cerevisiae? Nucleic Acids Res 40(5):2010–2019
7. Denis E, Sanchez S, Mairey B et al (2018) Extracting high molecular weight genomic DNA from saccharomyces cerevisiae. Research Square Preprint. https://doi.org/10.1038/Protex.2018.076
8. Li H (2018) Minimap2: pairwise alignment for nucleotide sequences. Bioinformatics 34(18): 3094–3100
9. Lengronne A, Pasero P, Bensimon A et al (2001) Monitoring S phase progression globally and locally using BrdU incorporation in TK (+) yeast strains. Nucleic Acids Res 29(7): 1433–1442
10. Vernis L, Piskur J, Diffley JF (2003) Reconstitution of an efficient thymidine salvage pathway in saccharomyces cerevisiae. Nucleic Acids Res 31(19):e120
11. Siow CC, Nieduszynska SR, Muller CA et al (2012) OriDB, the DNA replication origin database updated and extended. Nucleic Acids Res 40:682–686

Chapter 9

CUT&RUN Profiling of the Budding Yeast Epigenome

Sandipan Brahma and Steven Henikoff

Abstract

Mapping the epigenome is key to describe the relationship between chromatin landscapes and the control of DNA-based cellular processes such as transcription. Cleavage under targets and release using nuclease (CUT&RUN) is an in situ chromatin profiling strategy in which controlled cleavage by antibody-targeted Micrococcal Nuclease solubilizes specific protein-DNA complexes for paired-end DNA sequencing. When applied to budding yeast, CUT&RUN profiling yields precise genome-wide maps of histone modifications, histone variants, transcription factors, and ATP-dependent chromatin remodelers, while avoiding cross-linking and solubilization issues associated with the most commonly used chromatin profiling technique Chromatin Immunoprecipitation (ChIP). Furthermore, targeted chromatin complexes cleanly released by CUT&RUN can be used as input for a subsequent native immunoprecipitation step (CUT&RUN.ChIP) to simultaneously map two epitopes in single molecules genome-wide. The intrinsically low background and high resolution of CUT&RUN and CUT&RUN.ChIP allows for identification of transient genomic features such as dynamic nucleosome-remodeling intermediates. Starting from cells, one can perform CUT&RUN or CUT&RUN.ChIP and obtain purified DNA for sequencing library preparation in 2 days.

Key words Chromatin profiling, Genome-wide mapping, Chromatin dynamics, Nucleosomes, Transcription factors

1 Introduction

Chromatin organization, or how DNA is packaged inside the nucleus, determines cell-type-specific gene expression and cellular identity. Nucleosomes, composed of DNA wrapped around an octameric core of histone proteins are the fundamental repeating units of chromatin. Chromatin organization is highly dynamic, and functionally distinct chromatin states are outcomes of the interplay between multiple proteins and proteins complexes including, but not limited to, histone chaperones, histone modifying enzymes, ATP-dependent nucleosome remodelers, and transcription factors (TF) [1]. Mutations affecting the functions of these proteins are associated with a variety of developmental disorders and cancers.

Frédéric Devaux (ed.), *Yeast Functional Genomics: Methods and Protocols*,
Methods in Molecular Biology, vol. 2477, https://doi.org/10.1007/978-1-0716-2257-5_9,

The first step in understanding the mechanisms of chromatin organization is to accurately map the genomic locations and chromatin contexts where chromatin-regulatory proteins bind.

The most widely used technique for genome-wide mapping of chromatin-associated proteins is chromatin immunoprecipitation followed by deep sequencing (ChIP-seq) [2]. ChIP-seq generally involves formaldehyde fixation (cross-linking) of bulk chromatin within cells followed by mechanical shearing or enzymatic cleavage into smaller fragments and immunoprecipitation of the DNA cross-linked with the protein-factor of interest. Problems arise due to differential sensitivity of genomic regions to chromatin fragmentation approaches, the need to fragment the entire genome to solubilize chromatin resulting in high background noise, as well as epitope masking and false positives potentially due to chemical cross-linking [3–6]. Although successive technological and methodological advancements have greatly improved the resolution of ChIP-seq, these methods still rely on formaldehyde cross-linking to preserve the in vivo patterns [7–9]. Native ChIP minimizes epitope masking and cross-linking artifacts; however, problems still remain with incomplete extraction efficiency of protein-DNA complexes [10].

Chromatin-tethering of a variety of DNA enzymes including methyl transferases [11], nucleases [12–14], and transposases [15, 16] are attractive alternatives for chromatin profiling. Enzyme tethering methods fundamentally differ from ChIP-seq as they are carried out in vivo or in situ where the tethered enzyme acts upon the DNA in its local vicinity, and the modified DNA is directly extracted from permeabilized cells, thus eliminating the need to solubilize and recover chromatin. Cleavage Under Targets and Release Using Nuclease (CUT&RUN) is an enzyme-tethering method in which a protein A - Micrococcal Nuclease fusion (pA-MNase) is targeted to genomic regions via factor-specific antibodies in permeabilized native cells or nuclei (in the case of yeast) such that chromatin around the factor-binding sites is selectively cleaved [12] (Fig. 1). Micrococcal Nuclease (MNase) is an endo-exo nuclease from *Staphylococcus aureus* that has been used for decades for mapping nucleosomes. In the presence of calcium ions, MNase nibbles away unbound DNA to generate minimal protections of protein-DNA complexes. Subjecting DNA fragments obtained after MNase treatment of nuclei to paired-end short-read sequencing enables determination of the genome-wide footprints corresponding to both small proteins (e.g., transcription factors) as well as nucleosomes at base-pair resolution [17]. In CUT&RUN, MNase cleavage solubilizes factor-associated chromatin, which diffuse out cleanly into the supernatant, leaving the bulk of the undigested chromatin insoluble. As only the targeted fragments enter into solution and the vast majority of DNA is left behind, CUT&RUN has exceptionally high signal-to-noise and

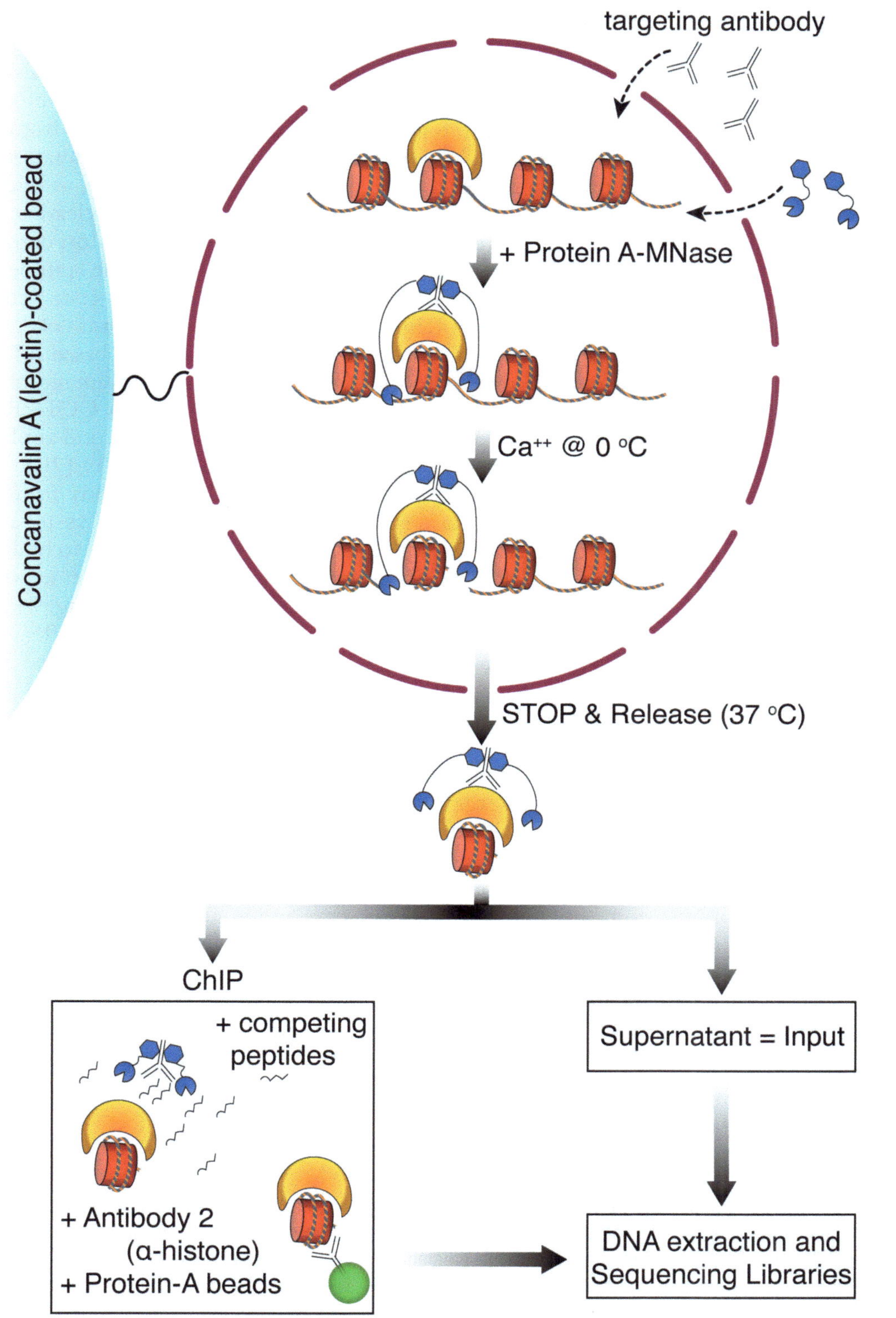

Fig. 1 Schematic of the CUT&RUN(.ChIP) approach. Permeabilized yeast nuclei attached to ConA beads are successively treated with the targeting antibody and pA-MNase fusion, which diffuse through the nuclear

data quality, requiring only ~1/tenth the sequencing depth as ChIP-seq [18, 19].

As an in situ method, CUT&RUN is inherently capable of probing the local chromatin environment such as the binding sites of TFs and the positions of flanking nucleosomes from the same experiment [12] (Fig. 2). This is possible by computationally filtering the sequence data based on fragment lengths, as nucleosomes typically protect ~150 bp of DNA from MNase cleavage and TFs usually protect short DNA fragments (<80 bp). Most chromatin profiling methods are capable of mapping only one epitope at a time and are therefore not able to directly test protein co-occupancies on chromatin that forms the basis of gene-regulatory mechanisms, while most conclusions are based on correlation. To overcome this limitation, we took advantage of the native chromatin complexes that are solubilized in CUT&RUN and used as input for immunoprecipitation with a second antibody under native conditions (CUT&RUN.ChIP) (Fig. 1) [20]. This required us to compete off the CUT&RUN antibody from the released particles using the immunizing peptide so as to avoid nonspecific enrichment in the ChIP step. When applied to budding yeast, CUT&RUN.ChIP mapped histone co-occupancies genome-wide with exceptional high efficiency and resolution (Fig. 3), and revealed that the ATP-dependent chromatin remodeler RSC and the general regulatory TFs Abf1 and Reb1 are associated with partially unwrapped dynamic nucleosomal intermediates at gene promoters [20].

In summary, CUT&RUN has several advantages for chromatin profiling in budding yeast: (1) The method is performed in situ in non–cross-linked nuclei and does not require chromatin fragmentation or solubilization; (2) The intrinsically low background allows low sequencing depth and identification of low-signal genomic features invisible to ChIP-seq; (3) The simple procedure can be completed in as little as two days and is suitable for robotic automation [21]; (4) Combining CUT&RUN with native ChIP allows mapping of protein co-occupancies on chromatin; (5) A simple spike-in strategy can be used for accurate quantitation of protein-DNA interactions. Although there is no published report of CUT&RUN profiling of the fission yeast *Schizosaccharomyces pombe* epigenome, the protocol described here starting from native nuclei should work as efficiently in principle.

Fig. 1 (continued) pores. Ca^{++} induces cleavage of target DNA by immobilized MNase. The reaction is done at 0 °C which limits diffusion of released fragments, thereby minimizing background cleavage. Because there is no cross-linking, cleaved chromatin particles are efficiently released into solution at 37 °C. In the ChIP step, the first antibody is competed off from the released particles, antibody for a second epitope is added, and immune-complexes are enriched by binding to protein A/G beads. DNA extracted from the CUT&RUN supernatant and from ChIP are used to prepare libraries for paired-end deep-sequencing

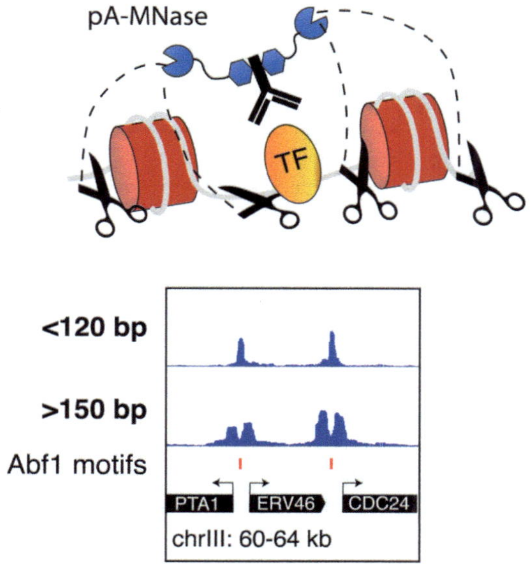

Fig. 2 Simultaneous mapping of TFs and flanking nucleosomes. Top: pA-MNase tethered to TFs bound at gene promoter nucleosome-depleted regions digests DNA around the bound TF as well as reach beyond the immediately flanking nucleosomes to solubilize both the TF- and flanking nucleosome-bound chromatin. Bottom: Separation of mapped DNA fragments by size from CUT&RUN targeting the general regulatory TF Abf1 shows crisp TF peaks (<120 bp size class) over Abf1 DNA-binding motifs (shown in red) [10] and the positions of the flanking nucleosomes (>150 bp size class) from the same experiment

2 Materials

Prepare all solutions using ultrapure water (prepared by purifying deionized water to attain a sensitivity of 18 MΩ-cm at 25 °C) and analytical grade reagents.

2.1 Yeast Culture and Nuclei Preparation

1. *Saccharomyces cerevisiae* cultured on Yeast Peptone Dextrose (YPD) agar.
2. Yeast Peptone Dextrose (YPD) media: 10 g/L yeast extract, 10 g/L bactopeptone, 20 g/L glucose. Autoclaved.
3. 1 M PIPES, pH 6.5: Mix 15.11 g PIPES in 30 mL water, add 10 N Sodium Hydroxide dropwise (takes about 5 mL) to adjust pH to 6.5, adjust final volume to 50 mL with water, filter-sterilize (0.2 μm), store at 4 °C.
4. Resuspension buffer (make fresh before use): 1.2 M Sorbitol, 100 mM Potassium Phosphate buffer, pH 7.5, 0.5 mM Calcium Chloride, 0.5 mM 2-Mercaptoethanol.

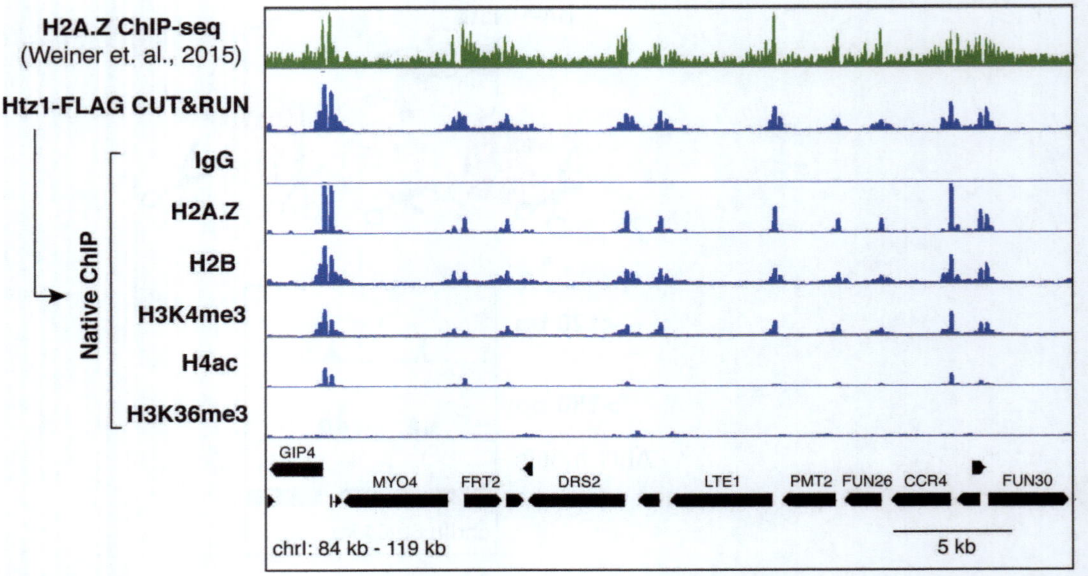

Fig. 3 CUT&RUN.ChIP maps histone co-occupancies. Blue tracks show spike-in normalized CUT&RUN.ChIP profiles (>150 bp or nucleosomal reads) of histone modifications in nucleosomes containing the histone variant H2A.Z. Supernatant from Htz1-FLAG CUT&RUN experiment was used as input for the subsequent native ChIP. IgG is a nonspecific negative control for ChIP. ChIP-seq of H2A.Z from cross-linked MNase-digested chromatin (green, top) is shown for comparison [22]. Results show that H2A.Z nucleosomes are present only at the beginning of genes and totally absent from gene bodies unlike profiles observed in conventional MNase-ChIP-seq. H2A.Z nucleosomes are enriched in histone H3 with trimethylation of lysine 4 (H3K4me3) and histone H4 with acetylated N-terminal tail (H4ac) but lack histone H3 trimethylated at lysine 36 (H3K36me3), commonly associated with gene-body nucleosomes. CUT&RUN.ChIP tracks were scaled to the same intensity (y-axes). The positions of the annotated yeast open reading frames are shown at the bottom with the gene names

5. Zymolyase: Dissolve 1 mg/mL Zymolyase in 1.0 M Sorbitol and 10 mM PIPES, pH 6.5, store aliquots at −20 °C.
6. SPC buffer: 1 M Sorbitol, 20 mM PIPES, pH 6.5, 0.1 mM Calcium Chloride. SPC + protease inhibitor: Add one Roche Protease inhibitor cocktail tablet per 50 mL buffer, and PMSF to 1 mM before use.
7. Ficoll buffer: 9% (w/v) Ficoll 400, 20 mM PIPES, pH 6.5, 0.5 mM calcium chloride. Add Ficoll gradually to half the final volume of water, stir at room temperature, bring up to final volume with water when Ficoll dissolves, filter-sterilize (0.2 μm).

2.2 CUT&RUN and Sequencing

1. Concanavalin A–coated (ConA) magnetic beads.
2. Binding buffer: 20 mM HEPES-KOH, pH 7.9, 10 mM Potassium Chloride, 1 mM Calcium Chloride, 1 mM Manganese Chloride.

3. 5% w/v Digitonin
4. Wash buffer: 20 mM Sodium HEPES, pH 7.5, 150 mM Sodium Chloride, 0.5 mM Spermidine, add one Roche Protease inhibitor cocktail tablet per 50 mL buffer. Wash buffer can be stored for up to 7 days at 4 °C, but spermidine might degrade over time. For Dig-Wash, add Digitonin to 0.05% final concentration (10 μL of 5% digitonin in water to 10 mL buffer).
5. Antibody buffer: Dig-Wash buffer supplemented with 2 mM ethylenediaminetetraacetic acid (EDTA).
6. Antibody to an epitope of interest, for example, rabbit anti-histone Htz1 (H2A.Z) antibody or mouse anti-FLAG M2 antibody (for FLAG-tagged proteins).
7. Secondary antibody, for example, guinea pig anti-rabbit IgG or rabbit anti-mouse IgG.
8. Isotype control IgG (negative control antibody).
9. Protein A-Micrococcal nuclease (pA-MNase) or Protein A-Protein G-Micrococcal nuclease (pAG-MNase) fusion proteins purified from overexpression in *Escherichia coli* (see [14] and [18] for protein purification protocols), or obtained commercially (e.g., CUTANA pAG-MNase, Epicypher; or CUT&RUN pAG-MNase, Cell Signaling Technology). pA-MNase and pAG-MNase expression plasmids can be obtained from Addgene.
10. Heterologous spike-in DNA, for example DNA purified from MNase-treated chromatin from *Drosophila* S2 cells (*see* refs. [12, 19]).
11. 2× STOP buffer: Dig-Wash buffer supplemented with 20 mM EDTA, 4 mM EGTA (ethylene glycol-bis(β-aminoethyl ether)-N,N,N′,N′-tetraacetic acid), 50 μg/ml RNase A (DNase free), 50 pg/mL heterologous spike-in DNA.
12. Immunizing peptide for competing off CUT&RUN antibody (only for CUT&RUN.ChIP), for example 3×-FLAG peptide.
13. Protein A- or Protein-G coated magnetic beads, for example Dynabeads Protein A.
14. 10% sodium dodecyl sulfate (SDS).
15. 20 mg/mL Proteinase K.
16. Phenol–chloroform–isoamyl alcohol 25:24:1.
17. Chloroform–isoamyl alcohol 24:1.
18. 2 mg/mL Glycogen.
19. Ethanol.
20. 10 M Tris–HCl, pH 8.0.

21. Fluorometric DNA quantification assay, for example, Qubit NGS equipment and Qubit dsDNA HS kit.
22. DNA capillary electrophoresis equipment and reagents, for example, TapeStation 4200 system and D1000 high-sensitivity Screen Tapes and reagents.
23. SPRI paramagnetic beads (e.g., Agencourt AMPure XP, Beckman Coulter).
24. NGS library preparation reagents, for example, KAPA HyperPrep kit.
25. Illumina TruSeq-compatible adapters and indexing primers.

2.3 General Equipment

1. Centrifuges, swinging bucket rotors that can accommodate 50 mL tubes at speeds up to 3000 × *g*, and tabletop refrigerated centrifuges that can accommodate 1.5 mL tubes at speeds up to 16,000 × *g*.
2. Microcentrifuge tubes (non–low-bind) and PCR tubes/96-well plates.
3. Magnetic separator that can hold 1.5 mL tubes and PCR tubes.
4. Nutating and end-over-end mixers.
5. Heating blocks or water baths at 37 °C.
6. Phase-lock tubes (e.g., Qiagen MaXtract High Density).

3 Method

3.1 Yeast Nuclei Preparation

1. Inoculate single *S. cerevisiae* colony into 10 mL YPD, grow overnight at 30 °C, use 5 mL–10 mL to inoculate 500 mL YPD next morning.
2. Grow 500 mL *S. cerevisiae* cultures to OD_{600} = 0.6–0.8 in YPD at 30 °C (expect about 5×10^8 cells).
3. Transfer culture to centrifugation bottles.
4. Centrifuge for 10 min at 2700 × *g* and 4 °C.
5. Discard the supernatant, wash cell pellets with 50 mL cold ultrapure H_2O, and transfer to 50 mL conical tubes.
6. Centrifuge for 10 min at 2700 × *g* and 4 °C.
7. Resuspend cells in 36 mL Resuspension buffer.
8. Incubate for 15 min in 37 °C water bath.
9. Add 500 μL 1 mg/mL Zymolase (freshly prepared in Resuspension buffer) and incubate for 10–20 min in 37 °C water bath (*see* **Note 1**).
10. Transfer to ice immediately once spheroplasting is complete and quickly proceed to next step.

11. Centrifuge for 5 min at 1000 × *g* and 4 °C.
12. Wash pellets (resuspend very gently with a 1 mL pipette) with 25 mL cold SPC buffer **with protease inhibitors**.
13. Centrifuge for 5 min at 1000 × *g* and 4 °C.
14. Repeat steps 10 and 11 once, resuspend pellets in 0.5 mL SPC buffer **without protease inhibitors** (***Pause point:*** If you are preparing nuclei from more than one yeast culture, it is okay to hold the spheroplasts in SPC buffer on ice for other samples to catch up).
15. Mix with cold 25 mL Ficoll buffer.
16. Transfer to 50 mL glass beaker and stir very gently (slowest setting) on a magnetic stirrer with a small (8–10 mm) magnetic bead at 4 °C for 15 min.
17. Transfer to 50 mL conical tubes and centrifuge for 10 min at 7500 × *g* rcf and 4 °C.
18. Wash pellet with 10 mL cold SPC buffer **with protease inhibitors**.
19. Centrifuge for 10 min at 4200 × *g* and 4 °C.
20. Repeat steps 18 and 19 once, resuspend each nuclei pellet in total 5 mL to 10 mL cold SPC buffer **with protease inhibitors**. Take out a small aliquot (~50 μL) and count nuclei concentration. Adjust volume of SPC to get 5×10^7 nuclei per mL. Slow-freeze nuclei in an isopropyl alcohol chamber at −80 °C. Nuclei can be stored at −80 °C (*see* **Note 2**).

3.2 CUT&RUN

This protocol is recommended for 5×10^7 yeast nuclei. We start with this high number so that the supernatant from CUT&RUN can be used for ChIP with a second antibody for up to 10 targets. Alternatively, for example for a time-course experiment, the starting material can be used for up to 10 time points. Scale down/up as necessary. Use freshly prepared or frozen nuclei. If using frozen nuclei, quick-thaw in a 37 °C water bath for 1–2 min until the frozen content starts to melt, and quickly transfer to ice to thaw completely. Avoid multiple freeze–thaw cycles, which are detrimental to nuclear integrity. Because it is crucial that DNA breakage is minimized throughout the protocol, we recommend that cavitation during resuspension and vigorous vortexing be avoided.

1. Equilibrate 100 μL Concanavalin A–coated (ConA) magnetic beads with 1 mL binding buffer in 1.5 or 1.7 mL microcentrifuge tubes (*see* **Note 3**). We recommend using 20 μL beads for 1×10^7 yeast nuclei.
2. Place on a magnet stand to clear and pull off the liquid.
3. Repeat steps 1 and 2 twice, and bring up beads in 100 μL binding buffer.

4. Add beads to thawed yeast nuclei (1 mL, 5×10^7 nuclei), rotate gently at room temperature for 10 min. Remove liquid from the cap and side with a quick spin on a microcentrifuge (no more than $100 \times g$ for 10 s). Place on a magnet stand to clear (<1 min) and pull off the liquid (*see* **Note 4**).
5. Wash beads (with the bound nuclei) with 1 mL Antibody buffer. Resuspend beads by inverting tubes a few times, or use a 1 mL pipette to break clumps very gently if necessary. Place on a magnet stand to clear, pull off the liquid, and resuspend beads in 1 mL Antibody buffer (*see* **Note 5**). From this step keep samples cold (on ice) as much as possible.
6. Add primary antibody against your target of interest (e.g., mouse anti-FLAG for a FLAG-tagged protein; recommended antibody dilution = 1:100 unless determined empirically; *see* **Note 6**), gently rotate for 1 h at room temperature or overnight at 4 °C (recommended, to obtain higher yields). To evaluate success of the procedure without requiring library preparation, include in parallel a positive control antibody (e.g., anti-H2A.Z or H3K4me3) and a negative control antibody (e.g., rabbit isotype-control IgG) (*see* **Note 7**). *Go to step 8 if not using a secondary antibody* (*see* **Note 8**).
7. Collect the contents of the tubes to the bottom with a quick spin on a microcentrifuge (no more than $100 \times g$ for 10 s). This will minimize carryover of reagents that could result in overall background cleavages during the digestion step.
8. Place tubes on a magnet stand to clear, pull off the liquid and wash beads with 1 mL cold Dig-Wash buffer. Place tubes on a magnet stand to clear, pull off the liquid, add 1 mL cold Dig-Wash buffer with secondary antibody (e.g., rabbit anti-mouse IgG) at 1:100 dilution. Rotate for 1 h at 4 °C.
9. Quick spin on a microcentrifuge (no more than $100 \times g$ for 10 s), place tubes on a magnet stand to clear, pull off the liquid and wash beads with 1 mL cold Dig-Wash buffer. Repeat once more.
10. Pull off the liquid on a magnet stand and resuspend beads in 1 mL cold Dig-Wash buffer with a final concentration of 500–1000 ng/mL pA-MNase or pAG-MNase (*see* **Note 9**, or follow manufacturer's recommendation for enzyme obtained from commercial sources), and rotate for 1 h at 4 °C.
11. Quick spin on a microcentrifuge (no more than $100 \times g$ for 10 s), place tubes on a magnet stand to clear, pull off the liquid and wash beads with 1 mL cold Dig-Wash buffer. Repeat twice more.

12. Pull off liquid on a magnet stand and resuspend beads in 1 mL cold Dig-Wash buffer. Hold tubes on a metallic tube-holder set an ice-water bath (0 °C) for 10 min.
13. Add 20 μL 100 mM Calcium Chloride (2 mM final concentration), mix quickly by placing tubes on a vortex mixer set at a very low speed. Immediately replace tubes to 0 °C. Incubate for 5–30 min for digestion (*see* **Notes 10** and **11**).
14. Stop digestion by adding 1 mL 2× STOP buffer containing 50 pg/mL *Drosophila melanogaster* spike-in DNA (*see* **Note 12**), mix by inverting tubes a few times. If there are multiple time points, remove 100 μL of reaction and add to 100 μL 2×STOP.
15. Incubate 20 min in 37 °C water bath to release CUT&RUN fragments from the insoluble nuclear chromatin.
16. Centrifuge 5 min at 4 °C and 16,000 × *g*, place on magnet stand and transfer liquid to a fresh tube.
17. If you are performing CUT&RUN.ChIP, jump to the next Subheading 3.3, or else proceed to DNA extraction (following steps).
18. To each sample or time point (200 μL) add 100 μL Wash buffer at room temperature, 3 μL 10% SDS (to 0.1%), and 2.5 μL 20 mg/mL Proteinase K. Mix by vortexing and incubate for 1 h at 50 °C.
19. Add 300 μL phenol–chloroform–isoamyl alcohol (25:24:1) and mix by full-speed vortexing ~2 s.
20. Transfer to a phase-lock tube (e.g., Qiagen MaXtract), and centrifuge 5 min at room temperature and 16,000 × *g*.
21. Add 300 μL chloroform–isoamyl alcohol (24:1) on top of the aqueous phase, invert ~10× to mix, and centrifuge for 5 min at room temperature and 16,000 × *g*.
22. Remove the aqueous phase by pipetting into a fresh tube containing 2 μL 10 mg/mL glycogen.
23. Add 750 μL chilled 100% ethanol, mix by inverting tubes, chill on ice for 10 min or longer at −20 °C, and centrifuge for 40 min at 16,000 × *g* and 4 °C (*see* **Note 13**).
24. Pour off the liquid and drain on a paper towel.
25. Rinse the pellet in 1 mL chilled 100% ethanol and centrifuge 5 min 16,000 × *g* and 4 °C.
26. Carefully pour off the liquid, drain on a paper towel, and let air dry by leaving the caps open.
27. When the pellet is dry, dissolve in 30 μL 10 mM Tris–HCl pH 8, and proceed to sequencing-library preparation (*see* Subheading 3.4).

3.3 Native ChIP with CUT&RUN Supernatant as Input (CUT&RUN.ChIP)

1. For CUT&RUN.ChIP, supplement 2× STOP buffer with 20 μg/mL of the immunizing/blocking peptide for the primary CUT&RUN antibody used in step 6 (e.g., FLAG peptide) (*see* **Note 14**).
2. Add 1 mL 2× STOP to 1 mL CUT&RUN reaction after 0 °C incubation for the desired duration (similar to **step 14** of Subheading 3.2), and mix by inverting tubes a few times.
3. Incubate 20 min in 37 °C water bath to release CUT&RUN fragments from the insoluble nuclear chromatin.
4. Centrifuge 5 min at 4 °C and 16,000 × *g*, place on magnet stand and transfer liquid to a fresh tube.
5. Remove 200 μL for each ChIP experiment, add 2 μL antibody against the epitope of interest, rotate 1 h at room temperature or overnight at 4 °C. Add an isotype control IgG to one sample as a negative control.
6. Save 200 μL CUT&RUN supernatant as the "ChIP input" sample at 4 °C.
7. Take 200 μL protein A– or protein G–coated magnetic beads (*see* **Note 15**), place on a magnet stand to clear and pull off the liquid.
8. Wash beads with 1 mL Wash buffer (without digitonin), place on a magnet stand to clear and pull off the liquid.
9. Repeat step 8 twice more, resuspend beads in 200 μL Wash buffer.
10. Add 20 μL beads to each ChIP experiment, except the "input" fraction.
11. Rotate for 30 min at 4 °C.
12. Quick spin on a microcentrifuge at 1000 × *g* for 10 s. Place on a magnet stand to clear and pull off the liquid.
13. Wash beads once with 1 mL Wash buffer, quick spin on a microcentrifuge, place on a magnet stand to clear and pull off the liquid.
14. Resuspend beads in 300 μL Wash buffer. Add 100 μL Wash buffer to the "input" fraction.
15. Proceed to DNA extraction (**steps 18–27** of Subheading 3.2) with each ChIP and input sample.

3.4 Library Preparation and Paired-End Illumina Sequencing

1. Optional: Use 1–2 μL of the DNA obtained from CUT&RUN or CUT&RUN.ChIP (as in **step 27** of Subheading 3.2) to quantify double-stranded (ds) DNA concentration, for example, using fluorescence detection with a Qubit instrument (*see* **Note 16**).

2. Optional: Evaluate the presence and size distribution of cleaved fragments by capillary electrophoresis with fluorescence detection, for example by using a TapeStation 4200 instrument with D1000 high-sensitivity reagents, following the manufacturer's instructions (*see* **Note 16**). CUT&RUN for a highly abundant chromatin-associated protein such as the histone variant H2A. Z (with an anti-FLAG antibody targeting Htz1-3XFLAG) typically shows a nucleosomal ladder; and DNA fragments corresponding to mononucleosomes get specifically enriched upon native ChIP for H2A.Z using an antibody targeting a different epitope on the protein (Fig. 4).
3. Prepare barcoded libraries for Illumina sequencing with Tru-Seq compatible adapters and indexing primers using a single-tube protocol, following the manufacturer's instructions (*also see published protocols* [12, 19] and **Note 16**). Good results have been obtained using the Hyper-prep kit from KAPA Biosystems (https://sequencing.roche.com/en-us/products-solutions/by-category/library-preparation/dna-library-preparation/kapa-hyperprep.html). Remove un-ligated adapters after the adapter ligation step by SPRI paramagnetic bead-based sequential cleanups (e.g., with AMPure XP beads) using 1.2× and 1.4× vol/vol ratios of sample to bead suspension. To minimize the contribution of large DNA fragments and to minimize PCR duplicates, PCR cycles should be between 8 and 14 cycles, preferably with a 10 s at 60 °C combined annealing/extension step (*see* Table 1 and **Note 17**). Remove excess PCR primers by two rounds of post-PCR-amplification cleanup using 1.4× vol/vol ratio of sample to bead suspension.
4. Quantify library yield using a dsDNA-specific assay as in step 1.
5. Determine the size distribution of libraries by capillary electrophoresis as in step 2, and mix libraries to achieve equal representation as desired aiming for a final concentration as recommended for the sequencing platform.
6. Perform paired-end 25 × 25 base pair Illumina sequencing on the barcoded libraries for 25 cycles using an Illumina HiSeq 2500 or another massively parallel DNA sequencer following the manufacturer's instructions (*also see published protocol* [12, 19] and **Note 18**).

3.5 Data Processing and Analysis

1. Align paired-end *Saccharomyces cerevisiae* reads to the sacCer3/V64 genome using Bowtie2 version 2.4.1 with parameters "--end-to-end --very-sensitive --no-mixed --no-discordant -q --phred33 -I 10 -X 700". Align spike-in fragments to repeat-masked FlyBase r6.06 (*Drosophila*) genome, with parameters "--end-to-end --very-sensitive --no-overlap --no-dovetail --no-mixed --no-discordant -q --phred33 -I 10 -X 700" to avoid cross-mapping of the experimental genome to that of the spike-

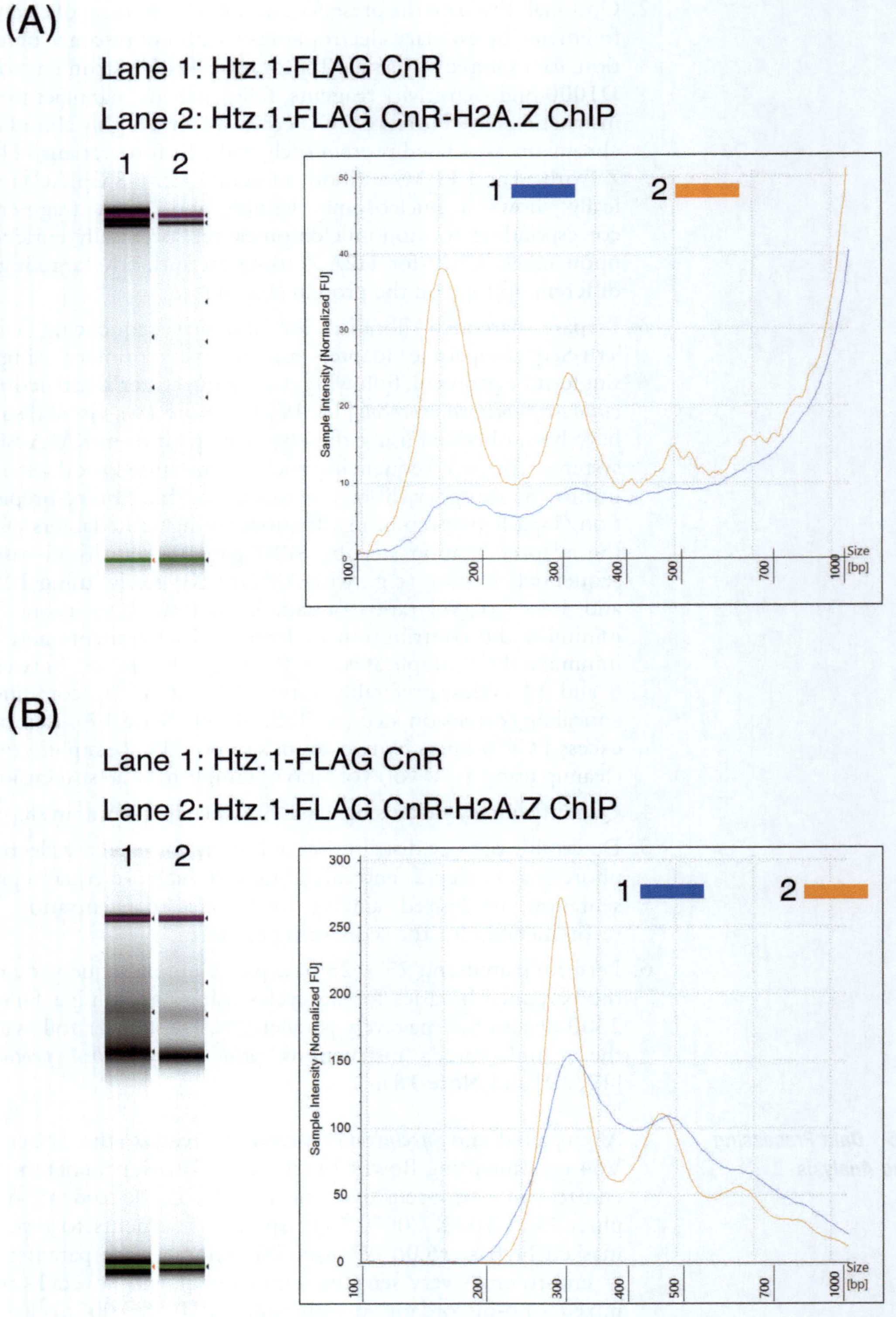

Fig. 4 TapeStation analysis of CUT&RUN(.ChIP) experiments. Example gel images (left) and densitometric scanning (right) show the fragment length profiles of DNA from Htz1-FLAG CUT&RUN (blue) and H2A.Z ChIP

in DNA (*see* **Note 19**). Extract properly paired reads from the alignments.

2. Make scaled fractional count or spike-in calibrated BedGraph for each sample using genomecov (with the "-bg" option) in BEDtools v2.30.0. For scaled fractional count, compute the fraction of all aligned bases at each base pair and scale it by the size of the sacCer3 genome (121,507,105 bp), so if the counts were uniformly distributed the value would be one at each base pair. For spike-in calibration compute the fraction of all aligned bases at each base pair, scale it by the total number of reads mapped to the spike-in genome, and divide by a large number (e.g., 10,000) (*see* **Note 19**). Unix-compatible scripts for processing CUT&RUN data and spike-in calibration are available from https://github.com/Henikoff/Cut-and-Run
3. To analyze TF-binding and the position of flanking nucleosomes from the same experiment, separate sacCer3 reads into $\leq$120 bp and $\geq$ 150 bp size classes for generating BedGraph (*see* **Note 20**).
4. Data shown in Figs. 2 and 3 are BedGraph files displayed as genomic tracks using Integrated Genome Browse (IGB) version 2.3.32.

4 Notes

1. The extent of spheroplasting can be monitored by removing 20 μL aliquots from the spheroplasting reaction into 1 mL of 1% SDS and measuring OD_{600} (vs. 1% SDS blank). Spheroplasting is typically complete when OD_{600} is 10–20% of the starting value (usually in 10–15 min at 37 °C). Spheroplasting should be carefully monitored, do not over- or under-digest. Work with one yeast pellet at a time.
2. We do not recommend flash freezing, as this can cause background DNA breakage that may impact final data quality.
3. Although low-retention pipette tips are preferred for accurate solution transfer, use only conventional (not low-binding) microcentrifuge tubes to avoid loss of beads between wash steps and buffer exchanges.
4. Take the supernatant after bead binding and count nuclei using a cell counter to determine the percent of nuclei bound to the

Fig. 4 (continued) targeting a separate epitope after Htz1-FLAG CUT&RUN (orange), before (panel A) and after (panel B) sequencing-library preparation. CUT&RUN produces a nucleosomal ladder, while mononucleosomal fragments are specifically enriched upon the subsequent ChIP. Note the increase in fragment sizes post library preparation due to adapter ligation (~120 bp)

Table 1
Thermocycler parameters for library amplification

Cycle number	Denature	Anneal + extend	Extend	Final
1	98 °C, 45 s	–	–	–
2–13	98 °C, 15 s	60 °C, 10 s	–	–
16	–	–	72 °C, 1 min	–
17	–	–	–	8 °C, hold

beads. Alternatively, take a small volume of the nuclei suspension before bead binding and the supernatant after bead binding, and compare nuclei counts under a microscope using a cell counting device such as a hemocytometer.

5. EDTA in antibody buffer removes excess divalent cations used to activate the ConA, because carryover of Ca^{++} from the beads can prematurely initiate strand cleavage after addition of pA-MNase. Chelation of divalent cations when cells are permeabilized also serves to quickly halt metabolic processes and prevent endogenous DNase activity. Washing out the EDTA before pA-MNase addition avoids inactivating the enzyme. Spermidine in the wash buffer is intended to compensate for removal of Mg^{++}, which might otherwise affect chromatin properties.
6. The primary antibody should be used at a dilution recommended for immunofluorescence-type experiments. If a working dilution is not known, we recommend 1:100 as a starting point. Although unbound antibodies are washed away in the subsequent washing steps, using excess antibodies can potentially lead to higher background and/or nonspecific signals.
7. Do not include a no-antibody control, as the lack of tethering may allow any unbound pA-MNase to act as a "time-bomb" and digest accessible DNA, resulting in a background of DNA-accessible sites.
8. The binding efficiency of Protein A to the primary antibody depends on host species and IgG isotype. For example, Protein A binds well to rabbit and guinea pig IgG but poorly to mouse and goat IgG, and so for these latter antibodies, a secondary antibody, such as rabbit anti-mouse IgG is recommended. The hybrid Protein A-Protein G-MNase fusion (pA/G-MNase) binds avidly to a broad range of antibody species and subtypes (e.g., mouse antibodies) [18]. Nevertheless, using a secondary antibody is likely to provide higher yield, and is therefore recommended.

9. Enzymatic activity of purified recombinant pA-MNase or pAG-MNase can be determined empirically by enzyme titration for in vitro digestion of plasmid DNA, or purified budding yeast chromatin to obtain mostly mononucleosomal fragments, and by CUT&RUN using an antibody against an abundant epitope. We found that when our in-house purified pAG-MNase was used in excess, there was no increase in release of background fragments, indicating that the washes are effective in removing excess unbound fusion protein [18].
10. MNase binds DNA but only cleaves when Ca^{++} is present, so that digestion is a zero-order reaction that seems to be less temperature-dependent than the subsequent diffusion of released pA-MNase-bound particles, which can digest accessible regions of the genome. Efficient cleavage and release of particles can be obtained at 0 °C while minimizing background cleavages attributable to diffusion. We found that digestion at ambient or higher temperatures result in unacceptable background cleavage levels.
11. Digestion time needs to be optimized for the factor of interest. For stable chromatin features such as histone posttranslational modifications on nucleosomes, longer digestion (up to 30 min at 0 °C) may increase yield without significantly increasing background cleavages. However, for proteins that weakly interact with DNA, such as certain transcription factors, a shorter digestion time might be more suitable as longer digestion with the tethered MNase may lead to their eviction from chromatin due to extensive digestion of the underlying DNA.
12. Heterologous spike-in DNA should be present in the 2× STOP to calibrate DNA amounts, for example to compare treatments or digestion time points. This is especially important for CUT&RUN as there is too little background cleavage for normalization of samples.
13. Long incubation at −20 °C and long centrifugation for DNA precipitation may result in better recovery of small DNA fragments.
14. It is extremely important to compete off bound antibodies from the released targets to prevent their recapture by Protein A– or Protein G–coated magnetic beads during the ChIP step. To evaluate successful competition of the antibodies used in CUT&RUN and specificity of the ChIP antibody, compare spike-in DNA-calibrated ChIP and IgG negative control samples versus CUT&RUN supernatant (used as ChIP input). *Also see* Fig. 3.
15. The choice of Protein A- or Protein-G coated magnetic beads depends on the host species and isotypes of ChIP antibodies.

16. Some long undigested DNA will leak through in CUT&RUN and dominate Qubit fluorescence, particularly in CUT&RUN of typical transcription factors. For these, the amount of targeted DNA recovered is too low to be detected, even by TapeStation analysis. For such samples, we often use the entire amount of recovered DNA for making sequencing libraries. We do not recommend size selection of DNA before library preparation.
17. Rapid PCR cycles favor exponential amplification of the desired CUT&RUN fragments over linear amplification of large DNA fragments that are too long for polymerase to complete. Higher than 14 PCR cycles may result in unacceptable proportions of PCR duplicates in the sequenced libraries. Higher ratio of beads during the cleanup steps ensure that small sub-nucleosomal size DNA fragments (such as those resulting from CUT&RUN of transcription factors) are retained in the library.
18. Paired-end sequencing is recommended for CUT&RUN, as single-end sequencing sacrifices resolution and discrimination between transcription factors and neighboring nucleosomes. Because of the very low background with CUT&RUN, typically one–three million paired-end reads suffices for transcription factors or nucleosome modifications in budding yeast.
19. We found that pAG-MNase purified over a nickel-agarose column using the protocol as described in Meers et al., 2019 [18] contains enough carryover *Escherichia coli* DNA which can be used for spike-in calibration, and shows close correspondence with adding heterologous spike-in DNA at the end of a CUT&RUN experiment. Align paired-end *E. coli* reads to NCBI Ecoli (Escherichia coli str. K-12 substr. MG1655 U00096.3) with parameters as described in Subheading 3.5, **step 1**, and use the data for spike-in calibration of SerCer3 reads as described in Subheading 3.5, **step 2**.
20. This analysis provides mapping of the local vicinity of a DNA-binding protein, but can vary depending on the steric access to the DNA by the tethered MNase.

Acknowledgments

We thank Christine Codomo for preparing Illumina sequencing libraries, the Fred Hutch Genomics Shared Resource for sequencing, and Jorja Henikoff for help with data analysis. Research reported in this publication was supported by the National Institute of General Medical Sciences of the National Institutes of Health under award number K99GM138920 (S.B.) and the Howard Hughes Medical Institute (S.H.).

References

1. Lai WKM, Pugh BF (2017) Understanding nucleosome dynamics and their links to gene expression and DNA replication. Nat Rev Mol Cell Biol 18:548–562
2. Barski A, Cuddapah S, Cui K et al (2007) High-resolution profiling of histone methylations in the human genome. Cell 129:823–837
3. Park D, Lee Y, Bhupindersingh G et al (2013) Widespread misinterpretable ChIP-seq bias in yeast. PLoS One 8:e83506
4. Teytelman L, Thurtle DM, Rine J et al (2013) Highly expressed loci are vulnerable to misleading ChIP localization of multiple unrelated proteins. Proc Natl Acad Sci U S A 110: 18602–18607
5. Teves SS, An L, Hansen AS et al (2016) A dynamic mode of mitotic bookmarking by transcription factors. elife 5:e22280
6. Jain D, Baldi S, Zabel A et al (2015) Active promoters give rise to false positive 'Phantom Peaks' in ChIP-seq experiments. Nucleic Acids Res 43:6959–6968
7. Rossi MJ, Kuntala PK, Lai WKM et al (2021) A high-resolution protein architecture of the budding yeast genome. Nature 592:309–314
8. He Q, Johnston J, Zeitlinger J (2015) ChIP-nexus enables improved detection of in vivo transcription factor binding footprints. Nat Biotechnol 33:395–401
9. Skene PJ, Henikoff S (2015) A simple method for generating high-resolution maps of genome-wide protein binding. elife 4:e09225
10. Kasinathan S, Orsi GA, Zentner GE et al (2014) High-resolution mapping of transcription factor binding sites on native chromatin. Nat Methods 11:203–209
11. Van Steensel B, Delrow J, Henikoff S (2001) Chromatin profiling using targeted DNA adenine methyltransferase. Nat Genet 27:304–308
12. Skene PJ, Henikoff S (2017) An efficient targeted nuclease strategy for high-resolution mapping of DNA binding sites. eLife 6:e21856
13. Grunberg S, Zentner GE (2017) Genome-wide mapping of protein-DNA interactions with ChEC-seq in saccharomyces cerevisiae. J Vis Exp 124:55836
14. Schmid M, Durussel T, Laemmli UK (2004) ChIC and ChEC; genomic mapping of chromatin proteins. Mol Cell 16:147–157
15. Carter B, Ku WL, Kang JY et al (2019) Mapping histone modifications in low cell number and single cells using antibody-guided chromatin tagmentation (ACT-seq). Nat Commun 10:3747
16. Kaya-Okur HS, Wu SJ, Codomo CA et al (2019) CUT&tag for efficient epigenomic profiling of small samples and single cells. Nat Commun 10:1930
17. Henikoff JG, Belsky JA, Krassovsky K et al (2011) Epigenome characterization at single base-pair resolution. Proc Natl Acad Sci U S A 108:18318–18323
18. Meers MP, Bryson TD, Henikoff JG et al (2019) Improved CUT&RUN chromatin profiling tools. Elife 8:e46314
19. Skene PJ, Henikoff JG, Henikoff S (2018) Targeted in situ genome-wide profiling with high efficiency for low cell numbers. Nat Protoc 13:1006–1019
20. Brahma S, Henikoff S (2019) RSC-associated subnucleosomes define MNase-sensitive promoters in yeast. Mol Cell 73(238–249):e233
21. Janssens DH, Wu SJ, Sarthy JF et al (2018) Automated in situ chromatin profiling efficiently resolves cell types and gene regulatory programs. Epigenetics Chromatin 11:74
22. Weiner A, Hsieh TH, Appleboim A et al (2015) High-resolution chromatin dynamics during a yeast stress response. Mol Cell 58:371–386

Chapter 10

ChIP-SICAP: A New Tool to Explore Gene-Regulatory Networks in *Candida albicans* and Other Yeasts

Lasse van Wijlick, Ansh Goyal, Sophie Bachellier-Bassi, and Christophe d'Enfert

Abstract

Chromatin immunoprecipitation followed by mass spectrometry (ChIP-MS) is a powerful method to identify protein interactions, and has long been used to gain insights into regulatory networks in relevant fungal species as well as many other organisms. In this chapter, we discuss a similar technique called ChIP-SICAP (chromatin immunoprecipitation with selective isolation of chromatin-associated proteins) that overcomes many of the traditional limitations of ChIP-MS, and describe a protocol that allows ChIP-SICAP to be applied to *Candida albicans* and other yeasts. Notably, the technique design permits stringent washing to remove contaminating proteins and antibodies before subsequent mass spectrometry processing, allows for genome-wide mapping of the bait protein by ChIP-seq after ChIP-SICAP from the same sample through a DNA recovery process, and specifically purifies and identifies proteins associating with chromatin. In the future, ChIP-SICAP will provide the yeast genomics research community an additional method to explore the complex dynamics of the gene-regulatory networks modulating morphology, metabolism and response to stress.

Key words Chromatin immunoprecipitation, Next-generation sequencing, Mass spectrometry, Gene-regulatory networks, *Candida albicans*, Yeast

1 Introduction

Chromatin immunoprecipitation (ChIP) was established over 30 years ago [1] and remains one of the most widely used techniques in modern molecular biology. ChIP is used to address the questions about the nature of DNA interactors, their exact binding location and, if specific conditions are applied, the timing of the interaction, within the native chromatin context of the cell. The fundamental ChIP technique is limited in many ways, but several techniques have been developed to overcome some of its limitations to broaden the questions scientists can ask. ChIP followed by microarray (ChIP-chip) or followed by next generation sequencing (ChIP-seq) permits exploration of transcriptional regulators

Frédéric Devaux (ed.), *Yeast Functional Genomics: Methods and Protocols*,
Methods in Molecular Biology, vol. 2477, https://doi.org/10.1007/978-1-0716-2257-5_10,

binding on a genome-wide scale and has been successfully used to gain insight into transcription factor networks operating together, for example, to establish biofilms in the human fungal pathogen *C. albicans* [2], a key virulence trait of the fungus, or how histones modulate gene expression [3]. Recently developed ChIP-seq techniques have been optimized to overcome technical limitations of classical ChIP, such as the need to cross-link protein–DNA complexes with formaldehyde or the need to shear chromatin by sonication, which are major sources of experimental bias. In the ChEC-seq (chromatin endogenous cleavage) approach, the bait transcriptional regulator is expressed as a fusion protein with MNase, a nuclease which introduces double stranded DNA breaks and is specifically activated by adding Ca^{2+} ions [4, 5]. ChEC-seq has also been recently established for *C. albicans* and other yeast species with a convenient set of plasmids which enables its use in wild-type backgrounds [6]. Similarly, in CUT&RUN, MNase fused with Protein A is tethered to the localization of the bait transcriptional regulator or histone by interacting with the Fc region of an antibody that is used against the bait protein, enabling robust mapping of genome-wide binding sites in situ at a high resolution [7].

However, even highly optimized protocols are fundamentally limited; for instance, they only give us insight into the target-gene repertoire of one single regulator, even though gene regulation requires coordinated interaction of protein complexes. To truly understand the highly dynamic process of gene regulation it is therefore necessary to characterize the activity of protein complexes, rather than isolated proteins. One approach that allows for capture of protein interactions on chromatin is ChIP followed by high-resolution mass spectrometry. Many variations have been developed, with ChIP-MS and RIME (Rapid Immunoprecipitation Mass spectrometry of Endogenous proteins) as the most popular [8, 9]. Similar to the basic ChIP-seq protocol, these techniques take advantage of formaldehyde to preserve protein–protein and protein–DNA interactions. After solubilizing the chromatin complexes by sonication, bait proteins are affinity-purified on Protein A- or Protein G-covered beads, and copurified proteins are digested with trypsin either on the beads or after elution for subsequent identification by mass spectrometry. One major technical challenge of these approaches is to remove background contaminants, because a stringent washing protocol cannot be applied. Another fundamental limitation of the techniques is that both soluble protein–protein complexes and chromatin–protein complexes are captured and enriched. It is therefore not possible to distinguish between interactions that have occurred at regulatory-sites of genes or between unbound proteins. A recently developed technique, ChIP-SICAP (selective isolation of chromatin-associated proteins) overcomes these limitations and combines ChIP-seq (to map genome-wide binding of a bait transcription factor) with a modified ChIP-MS procedure (to specifically identify proteins that are associated with the transcription factor occupying the same genomic sites) [10]. The method introduces a chromatin biotin-labeling step, which, due to the high specificity and high biotin-binding affinity of streptavidin ($K_d = 10^{-15}$ M) [11], permits stringent washing of the enriched chromatin–protein complexes coupled to streptavidin-coated beads to drastically reduce the protein background [10]. An alternative technique that makes use of the high affinity of biotin to streptavidin to enable stringent washing is BioTAP-XL, which relies on tandem purification of a bait protein modified with a Protein A tag fused to a biotin acceptor sequence; however, unlike ChIP-SICAP, this technique cannot discriminate between on- or off-chromatin interactions [12].

Here we adapted the ChIP-SICAP approach [10] to establish transcription factor interaction networks in the pathogenic yeast *C. albicans*. Our approach can easily be adjusted to other yeast species. The general protocol can be divided into 10 steps (Table 1): (1) All interactions are fixed with formaldehyde, and then the chromatin–protein complexes are (2) solubilized by sonication and (3) immunoprecipitated with antibodies coupled to Protein A- or Protein G-coated beads, targeting the epitope of the bait transcriptional regulator. The enriched chromatin is then (4) labeled with biotin by terminal deoxynucleotidyl transferase, (5) eluted from the beads by SDS and DTT treatment and subsequently (6) recaptured with streptavidin-coated beads (7). A stringent washing protocol consisting of highly concentrated detergent washes and salt washes followed by isopropanol and acetonitrile washes is applied to remove contaminants (8). The copurified proteins are digested by trypsin on the streptavidin-coated beads, separated from the beads and injected into LC-MS/MS. (9) The DNA is recovered from the beads and prepared for NGS. (10) After acquiring the raw data from the MS and NGS workflows, two distinct data analysis pipelines are used to identify potential coregulators of the bait transcriptional regulator and potential target genes. The two pipelines are also presented in our protocol. The method has been established with HA (Human influenza hemagglutinin)-tagged transcription factors and is straight-forward and highly reproducible. To keep the risk of proteolytic degradation of proteins to a minimum, only 12 samples were processed simultaneously, for example, corresponding to 3 transcriptional regulator bait proteins and 1 control strain, in biological triplicates. To overcome low-expression rates of some transcriptional regulators, we took advantage of our recently developed ORFeome collection, which enables Gateway™-mediated cloning of nearly every *C. albicans* ORF under the control of the inducible tetracycline-dependent promoter (TET-ON), along with C- or N-terminal epitope tags [13, 14].

Table 1
Estimated time for preparation of 12 ChIP-SICAP samples

	Timetable of the ChIP-SICAP workflow		
	Process	**Time**	**Day**
Step 1	Cross-linking and cell lysis	6 h	1
Step 2	Chromatin shearing	2–4 h	1
Step 3	Immunoprecipitation on magnetic beads and washing	Overnight +1 h	1/2
Step 4	Chromatin labelling with biotin	2–3 h	2
Step 5	Elution from the beads	0.5 h	2
Step 6	Enrichment of chromatin on streptavidin-coated beads	1 h	2
Step 7	Stringent washing to remove background contaminants	0.5 h	2
Step 8	Tryptic digestion of co-purified proteins on beads and LC-MS/MS injection	2 h + overnight	2/3
Step 9	Recovery of chromatin and library preparation for NGS	6–16 h	3/4
Step 10	Data analysis: MS- and NGS-pipelines	Several days	5

The data analysis can take several days and is often based on individual pipelines that are either provided by the sequencing platform or that have been specifically developed

2 Materials

Prior to preparation of buffers and solutions, clean all glassware with isopropanol and let them dry within a biosafety cabinet. Do not use autoclaved tubes. All buffers are prepared with MS grade water and subsequently filter-sterilized. If not specified, buffers and solutions can be stored and prepared at room temperature.

2.1 Strains and Growth Conditions

The protocol has been used with strains overproducing HA epitope-modified transcription factors from a tetracycline-inducible promoter [13], but can be applied to any other expression system, including native expression from the endogenous promoter. As a control, strains expressing unmodified transcription factors have been used. We recommend to use at least three biological replicates, ideally three independently transformed strains. *C. albicans* strains were grown in YPD medium (1% Yeast-Extract, 2% Bacto Peptone, 2% D-glucose) at 30 °C with 180 g agitation. For full induction of tetracycline-inducible promoters, the medium was supplemented to a final concentration of 6 μg/mL anhydrotetracycline from a 6 mg/mL stock solution dissolved in 100% EtOH. Expression rates can be adjusted by changing the final concentration of anhydrotetracycline in the growth medium (*see* **Note 1**).

2.2 Crosslinking and Cell Lysis

1. 30% formaldehyde solution, low-methanol (*see* **Note 2**).
2. 2.5 M glycine in 100 mL H_2O, freshly prepared.
3. Tris-buffered saline (TBS 10×): 0.2 M Tris–HCl, pH 7.5, 1.5 M NaCl in H_2O. Prepare a 1 L solution and store at 4 °C.
4. Liquid nitrogen.
5. Lysis I buffer: 50 mM HEPES–KOH, pH 7.5, 150 mM NaCl, 1 mM EDTA, 1% Triton X100, 0.1% Na-deoxycholate, in 500 mL H_2O, store at 4 °C.
6. Complete protease Inhibitor cocktail EDTA-free, store at 4 °C or prepare 100× stock by dissolving one tablet in 100 μL H_2O and store at −20 °C.
7. 100 mM Phenylmethylsulfonyl fluoride (PMSF), 100× stock in isopropanol, store at −20 °C.
8. FastPrep-24 instrument or equivalent.
9. 1.5 mL screw cap tubes compatible with the FastPrep-24 instrument or equivalent.
10. Glass beads, 0.5 mm diameter.
11. 23-G needles, 1 for each sample.
12. Round bottom tubes, 14 mL.
13. 50 mL Falcon tubes.
14. Photometer to measure absorbance at wavelength of 595–600 nm.
15. Cooled (tabletop) centrifuge for 1.5 mL, 2 mL, 16 mL and 50 mL tubes.
16. Tube rotator, suitable for 50 mL Falcon tubes.

2.3 Chromatin Fragmentation and Immunoprecipitation

1. Dynabeads™ Protein A, Protein G or IgG coated for Immunoprecipitation, store at 4 °C (*see* **Note 3**).
2. Magnetic stand suitable for 2 mL tubes.
3. Antibodies (*see* **Note 4**).
4. Adaptive Focused Acoustics (AFA, Covaris, E220 Focused-ultrasonicator), sonication bath (Diagenode Bioruptor Plus Sonication System with water cooler), with appropriate tubes, or probe sonicator with microtip (EpiShear Probe Sonicator V6) (*see* **Note 5**).
5. Lysis II buffer: 50 mM HEPES–KOH, pH 7.5, 360 mM NaCl, 1 mM EDTA, 1% Triton X100, 0.1% Na-deoxycholate, in 500 mL H_2O, store at 4 °C.
6. Wash III buffer: 10 mM Tris–HCl, pH 8.0, 1 mM EDTA, 250 mM LiCl, 0.5% NP40, 0.5% Na-deoxycholate, in 500 mL H_2O, store at 4 °C.
7. Tris buffer: 10 mM Tris–HCl, pH 8.0, in 10 mL H_2O.

8. Tube rotator, suitable for 2 mL tubes and temperatures of 4 °C–37 °C.
9. 2 mL low-retention reaction tubes.
10. Phosphate-buffered saline (10X PBS): 100 mM Na_2HPO_4, 18 mM KH_2PO_4, 1.37 M NaCl, 27 mM KCl in H_2O. Prepare a 1 L solution and store at 4 °C.
11. PBS-Tween (PBS-T): 0.05% Tween 20 in 1× PBS, 10 mL, freshly prepared.

2.4 Biotinylation and Capture of Coenriched Chromatin

1. Terminal deoxynucleotidyl Transferase (TdT) enzyme (20 U/μL), 50 μL, store at −20 °C.
2. TdT buffer (5 mL), store at −20 °C.
3. Biotin-11-ddUTP (1 mM, 125 μL, for example, Biotin-X-5-Propargylamino-ddUTP, Jena Bioscience), store at −20 °C.
4. TdT IP-mix (1×): 20 μL TdT buffer (5×), 5 μL biotin-ddUTP (1 mM stock), 2.5 μL TdT (50 U), 72.5 μL H_2O, freshly prepared.
5. Lysis I buffer (*see* Subheading 2.2), prechilled on ice.
6. Sodium dodecyl sulfate (SDS) 20%.
7. 1 M Dithiothreitol (DTT) in 1 mL H_2O, store at −20 °C.
8. Elution buffer: 10 mM Tris–HCl, pH 8.0, 1 mM EDTA, 5% SDS, 0.1 M DTT in H_2O, 2 mL, freshly prepared.
9. Streptavidin-coated magnetic beads, store at 4 °C.
10. Thermomixer.
11. Magnetic stand suitable for 2 mL tubes.

2.5 Stringent Washing Protocol

1. SDS-wash buffer: 10 mM Tris–HCl, pH 8.0, 1 mM EDTA, 0.1% SDS, 0.2 M NaCl in 50 mL H_2O.
2. Washing buffer II: 10 mM Tris–HCl, pH 8.0, 1 mM EDTA, 0.1% Triton X100, 2 M NaCl in 50 mL H_2O.
3. Isopropanol-wash solution: 15% 2-propanol in H_2O (v/v), 50 mL, freshly prepared.
4. Acetonitrile-wash solution: 25% acetonitrile (MS grade) in H_2O (v/v), 50 mL, freshly prepared.
5. Magnetic stand suitable for 2 mL tubes and 0.2 mL tubes or 96-well plates.

2.6 Protein Digestion and Sample Preparation for MS

1. Ammonium bicarbonate buffer: 25 mM ammonium bicarbonate, 10 mM DTT, in H_2O, 1 mL, freshly prepared.
2. 0.2 M DTT from 1 M stock (*see* Subheading 2.4), 1 mL in H_2O, freshly prepared.
3. 0.2 M Iodoacetamide, 1 mL in H_2O, freshly prepared.

4. Trypsin (MS grade), 20 μg.
5. Thermocycler.
6. Formic acid (MS grade).
7. ZipTip μ-C18 Pipette Tips.
8. Mass spectrometer (e.g., Q Exactive Plus Orbitrap Mass Spectrometer).

2.7 Chromatin Recovery and Preparation for NGS

1. TE buffer: 10 mM Tris–HCl, pH 8.0, 1 mM EDTA, in 10 mL H_2O.
2. 20 mg/mL Glycogen, store at −20 °C.
3. 10 mg/mL RNAse A, store at −20 °C.
4. TE-RNAse-mix (1×): 3 μL of 10 mg/mL RNAse A, 2 μL of 20 mg/mL glycogen, 35 μL TE buffer.
5. 20 mg/mL Proteinase K, store at 4 °C or − 20 °C.
6. Tris buffer: 10 mM Tris–HCl, pH 8.0, in 10 mL H_2O.
7. Agencourt Ampure XP beads, 5 mL, store at 4 °C (*see* **Note 6**).
8. 80% EtOH, 15 mL aliquots, freshly prepared from a new EtOH stock, or stored at −20 °C.
9. Qubit 2.0 Fluorometer and dsDNA high sensitivity assay kit.
10. High sensitivity DNA chip for Fragment-Analyzer and 2100 Bioanalyzer instrument.
11. Magnetic stand for 0.2 mL tubes or 96 well plates.
12. Thermomixer.
13. 96-well PCR plates and sealing film.
14. Microcentrifuge for 96-well PCR plates.
15. Library preparation kit for ChIP-seq (e.g., TruSeq ChIP Library Preparation Kit, Illumina).
16. Sequencing platform (e.g., Illumina NextSeq 550 sequencing system).
17. Appropriate flow cell and reagent kit for the sequencing platform (e.g., NextSeq 550 Mid Output Kit v2.5 with 150 cycles from Illumina).
18. 0.1 N NaOH, 1 mL, freshly prepared.
19. Tris–Tween buffer: 10 mM Tris–HCl, pH 8.5, 0.1% Tween 20, in 10 mL H_2O.
20. 0.2 M Tris–HCl, pH 7, in 1 mL H_2O.
21. HT1: Illumina hybridization buffer.

2.8 Data Processing and Analysis

1. PCs with Windows and Unix environments.

2.9 Software

1. SAV (https://support.illumina.com/sequencing/sequencing_software/sequencing_analysis_viewer_sav/downloads.html).
2. bl2fastq (https://support.illumina.com/sequencing/sequencing_software/bcl2fastq-conversion-software.html).
3. FASTQC (http://www.bioinformatics.babraham.ac.uk/projects/fastqc/).
4. HISAT2 (http://daehwankimlab.github.io/hisat2/).
5. SAMTOOLS (http://www.htslib.org/).
6. BEDTOOLS (https://bedtools.readthedocs.io/en/latest/content/tools/genomecov.html).
7. IGV (http://software.broadinstitute.org/software/igv/).
8. bPeaks (https://cran.r-project.org/web/packages/bPeaks/index.html).
9. Proteome Discoverer 2.2 (https://assets.thermofisher.com/TFS-Assets/CMD/manuals/Man-XCALI-97808-Proteome-Discoverer-User-ManXCALI97808-EN.pdf).
10. R studio (https://rstudio.com/).
11. RSAT (http://rsat.sb-roscoff.fr/).
12. MEME-ChIP (http://meme-suite.org/tools/meme-chip).

3 Methods

3.1 Culturing and Cross-Linking

Processing time: 6 h

1. Inoculate 5 mL YPD medium with a single colony and incubate overnight at 30 °C rotating or agitating at 180 × *g*.
2. Measure the optical density (OD_{600}) from a 1:20 diluted overnight culture and inspect the cells under the microscope for abnormalities. Inoculate 50 mL YPD medium supplemented with anhydrotetracycline in a 250 mL Erlenmeyer, to an OD_{600} of 0.2 and incubate for 4 h at 30 °C and 180 × *g* agitation. Cell density of OD_{600} 0.8–1.0, corresponding to approximately 2×10^7 yeast cells/mL, is optimal (*see* **Note 7**).
3. Prepare 50 mL Falcon tubes containing 1.7 mL of 30% formaldehyde (1 per sample). Fill the tubes with the cultures, up to the 50 mL graduated mark (1% final concentration). Rotate the tubes on a tube rotator at 25 °C for 20 min.
4. Cool down the centrifuge to 4 °C.

5. Quench formaldehyde by adding 5 mL of 2.5 M Glycine and continue rotating the tubes for 5 min and spin down the cells immediately at 3200 × *g* for 5 min.
6. Wash the cell pellet twice by resuspending the cells in 10 mL freezer-cold TBS (1×) and subsequent centrifugation at 3200 × *g* at 4 °C. Place the samples on ice in between wash cycles and keep processing time to a minimum.
7. Resuspend the cell pellet in 1 mL freezer-cold TBS (1×) and transfer to prechilled screw cap tubes. Spin down the cells at 3200 × *g*, remove the supernatant and snap freeze with liquid nitrogen. The cell pellets can be stored at −80 °C for 1 month (*see* **Note 8**).

3.2 Dynabead–Antibody Preparation

Processing time: 3–4 h, but only 30 min hands on sample time

1. Before starting with the cell lysis step (Subheading 3.3), prepare magnetic beads and the antibody used to immunoprecipitate the antigen of the epitope-modified bait protein. Take out the appropriate magnetic beads from 4 °C and vortex for 60 s until beads are well mixed (*see* **Notes 3** and **4**).
2. For each ChIP sample, corresponding to 10^9 yeast cells, 30 μL (0.9 mg) Dynabeads are pooled together in a 2 mL tube (e.g., for 12 samples, pool 360 μL Protein A Dynabeads).
3. Put the tube on the magnetic stand, wait until the beads are separated and remove the supernatant. Wash the beads once with 1 mL PBS-T, by rotating the beads on a tube rotator for 2 min at room temperature, separate the beads from the buffer as before and remove the supernatant.
4. Resuspend and dilute the beads in PBS-T in twice the starting volume (e.g., for 360 μL resuspend in 720 μL PBS-T) by pipetting (*see* **Note 9**).
5. Take out the antibody (4 °C or −20 °C) and add 2–5 μg per ChIP sample to the beads (e.g., 30 μL of anti-HA antibody (F-7) of a 2 μg/μL stock, corresponding to 60 μg antibody for 12 ChIP samples) (*see* **Note 10**).
6. Incubate the bead–antibody mixture on a tube rotator for 30 min at room temperature, then transfer in a cold room and incubate further for 2 h or until chromatin shearing is finished (Subheading 3.3).
7. Spin down the beads at room temperature for 10 s; put the tubes on the magnetic stand and remove the supernatant. Wash the antibody-coupled beads twice by adding 1 mL ice-cold PBS-T, vortexing and spinning them down again. Set the tubes back on the magnet and remove the supernatant.
8. Similarly, wash the antibody-coupled beads in 1 mL lysis I buffer, and remove the supernatant.

9. Resuspend the beads in an appropriate volume of lysis I buffer with 1/10 extra amount of the starting volume of the beads (e.g., for 360 μL of Protein A Dynabeads resuspend in 400 μL lysis I buffer).
10. Keep the antibody-coupled beads at 4 °C until chromatin shearing is complete (*see* **Note 11**).

3.3 Cell Lysis, Chromatin Shearing and Immunoprecipitation

Processing Time: 3–4 h

1. Prepare 10 mL lysis I buffer by adding 1 protease inhibitor cocktail tablet (1 per 10 mL) and 100 μL of the 100× PMSF stock and keep the lysis buffer on ice (*see* **Note 12**).
2. Transfer the screw cap tubes with the cell pellets from −80 °C (*see* Subheading 3.1, **step** 7) on ice and resuspend the pellets gently in 500 μL lysis I buffer, by pipetting up and down.
3. Add the equivalent of a 0.5 mL PCR tube of 0.5 mm glass beads to each sample and close the screw cap tubes firmly to prevent leakage during the subsequent bead beating step.
4. In a cold room, prepare total cell extracts by bead beating using a FastPrep-24 instrument with 6–8 cycles, each 40 s at 6.5 m/s and 3 min incubation on ice between cycles. Inspect the samples under the microscope to confirm complete cell lysis.
5. With a hot 23G needle, punch two holes into the bottom of the screw cap tube containing the total cell extract and quickly place it on a 10 mL round-bottom tube, prechilled on ice. Use a fresh needle for each sample (*see* **Note 13**).
6. Loosen the screw cap of the tubes and centrifuge at 1000 × *g* and 4 °C for 2 min. Inspect the glass beads for remaining cell lysate and if necessary repeat centrifugation.
7. Transfer the cell extract into a 1.5 mL tube and centrifuge at 21,000 × *g* and 4 °C for 10 min. This step is used to separate unbound cytosolic proteins from cross-linked protein–chromatin. The insoluble chromatin is in the pellet.
8. Carefully resuspend the chromatin pellet in 500 μL ice-cold lysis I buffer (with fresh protease inhibitors) and try to avoid generating bubbles. Transfer the sample containing the protein–chromatin complexes into a suitable tube for sonication. Transfer 10 μL of each sample for immunoblot analysis (label as A) and store at −80 °C (*see* **Note 14**).
9. Chromatin shearing is one of the most critical steps and requires prior optimization for individual experiments. The settings of the following steps have been successfully used to obtain homogeneous shearing of chromatin in the range of 200–600 bp: (*see* **Note 5**).

For a probe sonicator with microtip: 7–8 cycles of 20 s ON/40 s OFF, on high output (amplitude of 200 μm), with the tube on ice.

For sonication bath with cooling: transfer your sample into a 15 mL TPX tube and add 510 μL ice-cold lysis I buffer to each sample; the recommended minimal volume is 1 mL. Place 6 tubes into the rotor and use the following settings: 10 cycles of 30 s ON/30 s OFF, on high output. Take out the samples, vortex them for 30 s and place them on ice for 3 min. Replace the water in the bath with precooled water if it warmed up and repeat sonication for a total of 3 times.

For Covaris E220: transfer your sample into a 15 mL falcon tube and dilute the chromatin by adding 3.5 mL of lysis I buffer. Vortex well and aliquot the sample into 8 times 500 μL microtube-500 vessels (with AFA fiber and screw cap) and use the following settings: 5 min, 175 peak incident power, 20% duty factor, 200 cycles per burst. Pool the samples together after sonication is complete and proceed with the next sample (*see* **Note 15**).

10. Vortex the samples for 30 s and centrifuge at 21,000 × *g* and 4 °C for 10 min. Transfer the clear supernatant (containing the fragmented, soluble chromatin) into a new 2 mL tube and keep the samples on ice. The pellet should appear white. Grey pellets indicate overheating of the sample during the sonication step. Transfer 40 μL of the fragmented chromatin to a fresh 1.5 mL tube and label as input (INP), transfer 15 μL to a separate 1.5 mL tube for immunoblot analysis and label as B and store at −80 °C (*see* **Note 14**).
11. Start the immunoprecipitation by adding 30 μL of antibody-coupled beads (prepared in **step 10** of Subheading 3.2) to each fragmented chromatin sample and place the tubes into a tube rotator. Let the samples rotate overnight in a 4 °C cold room.

3.4 Biotinylation of Enriched Chromatin

Processing time: 4 h

Before starting with the next steps, incubate lysis I buffer, lysis II buffer, wash III buffer and Tris buffer at −20 °C for 10 min and keep them on ice onward. All subsequent washing steps are performed at room temperature with ice-cold buffers.

1. Pellet the beads by centrifugation and separate the beads on a magnetic stand. Transfer the supernatant into a fresh 1.5 mL tube, label as C and store at −80 °C. Take out samples labeled as input from −80 °C (INP) and keep on ice (*see* **Note 14**).
2. Wash beads twice with 1 mL of lysis I buffer, twice with 1 mL of lysis II buffer, twice with 1 mL of wash III buffer and twice with 1 mL of Tris buffer. Each washing step

consists of adding the respective wash buffer to the beads, placing the tubes on the tube rotator for 2 min and placing them back on the magnetic stand to separate the beads and removing the supernatant. It is usually not necessary to resuspend the beads by vortexing or pipetting.

3. After the final wash with Tris buffer, centrifuge the beads quickly, place them back on the magnetic stand and remove as much remaining liquid as possible.
4. Resuspend the beads immediately in 100 μL 1× TdT buffer by pipetting gently to equilibrate the beads. Incubate at room temperature for 2 min and separate the beads on a magnetic stand again. Remove the buffer and resuspend the beads in 100 μL TdT IP-mix and incubate for 60 min in a thermomixer at 37 °C with agitation (500 rpm).
5. Add 160 μL of the freshly prepared elution buffer to the 40 μL input samples, and incubate overnight at 65 °C to start the reverse cross-linking of the chromatin–protein complexes. It is not required to agitate the samples (*see* **Note 16**).
6. After the TdT-mediated biotinylation of the enriched chromatin, pellet the beads by centrifugation and separate them on a magnetic stand. Remove the supernatant and wash the beads 5 times by adding 1 mL of ice-cold lysis I buffer, placing the tubes on a tube rotator for 2 min and then back on the magnetic stand to separate the beads. Centrifuge the samples after the last washing step, set them back on the magnetic stand and remove as much remaining liquid as possible.
7. To elute the chromatin–protein complexes from the beads, resuspend the beads in 100 μL freshly prepared elution buffer and incubate the samples for 10 min at 37 °C with agitation (1000 × *g*) in a thermomixer; in the meantime, start with **step 8** of Subheading 3.4 just below. Afterward, vortex the samples for 30 s, quickly centrifuge and place the tubes on the magnetic stand. Transfer the supernatant into a fresh 2 mL tube and keep the tube on ice. Repeat the elution step once again and pool the two 100 μL samples together. Transfer 10 μL of the eluted immunoprecipitated samples into a fresh 1.5 mL tube and label as D (*see* **Notes 14** and **17**).
8. Prepare the streptavidin-coated magnetic beads by first pooling 50 μL beads per sample together in a new 2 mL tube (e.g., for 12 samples pool 600 μL of beads together). Centrifuge the beads and separate them on a magnetic stand. Remove the supernatant and add 1 mL of lysis I

buffer to the beads. Rotate the tube on a tube rotator for 5 min at room temperature, pellet the beads, remove the supernatant and repeat washing of the beads one more time. Resuspend the beads in an appropriate volume of lysis I buffer, for example, if the pooled volume was 600 μL, resuspend the beads in 650 μL and keep them at 4 °C until needed (*see* **Note 18**).

9. To capture the biotin-labeled chromatin, add 460 μL of lysis I buffer and 50 μL of the preequilibrated streptavidin beads to the eluted samples and vortex them for 10 s. Place the tubes on a tube rotator and incubate them at room temperature for 60 min rotating. Prepare all wash solutions described in Subheading 2.5 before proceeding with the next step of the protocol.

3.5 Removal of Protein Contaminations

Processing time: 0.5 h

1. Pellet the streptavidin-coated beads by centrifugation and place the tube on a magnetic stand. Remove the supernatant and start with the stringent washing of the bead bound chromatin–protein complexes to remove common protein contaminants.
2. Add 1 mL of SDS-wash buffer to each sample and resuspend the beads by vortexing the sample for 10 s, quickly centrifuge and separate the beads on a magnetic stand. Remove the supernatant and repeat this step twice.
3. Resuspend the beads in 1 mL wash buffer II by vortexing the sample for 10 s, quickly centrifuge and separate the beads on a magnetic stand and remove the supernatant.
4. Similarly wash twice with isopropanol-wash solution, and twice with acetonitrile-wash solution.
5. Resuspend the beads in 100 μL of acetonitrile-wash solution by gentle pipetting and transfer the beads into a new 0.2 mL PCR tube or a 96-well PCR plate if preferred. Repeat this step if not all beads have been transferred.
6. Separate the magnetic beads on a magnetic stand suitable for PCR tubes or 96-well PCR plate and remove the acetonitrile-wash solution. Centrifuge the samples quickly, place the tubes/plate back on the magnetic stand and remove as much of the remaining acetonitrile-wash solution as possible. Let the beads stand at room temperature for 2 min but do not let them dry.

3.6 On Bead Digestion of Copurified Proteins

Processing time: 2.5–3 h

1. Resuspend the beads in 20 μL of ammonium bicarbonate buffer and incubate the beads at 55 °C in a thermocycler for 60 min, to prepare the coimmunoprecipitated proteins for digestion (*see* **Note 19**).

2. To prevent disulfide-bond reformation add 2 μL of 0.2 M iodoacetamide to each sample and incubate at room temperature for 60 min in darkness.
3. Add 1 μL of 0.2 M DTT to each sample and incubate at room temperature for 10 min (*see* **Note 20**).
4. Centrifuge the samples and add 3 μL of 100 ng/μL trypsin to each sample. Resuspend the beads by pipetting and incubate overnight or for at least 12 h in a thermocycler at 37 °C.

3.7 Sample Preparation and LC-MS/MS Acquisition

Processing time: 1 h—pause step possible

1. Vortex the beads for 10 s, centrifuge and separate them on a magnetic stand. The digested peptides are in the supernatant and the biotin-labeled chromatin is still captured on the streptavidin-coated beads. Transfer the supernatant into a new 0.2 mL PCR tube and reverse the formaldehyde crosslinks by incubating the samples at 95 °C for 20 min. Recovery of the captured chromatin is explained in Subheading 3.9.
2. Add 1 μL of 1% formic acid to each sample. Samples can be stored at −20 °C or directly subjected to LC-MS/MS.
3. The resulting peptides are desalted using ZipTip μ-C18 pipette tips and analysed on an Orbitrap Q-Exactive Plus, coupled Nano-LC Proxeon 1000, equipped with an easy spray ion source (*see* **Notes 21** and **22**).

3.8 Qualitative LC-MS/MS Data Analyses

1. The resulting mass spectrometry spectra are processed on the Proteome Discoverer 2.2. The integrated Mascot software is used to identify proteins from a peptide sequence database (*see* **Note 23**).
2. Open the corresponding PDRESULT file in Proteome Discoverer and inspect the results carefully (*see* **Note 24**).
3. To identify enriched proteins in the immunoprecipitated samples over the control samples apply the FILTER function in the Proteome Discoverer. Select PROTEINS and filter for Score Mascot is greater or equal to a set baseline, for example, 200 to filter out proteins with low abundance, and filter for proteins absent from your control sample or proteins with Mascot Scores fivefold greater than in the control sample (*see* **Note 25**).
4. Export the filtered list of identified proteins for subsequent network analysis (Fig. 1).

3.9 Recovery of Chromatin Fragments and Preparation for NGS

Processing time: 3 h + time for library construction (4–12 h)

1. Resuspend the streptavidin beads from **step 1** of Subheading 3.7 in 40 μL of TE-RNase-mix and incubate at 37 °C for 30 min. Take out the input samples from **step 5** of Subheading 3.4 from 65 °C, and let them cool down to room temperature.

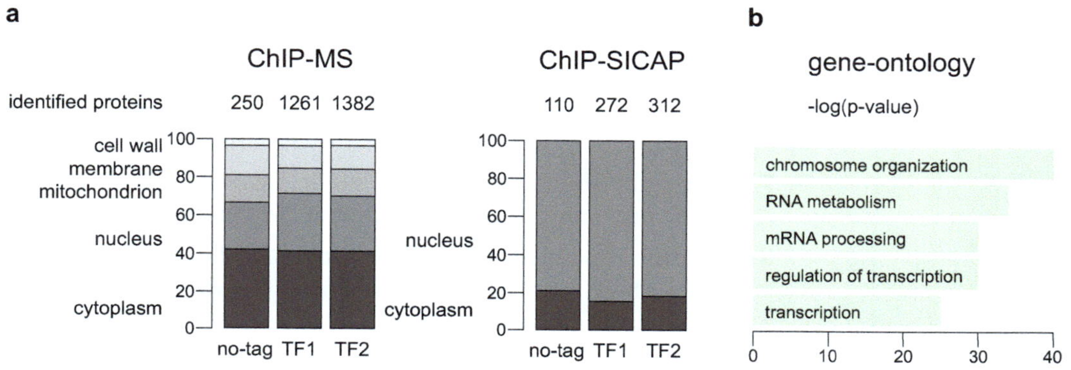

Fig. 1 ChIP-SICAP and ChIP-MS was applied to two transcription factors of *C. albicans* (TF1, TF2) and a control strain expressing unmodified transcription factors (no-tag). (**a**) Stacked bar plots showing the relative abundance of identified proteins and their cellular localization. (**b**) Enriched gene-ontology terms of identified proteins of the ChIP-SICAP approach

Add 5 μL of RNase A and 5 μL of glycogen to each sample, vortex the samples for 60 s, centrifuge quickly and incubate at 37 °C for 30 min.

2. After the RNase treatment add 5 μL of proteinase K and 5 μL of 20% SDS to each bead sample, and 20 μL of proteinase K and 20 μL of 20% SDS to each input sample. Vortex the samples quickly and incubate them at 55 °C for 60 min.
3. Centrifuge the bead samples, separate them on a magnetic stand and transfer the supernatant containing the DNA into a new tube.
4. Clean and concentrate the DNA with AMPure XP beads. Bring the beads to room temperature before use. Add 100 μL of the equilibrated beads to each 50 μL immunoprecipitated sample, or 200 μL of the equilibrated beads with 100 μL of each input sample (*see* **Notes 6** and **26**).
5. Vortex all samples for 10 s and incubate them at room temperature for 10 min. Centrifuge the tubes and separate the beads on a magnetic stand. Remove the supernatant and wash the beads by adding 200 μL of 80% EtOH, incubate for 30 s on the magnetic stand and remove the EtOH. Repeat the washing once and centrifuge the beads after the last wash step, place the tubes back on the magnetic stand and remove as much remaining EtOH as possible. Let the beads stand for 2 min but do not let them dry. Resuspend the immunoprecipitated-bead samples in 20 μL of Tris buffer and the input-bead samples in 100 μL of Tris buffer and incubate them at room temperature for 2 min. Separate the beads on a magnetic stand again and transfer the supernatant into a new tube.

6. Quantify the DNA using the Qubit 2.0 fluorometer along with a dsDNA high sensitivity assay kit. Evaluate the size distribution of your input samples on an Agilent bioanalyzer high sensitivity DNA chip, the peak should be around 300–400 base pairs.
7. Follow the library construction guidelines of the TruSeq ChIP library preparation kit from Illumina (*see* **Notes 27** and **28**).
8. Quantify the libraries using the Qubit 2.0 fluorometer with a dsDNA high sensitivity assay kit again. To estimate the average size of each individual library, run 1 μL of each sample on a bioanalyzer high sensitivity DNA chip. Use the following formula for the calculation:

$$\mathrm{nM} = (c(\mathrm{Qubit\,ng}/\mu\mathrm{L})/(660^{*}\mathrm{average\,size(bioanalyzer)})) * 10^{6}.$$

9. Normalize the concentration of each individual library to 10 nM by adjusting the volume with Tris–Tween buffer. The libraries can be stored at −20 °C.
10. Pool 4 μL of each normalized library together in a fresh 0.2 mL PCR tube or one well of a fresh 96-well plate (*see* **Note 29**).
11. For the NextSeq 550 sequencing format, it is required to dilute and denature the pooled libraries prior to cluster generation on the flow cell. Starting from a library concentration of 4 nM transfer 5 μL of the pooled library into a fresh 1.5 mL tube and add 5 μL of freshly prepared 0.2 N NaOH. Mix well by vortexing the tube for 5 s and centrifuge quickly. Incubate the tube at room temperature for 5 min.
12. Add 5 μL of 0.2 M Tris–HCl, pH 7 to the pooled library and mix by vortexing the tube for 5 s and centrifuge the tube quickly.
13. Dilute the denatured libraries to 20 pM by adding 985 μL of HT1 buffer to the tube of the libraries. Mix again by vortexing the tube and quick centrifugation.
14. To prepare the libraries to the loading concentration for the NextSeq flow cell, transfer 117 μL of the 20 pM library pool into a fresh tube and add 1183 μL of HT1 buffer. Invert the tube 10 times and quickly centrifuge. Inject the 1.3 mL into reservoir #10 of the reagent cartridge and follow the run setup instructions of the NextSeq 550 system using the NextSeq control software interface (*see* **Note 30**).
15. Start the run in single read mode with 151 cycles (*see* **Note 31**).

3.10 NGS Data Analyses

3.10.1 Quality Control Assessment and Adapter Trimming

1. Sequencing metrics are inspected with the Illumina Sequencing Analysis Viewer during or after the run is finished and a sample sheet is generated recording information about sample names and the corresponding index (*see* **Note 32**).
2. The sample sheet is used by the bcl2fastq2 conversion software in a Unix environment to convert the raw data BCL files into FASTQ files containing the raw reads of each sample. The software is also used for adapter trimming and demultiplexing (*see* **Note 33**).
3. The quality of the raw reads is assessed with the FastQC tool. Upload the FASTQ files and inspect the generated reports (in HTML format) carefully (*see* **Note 34**).

3.10.2 Mapping of the Reads to the Reference Genome

1. Use the *hisat2-build* function to generate the indexed genome file (*see* **Note 35**).
2. Use HISAT2 for the mapping of the reads to the indexed genome with default parameters and declare SAM format as the output format:

   hisat2 -x $reference_genome -U $fastq_file -S $fastq_file.sam --summary-file.
3. With SAMTOOLS convert the SAM formatted files into sorted BAM files:

   samtools view -b $fastq_file.sam > $fastq_file.bam
   samtools sort $fastq_file.bam -o $fastq_file.sorted.bam
   samtools index $fastq_file.sorted.bam.
4. Inspect the BAM files manually with the Integrative Genomics Viewer (IGV) and compare the location of reads along the genome between immunoprecipitated and corresponding input samples and control samples (*see* **Note 36**).

3.10.3 Peak-Detection

1. Use BEDTOOLS GENOMECOV to compute per-base genome coverage of the sorted BAM files (*see* **Note 37**).

   bedtools genomecov -ibam $fastq_file.sorted.bam -d >$fastq_file_gcb.txt.
2. Peak calling is performed with the bPeaks program in an R environment, for example, R studio, by comparing immunoprecipitated samples to corresponding input samples and control samples (*see* **Note 38**).
3. Inspect the generated BED files manually in IGV and confirm all identified peaks (*see* **Note 39**).

3.10.4 Motif Discovery

To identify sequence specific motifs of the investigated transcriptional regulator the MEME-ChIP and the RSAT tool can be used. Both programs require FASTA formatted sequence files as input which can be generated with the BEDTOOLS GETFASTA function based on the BED files (*see* **Note 40**).

bedtools getfasta -fi $reference_genome.fasta -bed bPeaks_$.bed -fo bPeaks_$fasta.fasta.

4 Notes

1. An overproduction approach has been used to overcome potential limitations of low expression rates of transcription factors. However, the TET-ON system can also be used to achieve native expression rates by adjusting the concentration of anhydrotetracycline. Alternatively, the tetracycline analog doxycycline can be used. A concentration of 100 μg/mL is used to achieve the same levels of induction as for 6 μg/mL anhydrotetracycline. Gene expression rates should be investigated by qRT-PCR and confirmed by immunoblotting prior to using this protocol. To achieve moderate but constitutive expression of the transcriptional regulator, all cultures are grown in the presence of 1 μg/mL anhydrotetracycline.
2. Instead of using a commercial 30% formaldehyde stock solution, a fresh paraformaldehyde stock solution, can also be used that has the advantage that no Methanol is required for stabilization, which might have a negative effect on protein complexes and cross-linking. However, there are low-methanol (<1.5%) formaldehyde solutions available. The final formaldehyde concentration is crucial for proper cross-linking of proteins and chromatin and protein–protein interactions. Only purchase formaldehyde in small aliquots (25 mL) and do not use formaldehyde older than 3 months.
3. Binding characteristics of immunoglobulins (Igs) can differ significantly between species and Ig subclasses. To determine if Protein A- or Protein G-coated beads are most appropriate for your antibody, compare the binding strength of the antibody's specific Ig with the binding properties of Protein A or Protein G, provided by the manufacturer of the beads. Assuming an anti-HA antibody (F-7, from Santa Cruz Biotechnology, monoclonal Mouse, IgG2a) is used, binding properties are equally strong, either Protein A- or Protein G-coated beads can be used. For TAP-tagged proteins (Protein A-epitope) we use PanMouse IgG-coated Dynabeads, as IgG has a very high affinity to Protein A.
4. We recommend to use ChIP-seq grade antibodies only. The antibody should provide a minimum signal to noise ratio as compared to the input chromatin control. We successfully used anti-HA antibody (F-7, Mouse monoclonal, from Santa Cruz Biotechnology) and anti-myc antibody (9E-11, Mouse monoclonal, from Abcam).
5. Shearing of chromatin is one of the most crucial steps for successful ChIP, and optimal conditions need to be established with test samples (cross-linked chromatin). If available, Adaptive Focused Acoustics (AFA) technology from Covaris should

be used, as it enables mechanical processing of samples through focused ultrasonication and prevents samples from overheating. In contrast to probe-sonicators, AFA and sonication baths are usually more reliable, produce less heat stress to the samples and result in more homogenous chromatin fragment-sizes across samples, thus less bias. The optimal fragment-size range for this protocol is between 200 and 600 bp and should be determined on DNA high sensitivity Bioanalyzer chips or equivalent.

6. Binding specificity of AMPure XP beads is greatly influenced by temperature. We recommend to take an aliquot of the beads in a separate tube and bring the beads to room temperature by rotating them for 15 min on a tube rotator. Always vortex the beads for 30 s before use.
7. With the overexpression approach that we are using, 10^9 cells is by far sufficient. The yield of protein and chromatin overexceed the required input amount for downstream MS and NGS approaches by 10- to 100-fold. However, for low expressed transcription factors or when specific growth conditions are used, upscaling of culture volume might be required. The protocol can easily be adjusted to 10^{10}–10^{11} yeast cells.
8. We usually prepare the cell samples on a Thursday or Friday and store the cross-linked samples over the weekend at −80 °C prior to cell lysis. This allows more time for sample processing during the following week. We never handle more than 12 ChIP samples at the same time to avoid proteolytic degradation of the samples.
9. If smaller volumes of Dynabeads are used, it is recommended to use 0.2 mL or 0.5 mL tubes, instead of the 2 mL tubes, to ensure a good ratio of bead–antibody volume in regard to the total volume of the tube to facilitate sufficient mixing.
10. 2–5 μg per ChIP sample are sufficient for ChIP-seq approved antibodies. However, the amount of antibody required depends on the abundance of the antigen in the sample and can be determined empirically by immunoblot analysis of different antibody–bead ratios, as compared to antigen left in the supernatant of the immunoprecipitation sample. Different binding capacities of Dynabeads also need to be considered when selecting the optimal ratio of antibody to beads, for example, 1 mg Protein A Dynabeads has a binding capacity of approximately 5 μg Mouse IgG2. We use 0.9 mg of beads and mix them with 5 μg of the anti-HA antibody.
11. If antibodies with lower affinity are used it is recommended to first incubate the antibody with the ChIP sample, and to add the beads afterward. Moreover, the incubation time for the antibody–antigen complex formation is flexible and can be adjusted.

12. PMSF is a serine protease inhibitor that crystallizes at temperatures below 4 °C. To solubilize vortex the tube and incubate at room temperature for several min before preparing the lysis buffer. Activity of PMSF decreases rapidly after addition to the lysis buffer.
13. We cut 2 mL tubes in half to remove the bottom part of the tube. This can be slipped over the screw cap tubes (1.5 mL) and the cap of the 2 mL tube prevents the screw cap tube from falling into the larger 14 mL round bottom tube (or 15 mL falcon tube).
14. There are several steps throughout the protocol where the integrity of the proteins should be investigated by immunoblot analysis. We take samples before performing sonication (A), after sonication (B), a small aliquot of the beads that are used for immunoprecipitation of the protein–chromatin complexes (C) and a sample of the supernatant after immunoprecipitation (D), to check if the protein was successfully captured. The immunoblot is performed once all samples have been collected.
15. Although Covaris AFA technology appears to be the most reliable technique in terms of control of fragment size with very low potential to create bias, the technique requires to buy additional rather costly consumables.
16. The TdT enzyme has a template-independent DNA polymerase function, extending 3′ ends of DNA with the supplied biotin-labeled ddUTPs. Continue with the input samples at **step 1** of Subheading 3.9 of the protocol.
17. If TAP-tagged bait proteins are used, a TEV-mediated cleavage step needs to be added to the elution protocol, before adding the elution buffer. In our experience this requires more cell material (minimum of 10^{10} cells). The combination of the ionic detergent SDS and the reducing agent DTT in the elution buffer is also used to remove all proteins that are not cross-linked to chromatin.
18. It is possible to chemically modify the streptavidin-coated beads to improve resistance to proteolytic cleavage by trypsin and LysC, thereby reducing streptavidin contaminations of the protein samples [15]. However, we had no problems with streptavidin contaminations in our experiments, but this possibility should be considered if only low peptide amounts are recovered after the trypsin digestion of the enriched proteins.
19. The following protocol can be used for the thermocycler: 60 °C—60 min, 18 °C—5 min, hold at 18 °C. Alternatively the samples could be incubated in a suitable thermo block or water bath, but the samples need to be brought to room temperature before proceeding with the next step of the protocol.

20. The treatment with first DTT and then iodoacetamide is required to fully reduce and alkylate available thiol groups of cysteines, in order to prevent the formation of disulfide bonds. A final addition of an excess of DTT is used to neutralize iodoacetamide.
21. Liquid chromatography with tandem mass spectrometry is usually performed by core facilities with highly standardized pipelines and therefore the protocol in LC-MS/MS acquisition can be variable; however, here are the details of the LC-MS/MS acquisition in our workflow. Peptides are loaded with an online preconcentration method and separated by chromatography using a Pepmap-RSLC C18 column (0.75 × 500 mm, 2 μm, 100 Å), equilibrated at 50 °C and operated at a flow rate of 300 nL/min. Peptides are eluted by a gradient of solvent A (H_2O, 0.1% FA) and solvent B (100% ACN, 0.1% FA), the column is first equilibrated 5 min with 95% of solvent A, then solvent B is raised to 35% in 93 min, and finally, the column is washed with 80% solvent B during 10 min and reequilibrated at 95% solvent A during 10 min. Peptides masses are analyzed in the Orbitrap cell in full ion scan mode at a resolution of 70,000 with a mass range of m/z 375–1500 and an AGC target of 3×10^6. MS/MS is performed in a Top 20 DDA mode. Peptides are selected for fragmentation by Higher-energy C-trap Dissociation (HCD) with a Normalized Collisional Energy of 27%, and a dynamic exclusion of 30 s. Fragment masses are measured in the Orbitrap cell at a resolution of 17,500, with an AGC target of 2×10^5. Monocharged peptides and unassigned charge states are excluded from the MS/MS acquisition. The maximum ion accumulation time is set to 50 ms for MS and 45 ms for MS/MS acquisitions respectively.
22. To minimize run-to-run and column-to-column bias, it is recommended to run all samples in one series.
23. The RAW files are imported into Proteome Discoverer 2.2, a MS data analysis platform provided by Thermo Fisher Scientific. The protein sequence information of the *C. albicans* proteome was obtained from the candida-genome-database (http://www.candidagenome.org/), and a custom protein database with default settings was created. The database should only include species-specific protein sequences from an organism with fully sequenced genome. The Mascot node [16] was specified to identify proteins based on a tryptic digestion profile with a tolerance of 2 missed cleavage sites. The fragment and precursor mass tolerance were set to 0.5 Da and 6 ppm, respectively. Post-translational modifications can also be specified as fixed or variable parameters. In our MS pipeline, the following modifications are set as variable: acetylation, oxidation and

phosphorylation. The integrated percolator node was used to filter the spectra to limit FDR rates to 0.1% (based on a decoy database search). In addition, a Mascot significance threshold for identity was applied.

24. The Proteome Discoverer results file contains several information, such as the number of identified proteins in the experiment, including all runs, the proteins false discovery rate (FDR) for every identified protein, the % coverage of the protein, sub columns with unique peptides for the identified protein, and the Mascot Score. A detailed guide is provided by Thermo Fisher Scientific. If defined, also post-translational modifications can be inspected and used for further biological interpretation.

25. We usually establish a baseline for the Mascot Scores based on obtained Mascot Scores of nonspecifically associated proteins like ribosomal proteins, which are always overrepresented. We consider proteins as enriched in our immunoprecipitated samples when they are absent from all control samples, or if the Mascot Score is at least fivefold higher in all biological replicates as compared to the control samples. If required, different Mascot Score cutoffs can be applied to limit the number of identified proteins to the most abundant. We further shorten the list of identified proteins using GO-term annotations, by filtering for transcriptional regulators, kinases and phosphatases. All results are inspected manually at every step of the workflow. A quantitative data analysis pipeline can be established with the MaxQuant software package, in combination with Perseus for statistical interpretation of label-free quantification MS approaches (https://www.maxquant.org/), or with the Progenesis QI Proteomics software (http://www.nonlinear.com/). Downstream analyses involve further GO-term analysis, pathway enrichment, and comprehensive interaction-network construction and analysis (e.g., with Cytoscape).

26. AMPure XP beads are solid phase reversible immobilisation paramagnetic beads immersed in a PEG/NaCl buffer which can efficiently bind DNA fragments ranging from 100 to 800 bp. The ratio of beads/PEG/NaCl to sample volume is important for the binding properties of the beads and can be used for size selection. The advantage is that DNA fragments of larger size are already excluded from the following DNA quantification and thereby reduce overestimating the concentration of DNA input material that is used for library construction. Alternatively, MinElute column-based purification kits can be used.

27. We use the TruSeq ChIP library preparation kit from Illumina. If the DNA yield of the immunoprecipitated samples is <5 ng,

15 PCR cycles are applied during the DNA enrichment step. For samples with >5 ng DNA only 12 cycles are used, to minimize PCR-based bias. If the concentration is <1 ng use the entire volume of the eluted DNA and use 1 ng of the input samples as starting material for the libraries. For removal of adapter–adapter dimers we use AMPure XP beads in a bead-to-sample ratio of 1.6× or 1.8× after the PCR step of the protocol.

28. Usually we multiplex 12 samples on one NextSeq 550 Mid Output flow cell from Illumina (*see* also Subheading 2.7, **step 17**) in single read mode (SR), which gives about 120–130 million reads in total, yielding eight to ten million reads per ChIP sample (or input sample) of up to 150 bp length. For small genomes <20 Mb, such as budding yeast *S. cerevisiae* or *C. albicans*, ten million reads are deep enough to map genome-wide binding of transcriptional regulators.

29. If the total volume of pooled libraries is greater than the desired final volume, AMPure XP beads can be used once more to bring the samples to the desired volume and molarity. The quantity and the size distribution of the pooled library should be determined again before proceeding to library dilution and denaturation.

30. Before starting with the denaturation of the pooled libraries, take out the flow cell (from 4 °C) and the reagent cartridge (from −20 °C) and bring them to room temperature. The reagent cartridge should be incubated in a water bath until thawed, but not submerged.

31. Each run on the NextSeq 550 with the mid output kit for 150 cycles takes about 26 h to complete and produces up to 20 GB of raw-data.

32. Important quality control metrics to inspect are %Q > 30, which refers to the percentage of bases with a quality score of 30 or higher, Density and Density PF, referring to the density of clusters passing the filter, %Phasing and %Prephasing, the rate of molecules in a given cluster for which sequencing is not synchronic anymore, Cluster and Cluster PF, referring to the number of clusters passing the filter. For detailed information and interpretation of the quality metrics produced by the NextSeq system see also the SAV documentation.

33. The BCL to FASTQ conversion process is used to convert base call files (BCL) into FASTQ files. Local Run Manager and MiSeq Systems automatically convert BCL files into FASTQ format. It is important to name the sample sheet "SampleSheet.csv", otherwise it will not be recognized by the software. The default location is the root output folder.

34. FastQC is widely used for NGS quality control assessment and provides a user-friendly web interface. The most useful plots are the Per Base Sequence Quality box-plot, which plots the quality of each sequence read per cycle, and the Per Base Sequence Content plot, which provides information about the proportion of each base (GATC) per cycle and can be used to detect GC bias and issues with the complexity and quality of the library.

35. We use the HISAT2 (hierarchical indexing for spliced alignment of transcripts 2) program in a Unix environment to build an indexed reference genome and to align the sequence reads to the indexed genome. The HISAT2 program benefits from a graph-based data structure and an alignment algorithm to enable fast and sensitive alignments of sequencing reads through its graph index [17]. The ***hisat2-build*** function requires the sequence information of the reference genome in FASTA format.

36. The manual inspection with IGV [18] enables the user to identify experimental problems and genomic regions that are overrepresented in the background input samples or the control samples. To visualize the localization of open reading frames a GFF (general feature format) file of the referenced genome is required. Be sure to use the same version of the reference genome for IGV which had been used for the creation of the indexed reference genome. In addition to open reading frames, the localization of nucleosomes or transcript start-sites can be annotated if the information is available.

37. The −d option of the BEDTOOLS GENOMECOV program is used to report the depth at each position in the genome with 1-based coordinates. The use of the program is only necessary if the bPeaks program (Subheading 3.10.3, **step 2**) is used for subsequent peak-calling which is able to resolve enrichment analyses at the single nucleotide level.

38. The bPeaks R package is optimized to detect transcription factor binding sites from ChIP-seq data derived from small eukaryotic genomes [19]. Enrichment of immunoprecipitated samples over input samples is calculated based on four different thresholds: (T1) high number of reads in the immunoprecipitated sample, (T2) low number of reads in the control sample, (T3) high log fold-change between immunoprecipitated and control sample and (T4) good sequencing coverage in both, the immunoprecipitated and the control sample. We always run the bPeaksAnalysis function in multiparameter mode, which uses a range of different parameters and reports a table of all peaks found for each condition respectively: IPcoeff = *c*(6,4,2), controlCoeff = *c*(2,4,6), log2FC = *c*(3, 2, 1.5),

averageQuartile = *c*(0.9, 0.7). Detailed information can be found in the R package description (https://rdrr.io/cran/bPeaks/man/bPeaksAnalysis.html). bPeaks is also available in a web-based user-interactive interface for scientist without R experience [20].

39. Although bPeaks is very sensitive in peak detection, it is still required to investigate the identified binding sites manually with IGV. We only consider binding sites as relevant if they have been detected in at least two biological replicates. In addition to the input sample we always include a control sample which is derived from a strain expressing the unmodified transcriptional regulator, without any epitope, but treated with the antibody. In our experience this control works best to detect and remove unspecific genomic regions. This control is also used for the background detection for the protein MS workflow.
40. One important aspect of ChIP-seq is the identification of *cis*-regulatory motifs. MEME-ChIP [21] and the RSAT tool for fungi [22] are used for motif discovery in parallel and predicted motifs are compared. Another motif discovery algorithm, and part of the MEME-ChIP suite, DREME, also reports core DNA-binding motifs of potential coregulators [23], identified within the supplied sequences obtained from bPeaks. This information is extremely useful and can be combined with identified coregulators of the protein MS workflow. Co-occurrence of motifs in regulatory sequences of identified target genes can also be investigated with the PathoYeastract database [24] and used for the establishment of the final interaction network.

Acknowledgments

This work was supported by an Agence Nationale de la Recherche grant CANDIHUB (ANR-14-CE-0018 to C. d'E.) and the French Government's Investissement d'Avenir program (Laboratoire d'Excellence Integrative Biology of Emerging Infectious Diseases; ANR10-LABX-62-IBEID). L.v.W. was the recipient of a postdoctoral fellowship under the Agence Nationale de la Recherche project CANDIHUB (ANR-14-CE-0018). A.G. received a fellowship of the Mitacs Globalink Research Award programme. We are also thankful to Thibaut Léger, Camille Garcia, and Samuel Terrier for supporting our work at the Plateforme Protéomique/Spectrométrie de masse de l'Institut Jacques-Monod, to Gaelle Lelandais for her help with the establishment of the ChIP-seq pipeline and to Juliana Pipoli Da Fonseca from the Biomics team at Institut Pasteur.

References

1. Gilmour DS, Lis JT (1984) Detecting protein-DNA interactions *in vivo*: distribution of RNA polymerase on specific bacterial genes. Proc Natl Acad Sci U S A 81(14):4275–4279. https://doi.org/10.1073/pnas.81.14.4275
2. Nobile CJ, Fox EP, Nett JE, Sorrells TR, Mitrovich QM, Hernday AD, Tuch BB, Andes DR, Johnson AD (2012) A recently evolved transcriptional network controls biofilm development in *Candida albicans*. Cell 148(1–2):126–138. https://doi.org/10.1016/j.cell.2011.10.048
3. Price RJ, Weindling E, Berman J, Buscaino A (2019) Chromatin profiling of the repetitive and nonrepetitive genomes of the human fungal pathogen *Candida albicans*. MBio 10(4). https://doi.org/10.1128/mBio.01376-19
4. Schmid M, Durussel T, Laemmli UK (2004) ChIC and ChEC; genomic mapping of chromatin proteins. Mol Cell 16(1):147–157. https://doi.org/10.1016/j.molcel.2004.09.007
5. Zentner GE, Kasinathan S, Xin B, Rohs R, Henikoff S (2015) ChEC-seq kinetics discriminates transcription factor binding sites by DNA sequence and shape *in vivo*. Nat Commun 6: 8733. https://doi.org/10.1038/ncomms9733
6. Tebbji F, Khemiri I, Sellam A (2020) High-resolution genome-wide occupancy in *Candida* spp. Using ChEC-seq. mSphere 5(5): e00646–e00620. https://doi.org/10.1128/mSphere.00646-20
7. Skene PJ, Henikoff S (2017) An efficient targeted nuclease strategy for high-resolution mapping of DNA binding sites. Elife 6: e21856. https://doi.org/10.7554/eLife.21856
8. Ji X, Dadon DB, Abraham BJ, Lee TI, Jaenisch R, Bradner JE, Young RA (2015) Chromatin proteomic profiling reveals novel proteins associated with histone-marked genomic regions. Proc Natl Acad Sci U S A 112(12): 3841–3846. https://doi.org/10.1073/pnas.1502971112
9. Mohammed H, Taylor C, Brown GD, Papachristou EK, Carroll JS, D'Santos CS (2016) Rapid immunoprecipitation mass spectrometry of endogenous proteins (RIME) for analysis of chromatin complexes. Nat Protoc 11(2): 316–326. https://doi.org/10.1038/nprot.2016.020
10. Rafiee MR, Girardot C, Sigismondo G, Krijgsveld J (2016) Expanding the circuitry of pluripotency by selective isolation of chromatin-associated proteins. Mol Cell 64(3):624–635. https://doi.org/10.1016/j.molcel.2016.09.019
11. Chaiet L, Wolf FJ (1964) The properties of streptavidin, a biotin-binding protein produced by *Streptomycetes*. Arch Biochem Biophys 106:1–5. https://doi.org/10.1016/0003-9861(64)90150-x
12. Alekseyenko AA, McElroy KA, Kang H, Zee BM, Kharchenko PV, Kuroda MI (2015) BioTAP-XL: cross-linking/tandem affinity purification to study DNA targets, RNA, and protein components of chromatin-associated complexes. Curr Protoc Mol Biol 109: 21.30.1–21.30.32. https://doi.org/10.1002/0471142727.mb2130s109
13. Legrand M, Bachellier-Bassi S, Lee KK, Chaudhari Y, Tournu H, Arbogast L, Boyer H, Chauvel M, Cabral V, Maufrais C, Nesseir A, Maslanka I, Permal E, Rossignol T, Walker LA, Zeidler U, Znaidi S, Schoeters F, Majgier C, Julien RA, Ma L, Tichit M, Bouchier C, Van Dijck P, Munro CA, d'Enfert C (2018) Generating genomic platforms to study *Candida albicans* pathogenesis. Nucleic Acids Res 46(14):6935–6949. https://doi.org/10.1093/nar/gky594
14. Chauvel M, Nesseir A, Cabral V, Znaidi S, Goyard S, Bachellier-Bassi S, Firon A, Legrand M, Diogo D, Naulleau C, Rossignol T, d'Enfert C (2012) A versatile overexpression strategy in the pathogenic yeast *Candida albicans*: identification of regulators of morphogenesis and fitness. PLoS One 7(9):e45912. https://doi.org/10.1371/journal.pone.0045912
15. Rafiee MR, Sigismondo G, Kalxdorf M, Forster L, Brugger B, Bethune J, Krijgsveld J (2020) Protease-resistant streptavidin for interaction proteomics. Mol Syst Biol 16(5):e9370. https://doi.org/10.15252/msb.20199370
16. Perkins DN, Pappin DJ, Creasy DM, Cottrell JS (1999) Probability-based protein identification by searching sequence databases using mass spectrometry data. Electrophoresis 20(18):3551–3567. https://doi.org/10.1002/(SICI)1522-2683(19991201)20:18<3551::AID-ELPS3551>3.0.CO;2-2
17. Kim D, Paggi JM, Park C, Bennett C, Salzberg SL (2019) Graph-based genome alignment and genotyping with HISAT2 and HISAT-genotype. Nat Biotechnol 37(8):907–915. https://doi.org/10.1038/s41587-019-0201-4

18. Robinson JT, Thorvaldsdottir H, Winckler W, Guttman M, Lander ES, Getz G, Mesirov JP (2011) Integrative genomics viewer. Nat Biotechnol 29(1):24–26. https://doi.org/10.1038/nbt.1754
19. Merhej J, Frigo A, Le Crom S, Camadro JM, Devaux F, Lelandais G (2014) bPeaks: a bioinformatics tool to detect transcription factor binding sites from ChIPseq data in yeasts and other organisms with small genomes. Yeast 31(10):375–391. https://doi.org/10.1002/yea.3031
20. Denecker T, Lelandais G (2018) Empowering the detection of ChIP-seq "basic peaks" (bPeaks) in small eukaryotic genomes with a web user-interactive interface. BMC Res Notes 11(1):698. https://doi.org/10.1186/s13104-018-3802-y
21. Bailey TL, Boden M, Buske FA, Frith M, Grant CE, Clementi L, Ren J, Li WW, Noble WS (2009) MEME SUITE: tools for motif discovery and searching. Nucleic Acids Res 37(web server issue):W202–W208. https://doi.org/10.1093/nar/gkp335
22. Thomas-Chollier M, Herrmann C, Defrance M, Sand O, Thieffry D, van Helden J (2012) RSAT peak-motifs: motif analysis in full-size ChIP-seq datasets. Nucleic Acids Res 40(4):e31. https://doi.org/10.1093/nar/gkr1104
23. Bailey TL (2011) DREME: motif discovery in transcription factor ChIP-seq data. Bioinformatics 27(12):1653–1659. https://doi.org/10.1093/bioinformatics/btr261
24. Monteiro PT, Oliveira J, Pais P, Antunes M, Palma M, Cavalheiro M, Galocha M, Godinho CP, Martins LC, Bourbon N, Mota MN, Ribeiro RA, Viana R, Sa-Correia I, Teixeira MC (2020) YEASTRACT+: a portal for cross-species comparative genomics of transcription regulation in yeasts. Nucleic Acids Res 48(D1):D642–D649. https://doi.org/10.1093/nar/gkz859

Part III

Translation Dynamics, Protein Complexes, and Proteomics

Chapter 11

Purification of Ribosome-Nascent-Chain Complex for Ribosome Profiling and Selective Ribosome Profiling

Hagit Bar-Yosef, Johannes Venezian, Kevin Klann, and Ayala Shiber

Abstract

Selective Ribosome Profiling (SeRP) is an emerging methodology, developed to capture cotranslational interactions in vivo. To date, SeRP is the only method that can directly capture, in near-codon resolution, ribosomes in action. Thus, SeRP allows us to study the mechanisms of protein synthesis and the network of protein–protein interactions that are formed already during synthesis. Here we report, in detail, the protocol for purification of ribosome- and Nascent-Chain associated factors, followed by isolation of ribosome-protected mRNA footprints, cDNA library generation and subsequent data analysis.

Key words Affinity purification, Ribosome, Nascent-chain, Cotranslational interactions, Ribosome occupancy

1 Introduction

The Selective Ribosome Profiling (SeRP) method combines biochemistry and deep sequencing approaches, to enable direct identification and characterization of the diverse range of ribosome associated interactions, in vivo. This breakthrough method was developed recently [1] and has opened up previously unreachable nascent chain biology to direct analysis (see examples in [2–4]). SeRP allows for identification of the entire in vivo landscape of nascent-chain interactors for each ribosome and/or nascent-chain associated factor [5]. Furthermore, SeRP captures local variation in ribosomes density, thus allowing for the identification of translation rate changes associated with cotranslational interactions, such as ribosome stalling or speed-up [3]. The principals of SeRP include rapid cell harvesting, within seconds, by filtering and then immediately flash freezing, to capture the cellular "snapshot" of all

Supplementary Information The online version contains supplementary material available at [https://doi.org/10.1007/978-1-0716-2257-5_11].

Frédéric Devaux (ed.), *Yeast Functional Genomics: Methods and Protocols*,
Methods in Molecular Biology, vol. 2477, https://doi.org/10.1007/978-1-0716-2257-5_11,

cotranslational interactions (Fig. 1). Lysis is done in frozen state. Then, the cells are treated with RNase to eliminate all mRNA in the cell except for ribosome protected mRNA footprints. This is done in the presence of a cocktail of translation inhibitors, targeting all the ribosome various conformations during translation to avoid any ribosomal run-off during the RNase treatment. The ribosome-nascent chain complexes (RNCs) associated with the targeted factor of interest are isolated by affinity purification of the tagged factor. Then, we continue to isolate the ribosome-protected mRNA footprints, generate a cDNA library and perform next generation sequencing.

2 Materials

1. 0.45-μm nitrocellulose blotting membrane
2. Teflon Grinding Jar (10 mL), equipped with a grinding ball (20 mm), fitting for TissueLyser II by QIAGEN, or the Mixer Mill MM400 by Retsch
3. RNase I.
4. Lysis Buffer:20 mM Tris–HCl pH 8.0, 140 mM KCl, 6 mM $MgCl_2$, 0.1% NP-40, 0.1 mg/mL cycloheximide (CHX), 1 mM PMSF, 2× protease inhibitors (e.g., Complete EDTA-free, Roche), 0.02 U/mL DNase I, 20 mg/mL leupeptin, 20 mg/mL aprotinin, 10 mg/mL E-64, 40 mg/mL Bestatin.
5. Wash buffer: same compositions as lysis buffer, replacing the NP-40 with 10% glycerol.
6. Acid phenol to 65 °C for each extraction.
7. 20% SDS
8. Chloroform.
9. 3 M NaOAc, pH 5.5
10. Isopropanol.
11. Ethanol.
12. GlycoBlue.
13. Tris–HCl, pH 7 and Tris–HCl, pH 8.
14. 15% TBE-Urea polyacrylamide gels
15. 10% TBE-Urea polyacrylamide gels
16. 8% TBE-Urea polyacrylamide gels
17. TBE running buffer.
18. 2× sample loading buffer (TBE-Urea)
19. SYBR Gold.
20. T4 polynucleotide kinase buffer without ATP.

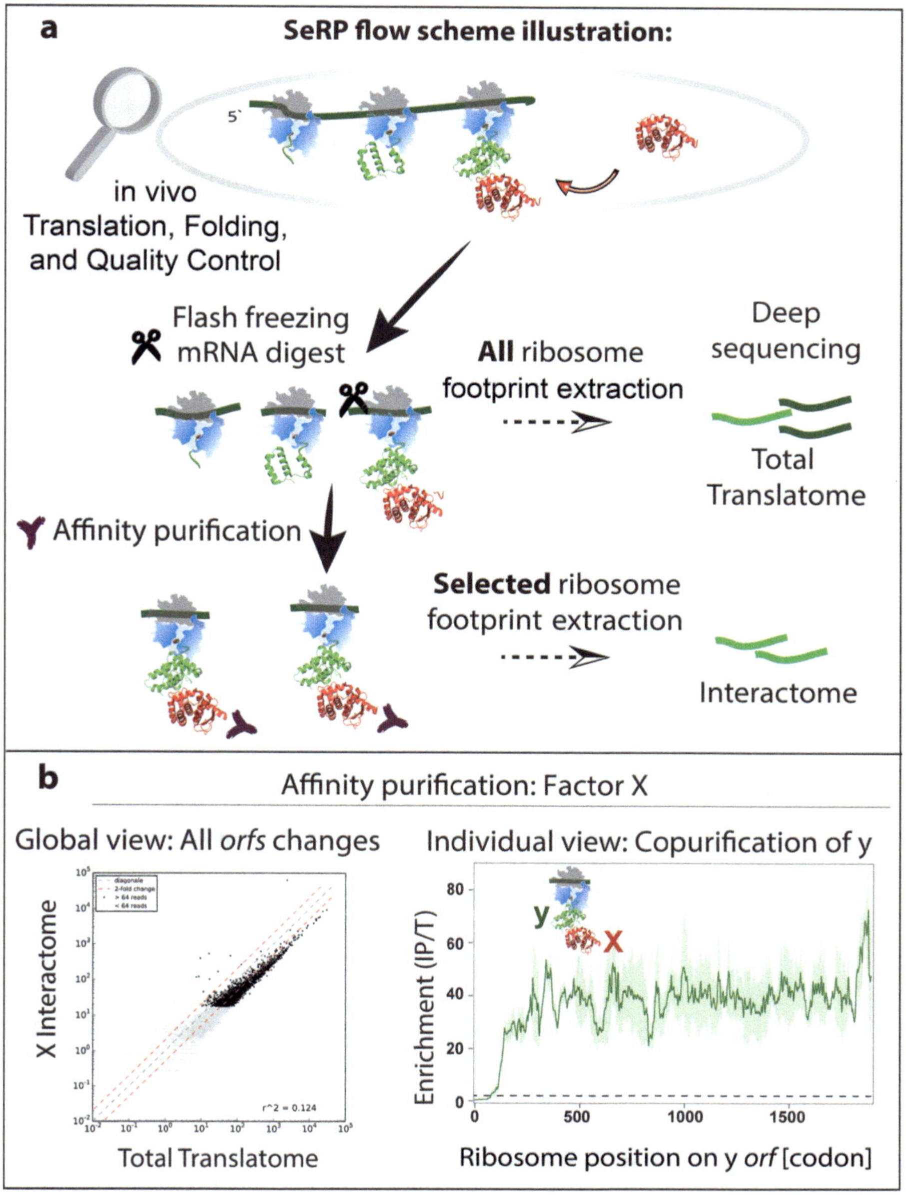

Fig. 1 Selective ribosome profiling (SeRP) method to capture and characterize cotranslational interactions. (**a**) SeRP flow scheme illustration. SeRP is a method for global profiling of interactions of any factor with translating ribosomes in vivo. The method relies on flash freezing of growing cells, preserving active translation. Cell lysates are then treated with RNase to digest all RNA species in the cell except ribosome-protected mRNA fragments (ribosome footprints). The sample is then split into two parts. One part is directly used for isolation of all ribosome footprints (total translatome). A second part is used for affinity purification of the specific subset of ribosomes associated with a factor of interest (e.g., a quality-control factor), also termed the interactome. Then, the protected mRNAs are extracted and used for cDNA library generation, followed by deep sequencing. (**b**) Comparative analysis of the total and interactome samples allows for identification of all *orfs* which associate with the factor (left) as well as characterization of each *orf* interaction profile at near-residue resolution (right). This profile reports on the precise engagement onset, dynamics as well as ribosome speed

21. SUPERaseIn stock (Ambion).
22. PEG MW 8000.
23. DMSO.
24. Superscript III, First strand buffer, and 0.1 M DTT.
25. CircLigase™, 1 mM ATP, and 50 mM $MnCl_2$ (EPICENTRE).
26. HF Phusion polymerase.
27. DEPC water.
28. RNase I.
29. Sucrose cushion.
30. High-speed centrifuge.
31. Bead-conjugated-GFP-binder slurry.
32. RNA Bioanalyzer chip (Agilent).
33. 1 N NaOH
34. Spin-X cellulose acetate column.
35. 5 M NaCl
36. 0.5 M EDTA
37. Oligonucleotide sequences

    o199-P5′_AUGUACACGGAGUCGACCCGCAACGCGA3phos_3′

    Linker-1: 5′_App/CTGTAGGCACCATCAAT/3ddC_3′

    o225-Link15′_5phos/GATCGTCGGACTGTAGAACTCTGAACCTGTCGGTGGTCGCCGTATCATT/iSp18/CACTCA/iSp18/CAAGCAGAAGACGGCATACGAATTGATGGTGCCTACAG_3′

    o230: 5′_AATGATACGGCGACCACCGA_3′

    o231: 5′_CAAGCAGAAGACGGCATACGA_3′

    o202: 5′_CGACAGGTTCAGAGTTCTACAGTCCGACGATC_3′

    o1055: 5′_5Biosg/TCATCTCCGGGGGTAGAGCACTGTTTCG_3′

    o1056: 5′_5Biosg/GGCTAAACCATGCACCGAAGCTGCGGCAG_3′

    o1057: 5′_5Biosg/AAGGCTGAGGCGTGATGACGAGGCACT_3′

    o1058: 5′_5Biosg/CGGTGCTGAAGCAACAAATGCCCTGCTT_3′.

3 Methods

3.1 Cell Growth, Collection, and Lysis

1. Grow ~200 mL of cell culture to OD_{600nm} of 0.5, at 30 °C, in appropriate medium (e.g., YPD or dropout media).

2. Filter cells in vacuum on 0.45-μm nitrocellulose (Amersham) blotting membrane and then flash-freeze in liquid nitrogen. Keep in −80 °C

   The next 3 steps are done in liquid nitrogen.

3. Scrape the cells into a grinding chamber that have been submerged in liquid nitrogen (e.g., use a Teflon grinding jar (10 mL), equipped with a grinding ball (20 mm)).
4. Add 600 μL of lysis buffer, prefrozen in droplets into liquid nitrogen.
5. Perform cryogenic grinding, each time for: 2 min, 30 Hz at least 3 times. After each interval submerge the grinding chamber in liquid nitrogen (*see* **Note 1**).

3.2 Total Translatome Extraction and Affinity Purification of GFP Tagged Subunits (See Note 2)

1. Thaw filtered and mixer milled cultures for 2 min 30 °C bath, then put on ice.
2. Spin lysates 2 min at 20,000 × *g*, 4 °C, to sediment cell debris.
3. Transfer lysate to new tube.
4. Measure RNA concentration of lysates with the nanodrop (1: 100 dilution) in DEPC-treated water.
5. From now on use nonstick tubes always (to decrease loss of sample material, *see* **Note 2**).

   Divide samples for Total translatome (around 200 μg) and AP (Affinity Purification)—the rest of the sample (around 5000 μg).

   For total translatome:

6. Digest lysates using RNase I (use 10u/A260 for 30 min at 4 °C), to generate ribosome protected footprints.
7. Load digested lysates onto the sucrose cushion.
   (a) Use small rotor and tubes (TLA120-rotor).
   (b) Fill tubes with 400 μL of sucrose-cushion and overlay carefully with 150 μL of the lysate.
   (c) Centrifuge for 90 min at 245,000 × *g* and 4 °C in order to pellet the monosomes.
   (d) Remove supernatant quickly and overlay pellets with buffer (use 100 μL of lysis buffer).
   (e) Resuspend pellets by shaking in cold room for 1 h (or overnight).
8. For Affinity Purification (AP):
9. Wash 100–500 μL of bead-conjugated-GFP-binder slurry (calibrate the amount of GFP binder required to saturate, according to protein abundance) 3 times, each time with 1 mL of your respective buffer (best to use lysis buffer without DNase I and the additional protease inhibitors): each wash is done for 1 min,

rotating at room temperature (to pellet beads spin for 1 min at 450 × *g*).

10. Remove buffer from beads as completely as possible and add lysate.
11. Digest lysates using RNase I. use 10u/A260 for 30 min at 4 °C. 1 μL = 100u. RNase I digestion of mRNA and affinity purification of ribosomes are performed at the same time but it is important to first add GFP binding slurry and then add RNAse I.
12. Incubate APs rotating for 25–30 min at 4 °C.
13. Wash beads 2 times in 1 mL of "lysis buffer" for 1 min (*see* **Note 3**).
14. Wash beads 2 times in 1 mL of "wash buffer" for 1 min.
15. 15. Wash beads 1 time in 1 mL of wash buffer for 5 min, rotating.
16. Put aside 1/10 of beads for protein elution with 2× sample loading buffer for analysis by western blotting the success of the affinity purification steps.
17. Use rest of beads for RNA extraction using hot phenol (*see* Subheading 3.3 below)
18. Freeze in liquid nitrogen after aspiration of all upper liquid.

3.3 Preparation of a Deep Sequencing Library, in RNA Safe Conditions (See Note 4)

3.3.1 Acid Phenol Extraction of RNA

Work in the hood—phenol evaporates quickly

1. Prewarm 0.75 mL acid phenol to 65 °C for each extraction.
2. Add 40 μL 20% SDS to 0.7 mL of "Total" RNA or "AP" elution.
3. Add 0.75 mL prewarmed acid phenol to samples.
4. Incubate at 65 °C for 5 min. Shake at 30 × *g* rpm in thermomixer (use safe-lock tubes, *see* **Note 5**).
5. Chill samples on ice for 5 min.
6. Spin samples for 2 min at 20,000 × *g*. Transfer top, aqueous layer to a fresh tube.
7. Add 0.7 mL acid phenol.
8. Incubate at room temperature for 5 min. Vortex occasionally.
9. Spin samples for 2 min at 20,000 × *g*. Transfer top, aqueous layer to a fresh tube.
10. Add 0.6 mL chloroform. Mix by vortex.
11. Spin samples for 1 min at 20,000 × *g*. Transfer top, aqueous layer to a fresh tube.

12. Precipitate nucleic acids: to each sample, add 78 μL 3 M NaOAc, pH 5.5 (1/9 equivalence volume), 0.75 mL isopropanol (1 equivalence volume).
13. Mix by vortex.
14. Chill samples at −20 °C overnight.
15. Spin samples at 20,000 × *g* for 30 min in 4 °C microfuge.
16. Aspirate supernatant by pipette.
17. Wash pellet with 0.75 mL ice-cold 80% ethanol. Invert tube for a thorough wash.
18. Pulse-spin to collect ethanol wash and aspirate remaining liquids by pipette.

    Then incubate in 65 °C for 5–10 min.
19. Resuspend samples in 10 μL (AP) or 20 μL (Total translatome) of prewarmed 10 mM Tris–HCl, pH 7.

3.3.2 Quantify Total RNA Concentration by Nanodrop

1. Prepare 1:10 dilution of acid phenol extracted total RNA in 10 mM Tris–HCl, pH 7.
2. Quantify by Nanodrop.
3. Dilute 50 μg of samples into 10 μL in 10 mM Tris–HCl, pH 7.
4. Do not measure AP samples; use everything for the next step (*see* **Note 6**).

3.3.3 Gel-Purify Ribosome Protected Footprint Fragments

1. Set up 15% TBE-Urea polyacrylamide gels in 1× TBE. Prerun for 1 h at 200 V.
2. Add either 15 (for 12 well gels) or 20 μL (for 10 well gels) 2× sample loading buffer to each sample.
3. Set up o199-P oligo in parallel: 2 μL 20 μM o199-P oligo (RNA control oligo), 8 μL 10 mM Tris 7, 10 μL 2× sample loading buffer.
4. Set up ladder in parallel: 1 μL 10 bp ladder, 9 μL 10 mM Tris–HCl, pH 8, 10 μL 2× sample loading buffer.
5. Denature all samples at 80 °C for 2 min. Return to ice.
6. Load samples onto gel. Run for 65 min (if gels are old: run shorter ~55 min) at 200 V.
7. Stain gel with SYBR Gold for 20 min. Dilute 6 μL SYBR Gold stock in 60 mL 1× TBE.
8. Photograph gel.
9. Excise desired bands with a sterile scalpel.
10. Recover ribosome protected footprint fragments. See appendix §A-RNA.
11. Photograph gel.

3.3.4 Dephosphorylate 3′ Ends of Gel-Purified Fragments

1. Resuspend samples in 15 μL 10 mM Tris–HCl, pH 7.
2. Transfer resuspended RNA to a fresh nonstick tube.
3. Add 2 μL 10× T4 polynucleotide kinase buffer without ATP, 1 μL SUPERaseIn stock to each sample. Make a 10× master mix of aforementioned reagents to reduce variability.
4. Add 2 (better indicate unit numbers) T4 polynucleotide kinase.
5. Mix thoroughly by pipette.
6. Incubate at 37 °C for 2 h.
7. Heat kill enzyme at 75 °C for 10 min.
8. Precipitate nucleic acids. To each sample, add 0.5 mL 10 mM Tris–HCl, pH 7, 55 μL 3 M NaOAc, pH 5.5 (1/9 equivalence vol), 2 μL GlycoBlue, 0.55 mL isopropanol (1 equivalence vol).
9. Mix by vortex.
10. Chill samples at −20 °C overnight.
11. Spin samples at 20,000 × *g* for 30 min in 4 °C microfuge.
12. Aspirate supernatant by pipette.
13. Wash pellet with 0.75 mL ice-cold 80% ethanol. Invert tube for a thorough wash.
14. Pulse-spin to collect ethanol wash and aspirate remaining liquids by pipette.
15. Resuspend samples in 11 μL 10 mM Tris–HCl, pH 7.

3.3.5 Quantify Concentration by Bioanalyzer/TapeStation

1. Make a 1:5 dilution by mixing 1 μL resuspended RNA in 4 μL 10 mM Tris–HCl, pH 7.
2. Run a small RNA bioanalyzer chip (a note describing your usual quality control criteria might be useful here).

3.3.6 Ligate 3′ End with Linker-1

1. Dilute 5 pmol of small RNA fragments to 5 μL in 10 mM Tris–HCl, pH 7.
2. Denature all samples at 80 °C for 2 min. Return to ice.
3. Add 8 μL 50% sterile filtered PEG MW 8000, 2 μL DMSO, 2 μL 10× T4 RNA Ligase 2 buffer, 1 μL SUPERaseIn or 1 μL RNase Inhibitor, Murine (40,000 units/mL, NEB M0314L), 1 μg/μL Linker-1 (Linker L1) to each sample. Make a 10× master mix of aforementioned reagents to reduce variability. Dilute Linker if less than 5 pmol of RNA are used.
4. Mix thoroughly by pipette.
5. Add 1 μL T4 RNA Ligase 2, truncated.
6. Mix thoroughly by pipette.
7. Incubate at 37 °C for 2.5 h.

8. Precipitate nucleic acids to concentrate. To each sample, add 0.5 mL 10 mM Tris–HCl, pH 7, 55 μL 3 M NaOAc, pH 5.5 (1/9 equivalence vol), 2 μL GlycoBlue, 0.55 mL isopropanol (1 equivalence vol).
9. Mix by vortex.
10. Chill samples at −20 °C overnight.
11. Spin samples at 20,000 × *g* for 30 min in 4 °C microfuge.
12. Aspirate supernatant by pipette.
13. Wash pellet with 0.75 mL ice-cold 80% ethanol. Invert tube for a thorough wash.
14. Pulse-spin to collect ethanol wash and aspirate remaining liquids by pipette.
15. Resuspend samples in 6 μL 10 mM Tris–HCl, pH 7.

3.3.7 Gel-Purify 3′ Linked Footprint Fragments

1. Set up a 12 well, 10% TBE-Urea polyacrylamide gel in 1× TBE. Prerun for 1 h at 200 V.
2. Add 6 μL 2× sample loading buffer to each sample.
3. Set up o199-P oligo in parallel: 1 μL 20 μM o199-P oligo (RNA control oligo), 5 μL 10 mM Tris–HCl, pH 7, 6 μL 2× sample loading buffer.
4. Set up ladder in parallel: 1 μL 10 bp ladder, 5 μL 10 mM Tris–HCl, pH 8, 6 μL 2× sample loading buffer.
5. Denature all samples at 70 °C for 2 min. Return to ice.
6. Load samples onto gel. Run for 50 min at 200 V.
7. Stain gel with SYBR Gold for 20 min. Dilute 6 μL SYBR Gold stock in 60 mL 1× TBE.
8. Photograph gel.
9. Excise desired bands with a sterile scalpel.

3.3.8 Recovering 3′ Linked Footprint RNA Fragments from Gel

1. Use 0.5 mL gel breaker tubes (The gel breaker tubes are a 0.5 mL tubes with 4 × 10 μm holes designed to fragment excised gel. The gel slice is placed inside the gel breaker tube which is then stacked into a 2 mL microfuge tube. Following high-speed centrifugation allows the gel to move through the holes, fragmenting the gel. Buffer is next added to elute out the RNA. Alternatively, manually smash the gel into several smaller pieces with a pipette.
2. Place the excised gel slice in the 0.5 mL tube and nest it in a 1.5 mL tube.
3. Spin at 20,000 × *g* for 3 min.
4. Collect and transfer remaining gel pieces.

5. Add 0.5 mL 10 mM Tris–HCl, pH 7.
6. Incubate at 70 °C for 10 min with max shaking.
7. Transfer the gel slurry to a Spin-X cellulose acetate column using a wide-bore pipette. Spin for 2 min at 20,000 g.
8. Transfer the flow through to a fresh nonstick tube.
9. Add 55 μL 3 M NaOAc, pH 5.5 (1/9 equivalence vol), 2 μL GlycoBlue, 0.55 mL isopropanol (1 equivalence vol).
10. Mix by vortex.
11. Chill samples at −20 °C overnight.
12. Spin samples at 20,000 × *g* for 30 min in a 4 °C microfuge.
13. Aspirate supernatant by pipette.
14. Wash pellet with 0.75 mL ice-cold 80% ethanol. Invert tube for a thorough wash.
15. Pulse-spin to collect ethanol wash and aspirate remaining liquids by pipette.
    Then incubate in 65 °C for 5–10 min.

3.3.9 Reverse Transcribe 3′ Linked Footprint Fragments to Generate ssDNA

1. Resuspend samples in 10 μL 10 mM Tris–HCl, pH 7.
2. Transfer resuspended RNA to a fresh nonstick tube.
3. Add 1 μL 10 mM dNTP, 1 μL 25 μM o225-Link1 (Linker L1′L2′, and 1.5 μL DEPC H20 to each sample. Make a 10× nucleic acids master mix of aforementioned reagents to reduce variability.
4. Incubate samples at 65 °C for 5 min to denature nucleic acids. Return to ice.
5. Add 4 μL 5× First Strand buffer, 1 μL SUPERaseIn stock, and 1 μL 0.1 M DTT to each sample. Make a 10× Reverse Transcription buffer master mix of aforementioned reagents to reduce variability.
6. Mix thoroughly by pipette.
7. Add 1 μL Superscript III to each sample.
8. Mix thoroughly by pipette.
9. Incubate at 50 °C for 30 min.
10. Add 2.3 μL 1 N NaOH to hydrolyze RNA to quench RT reaction. Incubate at 95 °C for 15 min. The samples turn pink in color.
11. Set up a 12 well, 10% TBE-Urea polyacrylamide gel in 1× TBE. Prerun for 30 min at 200 V.
12. Add 23 μL 2× sample loading buffer to each sample.
13. Set up ladder in parallel: 1 μL 10 bp ladder, 9 μL 10 mM Tris–HCl, pH 8, 10 μL 2× sample loading buffer.

14. Denature all samples at 70 °C for 2 min. Return to ice.
15. Load samples onto gel. Run for 70 min at 200 V.
16. Stain gel with SYBR Gold for 2 min. Dilute 6 μL SYBR Gold stock in 60 mL 1× TBE.
17. Photograph gel.
18. Excise desired bands with a sterile scalpel.

3.3.10 Recovering DNA Fragments from Gel

1. Use 0.5 mL gel breaker tubes.
2. Place the excised gel slice in the 0.5 mL tube and nest it in a 1.5 mL tube.
3. Spin at 20,000 × *g* for 3 min.
4. Collect and transfer remaining gel pieces.
5. Add 0.5 mL 10 mM Tris–HCl, pH 8.
6. Incubate at 70 °C for 10 min with max shaking.
7. Transfer the gel slurry to a Spin-X cellulose acetate column using a wide-bore pipette. Spin for 2 min at 20,000 × *g*.
8. Transfer the flow through to a fresh nonstick tube.
9. Add 32 μL 5 M NaCl (1/16 equivalence vol), 1 μL 0.5 M EDTA (1/500 equivalence vol), 2 μL GlycoBlue, 0.55 mL isopropanol (1 equivalence vol).
10. Mix by vortex.
11. Chill samples at −20 °C overnight.
12. Spin samples at 20,000 × *g* for 30 min in a 4 °C microfuge.
13. Aspirate supernatant by pipette.
14. Wash pellet with 0.75 mL ice-cold 80% ethanol. Invert tube for a thorough wash.
15. Pulse-spin to collect ethanol wash and aspirate remaining liquid by pipette.

    Then incubate in 65 °C for 5–10 min.
16. Return to main protocol.

3.3.11 Circularize ssDNA

1. Resuspend samples in 15 μL 10 mM Tris–HCl, pH 8.
2. Transfer resuspended ssDNA to a fresh nonstick tube.
3. Add 2 μL 10× CircLigase buffer, 1 μL 1 mM ATP, 1 μL 50 mM $MnCl_2$ to each sample. Make a 10× master mix of aforementioned reagents to reduce variability.
4. Add 1 μL CircLigase.
5. Incubate at 60 °C for 1 h.
6. Spike in additional 1 μL CircLigase to each sample.

7. Incubate at 60 °C for 1 h.
8. Incubate at 80 °C for 10 min to inactivate enzyme.
9. Return to ice.

3.3.12 PCR Amplify

1. Aliquot 5 μL circularized DNA to a fresh nonstick tube.
2. Add 16.7 μL 5× HF buffer, 1.7 μL 10 mM dNTPs, 0.4 μL 100 μM o230 (PCR L1′), 0.4 μL 100 μM o230 (barcode primer), 59.2 μL DEPC H20, 0.8 μL HF Phusion to each aliquot. Make a 10× master mix of aforementioned reagents to reduce variability.
3. Mix by vortex.
4. Aliquot 17 μL PCR mix to 4 separate PCR strips.
5. Perform the following PCR reaction steps. Repeat steps 2 through 4 ten times. Remove strips after cycles 7, 8, 9, 10.

Step	Time	Temp
1	30 s	98 °C
2	10 s	98 °C
3	10 s	60 °C
4	5 s	72 °C

6. Add 3.5 μL 6× DNA loading dye to each sample.
7. Set up ladder in parallel: 1 μL 10 bp ladder, 16 μL 10 mM Tris–HCl, pH 8, 3.5 μL 6× DNA loading dye.
8. Load samples directly a 12 well, 8% TBE polyacrylamide gel in 1× TBE. Run at 180 V for 45 min. Do not run gel longer than 45 min or samples will run off!
9. Stain gel with SYBR Gold for 20 min. Dilute 6 μl SYBR Gold stock in 60 mL 1× TBE.
10. Excise desired bands with a sterile scalpel.
11. Recover dsDNA. *See* Subheading 3.3.10.

3.3.13 Analysis and Quantification of the cDNA Library Products

1. Resuspend samples in 21 μL 10 mM Tris–HCl, pH 8.
2. Transfer resuspended dsDNA to a fresh nonstick tube.
3. Size distribution analysis: Run a high-sensitivity DNA Bioanalyzer/TapeStation chip to verify the dominant product size is the footprints, linker-ligated (~160 bp) and not only linkers (a range up to ~125 bp). A negative control can be a sample made without ribosome foot-prints.

4. High-precision concentration analysis by Qubit: evaluate each sample concentration by dsDNA high-sensitivity kit with Qubit, according to the manufacturer's recommendations.

3.3.14 Next-Generation Sequencing by HiSeq \NextSeq\NovaSeq

1. Multiplex your sample to the same stoichiometry (range: 2–10 nM each sample).
2. 50 μL of the pooled/multiplexed sample is enough for several lanes
3. Application type: Small RNA-seq.
4. Sequencing type: Single-Read 50 bp.

3.4 Data Analysis

Data analysis can be performed by the python script provided as supplementary material on line (titled pipeline v6.6).

This is a Linux Script, so path must be given in linux syntax.

1. This pipeline uses raw Deep Sequencing data (fastq files) and processes it as follows:
2. Trimming adaptor sequences using Cutadapt [6] in order to remove 3′-end adaptor sequences. Following processing, the reads are written into an output file, with each processed read ranging between 20 and 45 nt.
3. rRNA and other noncoding RNA sequences (such as anoRNA) **removal** using Bowtie2 [7], for fast gapped-read alignment. We have supplied a reference file containing all noncoding *S. cerevisiae* elements. This file can be updated, utilizing the *Saccharomyces* Genome Database (SGD) [8].
4. Genome Alignment using Tophat2 [9]. Tophat2 is a splice-aware aligner, allowing for mapping ribosome-footprint reads spanning introns (utilizing Bowtie2 [7]). As in the previous step, we have supplied a reference genome file containing all coding *S. cerevisiae* elements. This file can be updated, utilizing the Saccharomyces Genome Database (SGD) [8].
5. Next sequencing processing steps is assignment of ribosome positions, a specialized Python script. You can decide in this script which analysis mode do you want to use, center = center weighting, complete = complete distribution, 5-end = 5-end alignment, 3-end = 3-end alignment. In determining 3 nt periodicity of decoding, 5′ end is recommended, but can lead to loss of many reads (*see* **Note 6**). You can also decide whether you want a .txt file with read densities for each nucleotide in each chromosome or not, just set variable txt to yes or no.
6. Metagene profile and 64-threshold data is generated for each sample. Metagene profiles show the average ribosomal densities between all transcripts which are significantly detected above background (these transcripts are termed "included" in the script, meaning above the 64 reads per *orf* threshold,

ensuring reproducibility between biological replicates [10] where *orfs* below this threshold are termed "excluded"). Installer package is added with REFERENCE files and so on. References and scripts are copied to your installation path and are accessed directly, and there is no need to access the server.

4 Notes

1. Alternatively, cryogenic grinding can be performed with a mortar and pestle.
2. It is important to note that the affinity purification conditions need to be calibrated prior to SeRP, in order to achieve low background binding of rRNA and proteins. This can be determined by qPCR/western blotting/proteomics analysis of Mock AP compared to positive control. High background binding can be avoided by increasing washing steps and/or increasing detergents concentration in the washing buffer and/or increasing salt concentration in the washing buffer. Various Affinity purifications tags can also lead to increased rRNA background. For example, positively charged His tags followed by nickel-based affinity purification leads to high background binding of the negatively charged ribosomes. These tags should be avoided, as the enrichment analysis will be greatly impaired.
3. During washing procedure change tubes 2× (after wash 2 and 4 for example).
4. Use DNA/RNA nonstick tubes to avoid loss of RNA to the adherent tubes.
5. It is important to utilize safe-lock tubes or lid clamps to avoid samples loss during the extensive shaking.
6. Throughout the cDNA library preparation, it is important to work at RNA safe conditions (use gloves, filter tips, clean surfaces with RNase AWAY or 0.5 M NaOH).
7. Affinity Purifications samples (AP) are often very low in their RNA concertation, often below detection with Bioanalyzer/ TapeStation. This is expected as the purification steps are resulting in a smaller ribosome-nascent-chain population, compared to the total translatome. However, next-generation sequencing is sensitive enough to obtain results, if we mix samples with low RNA concentration with samples with high concentration—such as total translatome—in the Next generation Sequencing run.

   If Total Translatome RNA concentration is too low, you can upscale input material in Subheadings 3.1 and 3.2.

8. In determining 3 nt periodicity of decoding, 5′ end is recommended, but can lead to loss of many reads (as the ribosome footprint size becomes very important). For determination of cotranslational interactions, the more permissive center weighting is recommended.

References

1. Becker AH, Oh E, Weissman JS et al (2013) Selective ribosome profiling as a tool for studying the interaction of chaperones and targeting factors with nascent polypeptide chains and ribosomes. Nat Protoc 8:2212–2239
2. Shiber A, Döring K, Friedrich U et al (2018) Cotranslational assembly of protein complexes in eukaryotes revealed by ribosome profiling. Nature 561:268–272
3. Döring K, Ahmed N, Riemer T et al (2017) Profiling Ssb-nascent chain interactions reveals principles of Hsp70-assisted folding. Cell 170(2):298–311
4. Buschauer R, Matsuo Y, Sugiyama T et al (2020) The Ccr4-not complex monitors the translating ribosome for codon optimality. Science 368(6488):eaay6912
5. Galmozzi CV, Merker D, Friedrich UA et al (2019) Selective ribosome profiling to study interactions of translating ribosomes in yeast. Nat Protoc 14(8):2279–2317
6. Martin M (2011) Cutadapt removes adapter sequences from high-throughput sequencing reads. EMBnet J 17:1
7. Langmead B, Salzberg SL (2012) Fast gapped-read alignment with bowtie 2. Nat Methods 9: 357–359
8. Engel SR, Dietrich FS, Fisk DG et al (2014) The reference genome sequence of Saccharomyces cerevisiae: then and now. G3 Genes Genom Genet 4:389–398
9. Kim D, Pertea G, Trapnell C et al (2013) TopHat2: accurate alignment of transcriptomes in the presence of insertions, deletions and gene fusions. Genome Biol 14:R36
10. Ingolia NT, Lareau LF, Weissman JS (2011) Ribosome profiling of mouse embryonic stem cells reveals the complexity and dynamics of mammalian proteomes. Cell 147:789–802

Chapter 12

Single-Step Affinity Purification (ssAP) and Mass Spectrometry of Macromolecular Complexes in the Yeast *S. cerevisiae*

Christian Trahan and Marlene Oeffinger

Abstract

Cellular functions are mostly defined by the dynamic interactions of proteins within macromolecular networks. Deciphering the composition of macromolecular complexes and their dynamic rearrangements is the key to get a comprehensive picture of cellular behavior and to understand biological systems. In the past two decades, affinity purification coupled to mass spectrometry has become a powerful tool to comprehensively study interaction networks and their assemblies. To overcome initial limitations of the approach, in particular, the effect of protein and RNA degradation, loss of transient interactors, and poor overall yield of intact complexes from cell lysates, various modifications to affinity purification protocols have been devised over the years. In this chapter, we describe a rapid single-step affinity purification method for the efficient isolation of dynamic macromolecular complexes. The technique employs cell lysis by cryo-milling, which ensures nondegraded starting material in the submicron range, and magnetic beads, which allow for dense antibody-conjugation and thus rapid complex isolation, while avoiding loss of transient interactions. The method is epitope tag-independent, and overcomes many of the previous limitations to produce large interactomes with almost no contamination. The protocol as described here has been optimized for the yeast *S. cerevisiae*.

Key words Proteomics, Single-step affinity purification, Cell lysis, Cryo-milling, Yeast, Mass spectrometry

1 Introduction

In the past two decades, many researchers have put considerable effort into elucidating protein-protein interaction networks in order to better understand the interplay between proteins and gain a comprehensive picture of dynamic pathways that govern cellular processes [1–3]. The methods to study macromolecular complexes within these networks have improved greatly as part of the ever expanding proteomics field, and affinity purification—mass spectrometry (AP-MS) has become a powerful approach to study interactions within a variety of cellular complexes in many

Frédéric Devaux (ed.), *Yeast Functional Genomics: Methods and Protocols*,
Methods in Molecular Biology, vol. 2477, https://doi.org/10.1007/978-1-0716-2257-5_12,

organisms [1, 4–6]. However, determining the interactions between macromolecules in a cell is a formidable undertaking for several reasons. The number of interacting entities is large. In *Saccharomyces cerevisiae* alone, there are ~6000 open reading frames that code for proteins [7, 8]. Moreover, proteins do not work in isolation but within organized multiprotein complexes along often dynamic pathways, forming intricate information networks, and are present in a broad range of abundance (10^1–10^6 copies/cell) [9–11]. Finally, protein interactions have a wide range of affinities [12].

Many of the early affinity purification approaches came with a number of limitations that were not up to handling these many variations, which make up the very nature of dynamic macromolecular complexes and pathways. Heat-generating lysis methods, long incubation times due to slow affinity kinetics, and resins with limited pore dimensions all resulted in partial complex degradation, loss of interactors, association and identification of a wide range of contaminants, and overall poor yields and low signal-to-noise ratios [13–15]. Here, we describe a faster method, which employs cell lysis by cryo-milling followed by a rapid single-step affinity purification (ssAP) using solely one epitope tag and magnetic resin, which allow us to preserve the integrity of the isolated complexes without crosslinking.

The biggest challenge of any affinity purification experiment is to preserve the interactions within the targeted complex, or pool of complexes, with a high degree of fidelity. Cell lysis disrupts the cellular environment and causes the intermingling of components that are not normally exposed to one another, and thus the resultant possibility of aberrant molecular interactions—a major source of "nonspecific background." Another undesired result of this unnatural intermingling is the exposure to degradation enzymes, such as proteases, RNases, and DNases, that are normally kept at bay in a living cell. Contrary to other cell disruption methods, cryolysis better preserves the integrity of complexes as well as its transient and weak interactors as cell breakage occurs in liquid nitrogen, at a temperature below −196 °C, keeping cellular components in a solid state, thus preventing rearrangements of complexes during the step of cell lysis and subsequent disruption of the subcellular compartments. The method also protects complexes from the activity of nucleases and proteases released during lysis. Cryolysis is efficiently performed using a planetary ball mill, which yields a homogeneous cell grindate with particles of approximately 1–2 μm in diameter [16].

Besides many stable interactions, there are proteins, subcomplexes and macromolecules that constantly associate and dissociate constituting dynamic or transient interactions. In addition, while not directly interacting, there are many vicinal macromolecules—that is, macromolecules that are particular to the surrounding of the tagged macromolecule—which help to define the physical and chemical characteristics of its environment. After cell lysis, the goal is to isolate these complexes very rapidly, before many macromolecular assemblies degrade, and transient and dynamic interactions fall apart.

In the past, commonly used tandem affinity epitope tags have often required long incubation times and several steps of purification, which resulted in the loss of dynamic and transient interactions. Single-step affinity approach described in this chapter reduces the necessity for multistep purifications and thus time and handling, resulting in faster isolations and conservation of these labile interactions. While either Protein A (PrA), GFP, FLAG, HA, or myc are now commonly used, PrA tag in particular has high recovery rates due to the presence of multiple IgG-binding moieties, which enables faster kinetics during the binding step, and high affinity for rabbit IgG (~10 nM) [17]. In addition, affinity resins can also pose a limiting factor. Sepharose-based affinity isolations, still the most widely used in many laboratories, tend to be long (4–14 h) as diffusion rates through porous Sepharose resin are slow, and increase background. Moreover, the resin pore dimensions limit the upper size of any complexes to be isolated, which is a particular issue for affinity purification of macromolecular complexes. On the other hand, magnetic resin is small (1 nm to 2.8 μm in diameter) and nonporous, which allows the dense conjugation of antibody to its inert surface. This ensures a large surface-to-volume ratio, allowing for short incubation times (5–30 min) and efficient, >90%, complex recovery, regardless of the nature or cellular abundance of the bait protein or size of its associated complexes [16]. Short incubation times also have the advantage of minimizing nonspecific binding and loss of significant transient or weakly associated interactors [14, 16]. After complex isolation, the affinity-purified proteins are then either trypsin digested directly on the magnetic resin ("on-bead") to prepare them for mass spectrometry, or eluted from the resin under denaturing conditions for SDS-PAGE and western blotting analysis.

Sample preparation protocols and mass spectrometry (MS) methods have also advanced significantly over the past decade. In particular, a number of new proteolytic approaches for peptide digestion have emerged in the last few years. While for many years in-gel trypsin digestion was the standard for MS sample preparation, with the emergence of gel-free methods such as in-solution and, most recently, on-bead trypsin digest, sample handling has become much faster and simpler. In a gel electrophoresis-based approach, the proteins are separated in one or two dimensions (1D/2D) on an SDS-PAGE and enzymatic digestion is performed in-gel after excision of protein bands or spots, which is a time-consuming process [18]. During gel-free or in-solution based approaches, the proteins or peptides, or both, are separated chromatographically using online LC systems and the proteins are then digested in-solution [19]. More recent protocols use on-bead trypsin digest, wherein proteins are digested directly on the resin in lieu of elution [20]. The in-solution and on-bead based approaches tend to be the simplest in terms of sample handling and speed as well as recovery efficiency of preys.

In this chapter, we describe a step-by-step protocol for affinity-purifying epitope tagged dynamic macromolecular complexes using a single-step approach in the yeast *S. cerevisiae*. We first describe cell growth, harvesting cell material, buffer selection, and antibody conjugation to magnetic resin. We then provide a step-by-step guide of cryo-milling, affinity purification, on-bead trypsin digest, and mass spectrometry. Finally, we briefly describe aspects of data analysis.

2 Materials

2.1 Conjugation of Magnetic Dynabeads with Rabbit IgG (See Note 1)

1. Dynal Dynabeads M-270 Epoxy.
2. IgG from rabbit serum.
3. Polypropylene tubes: 4 × 15 mL, 2 × 50 mL.
4. Magnetic holder for 15 mL tubes.
5. Nutator or any slow rotating platform or wheel.
6. Millex-GP Syringe Filter Unit, 0.22 μm, polyethersulfone (EMD Millipore).
7. 20 mL syringe.
8. 1.5 mL polypropylene tubes.
9. Microcentrifuge (≥13 k rpm).
10. Vortex.
11. Parafilm.
12. M sodium phosphate, pH 7.4.
13. 3 M $(NH_4)_2SO_4$, pH 7.4.
14. 1× PBS pH 7.4: 10 mM sodium phosphate, pH 7.4, 150 mM NaCl.
15. 100 mM glycine pH 2.5.
16. 10 mM Tris pH 8.8, store up to 1 year at RT.
17. 100 mM triethylamine (make fresh) (*see* **Note 2**).
18. 1× PBS, 0.5% Triton X-100.
19. 10% NaN_3 (*see* **Note 2**).

2.2 Harvesting and Freezing Yeast Cells

1. Epitope tagged yeast strain (here, we use a W303 background) grown to mid-log phase OD_{600nm} ~ 0.8 (*see* **Note 3**).
2. Liquid N_2 (~1 L/strain).

3. Cryo-gloves, eye protection, and lab coat.
4. Spatula, cooled in liquid N_2.
5. 50 mL polypropylene tubes.
6. 20 mL syringe.
7. 1 M DTT.
8. 100× Solution P: 100 mM PMSF, 0.4 mg/mL Pepstatin A; store at −20 °C for up to 3 weeks.
9. Protease inhibitor cocktail.
10. Resuspension buffer: 20 mM Hepes, pH 7.4, 1.2% PVP-40, store at 4 °C.
11. Centrifuge(s) for 1 L and 50 mL tubes, capable of 3500 × *g*.
12. Clean and dry Styrofoam box big enough to fit a 9 × 50 mL tube rack.
13. 50 mL tube rack with open sides that can withstand liquid nitrogen.

2.3 Cryolysis (Cryomilling) (See Note 4)

1. Frozen yeast "noodles" from yeast pellet paste.
2. 10 L/strain of liquid N_2.
3. Cryo-gloves, eye protection, and lab coat.
4. PM-100 Planetary Ball Mill (Retsch).
5. For 20–50 mL of "noodles": 125 mL stainless steel jar and twelve 20 mm stainless steel balls. For less than 20 mL of "noodles": 50 mL stainless steel jar and four 20 mm stainless steel balls.
6. Spatula.
7. Clean and dry Styrofoam box deep enough to immerse a 9 × 50 mL tube rack, with lid, or covered with aluminum foil.
8. 50 mL tube rack that can be immersed in liquid N_2.
9. 50 mL polypropylene tube pierced with a spatula as handle (Fig. 1).
10. Long stainless-steel tweezers (25–30 cm).

2.4 Weighing Yeast Cell Grindate

1. 1 L Liquid N_2.
2. Protection equipment for N_2 as mentioned previously.
3. Yeast cell grindate (From Subheading 2.3) (*see* **Note 5**).
4. 50 mL polypropylene tubes (*see* **Note 5**).
5. Spatula.
6. Clean and dry Styrofoam box deep enough to immerse a 9 × 50 mL tube rack.
7. Two 9 × 50 mL tube racks that can be immersed in liquid N_2.

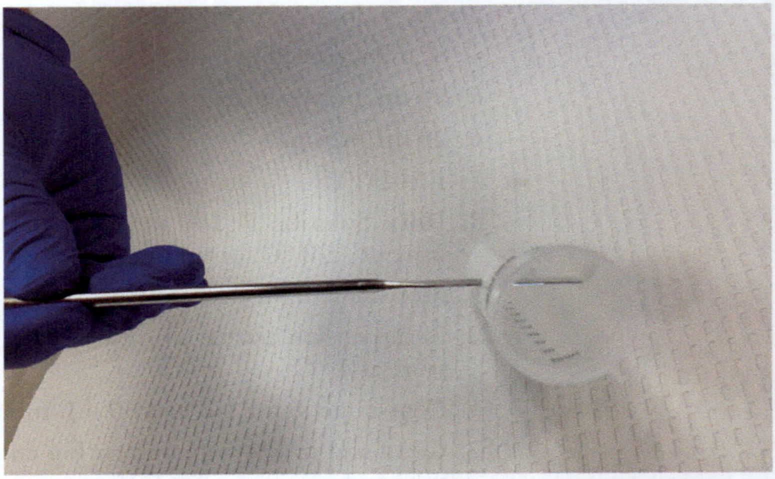

Fig. 1 Homemade liquid N_2 baster. A metal spatula with a straight end was inserted into the side of an empty 50 mL polypropylene tube. The tube side opening was created with a Kelly forceps-held razor blade that was heated under a Bunsen burner flame, and then used to pierce a straight line into the tube

2.5 Single-Step Affinity Purification

1. Preweighted yeast cell grindate.
2. Predetermined ssAP buffer prepared with MS grade water (*see* **Note 6**).
3. Vortex.
4. Ice and ice bucket.
5. Polytron homogenizer (Kinematica PT 1200 E) equipped with a 7 mm probe.
6. Centrifuge for 15 and 50 mL tubes capable of supporting 2700 × *g*.
7. Conjugated IgG magnetic beads (*see* **Note 1**) (From Subheading 2.1).
8. Protein LoBind 1.5 mL tubes (Eppendorf).
9. Timer.
10. Nutator or any slow rotating platform or wheel.
11. Cold room.
12. DynaMag-2, -15, and/or -50 Magnets.
13. Last Wash Buffer (Attention: make fresh just prior to use): 0.1 M NH_4OAc, 0.1 mM $MgCl_2$, 0.02% Tween 20 in MS grade water.
14. Last Wash Buffer without detergent (Attention: make fresh just prior to use): 0.1 M NH_4OAc, 0.1 mM $MgCl_2$ in MS grade water.

15. 20 mM Tris-HCl, pH 8.0.
16. Elution buffer (make fresh): 0.5 M NH_4OH, 0.5 mM EDTA.
17. Vacuum concentrator (SpeedVac) (*see* **Note 7**).
18. SDS-PAGE solution A: 0.5 M Tris-HCl, pH 8.0, 5% SDS.
19. SDS-Page solution B: 75% glycerol, 124.5 mM DTT, 0.05% Bromophenol Blue; store at 4 °C. DTT should be kept in frozen aliquots and should be added freshly to solution B needed.
20. Antifoam B Emulsion.

2.6 Trypsin Digest

1. Protein LoBind 1.5 mL tubes.
2. Sequencing Grade Modified Trypsin .
3. Thermomixer for 1.5 mL tubes.
4. LC-MS grade formic acid.

2.7 Mass Spectrometry

1. Tryptic peptides (from ssAP on-beads trypsin digestion).
2. LC-MS grade formic Acid.
3. LC-MS grade acetonitrile.
4. MS grade water.
5. C^{18} ZipTips with 5 μg capacity (e.g., Millipore Sigma).
6. MixMate (Eppendorf).
7. Ultrasonic bath (e.g., Branson Ultrasonics).
8. Easy-nLC II HPLC system (Proxeon Biosystems).
9. PicoFrit fused silica capillary column (17 cm × 75 μm i.d; New Objective), packed in-house with C-18 reverse-phase resin (Jupiter 5 μm particles, 300 Å pore size; Phenomenex) using a high-pressure packing cell.
10. Mass spectrometer. We use either the Orbitrap Q Exactive or Orbitrap Fusion Tribrid mass spectrometer.
11. Solvent A (0.2% formic acid in MS grade water).
12. Solvent B (99.8% acetonitrile, 0.2% formic acid).

2.8 Sample Analysis

1. LC-MS Raw files.
2. Protein fasta database library.
3. Proteome Discoverer (ThermoFisher Scientific).
4. Mascot (Matrix Sciences).
5. Percolator algorithm.
6. Scaffold (Proteome Software).

3 Methods

3.1 Conjugation of Magnetic Dynabeads with Rabbit IgG

1. *Resuspending and washing the magnetic beads:* Resuspend 300 mg of Dynabeads in 16 mL of 0.1 M Sodium phosphate, pH 7.4 (vortex, 30 s). Divide the mixture evenly into four 15 mL polypropylene tubes (4 mL/tube). Incubate the solution with slow rotation (RT, 10 min). Recover the beads using a magnetic holder and remove the buffer (*see* **Note 8**). Resuspend the beads once again in 4 mL of 0.1 M Sodium phosphate, pH 7.4 (vortex, 15 s).
2. *Reconstituting the antibodies:* Rabbit IgG is received lyophilized (100 mg). Solubilise the antibodies in ultrapure water (7 mL) to obtain a final concentration of ~14 mg/mL. Aliquot the solution (1 mL) into 1.5 mL tubes and store at −80 °C.
3. *Preparing the antibodies for conjugation:* Take 4 mL of antibodies prepared above, and centrifuge (10 min, V_{max}) using a microcentrifuge. In a 50 mL polypropylene tube, mix 3.525 mL of the subsequent supernatant with 9.85 mL of 0.1 M Sodium phosphate, pH 7.4 (*see* **Note 9**). Slowly add 6.65 mL of 3 M $(NH_4)_2SO_4$ drop-by-drop while vortexing the tube at the lowest speed to avoid local precipitation. After all $(NH_4)_2SO_4$ has been added, briefly vortex the tube for 2 s. Filter the solution into another 50 mL tube (0.22 μm filter fitted on a 20 mL syringe) to eliminate any precipitate. The antibody solution is now ready for conjugation.
4. Magnetically recover the beads, and wash them once more with 4 mL 0.1 M Sodium phosphate, pH 7.4 buffer (vortex, 15 s).
5. *Conjugation:* Magnetically recover the beads, remove the buffer and resuspend them in 5 mL of antibody solution. Close the tubes, seal them with parafilm, and incubate overnight (18–24 h, 30 °C, slow rotation).
6. Wash Successively the conjugated beads with the following solutions (*see* **Note 10**):
   (a) One quick wash with 4 mL of 100 mM Glycine, pH 2.5.
   (b) One wash with 4 mL of 10 mM Tris-HCl, pH 8.8.
   (c) One quick wash with 4 mL of 100 mM triethylamine.
   (d) One 5 min wash with 4 mL of 1× PBS, 0.5% Triton X-100, leaving the tube rocking or slowly rotating for 5 min.
   (e) One 15 min wash with 4 mL of 1× PBS, 0.5% Triton X-100, leaving the tube rocking or rotating, as in previous step.
7. *Final Dynabeads resuspension and storage:* Pool the beads into 1 mL of 1× PBS, 0.02% NaN_2 in a 5 mL tube. Use another

1 mL to recover any beads left behind in the tubes. The beads are now resuspended in 2 mL of 1× PBS, 0.02% NaN_3. Seal the tube with parafilm to prevent buffer evaporation and store at 4 °C.

The conjugation provides densely antibody-coated Dynabeads (160 μg of rabbit IgG per mg of Dynabeads) with a final concentration of ~0.15 μg of beads per μL of solution. The beads can be stored for several months at 4 °C without loss of activity (*see* **Note 11**).

3.2 Harvesting and Freezing of Cells

1. Grow yeast cells to no more than a density of 0.866–2.7 × 10^7 cells/mL (OD_{600} ~ 0.8–1.2; early log phase) (*see* **Note 3**).
2. Harvest cells by centrifugation (4 °C, 5–10 min, 3500 × *g*).
3. Resuspend and wash the cell pellet(s) with ultrapure water and transfer them into a 50 mL tube.
4. Pellet the cells by centrifugation (4 °C, 5 min, 2600 × *g*).
5. Wash the cells once more with an isovolume of ultrapure water and centrifuge once more.
6. While the cells are centrifuging, supplement an isovolume of cold resuspension buffer with DTT (1:1000), PIC (1:100) and solution P (1:100).
7. Remove the supernatant from **step 5**, resuspend the cell pellet in the supplemented Resuspension Buffer prepared in **step 6** and vortex.
8. Pellet the cells by centrifugation (4 °C, 5–15 min, 2600 × *g*), remove the supernatant and centrifuge the cells once again (4 °C, 5–15 min, 2600 × *g*) to eliminate as much supernatant as possible.
9. While the cells are centrifuging, place a 50 mL tube rack with an empty 50 mL polypropylene tube into the Styrofoam box. Fill both the Styrofoam box and the 50 mL polypropylene tube with liquid N_2. As the liquid nitrogen will rapidly evaporate from the tube during cooling, add more liquid N_2 until the temperature has equilibrated. Unplug the syringe and have a clean spatula ready.
10. After centrifugation, remove any remaining buffer from the top of the cell pellet. Use the spatula to fill a 20 mL syringe with the paste-like yeast pellet. Cool a fresh spatula in liquid N_2 for the next step.
11. Put the plunger back into the syringe and push the yeast pellet through the syringe into the liquid N_2-filled 50 mL tube, filling the tube to the rim. Yeast cells will freeze instantly. Using the cold spatula, push the noodles to the bottom of the tube and

break them (*see* **Note 12**). Repeat this step until there is no more pellet left in the syringe.

12. Using a 18G × 11/2 needle, punch a few holes into the center of a 50 mL tube screw cap and close the noodle-filled tube, before rapidly inverting the tube over the Styrofoam box to remove any remaining liquid N_2. Care should be taken at this step, as liquid N_2 will burst out from the tube when inverted. Invert the tube in a way that the cap rotates away from yourself. Wear eye protection, a lab coat, and cryo-gloves.
13. Tubes with frozen yeast noodles can then be transferred to a −80 °C freezer, and the punctured screw caps changed after 15–30 min to allow for complete evaporation of liquid N_2. This will prevent condensation from forming inside the tube during long term storage.

3.3 Cryogenic Cell Lysis (Cryo-Milling)

1. Prechill the metal jar and metal beads by filling them in liquid nitrogen inside a Styrofoam box Liquid N_2 levels in the Styrofoam box should not exceed half of the jar's height (*see* **Notes 13** and **14**).
2. Remove any liquid N_2 from inside the stainless steel jar, add your yeast noodles and the prechilled stainless steel balls until the jar is full, leaving just enough space to completely close the lid (*see* **Note 14**).
3. Weigh the jar containing the noodles and metal balls wearing cryo-gloves.
4. Put the closed jar back into the liquid N_2-filled Styrofoam box.
5. Weight the filled jar and match its weight with the appropriate counterbalance setting on the Retsch PM-100 planetary ball mill.
6. Use the following program on the ball mill to perform the following cycle: 400 rpm rotations for a total of 3 min per cycle, with intervals of 1 min between alternating rotation directions.
7. Secure the jar tightly on the ball mill with the provided clamp. Close the cover on the machine. Once "Start" is pressed, the machine will ask whether you have tightly secured the jar. Press "Yes" to start cycle. This is cycle 1 (*see* **Notes 15**, **16** and **17**).
8. Once the machine stops, cool the jar in liquid N_2 for at least 2 min or until the vigorous bubbling stops. The jar cover can be basted with liquid N_2 using a home-made N_2 basting tool as shown in Fig. 1 to accelerate the cooling process.
9. Repeat **steps** 7 and **8** until you have performed 8 full cycles.

10. On the last cycle, put a 50 mL rack into a clean Styrofoam box filled with liquid N_2, immerse a clean 50 mL tube labeled with the sample information and add a spatula in the rack to prechill.
11. Open the stainless-steel jar and place the balls into the liquid N_2 using tweezers. The cryo-lysed yeast cell grindate will be visible and compacted at the sides and bottom of the jar. To obtain a fluffy cell grindate, use the prechilled spatula to scrape the cell grindate off the side and bottom of about 1/4 of the jar, place five prechilled metal balls back into the jar, cool the jar and repeat **step 7** stopping the milling cycle after ~15 s. Make sure that the tips of the pair of tweezers and spatula have been left in liquid nitrogen to keep cold.
12. Remove the metal balls with the tweezers. Open the prepared 50 mL tube and make sure that there is no liquid N_2 in it. Wearing a cryo-glove, tilt the jar over the 50 mL tube. With the other hand, using the prechilled spatula, transfer the yeast cell grindate from the jar to the tube. Cool the tip of the spatula in liquid N_2 from time to time to avoid sticking of the cell grindate to the spatula. Close the tube (hand-tight) and store the cryo-lysed cell grindate at −80 °C for an indefinite time.

3.4 Single-Step Affinity Purification (See Note 18)

3.4.1 Weighing the Yeast Cell Grindate

1. Prepare the 50 mL tubes (*see* **Note 5**), and immerse them in liquid N_2. Prechill the tip of the spatula and place one rack into liquid N2, and the other on the balance.
2. Place the tube containing the cell grindate into liquid N_2.
3. Completely remove the liquid N_2 from one of the empty tubes, place it on the balance and zero the balance.
4. With the prechilled spatula, quickly transfer the required amount of grindate for the ssAP into the prechilled tube, and put the closed tubes back into liquid N_2. If there are many strains or a large quantity of cell grindate to weigh out, cool the sample in between. Dip the spatula in liquid nitrogen from time to time to prevent any sticking of the grindate. Use a fresh spatula for each strain. Move the tubes containing the yeast cell grindate stock back to −80 °C for storage, or continue to Subheading 3.4.2.

3.4.2 Affinity Purification

1. Leave the apportioned yeast cell grindate to warm up on ice until it has an ice cream-like look.
2. Add ice-cold ssAP buffer in a 9:1 mL of buffer per gram of grindate ratio.
3. Vortex at maximum speed for 30–60 s to resuspend the cell grindate.

4. Put the tube on ice and further homogenize the grindate using the polytron at 2/3 of the maximum speed for 30 s trying to avoid excess foaming.
5. Centrifuge the tubes at 4 °C, 2700 × *g*, 10 min.
6. While the lysate is clearing in the centrifuge, take an appropriate amount of IgG-conjugated magnetic beads per sample, transfer them into 1.5 mL LoBind tubes and add 1 mL of ssAP buffer; for example, use 50 μL of beads for 1 g of grindate for Mex67-PrA, or 10 μL for 0.2 g; adjust the quantities according to Fig. 4. Beads can be pooled for washes but should be divided in different tubes during the last wash in **step 8**.
7. Invert tubes until the beads are homogeneously dispersed.
8. Magnetically recover the beads and proceed with two more washes using the ssAP buffer. Leave the beads resuspended in the last ssAP buffer wash.
9. When the samples have finished clearing by centrifugation, transfer the supernatants into an appropriately sized tubes according to the amount of material. After placing the 1.5 mL microtubes on a magnet holder, aspirate the wash buffer from the beads, then remove the washed bead microtubes from the magnet and place them on a regular rack. Use ~300 μL cleared lysates to resuspend and transfer the beads to the tubes containing the cleared lysate and rinse the 1.5 mL washed-beads tubes and pipette tip thoroughly to recover all beads.
10. Incubate the tube containing lysate and beads for 30 min at 4 °C, rocking on the nutator or similar device, or a slow rotating wheel.
11. Prepare the Last Wash (with and without detergents) and Elution buffers during the 30 min incubation time.
12. The remaining steps of the ssAP are performed at room temperature. At room temperature, magnetically recover the beads on a magnetic rack and remove the Flow-Through (*see* **Note 8**).
13. Use 1 mL of fresh ssAP buffer (first wash) to recover and transfer the beads into a clean 1.5 mL tube sitting on a magnet. Wait for beads to accumulate toward the magnet and then reuse half of the cleared buffer to rinse and recover the beads left in the original tube and pipette tip.
14. Wash the beads with 1 mL of ssAP buffer. This can be done by slowly resuspending the beads with a P-1000 pipette, then putting the tube on the magnet to wait for the beads to clear the sample, and then rinsing the pipette tip with the cleared sample while the tube is till on the magnet. Alternatively, the

tube can be inverted several times to resuspend and wash the beads before putting it back on the magnet. If inversion is used to resuspend and wash the beads, after putting the tubes back on the magnet and buffer has cleared, the tubes should be slowly inverted again to recover all the beads located beneath the cap of the tube, and, when removing the wash buffer, it I important to remove any buffer remaining in the cap. Proceed to washing the beads two to three more times with 1 mL of ssAP buffer.

15. Incubate the beads for 5 min with 1 mL of Last Wash Buffer rocking on the nutator or similar device, or a slow rotating wheel.
16. Wash the beads quickly 3 more times with 1 mL of Last Wash Buffer without detergent.
17. Incubate beads with 1 mL of Last Wash Buffer without detergent for 5 min, rocking on the nutator or similar device, or a slow rotating wheel.
18. Make sure that the beads are homogeneously dispersed, and transfer 20% v/v (e.g., 200 μL of a 1 mL wash), corresponding to 10 μL of beads, to a new tube for elution and subsequent SDS-PAGE gel for silver staining and Western blotting (Subheading 3.6). The remaining 800 μL from the last wash, which corresponds to 40 μL of beads, will be used for MS analysis.
19. To equilibrate the samples for on-beads digestion, wash the beads with 1 mL of 20 mM Tris-HCl, pH 8.0, or alternatively, with 1 mL of 50 mM ammonium bicarbonate pH 8.0 (*see* **Note 19**).
20. Resuspend the beads in 50 μL of 20 mM Tris-HCl or 50 mM ammonium bicarbonate pH 8.0 and continue with the on-beads trypsin digestion.

3.5 On-Beads Trypsin Digestion

3.5.1 Preparations

1. Solubilize 1 vial (20 μg) of lyophilized trypsin in 40 μL 20 mM Tris-HCl, pH 8.0 and aliquot them in 2.5–5 μL aliquots at a concentration of 0.5 μg/μL. Store the aliquots at −80 °C. One μg is needed to digest one sample. If many aliquots are needed to digest all the samples, pool the aliquots and mix by pipetting prior to dispensing them to ensure uniform enzyme distribution and activity across samples. Avoid freeze-thaw cycles of the enzyme as this affects its activity.

3.5.2 On-Bead Digestion

1. Trypsin digestion (*see* **Notes 20** and **21**):

   Over day then overnight.

   (a) Spike the beads with 500 ng of trypsin and incubate 4 h at 37 °C with gentle agitation (350 rpm).

(b) Transfer the supernatant into another tube and spike it with another 500 ng of trypsin. Incubate overnight (12–16 h) at 37 °C with gentle agitation.

Overnight then over day

(a) Spike the beads with 750 ng of trypsin and incubate overnight (12–16 h) at 37 °C with gentle agitation (350 rpm).

(b) Transfer the supernatant into another tube and spike it again with 250 ng trypsin. Incubate for 3 h at 37 °C with gentle agitation.

2. Quench the trypsin reaction by adding formic acid to a final concentration of 2%.
3. In order to eliminate all residual Dynabeads that could clog the HPLC system and capillary column, first centrifuge the samples at 13,000 × *g* for 1–2 min. Then place the tubes on a proper magnet holder and slowly aspirate the sample taking care not to place the pipette tip orifice where the beads pellet should be. For safety, leave 1–2 μL of sample behind. The volume left behind should be identical in all samples. Transfer the supernatant into a new 1.5 mL microtube. While the samples can be stored at −20 °C, it is preferable to analyse them directly by mass spectrometry. When it is not possible to proceed immediately with LC-MS, we routinely store our digested samples at −80 °C until a time slot is available for further sample cleaning/processing and LC-MS injections. If the digestion was carried out in ammonium bicarbonate buffer, the samples should be lyophilized using a SpeedVac before being either stored at −80 °C or resolubilized immediately for direct LC-MS analysis (Subheading 3.7.2). For trypsin digestion carried out in Tris buffer, split the peptide digest into two tubes of equal volume and either store both at −80 °C or proceed to Subheading 3.7.1 for C18 ZipTip sample cleanup for one of the tubes.

3.6 Protein Elution for SDS-PAGE Staining and Western Blotting

1. After the last wash, elute the purified complexes from the beads (from Subheading 3.4.2, **step 18**) with 500 μL of elution buffer (20 min, RT, slow rotation).
2. Transfer the eluate in a new tube and keep on ice.
3. Proceed with a second elution as above (500 μL Elution Buffer, 20 min, RT, slow rotation). A third elution might be necessary.
4. After all elutions, place the tubes containing the beads in a magnetic rack, and resuspend the beads in 10 μL of Solution A; then add 10 μL of Solution B supplemented with fresh DTT.
5. Boil the bead samples at 95 °C, for 5–10 min and store at −20 °C.

6. Pool the eluates and speed-vac the samples overnight at room temperature with the tubes open, lids pointing outward. Turn off the heating function of the SpeedVac. If three elutions were carried out, split the total volume between two tubes.
7. The next morning, resolubilize the lyophilized proteins by carefully rinsing the inside wall of each 1.5 mL tube with 10 μL of Solution A. For samples that were split into two tubes, resolubilize peptides from each tube successively with the same 10 μL of solution A.
8. Quick-spin for 15 s at V_{max}.
9. Place the tubes containing the beads after elution in a magnetic rack, transfer the supernatants into new tubes and add 10 μL of Solution B.
10. Denature your sample at 70 °C, for 10–20 min.
11. Load half of the eluted sample per well on SDS-PAGE gel for Silver or Coomassie staining. While Coomassie blue R-250 has good detection range (10–100 ng/μL depending on protocols), silver staining is the method of choice since it has sub-nanogram sensitivity.
12. Load 1/10th to 1/4th of your eluted sample on an SDS-PAGE gel for subsequent western blotting. This will depend on the sensitivity of the antibodies used for detection as well as the method used for visualization (i.e., film, fluorescent or chemiluminescent imagers).

3.7 Mass Spectrometry

3.7.1 C18 ZipTip Sample Cleanup

If Tris was used for on-beads trypsin digestion, clean half of the sample (one tube) using a C^{18} ZipTip to efficiently remove the Tris (*see* **Note 22**). The remaining half can be kept at −80 °C as a backup.

1. Make sure that the sample pH is around 3 and adjust the pH by adding more formic acid up to 1% if needed. If ever the sample has been lyophilized, resuspend the peptides as described in Subheading 3.7.2 first, and then proceed to **step 2** below. Hydrate the ZipTip by pipetting 20 μL of a 50% acetonitrile 0.2% formic acid solution and discarding the solution into a waste receptacle. Repeat 3 times.
2. Equilibrate the ZipTip with 20 μL of H_2O 0.2% formic acid solution and discard the solution. Repeat 10 times.
3. Bind the sample on the ZipTip by pipetting up-and-down 40 times through the sample, with the pipette volume set to 20 μL.
4. Wash the ZipTip-bound sample with 20 μL of 0.2% formic acid in H_2O and discard the solution. Repeat 3 times.

5. Elute the peptides with 15 μL of 80% acetonitrile 0.2% formic acid solution. To efficiently elute the peptide, pipette 12 times up-and-down using the same 15 μL.
6. Finally, dry the 15 μL of eluted peptides using a SpeedVac.

3.7.2 Resolubilizing the Dried Peptides for LC-MS Injection

1. Dried peptides from C^{18} ZipTip cleanups and ammonium bicarbonate trypsin digests are resolubilized by adding 12 μL of 1% formic acid and 2% acetonitrile in MS grade water per sample, by first pipetting 15 times up-and-down. Samples are then mixed for 15 min on a MixMate at 1200 rpm.
2. To ensure maximum peptide solubilization, the samples are sonicated in a water bath sonicator for 5 min at room temperature.
3. Samples are finally centrifuged to remove any air bubbles. Samples are now ready for LC-MS injection.

3.7.3 Sample QC and Sample Quantity Determination for LC-MS Injections

1. A saved MS method file (according to parameters in Subheading 3.7.4), should be used as template to scale down your elution gradient to about 20 min and save it for a short MS test run.
2. Make a 1/5th dilution with 1 μL of your positive ssAP sample and inject 1 μL of this dilution using the LC-MS method file with a short gradient as mentioned above. After the short LC-MS test run, open the raw file with Xcalibur Qual Browser, and look at the ion map. If the map does not show any excessive presence of detergents or polymers in the sample, use the total ion current to estimate the volume of your positive sample to be injected to achieve near LC or MS signal saturation. LC chromatogram peaks that are highly protuberant must also be considered to avoid them oversaturating the system. The same volume determined to inject your positive ssAP sample in the next step for near signal saturation will be used for the injections of the negative control samples.

3.7.4 LC-MS Method and Sample Injection

Each sample should be injected at least in triplicate. The same liquid chromatography (LC) column should be used for all samples for consistency across samples. In order to reduce variations between samples, the LC column should be conditioned with at least 2–3 LC-MS injections in order for the column to stabilize. Ideally, a small amount of reserved positive sample, a similar sample, or at total protein extracts from the same species are used for column conditioning.

1. *Injecting your sample:* Samples are first loaded at higher flow rate (400–600 nL/min) in 2% solvent B (98% solvent A), before reducing the flow rate to 250 nL/min.

2. *LC gradient:* At the same flow rate, peptides are eluted on a 3-slope gradient using solvent A and B mixtures. First, increase solvent B from 2% to 25% over a period of 20 min, then increase to 45% over 40 min, and lastly to 80% over 10 min. The gradient may be adjusted depending on the sample complexity, composition, and presence of detergents or polymers (*see* **Note 22**).
3. *Suggested parameters to be used in your method:*

   The data is acquired in data dependent acquisition (DDA) mode, using a top N ions method on both the Orbitrap Fusion Tribrid and Q Exactive mass spectrometers. A top 20 to 25 ion range is usually selected for the Fusion compared to a top 12 to 16 ion range for the Q Exactive, given that the latter operates on a slower duty cycle. These values are based on complexity of the sample: the more complex a sample, the higher top N ions range should be selected in the method. For a highly complex sample, extending the gradient would enable reduction of sample complexity during acquisition, and the selection of a lower top N ion range to achieve a higher duty cycle. Tip: use the short LC-MS run in Subheading 3.7.3 to evaluate the complexity of your sample along with peptide abundance and contaminants.

   (a) *Orbitrap Fusion Tribrid mass spectrometer:* an MS1 full scan range of 360–1560 m/z at a 120 K orbitrap resolution with an automatic gain control (AGC) target of 3×10^6 coupled to a maximum injection time of 50 ms should be used. The dynamic exclusion criteria are set to 2 scans within 14 s, with an exclusion time of 15 s. Mass tolerance of precursors is set to ± 10 ppm. The Quadrupole precursor isolation window is set to 1.6, and precursor ions are fragmented by HCD with a collision energy of 28%. MS2 fragmented ions are detected in the ion trap at a rapid scan rate with a first mass-to-charge set to 110 m/z. MS2 AGC target and maximum injection time are set to 2×10^4 and 50 ms, respectively.

   (b) *Orbitrap Q Exactive mass spectrometer:* an MS1 full scan range of 360–2000 m/z at a 70 K orbitrap resolution, with the same injection time and AGC target as on the Fusion should be used. Precursors mass tolerance is also set to ± 10 ppm and precursors are isolated in the Quadrupole with an isolation window of 2 m/z. Precursors are fragmented by HCD with an NCE of 28, and MS2 scans are acquired at a 17.5 K orbitrap resolution, with a scan range of 200–2000 m/z. MS2 AGC target and maximum injection time are set to 1×10^5 and 50 ms, respectively. Dynamic exclusion is set to 9 s.

3.8 Data Processing and Analysis

3.8.1 Data Processing

The raw data can be analyzed using one of several search engine algorithms: Mascot (Matrix Science), MSFragger, Seaquest or Tide, X!Tandem Comet, Andromeda (part of MaxQuant) amongst others [21–35] (*see* **Note 23**). Search engines will require additional tools to provide peptides confidence measurements as well as to support data analysis, while others software offer a more complete solution like ProHits, MaxQuant, the Trans Proteomic Pipeline (TPP), and others, including commercial software suits that require licenses like Scaffold (Proteome Software). Here, we describe the pipeline used in our lab.

1. *Processing the RAW Data:* The raw data files are first processed using Proteome Discoverer, which is interfaced with the Mascot search engine and Percolator [36–39]. The Mascot and Percolator search parameters are entered in Proteome Discoverer, and the Mascot searches are sent through Percolator to compute confidence measures for peptide identification prior to final export as MSF file format. We normally use reference proteome fasta files from Uniprot.
2. *Ions Tolerance and Charges and Missed Cleavages:* For MS acquisitions from the Fusion Tribrid mass spectrometer, parental ions errors are set to ±10 ppm while 0.6 Da is used for MS2 fragments detected in the linear trap. For the Q Exactive, ±10 ppm is also used for precursor ions, while 0.05 Da is used for fragmented ions. Only peptides of ≥2 positive charges that have a maximum of 2 missed cleavages are considered.
3. *Variable Modifications:* Since cysteines are not reduced and alkylated prior to trypsin digestion in this protocol, fixed carbamidomethyl modification of cysteines should be included in the search. Routinely, methionine oxidation and protein N-terminal acetylation are added as variable modifications to search parameters, and less frequently, phosphorylation of STY.
4. *Creating the Scaffold File:* When MSF files for all samples have been created, they can be combined and visually ordered in the Scaffold software (Proteome Software). The same fasta database that was used in Proteome Discoverer for the Mascot search must be used in Scaffold. To further improve peptide confidence levels, Scaffold makes use of Peptide Prophet. After creating a Scaffold file with all the ordered sample data associated with the correct fasta database, the file can be saved and then reopened in Scaffold Viewer, which does not require a license. It is worth noting that Scaffold supports Mascot, SeaQuest, X!Tandem, MSFragger, and MaxQuant data output formats.
5. *Scaffold Viewer:* While Scaffold requires a license, any user can then analyze their data on any computer using Scaffold viewer without the need for a licence.

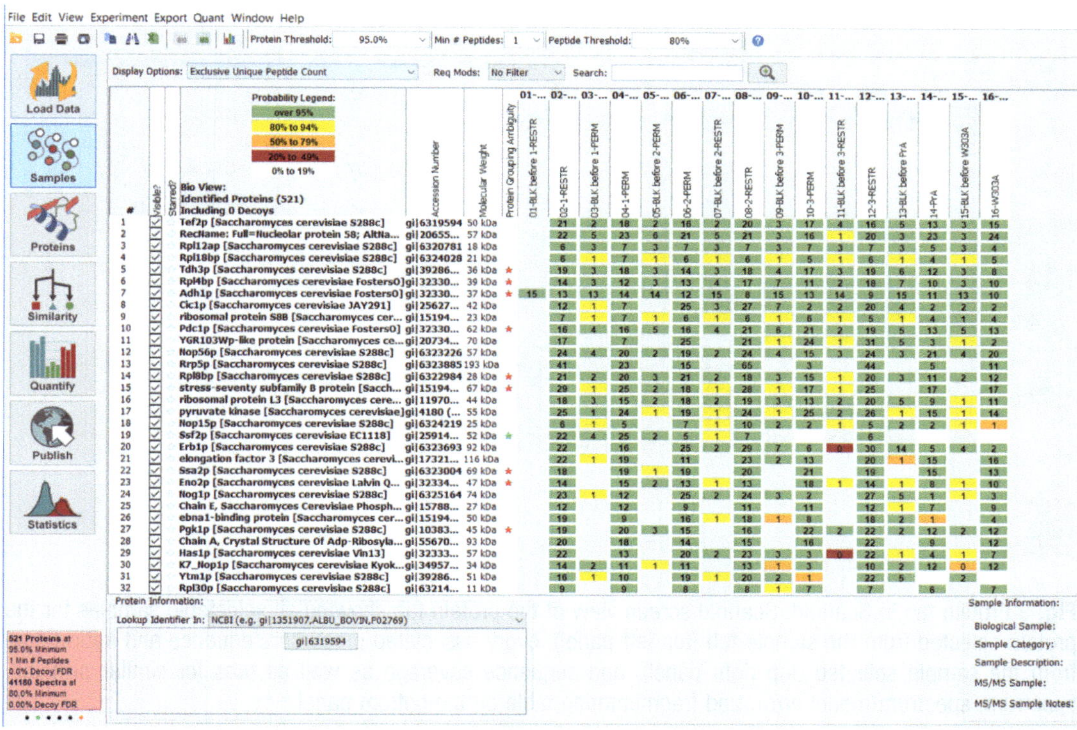

Fig. 2 Sample tab in Scaffold. Scaffold screen view of the sample tab showing exclusive unique spectrum counts for several samples including the two controls and blank runs between samples to evaluate carryover of abundant peptides in between sample injections

3.8.2 Data Analysis

1. Scaffold Viewer is intuitive and offers many useful visual and integrated filtering tools. As visual tools, protein identification probability, false discovery rate (FDR), protein coverage, exclusive and unique peptide or spectrum counts and peptide precursor information as well as spectrum with fragmentation table and interactive Venn diagram are available. Filtering tools include requested post-translational modifications from the search engine, protein and peptide confidence thresholds as well as minimum number of peptides per protein that can be changed in the top panel bar, causing the FDR for both protein and peptide levels to be promptly updated on the bottom left corner of the window. Both proteins and peptides can however be selected based on FDR values (Fig. 2).
2. As many end-users still use spectral counting to quantify their MS data [40], for this protocol spectral counting is carried out as follows. First, exclusive unique spectrum count (EUSC) with the default protein and threshold level are selected. Attention: by selecting exclusive unique spectrum counting, peptides that are shared between proteins will no longer appear in the sample panel list. If the FDR values for proteins and peptides are too

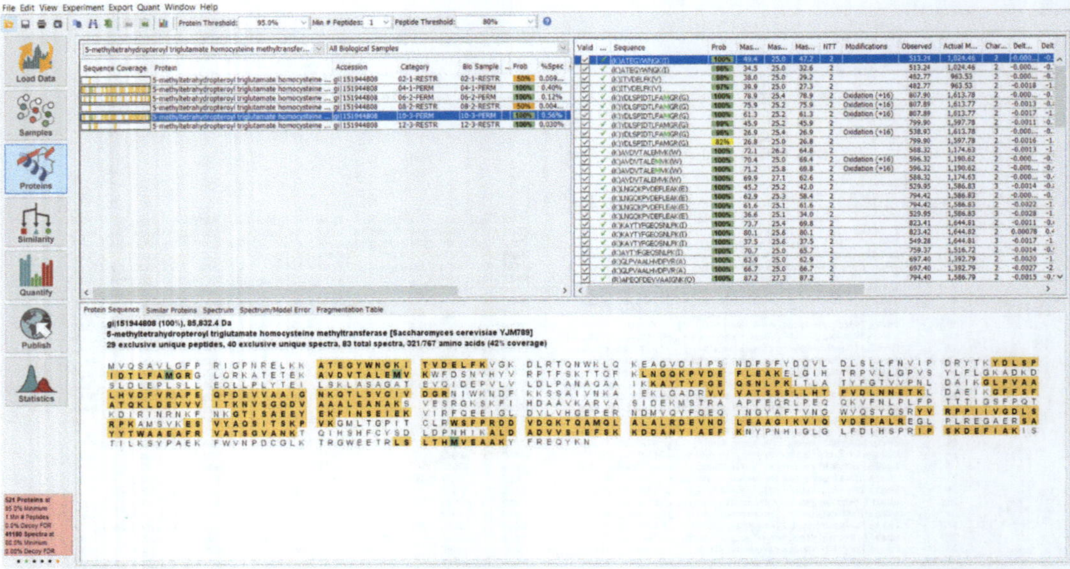

Fig. 3 Protein tab in Scaffold. Scaffold screen view of the protein tab showing all selectable samples for the protein selected from the sample tab (top left panel), every associated precursor sequence and information from the sample selected (top right panel), and sequence coverage as well as tabs for similar proteins, spectrum, spectrum/model error, and fragmentation table on the bottom panel

high, adjust the protein or peptide thresholds accordingly, or increase the minimum number of peptides to two in order to reduce the FDR to more acceptable values (Fig. 2).

3. By selecting a protein in the sample tab and switching to the protein tab, every exclusive unique spectrum (EUS) for the selected protein can be further compared across all samples. The upper left panel enables the user to select the sample for which all EUS sequences and information are displayed in the top right panel. Unmarking a spectrum will remove it from the results in the sample tab. The bottom panel shows the EUS coverage and its alignment to the protein sequence, with displays information upon cursor hovering. This panel also has specific tabs that are useful to manually verify the validity of any spectrum; Similar Proteins, Spectrum, Spectrum/Model Error, and the fragmentation table (Fig. 3).
4. Once all protein and peptide ambiguity has been addressed and lower confidence peptides have been verified or discarded, export the table from the sample tab to filter and transform the data manually. First, all proteins that do not display at least 2 EUSC in all ssAP positive samples are filtered out. While we process all the data for which at least 2 EUSC are displayed in all positive ssAP samples, only proteins of >10 EUSC are considered as high-confidence preys. We attribute a medium confidence level to preys with EUSC between 5 and 9, and a low confidence level for preys with EUSC between 2 and 5.

5. We then calculate EUSC average values of each protein triplicate along with the standard deviations and variation coefficient. Any protein for which the EUSC is at least twofold enriched compared to the negative control is considered above background. It is worth looking at proteins having high variation coefficients to decide if they should be excluded based on EUSC outliers and background levels from the negative control as well as low EUSC. A threefold increase filter criterion can also be applied to proteins from a positive ssAP having between 5–9 EUSC to strengthen the confidence of such prey proteins. Proteins identified with low peptide numbers should be considered only if they have been associated with exclusive unique and consistent ion fragmentation patterns (Fig. 3; *see* **Note 24**).
6. Lists of common contaminants can be found on https://reprint-apms.org/ website according to protocols, epitope tag, cell line, antibody, types of beads, and cell fraction [41].
7. To compensate for protein size, each averaged EUSC is divided by the total theoretical number of peptides per single protein generated by trypsin digestion. To obtain these theoretical numbers for each protein, MS-Digest from Protein Prospector (https://prospector.ucsf.edu/) should be used; all proteins are entered as a list. The *in-silico* digest is performed with 0 missed cleavages, no variable or fixed modification, and the peptide mass range used should be set between 720 and 6000 Da. Tab delimited text should be selected as output and the "hits to file" option marked.
8. Once the values are normalized for protein size, divide the new values over the bait value to normalize the data across all runs.

4 Notes

1. Our laboratory uses rabbit IgG from total serum as it has high binding affinity for ProteinA (PrA). Guinea pig IgG from serum could be used with similar results [42]. Subclasses of IgG which could be used with similar results include human IgG_1, IgG_2, and IgG_4 as well as mouse IgG_{2a} [42]. IgG from other species should not be used as they bind PrA with lower affinities. The IgG from goat, rat and chicken should be absolutely avoided as they have a low affinity for PrA [42]. Once conjugated, IgG-coupled magnetic beads last for up to 6 months at 4 °C. Caution should be taken with other antibodies like anti-GFP for example, as they only maintain a relatively good efficiency for about 2 weeks after conjugation.

2. Refer to MSDS specifications sheet and handle these reagents with caution.
3. For these experiments, usually 6 L (3 × 2L) of culture are grown to OD_{600} ~ 0.8–1.2. This will produce a reasonable amount of material to work with. A strain expressing the PrA-tag alone from an endogenous promoter is best used as negative control for PrA-tagged bait proteins, and, ideally, expression levels should be similar. Unconjugated beads should be used as negative control if an antibody against an endogenous protein is used for the ssAP.
4. As alternatives to planetary ball mill grinding, cell cryolysis can also be achieved using a swing-mill or mortar or a grinder under cryo-freeze conditions; however, for the latter, the efficiency of cell breakage will be reduced and less homogeneous.
5. *Sample Amount:* For small scale ssAP used for buffer and additives optimization experiments (*see* **Note 6**), 0.2 g of Mex67-PrA grindate and 10 μL of conjugated Dynabeads are sufficient to perform a silver-stained SDS-PAGE and western blot. In this case, do a scale up of the amount of grindate according to the number of small-scale ssAP samples needed in one resuspended lysate, then dispense the proper amount of cleared lysate into proper size tubes. For MS analysis, five times more grindate and beads are required compared to the small scale. While the volume of beads remains fixed, that is, 10 μL for small scale and 50 μL for MS analysis scale, the amount of grindate will vary according to protein copy number per cell. One gram of Mex67-PrA at 3300 copy per cell purified from 50 μL of beads is our reference to determine amounts of grindate needed for different baits (Fig. 4). For 0.2 g of cell grindate and lower, use a 2 mL microtube and skip the Polytron step. For more than 0.2 g up to 1 g, use a 15 mL polypropylene tube. When 1.5–2 g of cell grindate is needed, divide the cell grindate into two 15 mL tubes. If 15 g of cell grindate and more are required, divide into three or more 50 mL polypropylene tubes.

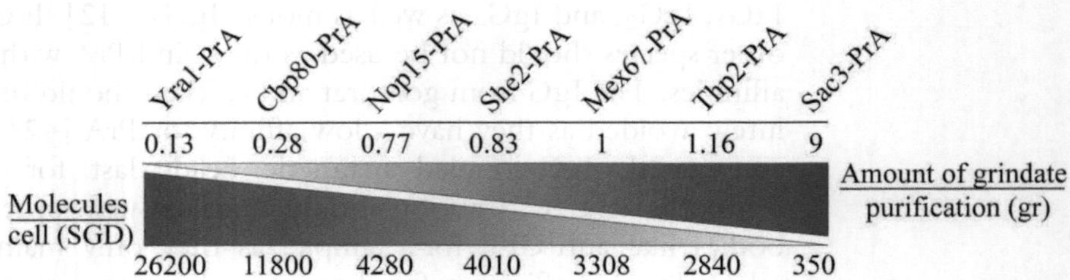

Fig. 4 Sample amounts. The amount of yeast cell grindate required to recover similar amounts of bait protein per 50 μL Dynabeads depends on the cellular abundance of the bait protein

6. *Buffer Selection*: The buffers used for ssAPs vary and need to be determined empirically as their choice mainly depends on the nature of the bait protein and the stability of its associated complexes. For example, our usual buffers contain 20 mM HEPES-KOH, pH 7.4, a salt, and two types of detergents. Salt concentrations can vary from 0 mM-1 M. Different salts can be used according to their stabilisation properties as determined by the Hofmeister Series: KOAc > NH_4OAc > NaCl > NaCitrate [43–45]. Standard buffers usually contain 0.5% Triton X-100 and 0.1% Tween 20, 1:100 Solution P, 1:5000 antifoam. A buffer commonly used in our laboratory for complex affinity purification is 20 mM HEPES–KOH, pH 7.4, 110 mM KOAc, 100 mM to 300 mM NaCl or 250 mM NaCitrate, 0.5% Triton X-100, 0.1% Tween 20, 1:100 Solution P and 1:5000 antifoam.
7. The SpeedVac must be compatible with organic solvents or else it will corrode over time.
8. *Buffer Removal* should be performed slowly with a pipette, or, alternatively, using either a 2 mL filter-free or micropipette tip, or clean Pasteur pipette connected to a vacuum source. Aspirate the liquid from the side of the tube opposite to the magnetized beads in a slow movement to ensure no to negligible loss of beads and efficient buffer removal. Pasteur pipettes and micropipette tips should be changed between samples to avoid cross-contamination. Tip: when using a vacuum system, a 1 mL filter-free pipette tip connected to the vacuum can be used to pick up clean 200 μL filter-free pipette tips from the tip box between samples for 1.5 mL microtube format.
9. *Determining Sodium phosphate amount for antibody-buffer-salt mixture*: If an antibody/tag combination other than IgG/PrA is used, the antibody concentration may differ and be more diluted in some instances. Since the stock concentration of ammonium sulfate and its volume added to the mixture should not be changed, the volume remaining for the addition of sodium phosphate may become limited compared to the original protocol. As the total volume of the IgG + sodium phosphate + ammonium sulfate mixture is 20 mL per 300 mg of Dynabeads, the remaining volume allowed for sodium phosphate should be used to calculate the concentration of the stock solution required to keep the concentration of every component constant. If less than 300 mg of Dynabeads are conjugated, then the entire conjugation protocol has to be scaled down accordingly.
10. During the washes, the beads should be completely resuspended in solution.

11. Special care should be taken in ensuring that the beads are covered by buffer. If beads dry out, the efficiency of the affinity purification will not be optimal.
12. The 50 mL tube should always be nearly completely filled with liquid nitrogen during this step. Do not place the syringe too close to the liquid nitrogen or the cell paste will freeze inside the syringe tip and clog it. If this happens, move the syringe further away from the tube, wait for a moment, then apply pressure on the piston while aiming at the tube opening until it unclogs. The yeast "noodle" coming out of the syringe should always be immersed in liquid nitrogen. If there is not enough liquid nitrogen in the tube, fill it immediately, and break the noodles with the cooled spatula if there is no more space in the 50 mL tube.
13. *Handling liquid nitrogen*. The temperature of liquid nitrogen is −196 °C. Since this procedure requires handling material prechilled in liquid nitrogen, it is preferable to work in a chemical hood, wear protective equipment as cryo-gloves, eye protection glasses and a lab coat. Any material immersed in liquid nitrogen will "boil" vigorously until the material reaches liquid nitrogen temperatures.
14. *Tips for cryo-grinding:* The jar should not be completely immersed in liquid nitrogen for it to be appropriately cooled with your sample in it. Liquid nitrogen levels should not reach higher than halfway up the jar. Once the jar is closed with your sample and stainless steel balls inside, it is not recommended to open it unless the balls are stuck (no more rattling sound can be heard; *see* **Note 17**). Avoid keeping the jar open for prolonged periods as condensation can occur as a result, which may mix with your yeast cell grindate. The precooled jar can be left at RT for several minutes before it reaches temperatures that will melt your sample. Your samples will not melt during a normal grinding cycle, but it is important to cool the jar enough between cycles. To accelerate the cooling process, the jar can be basted with a tool that can withstand liquid nitrogen, or with a simple homemade tool as shown in Fig. 1. Caution: if the cell grindate starts to look like "ice-cream," the sample is no longer usable and needs to be discarded.
15. *Milling Safety*: The planetary ball mill has a metal holder that perfectly fits the bottom of the jar to hold it securely in place during the run. The top of the jar is held in place by a clamp that must be manually secured with a force of ~10 Nm. The movement of the clamp is secured by a red ring, which needs to be pushed up to unlock the mechanism. As the clamp secures the jar, tightening will become more difficult. No more than two turns are necessary at this point to get to ~10 Nm. At the

end, the red ring should be clamped down and it may be necessary to turn the clamp back by approximately an eighth of a turn, resulting in a small clicking sound when locking occurs. Be careful not to overtighten the jar, as it may be difficult to remove afterward.

16. It is important to ensure the counterbalance has been appropriately set.
17. *Requirements for a full milling cycle.* The peculiar rattling sound of produced during milling indicates that the balls are moving. Sometimes the rattling can stop, and then start again after a few seconds. However, if the rattling is not heard for more than 10 consecutive seconds, it indicates that the balls are stuck. If this happens, stop the cycle and put the jar back in liquid nitrogen. While the jar cools down to prevent condensation upon opening, cool the tip of a long pair of tweezers. Once the jar is completely cooled, open the lid and unstuck/remove or add balls with the tweezers. It may be necessary to add or remove balls from the jar to achieve the correct ratio of balls to grindated volume to yield optimal lysis. Secure the closed jar back on the machine as usual (*see* **Note 15**) and resume the cycle. It is important to pay attention to the rattling sound because if the balls become continuously stuck for more than 20 s during a cycle, the cycle should be repeated.
18. *ssAP*: Before starting, clean your bench and other materials you will use. We use Windex to clean our surfaces, and then wash off residues left by the Windex with 70% ethanol. Avoid working in a chemical hood as these are meant to protect the worker, not the samples. Wear a clean lab coat and gloves throughout. Keratin is a usual contaminant in MS. It is however important to minimize contamination of keratin as it can reduce the number of peptide precursors of interest selected for MS2 if present in excess quantity. Keratin will not be detected in any MS analysis using a yeast database. Keep your tubes, bottles, and pipette tip boxes closed as much as possible. Use low binding pipette filtered tips for ssAP and MS work.
19. *Choice of on-beads digestion buffer and implication.* Using Tris-HCl for on-beads digestion will require a cleanup step of the tryptic peptides after the digestion. Alternatively, 50 mM ammonium bicarbonate pH 8.0 can be used instead of Tris-HCl, which allows to proceed directly with MS injections of the samples without any further peptide cleanup.
20. *Trypsin digest*: Trypsin is the most commonly used protease for bottom-up MS. It cleaves carboxyterminal of arginines and lysine, leaving a positive charge at the peptide C-terminus, which is advantageous for MS. Moreover, the fragments generated are of a suitable mass range for efficient MS2 fragmentation and identification with high specificity [46].

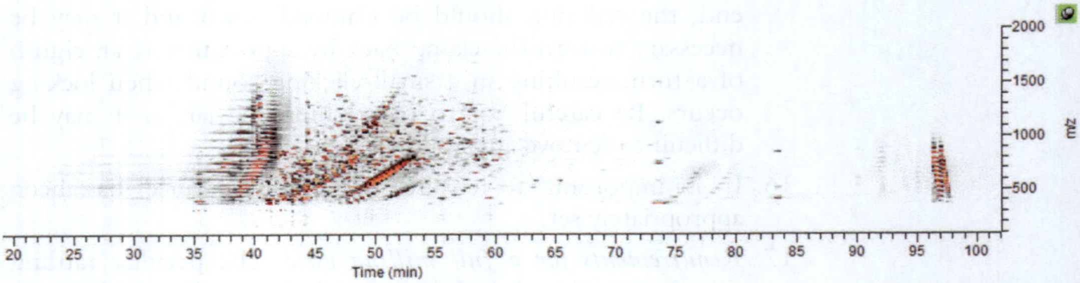

Fig. 5 Detergents and polymer contamination. Screen view of Xcalibur Qual Browser ion map showing detergents/polymers. Note the pattern observed between 35 and 42 min, as well the pattern of more abundant detergent/polymers ions throughout 41–54 min (orange, red, and some yellow). Some of them are less abundant (black and grey) but display a similar trail pattern (see 75 to 80 min for example)

21. *Cysteine reduction and alkylation:* Many MS samples are alkylated to avoid disulphide bridge formation between cysteines as most programs used to analyse MS spectra do not recognize peptides linked by disulphide bonds. This is not done in our lab for ssAP as yeast proteins contain overall only ~1.2% cysteines, and we found alkylation of cysteines not necessary to obtain better spectra [47].

22. *Sample cleanup*: Depending on the peptide sample buffer, salts, detergent, and polymer composition, it is advised to get remove these contaminants if their presence is too abundant prior to injecting your sample in LC-MS. Different methods can be used to remove different contaminants such as MCX (Mixed Cation Exchange), SCX (Strong Cation Exchange), or C^{18}. It may be difficult to detect the presence of polymers and detergents just by looking at the HPLC chromatogram. The ion map is more useful to detect these contaminants given that they most often show some characteristic patterns (Fig. 5). It is usual to see traces of detergent or polymers in MS samples; however, when these becomes abundant in any sample, they can create ion suppression which will result in lower numbers of identified peptides and proteins. To minimize ion suppression from detergents or polymers without further sample cleanup, less material can be injected in LC-MS and the gradient can be modified or stretched. However, when detergents and polymers are abundant, it is best to remove them from your sample by using a solid phase extraction method other than C^{18}, or to just simply repeat the ssAP, while making sure that all guidelines have been followed. Microtubes of lower quality than protein LoBind tubes from Eppendorf usually releases polymers from the internal plastic surface into solution. If other tubes than LoBind tubes are used, it is highly recommended to wash them three times with 1 mL of 70% acetonitrile, vortexing for 5 min for each wash, and drying/

autoclaving them prior to use. Samples produced using the described protocol normally show negligeable amounts of detergents and polymers in our hands, and clean up using C^{18} solid phase extraction support like ZipTips provides the best reproducibility.

23. File format requirements may vary depending on selected search engine. Although some will accept RAW files from ThermoFisher Scientific mass spectrometers, some will require other file formats: .MGF, .mzML, mzXML, for example. To convert any .RAW file to the formats indicated above, use the MSConvert tool from ProteoWizard (http://proteowizard.sourceforge.net/). For SeaQuest input file format conversion, a tool called RawExtractor is available on the Yates laboratory website (http://fields.scripps.edu/yates/wp/) in the resources download section.
24. *Matching peptides:* The expected fragmentation of the b and y ions are known for a given peptide as well as the expected fragmentation of its protonated ($^{+2}H/^{+3}H$), deaminated ($-NH_3$) and dehydrated ($-H_2O$) forms. The algorithm matches experimental fragmentation patterns according to a *best-fit* approach [48] and fragment ions experimentally found for this peptide are highlighted for all experimentally identified peptides in red, blue and green in Scaffold Viewer (Fig. 3).

Acknowledgments

C.T is supported by funding awarded to M.O. from the Canadian Institutes for Health Research (PJT153313).

References

1. Gavin A, Bosche M, Krause R et al (2002) Functional organization of the yeast proteome by systematic analysis of protein complexes. Nature 415:141–147
2. Goh K-I, Cusick ME, Valle D et al (2007) The human disease network. Proc Natl Acad Sci U S A 104:8685–8690. https://doi.org/10.1073/pnas.0701361104
3. Taylor IW, Linding R, Warde-Farley D et al (2009) Dynamic modularity in protein interaction networks predicts breast cancer outcome. Nat Biotechnol 27:199–204. https://doi.org/10.1038/nbt.1522
4. Ho Y, Gruhler A, Heilbut A et al (2002) Systematic identification of protein complexes in Saccharomyces cerevisiae by mass spectrometry. Nature 415:180–183
5. Gavin A, Aloy P, Grandi P et al (2006) Proteome survey reveals modularity of the yeast cell machinery. Nature 440:631–636
6. Krogan N, Cagney G, Yu H et al (2006) Global landscape of protein complexes in the yeast Saccharomyces cerevisiae. Nature 440:637–643
7. Costanzo MC, Hogan JD, Cusick ME et al (2000) The yeast proteome database (YPD) and Caenorhabditis elegans proteome database (WormPD): comprehensive resources for the organization and comparison of model organism protein information. Nucleic Acids Res 28:73–76
8. Lin D, Yin X, Wang X et al (2013) Re-annotation of protein-coding genes in the genome of Saccharomyces cerevisiae based on

support vector machines. PLoS One 8(7): e64477. https://doi.org/10.1371/journal.pone.0064477

9. Grigoriev A (2003) On the number of protein-protein interactions in the yeast proteome. Nucleic Acids Res 31:4157–4161. https://doi.org/10.1093/nar/gkg466
10. Collins SR, Kemmeren P, Zhao X-C et al (2007) Toward a comprehensive atlas of the physical interactome of Saccharomyces cerevisiae. Mol Cell Proteomics 6:439–450. https://doi.org/10.1074/mcp.M600381-MCP200
11. Johnson ME, Hummer G (2011) Nonspecific binding limits the number of proteins in a cell and shapes their interaction networks. Proc Natl Acad Sci U S A 108:603–608. https://doi.org/10.1073/pnas.1010954108
12. Picotti P, Bodenmiller B, Mueller LN et al (2009) Full dynamic range proteome analysis of S. cerevisiae by targeted proteomics. Cell 138(4):795–806. https://doi.org/10.1016/j.cell.2009.05.051
13. Oeffinger M (2012) Two steps forward-one step back: advances in affinity purification mass spectrometry of macromolecular complexes. Proteomics 12(10):1591–1608. https://doi.org/10.1002/pmic.201100509
14. Cristea I, Williams R, Chait B, Rout M (2005) Fluorescent proteins as proteomic probes. Mol Cell Proteomics 4:1933–1941
15. Rigaut G, Shevchenko A, Rutz B et al (1999) A generic protein purification method for protein complex characterization and proteome exploration. Nat Biotechnol 17:1030–1032
16. Oeffinger M, Wei KE, Rogers R et al (2007) Comprehensive analysis of diverse ribonucleoprotein complexes. Nat Methods 4:951–956. https://doi.org/10.1038/nmeth1101
17. Karlsson R, Jendeberg L, Nilsson B et al (1995) Direct and competitive kinetic analysis of the interaction between human IgG1 and a one domain analogue of protein a. J Immunol Methods 183:43–49
18. López-Ferrer D, Ramos-Fernández A, Martínez-Bartolomé S et al (2006) Quantitative proteomics using 16O/18O labeling and linear ion trap mass spectrometry. Proteomics 6(1): S4–S11. https://doi.org/10.1002/pmic.200500375
19. Capelo JL, Carreira R, Diniz M et al (2009) Overview on modern approaches to speed up protein identification workflows relying on enzymatic cleavage and mass spectrometry-based techniques. Anal Chim Acta 650:151–159. https://doi.org/10.1016/j.aca.2009.07.034
20. Belozerov VE, Lin Z-Y, Gingras A-C et al (2012) High-resolution protein interaction map of the Drosophila melanogaster p38 mitogen-activated protein kinases reveals limited functional redundancy. Mol Cell Biol 32: 3695–3706. https://doi.org/10.1128/MCB.00232-12
21. Kong AT, Leprevost FV, Avtonomov DM et al (2017) MSFragger: ultrafast and comprehensive peptide identification in mass spectrometry–based proteomics. Nat Methods 14(5):513–520
22. Yu F, Teo GC, Kong AT et al (2020) Identification of modified peptides using localization-aware open search. Nat Commun 11(1):1–9
23. Polasky DA, Yu F, Teo GC et al (2020) Fast and comprehensive N-and O-glycoproteomics analysis with MSFragger-Glyco. Nat Methods 17:1125–1132
24. Diament BJ, Noble WS (2011) Faster SEQUEST searching for peptide identification from tandem mass spectra. J Proteome Res 10(9):3871–3879. https://doi.org/10.1021/pr101196n
25. Bjornson RD, Carriero NJ, Colangelo C et al (2008) X!!Tandem, an improved method for running X!Tandem in parallel on collections of commodity computers. J Proteome Res 7(1): 293–299. https://doi.org/10.1021/pr0701198
26. Eng JK, Hoopmann MR, Jahan TA et al (2015) A deeper look into comet—implementation and features. J Am Soc Mass Spectrom 26(11):1865–1874. https://doi.org/10.1007/s13361-015-1179-x
27. Eng JK, Jahan TA, Hoopmann MR (2012) Comet: an open source tandem mass spectrometry sequence database search tool. Proteomics 13(1):22–24. https://doi.org/10.1002/pmic.201200439
28. Cox J, Mann M (2008) MaxQuant enables high peptide identification rates, individualized p.p.b.-range mass accuracies and proteome-wide protein quantification. Nat Biotechnol 26:1367–1372
29. Cox J, Michalski A, Mann M (2011) Software lock mass by two-dimensional minimization of peptide mass errors. J Am Soc Mass Spectrom 22:1373–1380
30. Schaab C, Geiger T, Stoehr G et al (2012) Analysis of high accuracy, quantitative proteomics data in the MaxQB database. Mol Cell Proteomics 11:M111.014068
31. Cox J, Hein MY, Luber CA et al (2014) Accurate proteome-wide label-free quantification by delayed normalization and maximal peptide

ratio extraction, termed MaxLFQ. Mol Cell Proteomics 13:2513–2526

32. Tyanova S, Temu T, Carlson A et al (2015) Visualization of LC-MS/MS proteomics data in MaxQuan. Proteomics 15:1453–1456
33. Tyanova S, Temu T, Cox J (2016) The MaxQuant computational platform for mass spectrometry-based shotgun proteomics. Nat Protoc 11:2301–2319
34. Liu G, Zhang JP, Larsen B et al (2010) ProHits: an integrated software platform for mass spectrometry-based interaction proteomics. Nat Biotechnol 28:1015–1017
35. Deutsch EW, Mendoza L, Shteynberg D et al (2010) A guided tour of the trans-proteomic pipeline. Proteomics 10(6):1150–1159. https://doi.org/10.1002/pmic.200900375
36. Käll L, Canterbury J, Weston J et al (2007) Semi-supervised learning for peptide identification from shotgun proteomics datasets. Nat Methods 4:923–925
37. Käll L, Storey JD, MacCoss MJ, Noble WS (2008) Assigning confidence measures to peptides identified by tandem mass spectrometry. J Proteome Res 7(1):29–34
38. Käll L, Storey JD, Noble WS (2008) Nonparametric estimation of posterior error probabilities associated with peptides identified by tandem mass spectrometry. Bioinformatics 24(16):i42–i48
39. The M, Noble WS, MacCoss MJ, Käll L (2016) Fast and accurate protein false discovery rates on large-scale proteomics data sets with percolator 3.0. J Am Soc Mass Spectrom 27:1719–1727
40. Arike L, Peil L (2014) Spectral counting label-free proteomics. Methods Mol Biol 1156:213–222. https://doi.org/10.1007/978-1-4939-0685-7_14
41. Mellacheruvu D, Wright Z, Couzens AL et al (2013) The CRAPome: a contaminant repository for affinity purification-mass spectrometry data. Nat Methods 10:730–736. https://doi.org/10.1038/nmeth.2557
42. Sambrook J, Russell DW (2001) Molecular cloning: a laboratory manual, 3rd edn. Cold Spring Harbor Laboratory Press, Cold Spring Harbor, NY
43. Bondos SE, Bicknell A (2003) Detection and prevention of protein aggregation before, during, and after purification. Anal Biochem 316: 223–231
44. Zhang Y, Cremer PS (2006) Interactions between macromolecules and ions: the Hofmeister series. Curr Opin Chem Biol 10:658–663. https://doi.org/10.1016/j.cbpa.2006.09.020
45. Damodaran S, Kinsella JE (1983) Dissociation of nucleoprotein complexes by chaotropic salts. FEBS Lett 158:53–57
46. Westermeier R, Naven T (2002) Proteomics in practice: a laboratory manual of proteome analysis. Wiley-VCH, Weinheim
47. Miseta A, Csutora P (2000) Relationship between the occurrence of cysteine in proteins and the complexity of organisms. Mol Biol Evol 17:1232–1239
48. O'Connor CD, Hames BD (2008) Proteomics. Scion Publishing Limited

Chapter 13

Composition and Dynamics of Protein Complexes Measured by Quantitative Mass Spectrometry of Affinity-Purified Samples

Abdelkader Namane and Cosmin Saveanu

Abstract

Multiple protein complexes are fundamental parts of living systems. Identification of the components of these complexes and characterization of the molecular mechanisms that allow their formation, function, and regulation can be done by affinity purification of proteins and associated factors followed by mass spectrometry of peptides. Speed and specificity for the isolation of complexes from whole cell extracts improved over time, together with the reliable identification and quantification of proteins by mass spectrometry. Relative quantification of proteins in such samples can now be done to characterize even relatively nonabundant complexes. We describe here our experience with proteins fused with the Z domain, derived from staphylococcal protein A, and IgG affinity purification for the analysis of protein complexes involved in RNA metabolism in the budding yeast *Saccharomyces cerevisiae*. We illustrate the use of enrichment calculations for proteins in purified samples as a way to robust identification of protein partners. While the protocols presented here are specific for yeast, their principles can be applied to the study of protein complexes in any other organism.

Key words Mass spectrometry, Affinity purification, Protein complexes, Complexes' dynamics

1 Introduction

Affinity purification of protein complexes originated in immunoprecipitation assays using antibodies against one of the complex components. Generic strategies for protein complexes characterization became available through the development of tags which, when fused with proteins of interest, are used for affinity purification. One of the most successful affinity isolation strategy involved the use of two purification steps in the tandem affinity purification method [1]. The isolated complexes can be used for functional and structural studies (examples for preribosomal intermediates reviewed in [2]). To identify the components of the purified samples, they are subjected to trypsin (Lys-C or other protease)

Frédéric Devaux (ed.), *Yeast Functional Genomics: Methods and Protocols*,
Methods in Molecular Biology, vol. 2477, https://doi.org/10.1007/978-1-0716-2257-5_13,

digestion, followed by the analysis of the obtained peptide mix through liquid chromatography coupled with high resolution mass spectrometry. Technical developments steadily increased the number of proteins identified from purified samples or whole cell extracts and also increased the peptide coverage of the detected proteins. Improvements in data analysis strategies allowed the combination of peptide signals for a given protein in a sample and comparisons with peptide signals for the same protein in other purified complexes or under modified conditions.

The introduction of superparamagnetic beads coated with affinity reagents led to the improvement in the recovery of very large complexes, such as polysomes [3] and allowed shorter incubation times [4]. The original TAP tag contains two 58 amino acid long Z domains, engineered based on staphylococcal protein A structure when bound to the Fc region of human IgG1 [5]. Using beads covalently bound to IgGs leads to much lower background compared with Sepharose or agarose-based IgG formulations. Thus, for most applications, we only use a single step purification, based on the binding of the two Z domains of a fusion protein of interest to rabbit IgG on superparamagnetic beads.

Mass spectrometry fundamental improvements, such as the development of the Orbitrap technology [6], together with better separation systems for peptide separation at very low flows, allow the identification and quantification of thousands to dozens of thousands of peptides in a single run [7]. Finally, crucial to the analysis of results is the ability to compare many independent runs and combine the results for visualization and further exploration. We found, for example, Perseus [8] to be efficient for manipulation of mass spectrometry data and interactive visualization of the results, including their annotations. Other options for the analysis of MaxQuant [9] results, based on the R language and libraries are Proteus [10] and the Bioconductor packages artMS [11] or DEP [12]. As soon as a corpus of mass spectrometry results is generated, it can be useful to present these data in a web-accessible interactive interface. For example, we included our published results on yeast nonsense-mediated mRNA decay (NMD) and NMD-associated complexes [13] in a web interface for rapid and interactive visualization of the results: hub05.hosting.pasteur.fr/NMD_complexes/. This interface presents both comparisons of intensity of signal for a given protein in two conditions, but also a derived measure that takes into account the relative abundance of proteins and adjust the intensity values accordingly. The "enrichment" value calculated this way allows a better discrimination of specific from nonspecific associations (for an example, *see* Fig. 1a).

Even small protocol changes can have a dramatic effect on the result of an affinity purification experiment. For example, we had several failed experiments due to decreased specific activity of the protease used to elute complexes under mild conditions. Thus, we

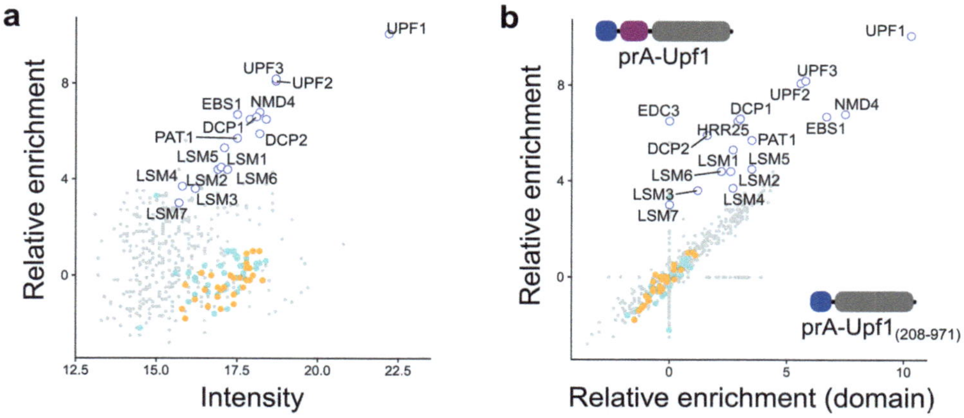

Fig. 1 Examples of data visualization for quantitative comparisons of proteins in purified samples. Data are from [13]. (**a**) Relative enrichment in the purified sample, vertical axis, corrects the obtained intensity values, horizontal axis, based on the abundance of each protein in a total extract. Labeled proteins are components of NMD complexes. Ribosomal proteins of the large subunit (cyan) and of the small subunit (orange) are also shown. (**b**) Comparison of enrichment to assess the involvement of a protein domain in the association of partner proteins in NMD complexes. Two variants of tagged Upf1 were used, one lacking the cysteine, histidine rich (CH) N-terminal domain, for purification. The tag used for the purification is depicted in blue, with the CH domain in purple. The values on the axes were log2 transformed (a one-unit difference corresponds to a factor of 2). Proteins showing a displacement of signal toward the upper left quadrant are likely to depend on the missing domain for their association with the RNA helicase Upf1

favor more robust elution protocols. High concentrations of urea or guanidine were relatively inefficient in eluting complexes. However, high salt concentration treatment using 2 to 4 M $MgCl_2$ or denaturing agents, such as SDS coupled with heat, were effective. These methods are both safer and cheaper than using a specific protease, but cannot be used to elute complexes for further purifications or biochemical tests. Elution conditions will affect subsequent steps of protein digestion and liquid chromatography. To remove excess SDS from eluates, for example, one can use a methanol-chloroform precipitation, which strongly reduces the amount of detergent in the sample [14].

Affinity purification of complexes is especially informative when performed under a variety of conditions that test the constitution or dynamics of the studied multiprotein assembly. For example, the importance of RNA to an RNA-protein complex can be tested by adding an RNase, such as the micrococcal nuclease (see for instance [15]), during one of the incubation steps. Complex formation can be dissected by performing the purification in strains or cell lines in which one of the complex components is absent. Finally, the relative importance of protein domains for interactions can be deduced from purifications performed with truncated proteins, as illustrated in Fig. 1b.

The following sections provide detailed protocols for the purification of proteins fused with the original TAP tag [1], although a single purification step is used. They include the preparation of IgG-coupled magnetic beads, yeast cells lysis, the purification procedure, elution and eluate treatment protocols. Digestion of the proteins to peptides and an example of LC-MS/MS protocol is also provided.

2 Materials

2.1 Yeast Growth Medium

1. YPD medium: 2% D-glucose, 1% Difco yeast extract, 1% Difco Bacto peptone. Autoclave at 110 °C for 20 min.

2.2 Magnetic Beads Preparation

1. Dynabeads M-270 epoxy (Dynal).
2. Rabbit IgGs.
3. 0.1 M sodium phosphate, pH 7.4 (for 100 mL of 1 M stock: 2.62 g $NaH_2PO_4*H_2O$ and 14.42 g $Na_2HPO_4*2\ H_2O$).
4. Ammonium sulfate solution (19.82 g of $(NH_4)_2SO_4$ in 50 mL 0.1 M sodium phosphate buffer, pH 7.4).
5. Glycine-HCl (3.75 g glycine, 4.3 mL of 37% HCl, for 500 mL).
6. PBS—phosphate buffered saline (137 mM NaCl, 2.7 mM KCl, 10 mM Na_2HPO_4, 1.8 mM KH_2PO_4).
7. 0.1 M Triethanolamine, pH 8.8.

2.3 Cell Lysis and Purification

1. Lysis buffer: 20 mM HEPES K, 100 mM potassium acetate, 5 mM $MgCl_2$, protease inhibitor (e.g., Roche complete EDTA-free), pH 7.4.
2. Wash buffer: 20 mM HEPES K, 100 mM potassium acetate, 5 mM $MgCl_2$, 0.5% Triton X-100, pH 7.4.
3. Acid-washed glass beads, 425–600 μm.
4. SDS elution buffer: 10 mM Tris-HCl, 1 mM EDTA, 2% SDS, pH 8.
5. Methanol (mass-spectrometry grade).
6. Chloroform (mass-spectrometry grade).
7. FastPrep-24 or equivalent bead beater vortex with adapter for 50 mL tubes.
8. Magnetic stand for 1.5 mL Eppendorf tubes.

2.4 Chromatography and Mass Spectrometry

1. 0.1 M Tris-HCl, pH 8, with 8 M urea.
2. 0.5 M TCEP (tris(2-carboxyethyl)phosphine): Dilute 1/5 in 0.1 M Tris-HCl, pH 8, 8 M Urea, as working solution.

3. 0.5 M Iodoacetamide (IAA) or chloroacetamide (ClAA) solutions (to be prepared freshly in 0.1 M Tris-HCl pH 8, 8 M urea).
4. 0.5 M $CaCl_2$ stock Solution in 0.1 M Tris-HCl pH 8.
5. Endoproteinase Lys-C: 20 μg are diluted in 100 μL of 0.1 M Tris-HCl pH 8, 8 M Urea (0.2 μg/μL). 20 μL aliquots are stored at −80 °C.
6. Trypsin Gold MS Grade: 100 μg are diluted in 500 μL 0.1 M Tris-HCl pH 8 (0.2 μg/μL). 20 μL aliquots are stored at −80 °C.
7. 0.1% trifluoroacetic acid (TFA).
8. Chromatography solvent formulation A: 0.1% formic acid/ 99.9% water (v/v).
9. Chromatography solvent formulation B: 0.1% formic acid/ 80% acetonitrile/19.9% water (v/v/v).
10. SpeedVac, Thermomixer or equivalent devices.
11. Omics C18 100 μL tips (Agilent).
12. EASY-Spray column (C18, particle size 2 μm, length 150 mm, pore size 100°A, diameter 75 μm), or equivalent.
13. EASY-nLC 1000 chromatography column, or equivalent.

3 Methods

3.1 Preparation of IgG Coupled Superparamagnetic B-eads

1. Wash 300 mg beads (about 2 × 10^{10} beads) with 20 mL of 0.1 M sodium phosphate buffer, pH 7.4. Vortex 30 s and incubate 10 min at room temperature. Recover the beads with a magnetic support (about 2 min). Repeat the washing. Suspend the beads in 4.5 mL sodium phosphate buffer, pH 7.4.
2. Thaw 1.5 mL rabbit IgGs (*see* **Note 1**) stock solution (10 mg/ mL in PBS) and add it to the 4.5 mL of magnetic beads suspension. Slowly add 3 mL of 3 M ammonium sulfate solution. Incubate on a spinning wheel at 30 °C for 20 h.
3. The next day, recover the beads on a magnetic support. Wash with 7.5 mL PBS, followed by 1.5 mL glycine-HCl 0.1 M, pH 2.5 buffer.
4. Wash sequentially with 7.5 mL each of the following solutions: PBS; 0.1 M Triethanolamine, pH 8.8; PBS; PBS with 0.5% Triton X100; PBS.
5. Store the IgG-beads preparation as a suspension in 1.5 mL PBS containing 0.01% Thimerosal at 4 °C. This corresponds to an approximate concentration of beads of 0.2 mg/mL. This preparation should be sufficient for about 70 purifications (*see* **Note 2**).

3.2 Preparation of a Total Yeast Extract

1. Prepare an initial culture of yeast expressing the protein of interest in fusion with the TAP tag by seeding the appropriate strain in the evening in 200 mL of YPD medium. In the next morning, dilute the saturated culture, grown at 30 °C with agitation, to an approximate optical density at 600 nm of 0.4 in 2 to 4 L of YPD. Allow to grow for 4 to 6 h until the optical density reaches an approximate value of 2.
2. Recover cells by centrifugation for 10 min at 3500 × *g* in a large volume centrifuge. Wash the cells with 30 mL cold water and transfer to a 50 mL tube. Centrifuge again for 10 min at 3500 × *g*, at 4 °C, and remove the supernatant. Cells can be stored at −80 °C until lysis or used directly. Typical yields are 3 to 5 g cells.
3. Suspend cells in an equal volume of lysis buffer by thorough pipetting. Transfer to a Falcon tube containing 20 mL of acid-washed glass beads. Perform a series of three 40 s bead beating in the FastPrep (6 m/s), with chilling intervals of 2 min in between.
4. Recover the cell lysate by centrifugation for 20 min at 3500 × *g*. Its appearance should be pale yellow. Keep a small sample of total extract at −20 °C, to be able to control the presence of the tagged protein and the total protein concentration.

3.3 Purification of Tagged Proteins

1. Prepare IgG-magnetic beads for purification by washing, for each purification, 20 μL of stock beads suspension, with 200 μL lysis buffer. Resuspend in lysis buffer.
2. Mix the total cell lysate that contains a TAP-tagged protein with the IgG-magnetic beads suspension. Allow binding to proceed between 20 and 120 min on a rotating wheel at 4 °C. Preliminary tests might be required to establish the optimal incubation period.
3. Remove the magnetic beads from solution using a magnetic stand. Suspend the beads in 1 mL lysis buffer and transfer to 1.5 mL Eppendorf tubes. Remove the supernatant.
4. Wash the beads five times with 1 mL each of wash buffer. Each time, resuspend the beads in the buffer.
5. Wash again the beads with 1 mL buffer without detergent. Add 50 μL SDS elution buffer (*see* **Note 3**) and resuspend the beads. Incubate 10 min at 65 °C. Carefully recover the eluate and proceed to performing controls of purification efficiency by denaturing polyacrylamide gels, immunoblots or follow up with precipitation for mass spectrometry analysis.

3.4 Protein Precipitation with Methanol-Chloroform (See Note 4)

1. Adjust the eluate volume to 100 μL with water. Add 400 μL high purity methanol and vortex thoroughly.
2. Add 100 μL chloroform and vortex.
3. Add 300 μL water. Mix thoroughly for 1 min. Mixing well at this step is important for phase separation. The solution becomes turbid.
4. Centrifuge 30 min at 14,000 × *g* at 4 °C. For eluates rich in protein a white precipitate ring can be observed at the interface of the two phases.
5. Remove upper phase (methanol and water) and leave in the tube the lower phase (chloroform). Pay attention to leave a small amount of the upper phase and not to remove the precipitate.
6. Add 300 μL methanol and mix thoroughly. Leave 10 min at −20 °C. Centrifuge for 20 min at 14,000 × *g* and 4 °C. Remove the supernatant. Sometimes, a white precipitate can be observed at the bottom of the tube or on the tube wall.
7. Leave for about 5 min at room temperature to allow evaporation of trace amounts of chloroform and methanol. Store at −20 °C until processing for mass spectrometry.

3.5 Proteolytic Lysis of Proteins for Mass-Spectrometry

3.5.1 Reduction and Alkylation

1. Dilute each sample in 30 μL of 0.1 M Tris-HCl, pH 8; 8 M Urea. Vortex.
2. Add 1.5 μL 0.1 M TCEP and leave at room temperature (20–22 °C) for 30 min.
3. Add 0.6 μL of freshly prepared 0.5 M IAA (or ClAA). Leave at room temperature for 30 min in the dark.

3.5.2 Protein Digestion

1. For each sample, add the endoproteinase Lys-C (5 μL) at 0.2 μg/μL in 0.1 M Tris-HCl, pH 8; 8 M urea. Incubate at 37 °C for at least 6 h.
2. Add to each sample 94 μL 100 mM Tris-HCl, pH 8, 0.5 μL of $CaCl_2$ (0.5 M stock solution) and 2.5 μL of stock trypsin at 0.2 μg/μL. Incubate at 37 °C under agitation overnight, with a Thermomixer device.
3. Repeat the addition of $CaCl_2$ and trypsin and incubate at 37 °C with stirring, using a Thermomixer, for 5 h.
4. Add 6 μL of formic acid for a final concentration of 5%.

3.5.3 Peptides Desalting

1. Desalting uses OMIX C18 100 μL tips. Perform 5 washes of the OMIX support with pure acetonitrile followed by 5 washes of the OMIX support with 0.1% TFA solution.
2. Load each sample by performing 20 back and forth pipetting through the tips.

3. Perform 7 washes of each sample with 150 μL 0.1% TFA solution, each.
4. To elute peptides, perform 20 back and forth pipetting in each tip with 150 μL of a mix in equal volumes of acetonitrile and 0.1% TFA.
5. Finally, rinse the tip with 150 μL of a 3/1 volume mix of acetonitrile and 0.1% TFA solution. Further processing requires dried material, that can be obtained with a SpeedVac. Store the eluted peptides at −20 °C.

3.6 Liquid Chromatography Coupled with Mass Spectrometry

1. Add 12 μL LC-MS buffer: water/acetonitrile/TFA (96/4/0.1, volume to volume mix). Mix well. Centrifuge at maximum speed for 10 min at 4 °C. Transfer 10 μL to the LC-MS injection vial. Store at 4 ° C until the start of the analysis.
2. Liquid chromatography analysis is performed with a specific capillary column, such as the EASY-nLC 1000. For each sample, 3 μL of peptide mixtures are loaded on an EASY-Spray column. Use a column temperature of 50 °C at a flow rate of 250 nL/min.
3. A linear gradient over 120 min is used for gradual peptide elution during each run. An example of how the gradient can be set up is the following:

Time (min)	Duration (min)	Mixture [%B]
0	0	4
70	70	25
95	25	45
100	5	95
105	5	95
105	0	4
120	15	4

4. For LTQ-Orbitrap Velos acquisition, full MS scans are acquired in the Orbitrap mass analyzer over the m/z 300–1700 range with resolution 60,000 (m/z 400). The target value is 5×10^5. The LTQ Orbitrap is operated in a data-dependent mode, switching automatically between one full-scan and subsequent MS/MS scans of the 20 most abundant peaks (Top20 method by CID). Ions with charge state ≥2 are selected for sequencing and fragmented in the ion trap with normalized collision energy of 35%, activation $Q = 0.25$, activation time of 10 ms,

and one microscan. The target value is 10^4. The ion selection threshold is 500 counts. Dynamic exclusion is enabled; with an exclusion list size of 500 and an exclusion duration of 30 s. Standard MS parameters are set as follows: 1.9 kV spray voltage and 275 °C heated capillary temperature.

3.7 Identification of Proteins and Raw Quantitation of Results

1. Processing of RAW output from the mass spectrometer can be done with MaxQuant with default parameters unless otherwise specified (http://www.maxquant.org). Database searches are performed with the built-in *Andromeda* search engine against a reference proteome (in our case the *S. cerevisiae* one recovered from Uniprot). MaxQuant provides a "contaminants. fasta" file including common laboratory contaminant proteins that is automatically added to the list of proteins for the in silico digestion.
2. Precursor mass tolerance is set to 6 ppm in the main search, and fragment mass tolerance is set to 0.5 Da (these parameters depend on the specific mass spectrometer). Digestion enzyme specificity is set to trypsin with a maximum of two missed cleavages. A minimum peptide length of 7 residues is required for identification. Up to five modifications per peptide are allowed; acetylation (protein N-terminal), and oxidation (Met) are set as variable modifications. Carbamidomethylation (Cys) is set as a fixed modification. Peptide and protein false discovery rates (FDR) are both set to 1%.
3. Relative, label-free quantification of proteins can be done using the MaxLFQ algorithm integrated into MaxQuant. Typical parameters are as follows: Minimum ratio count set to 2, and the FastLFQ option enabled. Proteins that shared same identified peptides are combined into a single protein group (*see* **Note 5**).

3.8 Interpretation of Quantitative MS Results

1. To identify proteins that interact with a tagged protein of interest in the presence of a vast number of background binding proteins, replicates of affinity-enriched bait samples are compared to a set of negative control samples. The tools used to perform these comparisons range from spreadsheet calculations to R packages or specific visualization and data analysis programs, such as the Perseus environment. Here, we describe a sequence of operations, using Perseus, for the analysis of MaxQuant output.
2. The "proteingroups.txt" file from MaxQuant is loaded and then entries in the for "Reverse," "Only identified by site modification," and "Potential contaminants" are removed before further processing. Protein groups identified by a single "razor and unique peptide" are removed from the data set.

3. Protein group LFQ values are log2 transformed. Missing LFQ values are assumed to represent mostly low abundance proteins that are below the MS detection limit. Imputation of these missing values is performed separately for each sample and missing values are replaced with random low level values. Alternatively, imputation can be done by using a constant value equal to the lowest LFQ value for each run.
4. In the presence of at least three replicate experiments, differences in relative abundance of proteins in different conditions are evaluated using a *t*-test. *T*-test results can be visualized with volcano plots, in which the difference between two samples is represented on the horizontal axis and the *p*-value of the *t*-test for difference for each protein between sample and control is represented on an inverse log scale. Enriched interactors appear in the upper right section of the plot (*see* **Note 6**).

4 Notes

1. The specific use of rabbit IgG is important here for proteins fused with the Z domain of protein A, which is present in the original TAP tag. Binding of this domain to the Fc region of antibodies is type and species specific. Thus, human IgG1, IgG2 and IgG4 bind protein A, while goat IgGs, for example, have low affinity for protein A.
2. Beads can also be produced in the laboratory, by using mixtures of $FeCl_2/FeCl_3$ and silica coating and chemical modification, for price reduction and customization [16].
3. Elution can also be done in the presence of high salt concentrations, for example with 200 μL of 2 M or 4 M $MgCl_2$. For specific purification types, such as the streptavidin-biotin purification, elution might not be necessary if streptavidin was rendered resistant to trypsin and Lys-C protease digestion [17].
4. Methanol-chloroform precipitation was developed to allow extraction of proteins and their separation from lipids and detergent [18]. It is also effective in the presence of high salt concentrations, such as 2 or 4 M $MgCl_2$. However, precipitation methods can lead to irreversible protein aggregation or protein loss. Such events can be avoided by using other preparation methods for samples that contain SDS, such as eFASP [19], one of the variants of the filter aided sample preparation protocol.
5. MaxLFQ is only one of several scores that have been proposed and used to compare relative amounts of the same protein in different samples. We obtained similar results using a variant of

the Top3 method [20], in which only the three most intense peptides are used to compare protein amounts. In light of the continuous improvement of coverage and sensitivity of mass spectrometers, it is likely that the label-free quantification scores will also further evolve.

6. While volcano plots are frequently used for the visualization of quantification results based on mass spectrometry results, one should be aware that p-values themselves are random variables with a variance that can be important, especially if only three replicates have been performed. While a low variability of quantitative values obtained in several replicates for a given protein correlates with low p values (high on the vertical axis of usual volcano plot), the difference between sample and control should be the most important information extracted from a volcano plot. As stated in [21] "no p-value can reveal the plausibility, presence, truth, or importance of an association or effect." This analysis of the use of p-values also suggest that the term "statistically significant" should be banned from scientific communication.

Acknowledgments

We are grateful to the Pasteur Institute Proteomic platform for providing access to a LC-MS/MS system. We thank Laurence Decourty and Marine Dehecq who performed the purifications for the results presented here and were involved in setting up the experimental conditions. Large-scale developments were made possible by the ANR funding of the CLEANMD and DEFineNMD grants (ANR-14-CE10-0014, ANR-18-CE11-0003).

References

1. Rigaut G, Shevchenko A, Rutz B et al (1999) A generic protein purification method for protein complex characterization and proteome exploration. Nat Biotechnol 17:1030–1032. https://doi.org/10.1038/13732
2. Biedka S, Wu S, LaPeruta AJ et al (2017) Insights into remodeling events during eukaryotic large ribosomal subunit assembly provided by high resolution cryo-EM structures. RNA Biol 14:1306–1313. https://doi.org/10.1080/15476286.2017.1297914
3. Halbeisen RE, Scherrer T, Gerber AP (2009) Affinity purification of ribosomes to access the translatome. Methods 48:306–310. https://doi.org/10.1016/j.ymeth.2009.04.003
4. Oeffinger M, Wei KE, Rogers R et al (2007) Comprehensive analysis of diverse ribonucleoprotein complexes. Nat Methods 4: 951–956. https://doi.org/10.1038/nmeth1101
5. Nilsson B, Moks T, Jansson B et al (1987) A synthetic IgG-binding domain based on staphylococcal protein A. Protein Eng 1:107–113. https://doi.org/10.1093/protein/1.2.107
6. Makarov A (2000) Electrostatic axially harmonic orbital trapping: a high-performance technique of mass analysis. Anal Chem 72: 1156–1162. https://doi.org/10.1021/ac991131p
7. Hebert AS, Richards AL, Bailey DJ et al (2013) The one hour yeast proteome. Mol Cell Proteomics 13:339–347. https://doi.org/10.1074/mcp.M113.034769

8. Tyanova S, Temu T, Sinitcyn P et al (2016) The Perseus computational platform for comprehensive analysis of (prote)omics data. Nat Methods 13:731–740. https://doi.org/10.1038/nmeth.3901
9. Cox J, Mann M (2008) MaxQuant enables high peptide identification rates, individualized p.p.b.-range mass accuracies and proteome-wide protein quantification. Nat Biotechnol 26:1367–1372. https://doi.org/10.1038/nbt.1511
10. Gierlinski M, Gastaldello F, Cole C, Barton GJ (2018) Proteus: an R package for downstream analysis of MaxQuant output. In: bioRxiv, p 416511. https://doi.org/10.1101/416511
11. Jimenez-Morales D, Rosa Campos A, Von Dollen J et al (2021) artMS: analytical R tools for mass spectrometry. R package version 1.12.0, http://artms.org.
12. Zhang X, Smits AH, van Tilburg GB et al (2018) Proteome-wide identification of ubiquitin interactions using UbIA-MS. Nat Protoc 13:530–550. https://doi.org/10.1038/nprot.2017.147
13. Dehecq M, Decourty L, Namane A et al (2018) Nonsense-mediated mRNA decay involves two distinct Upf1-bound complexes. EMBO J 37:e99278. https://doi.org/10.15252/embj.201899278
14. Arand M, Friedberg T, Oesch F (1992) Colorimetric quantitation of trace amounts of sodium lauryl sulfate in the presence of nucleic acids and proteins. Anal Biochem 207:73–75. https://doi.org/10.1016/0003-2697(92)90502-x
15. Choi YD, Dreyfuss G (1984) Isolation of the heterogeneous nuclear RNA-ribonucleoprotein complex (hnRNP): a unique supramolecular assembly. Proc Natl Acad Sci 81:7471–7475. https://doi.org/10.1073/pnas.81.23.7471
16. Oberacker P, Stepper P, Bond DM et al (2019) Bio-on-magnetic-beads (BOMB): open platform for high-throughput nucleic acid extraction and manipulation. PLoS Biol 17: e3000107. https://doi.org/10.1371/journal.pbio.3000107
17. Rafiee M-R, Sigismondo G, Kalxdorf M et al (2020) Protease-resistant streptavidin for interaction proteomics. Mol Syst Biol 16:e9370. https://doi.org/10.15252/msb.20199370
18. Wessel D, Flügge UI (1984) A method for the quantitative recovery of protein in dilute solution in the presence of detergents and lipids. Anal Biochem 138:141–143. https://doi.org/10.1016/0003-2697(84)90782-6
19. Erde J, Loo RRO, Loo JA (2014) Enhanced FASP (eFASP) to increase proteome coverage and sample recovery for quantitative proteomic experiments. J Proteome Res 13:1885–1895. https://doi.org/10.1021/pr4010019
20. Ahrné E, Molzahn L, Glatter T, Schmidt A (2013) Critical assessment of proteome-wide label-free absolute abundance estimation strategies. Proteomics 13:2567–2578. https://doi.org/10.1002/pmic.201300135
21. Wasserstein RL, Schirm AL, Lazar NA (2019) Moving to a World Beyond "p < 0.05.". Am Stat 73:1–19. https://doi.org/10.1080/00031305.2019.1583913

Chapter 14

Deep Mutational Scanning of Protein–Protein Interactions Between Partners Expressed from Their Endogenous Loci In Vivo

Alexandre K. Dubé, Rohan Dandage, Soham Dibyachintan, Ugo Dionne, Philippe C. Després, and Christian R. Landry

Abstract

Deep mutational scanning (DMS) generates mutants of a protein of interest in a comprehensive manner. CRISPR-Cas9 technology enables large-scale genome editing with high efficiency. Using both DMS and CRISPR-Cas9 therefore allows us to investigate the effects of thousands of mutations inserted directly in the genome. Combined with protein-fragment complementation assay (PCA), which enables the quantitative measurement of protein–protein interactions (PPIs) in vivo, these methods allow for the systematic assessment of the effects of mutations on PPIs in living cells. Here, we describe a method leveraging DMS, CRISPR-Cas9, and PCA to study the effect of point mutations on PPIs mediated by protein domains in yeast.

Key words Deep mutagenesis scanning, CRISPR-Cas9, Protein-fragment complementation assay, Protein–protein interaction, Yeast

1 Introduction

Cells are complex assemblies of smaller parts interacting with each other. Tremendous resources have been invested in trying to understand how these parts, for instance proteins, interact together. Such efforts have led to the mapping of interactomes for different model organisms [1–4]. These reference interactomes allow us to further assess how mutations can perturb protein–protein interactions (PPIs). Such perturbations, also referred to as edgetic mutations [5], are a fundamental link between genotypes and phenotypes [6] and therefore inform us on their potential causal role in many human genetic disorders [5, 7, 8]. Understanding the effects of mutations on PPI can lead to new ways to tackle pathologies. Such

Frédéric Devaux (ed.), *Yeast Functional Genomics: Methods and Protocols*,
Methods in Molecular Biology, vol. 2477, https://doi.org/10.1007/978-1-0716-2257-5_14,

efforts could also potentially have important implications in the field of synthetic biology. For example, binding affinity could be tunable based on mutant effects, leading to an improved version of signaling pathways [9].

Deep mutational scanning (DMS) is a family of approaches used to generate all possible variants of a protein fragment or of a full protein of interest. Studies have used DMS to assess the impact of variants on various molecular properties, such as protein activity [10], abundance [11], PPIs [12, 13], and other molecular phenotypes [14]. One potential limitation of many studies is the use of a plasmid expression system. For example, plasmid copy number is heterogeneous in a cell population and as such can affect data interpretation [15]. One way to circumvent this problem is by genomic integration of the mutant library. This can be efficiently carried out by the CRISPR-Cas9 system, which has been shown to be an powerful tool to perform yeast genome editing. Different protocols for CRISPR-Cas9 based genomic integration of mutant libraries in yeast have been developed so far [16–18]. Mutant donor DNA integration is often used to remove the CRISPR-Cas9 PAM site, thus preventing further double-strand break. With a complex library of DMS mutants, removal of the PAM site is impossible to achieve consistently without introducing additional variants. Another approach is to target an artificial sequence (stuffer) inserted in your gene of interest [19].

Here, we combine these different methodologies to test the effect of mutation on PPIs [10]. We first generate a receptor strain (stuffed strain) in which we integrate a DMS library directly in the yeast genome using CRISPR-Cas9. In parallel, we generate a DMS library in bacteria using the megaprimer method [20]. After integration in the yeast genome, we perform a protein-fragment complementation assay (PCA) in bulk to evaluate the effect of mutations on a specific PPI [4, 21]. Overall, the method we describe can be performed easily and rapidly by one person for a sequence up to 180 base pairs generating 1200 protein mutants.

2 Materials

Unless stated, water use for media and solutions preparation is distilled, deionized water. All autoclave steps are for 20 min at 121 °C. Methotrexate is a controlled chemical; make sure to verify with your institution if and how you can access it.

2.1 Strain Preparation (Stuffing/Gene Deletion/Gene Tagging) (See Note 1)

1. pCAS plasmid (Addgene # 60847) [17].
2. KAPA HiFi HotStart polymerase.
3. Mutagenesis primers.
4. Bacteria (*Escherichia coli*) competent cells.

5. 2YT: 10 g/L yeast extract, 16 g/L tryptone, 2 g/L glucose, and 5 g/L NaCl. Autoclaved. For solid 2YT media, add 20 g/L agar before autoclaving. For 2YT + KAN media, dilute kanamycin to reach a final concentration of 50 μg/mL. Kanamycin is added after the autoclave run (solid media) when media is tempered or just before inoculation of the media (liquid).
6. 50 mg/mL kanamycin. Filter-sterilize, aliquot in 1 mL tube and store at −20 °C.
7. DpnI.
8. Miniprep columns.
9. Stuffing primers; 2 forward, 2 reverse.
10. Yeast strain from the Yeast Protein Interactome (Horizon Discovery) (*see* **Note 2**).
11. YPD: 10 g/L yeast extract, 20 g/L tryptone, 20 g/L glucose. Autoclaved. For solid media, add 20 g/L agar before autoclaving. For YPD + NAT media, dilute Nourseothricin to reach a final concentration of 100 μg/mL. For YPD + G418 media, dilute G418 to reach a final concentration of 200 μg/mL. Nourseothricin or G418 is added after autoclave run (solid media) when media is tempered or just before inoculation of the media (liquid).
12. 100 mg/mL Nourseothricin. Filter-sterilize, aliquot in 1 mL tube and store at −20 °C.
13. 200 mg/mL G418. Filter-sterilize, aliquot in 1 mL tube and store at −20 °C.
14. 10× TE buffer: 100 mM Tris base, 10 mM EDTA, pH 8.0
15. LATE: 0.1 M Lithium Acetate, 1× TE buffer. Store at room temperature for up to 6 months [22].
16. PLATE: 40% polyethylene glycol (PEG) 3350, 0.1 M lithium acetate, 1X TE buffer. Prepare fresh before transformation [22].
17. 50% glycerol, autoclave to sterilize.
18. Carrier DNA: 10 mg/mL salmon sperm DNA, 1× TE buffer (*see* **Note 3**). With a syringe, pass 10 times (10 in and 10 out) the solution in a 23G needle. Boil for 10 min, aliquot in 1 mL tube and store at −20 °C. Before use, DNA carrier is boiled for 5 min and incubated on ice another 5 min.
19. 1.5% agarose gel.

2.2 Library Preparation

For KAPA HiFi HotStart Polymerase, Agarose gel, bacteria competent cells (*see* Subheading 2.1).

1. Row-Column primers.
2. Plate primers.

3. Magnetic beads.
4. 10 mM Tris-HCl, pH 8.0, Autoclaved

2.3 Yeast CRISPR-Cas9 Genomic Integration

For LATE, PLATE, 50% glycerol, YPD, G418, nourseothricin, and DNA carrier 10 mg/mL (*see* Subheading 2.1).

2.4 Liquid PCA

For YPD and nourseothricin (*see* Subheading 2.1).

1. 100 mg/mL Hygromycin B.
2. YPD + 1.8× NAT + 1.8× HPH: YPD, 180 μg/mL nourseothricin, 450 μg/mL hygromycin B.
3. SC complete (MSG), pH 6.0: 1.74 g/L yeast nitrogen base without ammonium sulfate without amino acids, 1 g/L monosodium glutamate salt (MSG), 20 g/L glucose, 10 g/L succinic acid, 6 g/L NaOH, 1.34 g/L in house drop-out complete. Autoclave.
4. SC complete (MSG) pH 6.0 + 1.1× NAT + 1.1× HPH media (*see* **Note 4**): SC complete (MSG) pH 6.0, 110 μg/mL nourseothricin, 275 μg/mL hygromycin B. Nourseothricin, and Hygromycin B are added just before inoculation of the media (liquid).
5. Drop-out complete: Combine in a mortar 0.5 g adenine, 2 g L-arginine, 2 g L-aspartic acid, 2 g L-glutamic acid, 2 g L-histidine, 10 g L-leucine, 2 g L-lysine, 2 g L-methionine, 2 g L-phenylalanine, 2 g L-serine, 2 g L-threonine, 2 g L-tryptophan, 2 g L-tyrosine, 2 g L-valine, and 2 g uracil. Mix with the pestle to homogenize the powder and store in a 50 mL tube. Use at 1.34 g/L for SC complete media.
6. 10 mg/mL Methotrexate in DMSO (*see* **Note 5**). Rinse the bottle with an aliquot of DMSO to ensure getting all the methotrexate powder dissolved.
7. 10× Liquid drop-out -ade-lys-met: 0.4 g/L L-tryptophan, 0.3 g/L L-tyrosine, 0.5 g/L L-phenylalanine, 1.0 g/L L-glutamic acid (monosodium salt), 1.0 g/L L-asparagine, 1.5 g/L L-valine, 2.0 g/L L-threonine, 3.75 g/L L-serine, 0.2 g/L Uracil, 0.2 g/L L-histidine HCl, 0.2 g/L L-arginine HCl, 0.6 g/L L-leucine. Filter-sterilize and store in a bottle at 4 °C (*see* **Note 6**).
8. 40% Glucose (*see* **Note 7**). Heating the solution is necessary to reach complete dissolution. Autoclave.
9. PCA -ade-lys-met +200 μg/mL methotrexate: 6.69 g/L yeast nitrogen base without ammonium sulfate without amino acids, 1× Liquid drop-out -ade-met-lys, 200 μg/mL methotrexate. Autoclave first the Yeast nitrogen base solution (83% of the desired final volume). Once solution is tempered, add 1/10th

final volume of 10× Liquid drop-out -ade-met-lys, 1/20th final volume of 40% Glucose and 1/50th final volume 10 mg/mL methotrexate (*see* **Note 8**).

3 Methods

3.1 Strain Preparation

1. The mutagenesis primer is designed as follows; the gRNA of interest is flanked in 5′ with CGGGTGGCGAATGGGACTTT and in 3′ with GTTTTAGAGCTAGAAATAGC (*see* **Notes 9** and **10**). The reverse primer is simply the reverse complement of the mutagenesis primer.
2. Set up the following PCR reaction (*see* **Note 11**).

Template DNA 10 ng/μL	0.75 μL
5× Buffer	5 μL
10 mM dNTPs	0.75 μL
10 μM Forward primer	0.75 μL
10 μM Reverse primer	0.75 μL
KAPA polymerase 1 U/μL	0.5 μL
PCR grade water	16.5 μL

3. Run the following PCR cycle.
   (a) 3 min 98 °C
   (b) 20 cycles
   (c) 20 s 98 °C
   (d) 15 s 61 °C
   (e) 10 min 72 °C
   (f) Final elongation.
   (g) 15 min 72 °C.
4. Remove template DNA from your PCR product by adding 0.5 μL DpnI (20,000 units/mL) directly in the PCR mix (*see* **Note 12**). Incubate in a water bath at 37 °C for 90 min.
5. Transform 5 μL of the digested PCR product in competent *E. coli* cells (*see* **Note 13**) and spread on 2YT + KAN petri dish. Incubate overnight at 37 °C.
6. Start four precultures in 5 mL 2YT + KAN using independent colonies (*see* **Note 14**). Cultures are incubated at 37 °C with 250 rpm for 16 h.

7. Extract plasmid from each culture and Sanger sequence the gRNA region with the following primer: CGGAATAG GAACTTCAAAGCG (*see* **Note 15**).
8. Set up the following PCR reaction twice.

Template DNA 10 ng/μL	0.75 μL (*see* **Note 16**)
5× Buffer	5 μL
10 mM dNTPs	0.75 μL
10 μM Forward primer	0.75 μL
10 μM Reverse primer	0.75 μL
KAPA polymerase 1 U/μL	0.5 μL.
PCR grade water	16.5 μL.

In the first reaction, the reverse primer has an overhang containing ACCACCACCTCCAGAACTTCCGCC sequence and in the second reaction, the forward primer has the overhang containing GGCGGAAGTTCTGGAGGTGGTGGT sequence (Fig. 1).

9. Run the following PCR cycle.
   (a) 3 min 98 °C
   (b) 35 cycles
   (c) 20 s 98 °C
   (d) 15 s 60 °C
   (e) 30 s 72 °C
   (f) Final elongation.
   (g) 1 min 72 °C.
10. Run PCR product on a 1.5% agarose gel. Usually 1 μL of PCR product is enough to produce bright bands. The size of the PCR will depend on the primer design.
11. Dilute both PCR products 1:20 in the same tube (1 μL PCR1: 1 μL PCR2:18 μL water).
12. Set up the following PCR reaction.

PCR dilution 1:20	0.75 μL
5× Buffer	5 μL
10 mM dNTPs	0.75 μL
10 μM Forward primer	0.75 μL
10 μM Reverse primer	0.75 μL
KAPA polymerase 1 U/μL	0.5 μL
PCR grade water	16.5 μL

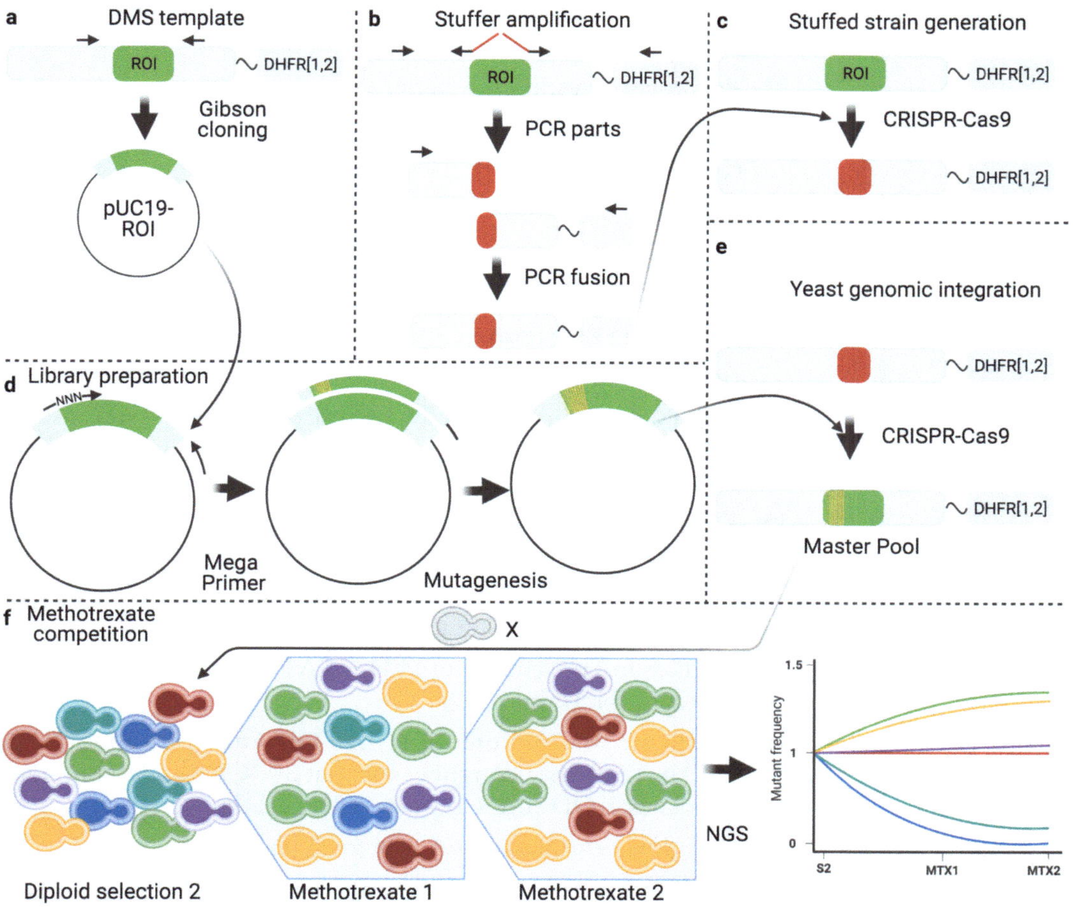

Fig. 1 Overview of the protocol for the integration of a library of mutant fragments into the yeast genome. (**a**) Deep mutational scanning (DMS) template generation. The region of interest (ROI) is amplified from the genome and cloned in a plasmid for bacterial cloning, here pUC19. (**b**) Stuffer amplification to remove ROI. Two fragments are amplified from the genome. The primers flanking ROI containing an overhang composed of the stuffer (GGCGGAAGTTCTGGAGGTGGTGGT) produce fragments that are fused by PCR. The resulting product contains long homology regions on each side of the ROI. (**c**) The ROI is replaced by a stuffer in the yeast genome using CRISPR-Cas9. The PCR fusion generated in (b) is used as a repair template to remove the ROI. (**d**) Library preparation using the megaprimer method. First, mutagenized PCR products are generated using a degenerate primer. A second step PCR is performed to integrate the mutations in the vector. Finally, plasmids containing the mutagenized fragments are recovered from bacteria and used to amplify the DMS library for yeast integration. (**e**) Libraries amplified in (d) are integrated in the stuffed strain generated in (c) using CRISPR-Cas9. The yeast colonies recovered at this step are pooled together to obtain the master pool. (**f**) The master pool is mated with prey of interest. Diploid cells then compete for two rounds in selective methotrexate (MTX) media. Growth in MTX media is the result of a protein–protein interaction between the bait and the prey. Mutation in your ROI stabilizing the interaction will increase in frequency and those destabilizing it will be eliminated from the population. Genomic DNA is extracted from diploid selection and the second MTX growth cycle. Genomic DNA is used to prepare next-generation sequencing libraries and sequenced on an Illumina platform. Mutant frequency is then compared between the last diploid selection and the last competition growth in MTX media

The primers are the forward and reverse from **step 3.1.8**, that do not have an overhang (Fig. 1).

13. Run the following PCR cycle.
    (a) 3 min 98 °C
    (b) 35 cycles
    (c) 20 s 98 °C
    (d) 15 s 60 °C
    (e) 30 s 72 °C
    (f) Final elongation.
    (g) 1 min 72 °C.
14. Run PCR product on a 1.5% agarose gel. Usually 1 μL of PCR product is enough to produce bright bands. The size of the PCR should be the addition of the first two PCRs.
15. Keep PCR product at −20 °C.
16. Retrieve from Yeast interaction collection the bait strain (XXX-DHFR [1, 2]) of interest and strike it on a YPD + NAT plate (*see* **Note 17**). Let grow for 3 days at 30 °C.
17. Start a 5 mL pre culture in YPD + NAT and grow at 30 °C with 250 rpm for 16 h.
18. Dilute the strain of interest at 0.15 OD/mL in 25 mL of fresh YPD in a 50 mL falcon tube. Grow at 30 °C with 250 rpm until it reaches 1 OD/mL (*see* **Note 18**).
19. Spin down the cells at 900 × *g* for 2 min.
20. Remove supernatant, resuspend the pellet in 500 μL of LATE solution and transfer to a 1.5 mL tube.
21. Spin down the cells at 9300 × *g* for 1 min.
22. Remove supernatant and resuspend pellet in 150 μL of LATE solution.
23. Add 150 μL of 50% glycerol, vortex and freeze at −80 °C (*see* **Note 19**).
24. In a 1.5 mL tube combine 90 μL of competent cells, 8 μL of stuffer PCR, 10 μL of DNA carrier 10 mg/mL, 500 ng of pCAS-gRNA and 900 μL of PLATE solution, mix with a vortex (*see* **Note 20**).
25. Incubate 30 min in a water bath set at 30 °C.
26. Heat shock the cells for 15 min in a water bath set at 42 °C.
27. Spin down the cell as in **step 3.1.21**.
28. Remove supernatant and resuspend the cells in 250 μL of YPD. Incubate at 30 °C for 4 h.
29. Spread 75 μL on a YPD + NAT + G418 Petri dish and let it grow for 3 days at 30 °C (*see* **Note 21**).

30. Grow 4 independent colonies in 200 μL of YPD for 16 h without shaking (*see* **Note 22**).
31. Streak on a YPD Petri dish 5 μL of each culture and grow for 2 days at 30 °C.
32. Select 2 colonies in each striation and resuspend them in sterile water. Spot 4 μL of each cell resuspension on YPD, YPD + NAT and YPD + G418. Grow for 2 days at 30 °C (*see* **Note 23**).
33. Use colony PCR to confirm stuffing of the region of interest. Also, Sanger sequence the PCR product to make sure nothing else has happened in your gene of interest.

3.2 Library Preparation

1. The primers to generate the megaprimer (*see* **Note 24**) are designed as follows. The forward mutagenic primer is composed of 13 bp before the codon of interest, the codon of interest replaced by NNN, and 14 bp after the codon of interest for a total length of 30 bp. The reverse primer is a generic primer that can be used with all forward mutagenic primers, usually annealing in your backbone.
2. Set up the following PCR reaction.

Plasmid 15 ng/μL	0.45 μL
5× buffer	3 μL
10 mM dNTPs	0.45 μL
10 μM forward primer	0.45 μL
10 μM reverse primer	0.45 μL
KAPA polymerase 1U/μL	0.1 μL
PCR grade water	10.1 μL

3. Run the following PCR cycle.
   (a) 3 min 98 °C
   (b) 35 cycles
   (c) 20 s 98 °C
   (d) 15 s 60 °C
   (e) 30 s 72 °C
   (f) Final elongation.
   (g) 1 min 72 °C.
4. Run PCR product on a 1.5% agarose gel. Usually 1 μL of PCR product is enough to produce bright bands. When PCR are loaded sequentially starting from codon 1, you should see a decrease in PCR product sizes.

5. For mutagenesis, set up the following PCR reaction:

Plasmid 15 ng/μL	0.45 μL
5× buffer	3 μL
10 mM dNTPs	0.45 μL
Megaprimer	0.45 μL (*see* **Note 25**)
KAPA polymerase 1 U/μL	0.1 μL
PCR grade water	10.55 μL

You should set at least one reaction in which you won't put any megaprimer as a negative control.

6. Run the following PCR cycle:
   (a) 3 min 98 °C.
   (b) 20 cycles.
   (c) 20 s 98 °C.
   (d) 15 s 68 °C.
   (e) 3 min 30 s 72 °C.
   (f) Final elongation.
   (g) 5 min 72 °C.
7. After the PCR, add 0.35 μL of DpnI (20,000 units/mL) in each of your mutagenesis PCR. Make sure to also add it into your negative control. Put back the PCR + DpnI in the thermocycler for 90 min at 37 °C.
8. Transform 7.5 μL of PCR + DpnI in *E. coli* competent cells using your favorite transformation protocol (*see* **Note 26**).
9. Incubate for 16 h at 37 °C on the appropriate selection media (*see* **Note 27**).
10. Once colonies have grown, resuspend the cells by spreading 5 mL 2YT media on the Petri dish using a sterile glass rake. Transfer the cell suspension in a 15 mL tube (*see* **Note 28**).
11. Extract plasmid DNA with your favorite miniprep kit and quantify all extractions using a NanoDrop.
12. In parallel, prepare a glycerol stock of each position by mixing 810 μL cell suspension and 190 μL 80% Glycerol and freeze at −80 °C.
13. For deep sequencing [23], primers to add Primer-site (PS) are designed as follows. The forward primer is composed of CCA TACGAGCACATTACGGG-18 bp sequence just before the region of interest and the reverse primer is CTTGACTGAGCGACTGAGG-reverse complement of the 18 bp sequence just after your region of interest (*see* **Note 29**) (Fig. 2).

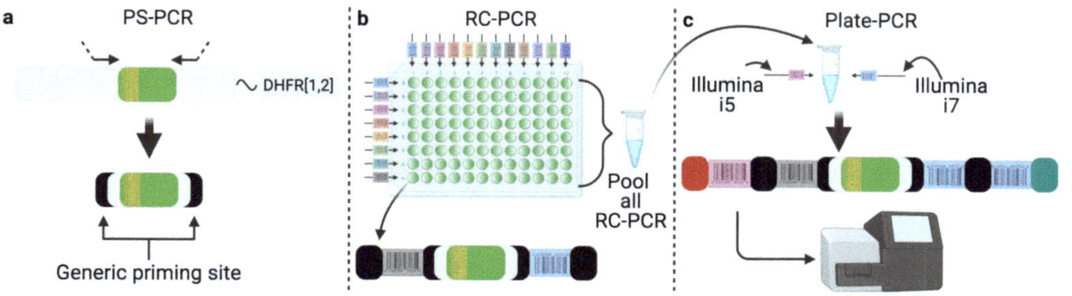

Fig. 2 Next-generation sequencing of the region of interest using the Row-Column PCR protocol. (**a**) The region of interest (ROI) is amplified using flanking primers. Each primer contains an overhang with generic priming sites (PS). (**b**) Primers with a first series of barcodes are used to amplify ROI from (a). Combination of 12 column primers and 8 row primers allows for the barcoding of a full 96-well plate, that is, 96 different samples. The 96 PCR products are then mixed uniformly and purified on magnetic beads. (**c**) Primers with a second series of barcodes are used to amplify the PCR pool generated in (b). These primers also contain the Illumina adaptor to perform next-generation sequencing. For instance, combinations of three forward primers and three reverse primers will allow to barcode nine 96-well plates, for a total of 864 different samples

14. Dilute each miniprep at a final concentration of 1 ng/μL using PCR grade water (*see* **Note 30**).
15. Set up the following PCR reaction:

Plasmid 1 ng/μL	4.5 μL
5× buffer	3 μL
10 mM dNTPs	0.45 μL
10 μM forward PS primer	0.45 μL
10 μM reverse PS primer	0.45 μL
KAPA polymerase 1 U/μL	0.1 μL
PCR grade water	6.05 μL

16. Run the following PCR cycle:
    (a) 3 min 98 C
    (b) 20 cycles
    (c) 30 s 98 °C
    (d) 15 s 60 °C
    (e) 30 s 72 °C
    (f) Final elongation.
    (g) 1 min 72 °C.
17. Run PS-PCR product on a 1.5% agarose gel. Usually 1 μL of PS-PCR product is enough to give a bright signal.
18. Dilute PS-PCR 1/2500 in PCR grade water.

19. Set up the following PCR reaction:

PS-PCR 1/2500	4.5 μL
5× Buffer	3 μL
10 mM dNTPs	0.45 μL
10 μM Row primer	0.45 μL
10 μM Column primer	0.45 μL
KAPA polymerase 1 U/μL	0.1 μL
PCR grade water	6.05 μL

See Fig. 2 for primer description. A list of the Row and Column primers can be found in Table 1.

20. Run the following PCR cycle:
    (a) 3 min 98 °C
    (b) 18 cycles
    (c) 30 s 98 °C
    (d) 15 s 60 °C
    (e) 30 s 72 °C
    (f) Final elongation.
    (g) 1 min 72 °C.
21. Run RC-PCR product on a 1.5% agarose gel. Usually 1 μL of RC-PCR product is enough to produce bright bands. RC-PCR products are quantified according to their band intensity.
22. Pool RC-PCR based on the quantification done on your agarose gel.
23. Purify RC-PCR pool on magnetic beads and elute in 10 mM Tris-HCl, pH 8.0 (*see* **Note 31**). Quantify your purified RC-PCR pool with a NanoDrop.
24. Dilute the purified RC-PCR pool at 0.01 ng/μL in PCR grade water.
25. Set up the following PCR reaction:

RC-PCR 0.01 ng/μL	1.8 μL
5× buffer	12 μL
10 mM dNTPs	1.8 μL
10 μM forward plate primer	1.8 μL
10 μM reverse plate primer	1.8 μL
KAPA polymerase 1 U/μL	0.4 μL
PCR grade water	40.4 μL

Table 1
The Row and Column primers

Primer name	Sequence
DB_BC_Row_1	TAACTTACGGAGTCGCTCTACGCAAGTGTTCCCATACGAGCACATTACGGG
DB_BC_Row_2	TAACTTACGGAGTCGCTCTACGAGGACATTCCCATACGAGCACATTACGGG
DB_BC_Row_3	TAACTTACGGAGTCGCTCTACGCACTAATGGCCATACGAGCACATTACGGG
DB_BC_Row_4	TAACTTACGGAGTCGCTCTACGAGCCTGATGCCATACGAGCACATTACGGG
DB_BC_Row_5	TAACTTACGGAGTCGCTCTACGTTACGCTAACCATACGAGCACATTACGGG
DB_BC_Row_6	TAACTTACGGAGTCGCTCTACGACTCTCCGTCCATACGAGCACATTACGGG
DB_BC_Row_7	TAACTTACGGAGTCGCTCTACGGTCGATGCACCATACGAGCACATTACGGG
DB_BC_Row_8	TAACTTACGGAGTCGCTCTACGACGGGAATTCCATACGAGCACATTACGGG
DB_BC_Col_1	GGATGGGATTCTTTAGGTCCTGCAAGTGTTCCTTGACTGAGCGACTGAGG
DB_BC_Col_2	GGATGGGATTCTTTAGGTCCTGAGGACATTCCTTGACTGAGCGACTGAGG
DB_BC_Col_3	GGATGGGATTCTTTAGGTCCTGCACTAATGGCTTGACTGAGCGACTGAGG
DB_BC_Col_4	GGATGGGATTCTTTAGGTCCTGAGCCTGATGCTTGACTGAGCGACTGAGG
DB_BC_Col_5	GGATGGGATTCTTTAGGTCCTGTTACGCTAACTTGACTGAGCGACTGAGG
DB_BC_Col_6	GGATGGGATTCTTTAGGTCCTGACTCTCCGTCTTGACTGAGCGACTGAGG
DB_BC_Col_7	GGATGGGATTCTTTAGGTCCTGGTCGATGCACTTGACTGAGCGACTGAGG
DB_BC_Col_8	GGATGGGATTCTTTAGGTCCTGACGGGAATTCTTGACTGAGCGACTGAGG
DB_BC_Col_9	GGATGGGATTCTTTAGGTCCTGCGCGCCCAGCTTGACTGAGCGACTGAGG
DB_BC_Col_10	GGATGGGATTCTTTAGGTCCTGACTAGTTTGCTTGACTGAGCGACTGAGG
DB_BC_Col_11	GGATGGGATTCTTTAGGTCCTGAGGTTGGGTCTTGACTGAGCGACTGAGG
DB_BC_Col_12	GGATGGGATTCTTTAGGTCCTGGTGAACCGACTTGACTGAGCGACTGAGG

Barcode in each primer is underlined [23]

Table 2
The Forward and Reverse Plate primers

Primer name	Sequence
P01_PE1.0_Forward	AATGATACGGCGACCACCGAGATCTACACTCTTTCCCTACACGACGC TCTTCCGATCTNNNNNCAAGTGTTCTAACTTACGGAGTCGCTC TACG
P02_PE1.0_Forward	AATGATACGGCGACCACCGAGATCTACACTCTTTCCCTACACGACGC TCTTCCGATCTNNNNNAGGACATTCTAACTTACGGAGTCGCTC TACG
P03_PE1.0_Forward	AATGATACGGCGACCACCGAGATCTACACTCTTTCCCTACACGACGC TCTTCCGATCTNNNNNCACTAATGGTAACTTACGGAGTCGCTC TACG
P01_PE2.0_Reverse	CAAGCAGAAGACGGCATACGAGATCGGTCTCGGCATTCCTGC TGAACCGCTCTTCCGATCTNNNNNCAAGTGTTCGGATGGGATTC TTTAGGTCCTG
P02_PE2.0_Reverse	CAAGCAGAAGACGGCATACGAGATCGGTCTCGGCATTCCTGC TGAACCGCTCTTCCGATCTNNNNNAGGACATTCGGATGGGATTC TTTAGGTCCTG
P03_PE2.0_Reverse	CAAGCAGAAGACGGCATACGAGATCGGTCTCGGCATTCCTGC TGAACCGCTCTTCCGATCTNNNNNCACTAATGGGGATGGGATTC TTTAGGTCCTG

Barcode in each primer is underlined [23]

Split in 15 μL in four PCR tubes.

See Fig. 2 for primer description. A list of the Forward and Reverse Plate primers can be found in Table 2.

26. Run the following PCR cycle:
    (a) 3 min 98 °C
    (b) 18 cycles
    (c) 30 s 98 °C
    (d) 15 s 61 °C
    (e) 35 s 72 °C
    (f) Final elongation.
    (g) 1 min 72 °C.
27. Run Plate-PCR product on a 1.5% agarose gel. Usually 1 μL of Plate-PCR product is enough to produce observable bands.
28. Pool the four reactions and purify on magnetic beads. Elute from the beads with 10 mM Tris-HCl, pH8.0.
29. Sequence the resulting PCR product with Illumina technology (*see* **Note 32**).

3.3 Yeast Integration

1. For library amplification, set up the following PCR reaction:

Template DNA 1 ng/μL	7.5 μL
5× buffer	5 μL
10 mM dNTPs	0.75 μL
10 μM forward primer	0.75 μL
10 μM reverse primer	0.75 μL
KAPA polymerase 1 U/μL	0.3 μL
PCR grade water	9.95 μL

2. Run the following PCR cycle:
   (a) 3 min 98 °C
   (b) 25 cycles
   (c) 20 s 98 °C
   (d) 15 s 60 °C
   (e) 45 s 72 °C
   (f) Final elongation.
   (g) 1 min 72 °C.
3. Run the PCR amplification on a 1.5% agarose gel. Usually 1 μL is enough to produce observable bands.
4. Retrieve strain validated at **step 33** in Subheading 3.1.
5. Perform **steps 17** to **29** in Subheading 3.1, with the following modification.
   (a) **Step 18**, the culture volume should be 100 mL in a 500 mL flask.
   (b) **Step 19**, the 100 mL culture is split in two 50 mL tubes. All subsequent volumes refer to each 50 mL tube.
   (c) **Steps 22** and **23**, the LATE solution and 50% Glycerol volume is now 300 μL.
   (d) **Step 24**, instead of a 1.5 mL tube use a PCR 96-well plate. In each well, combine 20 μL of competent cells, 8 μL of PCR library, 2 μL of DNA carrier 10 mg/mL, 250 ng of pCAS-gRNA and 200 μL of PLATE solution, mix by pipetting up and down.
   (e) **Step 28**, use 75 μL YPD instead of 250 μL.
   (f) **Step 29**, plate the full 75 μL.

6. Once colonies have grown, resuspend all the colonies by spreading 5 mL YPD media on the Petri dish. Transfer the cell suspension in a 15 mL tube (*see* **Note 33**).
7. Read optical density (OD) for each of your cell suspension (*see* **Note 34**).
8. Combine 5 units (U) of OD of each codon-specific library in a 50 mL tube (master pool).
9. Read OD for the master pool.
10. Prepare two cell pellets by centrifuging 500 μL of the master pool at 16,100 × *g* for 1 min. Remove the supernatant and freeze the pellet at –80 °C.
11. Prepare a glycerol stock for each library. For the master pool, prepare 6 different glycerol stocks.
12. Extract genomic DNA from one of the master pool pellets (*see* **Note 35**).
13. For Next Gen sequencing, perform **steps 13** to **29** of Subheading 3.2 completely with these modifications.
    (a) **Step 14**, dilute your genomic DNA at 30 ng/μL.
    (b) **Step 15**, use 3 μL of your genomic DNA at 30 ng/μL instead of 4.5 μL of plasmid. Don't forget to use 7.55 μL of PCR grade water instead of 6.05 μL.

3.4 Liquid Dihydrofolate Reductase Protein-Fragment Complementation Assay (DHFR-PCA) (See Fig. 3)

1. For mating and diploid selection, start a culture of your yeast library by transferring 100 μL of your master pool glycerol stock in 5 mL SC complete (MSG) pH 6.0 (*see* **Note 36**). Incubate with 250 rpm agitation at 30 °C for 16 h. At the same time, inoculate 5 mL YPD + HPH with your prey of interest from an isolated colony.
2. Combine in a 15 mL tube, 1.2 mL fresh YPD, 300 μL of master pool culture and 150 μL of your prey of interest for the mating step. Incubate 8 h with 250 rpm at 30 °C.
3. Start first diploid selection by combining 1.25 mL of the mating step with 1.75 mL of YPD + 1.8× NAT + 1.8× HPH in a 15 mL tube. Incubate 16 h with 250 rpm at 30 °C.
4. Transfer 300 μL of the first diploid selection in 2.7 mL of fresh SC complete (MSG) + 1.1× NAT + 1.1× HPH to perform second diploid selection (*see* **Note 4**). Incubate for 24 h with 250 rpm at 30 °C.
5. Centrifuge 5 U OD of diploid selection 2, remove supernatant and freeze pellet at −80 °C.
6. For PCA, dilute diploid selection 2 at 0.1 OD/mL in 10 mL PCA media (PCA -ade-lys-met +200 μg/mL methotrexate) to generate MTX1 (*see* **Note 37**). Incubate in the dark 72 h with 250 rpm at 30 °C.

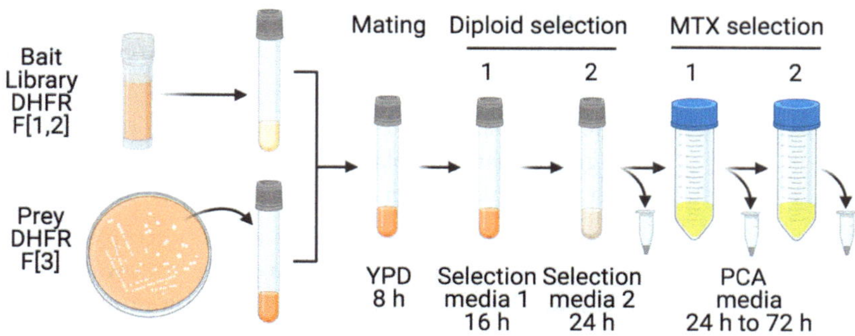

Fig. 3 Dihydrofolate reductase (DHFR) protein-fragment complementation assay (PCA) workflow in liquid medium. First, bait and prey strains are inoculated and grown separately in Synthetic or YPD media for 24 h at 30 °C. Next morning, the two strains are mixed together and allowed to mate for 8 h at 30 °C in nonselective media (Mating). The cells are then transferred to YPD + NAT + HPH media for 16 h at 30 °C to select for the newly formed diploid cells (Diploid selection 1). Diploid cells are selected for a second round, but this time in Synthetic Media (MSG) pH 6.0 + NAT + HPH for 24 h at 30 °C (Diploid selection 2). Following the Diploid selection 2, diploid cells are competed in PCA media at 30 °C between 24 h and 72 h. A cell pellet is conserved for these three steps: Diploid selection 2, MTX selection 1 and MTX selection 2

7. Centrifuge 5 U OD of MTX1, remove supernatant and freeze pellet at −80 °C (*see* **Note 38**).
8. Dilute MTX1 culture at 0.1 OD/mL in 10 mL of fresh PCA media (PCA -ade-lys-met +200 μg/mL methotrexate) to generate MTX2. Incubate in the dark 72 h with 250 rpm at 30 °C.
9. Centrifuge 5 U OD of MTX2, remove supernatant and freeze pellet at −80 °C.
10. For Next Gen Sequencing, perform **steps 12** and **13** of Subheading 3.3 with the pellet from diploid selection 2 and MTX2 (*see* **Notes 39**, **40**, **41** and **42**).

4 Notes

1. If your experiment is performed in a diploid yeast strain, you may want to consider removing your gene of interest from the prey strain. The primers used to amplify your library for next-generation sequencing will amplify both alleles. Removing your gene of interest from the prey strain will focus your next-generation sequencing on your library.
2. Yeast strains from the Yeast Protein Interactome are referred to as baits and preys. A bait strain contains an ORF tagged in C-terminal with the first fragment of the DHFR (DHFR F [1, 2]) and is *MAT***a**. A prey strain contains an ORF tagged in C-terminal with the second fragment of the DHFR (DHFR F

[3]) and is *MATα*. If you mate one bait and one prey you will recover a diploid able to grow in methotrexate media if the two proteins interact.

3. When preparing DNA carrier 10 mg/mL, agitation with a magnetic bar and plate stirrer will take a long time. You may also want to consider heating at 70 °C to accelerate the process.
4. When adding Nourseothricin and Hygromycin B to Synthetic media you need to make sure you follow those rules. First, Nourseothricin activity is reduced by ammonium sulfate, so you need to replace it by monosodium glutamate salt. Second, Hygromycin B activity is reduced by low pH, this is why we buffer our Synthetic media at pH 6.0 with succinic acid and NaOH.
5. Methotrexate is a harmful chemical. Always make sure to manipulate the powder under a chemical hood. We usually do not weigh the powder, we dissolve all the powder in one bottle and store it at −20 °C in small aliquots. Once an aliquot is thawed, it cannot be refrozen. Make sure you dispose of the waste according to your Institution's instructions.
6. Liquid drop-out is light sensitive. Wrap in foil the bottle containing your drop-out to prevent light exposure when stored at 4 °C.
7. When preparing 40% Glucose make sure to put water first in the beaker and then add the glucose powder when stirring. If you do the opposite, glucose will create a "rock" and it will be harder to dissolve. You can add water during the dissolution but always keep in mind that adding that much powder in water inflates the volume.
8. PCA media is not stable for a long time at room temperature and cannot be stored at 4 °C since DMSO freezes at 4 °C. Best thing to do is to prepare the quantity you need just before you need it.
9. As explained in Ryan et al. 2016 [22], avoid T stretches (>4 consecutive T) in your gRNA. Stretches of T could serve as RNA polymerase III terminators and impact gRNA efficiency.
10. Not all gRNAs will work the same to produce stuffed strains with high efficiency. To avoid losing precious time, order at least 2 gRNA for each region of interest to increase your chances of having at least one that works properly.
11. When setting up a PCR mutagenesis reaction, always prepare a reaction without mutagenesis primers. The reaction without primer will be the control for the rest of the mutagenesis process.
12. DpnI is used to remove parental DNA because it cleaves methylated DNA. Since DNA produced by PCR is

unmethylated, only mutated plasmids will stay intact and be transformed in bacteria.

13. We use competent cells prepared in house with a RuCl2 protocol [24]. Usually, the transformation efficiency of the cells is in the range of 2.5×10^7 CFU/μg of DNA.
14. Instead of picking four random colonies, you could do a colony PCR [25] on different colonies with these two primers: CGGAATAGGAACTTCAAAGCG and CTGCAGCGAGGAGCCGTAAT. The PCR product is then digested with a restriction enzyme cutting only the new gRNA. Colonies with the new digestion pattern are sent for sequencing. Colony PCR is the easiest way to reduce time to confirm mutagenesis.
15. The proposed mutagenesis protocol goes all around the pCAS plasmid. If elongation time is too long in the PCR protocol, you will see primer concatenation in your final plasmid. If you see concatenation in all your clones, you can try to shorten the extension time.
16. Template DNA to produce the stuffer can be either genomic DNA or plasmid. If you are using genomic DNA, dilute it to 15 ng/μL and use 6 μL in your PCR. Do not forget to reduce water volume to 11.25 μL.
17. The integration of the DHFR fragment at the proper loci was verified by PCR for strains originating from the DFHR collection. Before using a strain from the collection, we suggest to sequence the fusion between the gene and the DHFR [1, 2] fragment. By doing so, you will validate the fusion and therefore increase the chance of having an optimal PCA signal.
18. Growth between 0.15 and 1 OD/mL usually takes around 6 h. We suggest you monitor OD after 5 h, so that you can adjust time needed to reach the required OD.
19. Competent cells remain frozen at least overnight.
20. The protocol for CRISPR-Cas9 transformation is adapted/modified from Ryan et al. 2016 [22]. DNA carrier 10 mg/mL is boiled for 5 min and cooled down for 5 min on ice before use. Competent cells are thawed for 10 min at room temperature before use.
21. Yeast cells are very efficient at homologous recombination and it is why you should plate your transformation on a Petri dish containing antibiotics for all your markers. Since the vast majority of antibiotic markers have the same promoter and terminator, recombination can happen between the genomic marker and the plasmid marker. If that is the case, you won't edit the genome and will instead recover a DHFR [1, 2] strain that is now G418 resistant.

22. Colonies from CRISPR-Cas9 transformation show various textures. If all colonies are tiny and/or round, it could indicate a transformation problem or an ineffective gRNA. You should select colonies that look wrinkled.
23. When testing for pCAS loss in your strain, test for the presence of all markers. As stated in note 20 you need to be really careful with yeast's ability to recombine.
24. The complete protocol for library preparation requires a plasmid with your region of interest. We used the pUC19 backbone and inserted our sequence of interest using Gibson cloning. We cloned our sequence of interest with 70–80 bp of homology on each side to be able to perform yeast integration more easily.
25. The volume of megaprimer can be increased up to 0.75 μL if you notice a low colony number for your mutagenesis transformation. Also, pipetting extremely small volumes such as 0.45 μL can be less accurate, therefore increasing the volume to 0.75 μL can help increase uniformity across all reactions.
26. For this step, we do separate transformations for each codon of our sequence of interest. Keeping libraries separate at this step helps to recover a more uniform library. Also, if there is a problem with one codon, it gives you the opportunity to redo only the problematic one and not all the libraries.
27. In our case, appropriate media was 2YT + AMP, since we used pUC19 as backbone for our ROI.
28. Again, as in **Note 26**, keep every codon separated for glycerol stock and miniprep.
29. Test the primers to add the PS before you need them. If they do not work, order new ones.
30. We used 1 ng/μL to be able to work easily with a multichannel pipette. Pipetting 4.5 μL also helps with visual verification of your pipetting.
31. For magnetic beads purification follow the manufacturer's protocol. In our case, we use a ratio 1:1 (beads–PCR products).
32. Illumina Paired-end technology is necessary to take advantage of the RC-PCR protocol described here. When selecting the length of Illumina read, make sure the forward read and reverse read will overlap. Keep in mind the RC-PCR protocol adds 83 bp on the 5′ end of your ROI and 82 bp on the 3′ end. Also, RC-PCR does not use standard Illumina Indexes so all your reads will go in the Unassigned directory. Make sure you inform the person in charge of your sequencing to avoid confusion. The protocole has been optimized in our hands for the MiSeq technology.

33. Again, as in **Note 26** and **27**, keep all codons separated. If something is off with one codon, you will only need to redo the problematic one.
34. We usually perform a 1:40 dilution to have an accurate OD reading. Don't forget to convert to real OD based on your dilution.
35. We extracted genomic DNA using a phenol/chloroform/isoamyl alcohol protocol [26]. You could use your preferred extraction protocol, but you need to have high quality DNA at the end.
36. Start three independent precultures of your master pool to have biological replicates in your experiment.
37. You can perform the same experiment but you replace the MTX condition with a DMSO only condition. The DMSO condition will be without selection for the protein–protein interaction so you should not see an increase or decrease in the abundance of any mutant. If you see mutant abundance variation, it could mean your strain is sensitive to DMSO or has a fitness defect. So before starting your experiment, you could test your stuffed strain and the complete deletion of your gene in DMSO condition to see if it is sensitive. If yes, you need to include a DMSO control condition as a control of your PCA experiment.
38. We keep the intermediate time point, but only sequence the start and finish. Preliminary tests have shown that signals start to appear in MTX1, but it is still weak.
39. DMS libraries can be built using many established methods in-house or optionally be synthesized by commercial vendors.
40. If the region of interest is larger than 60 amino acids (180 bp), it will be difficult to sequence from start to finish one paired-end read with Illumina PE-300 chemistry, with uniform read quality. You could think about splitting your region in smaller parts with overlapping regions to help with normalization between the regions. You could also associate a barcode with each mutation and instead of sequencing your region of interest you sequence the barcode variation [27].
41. Sequencing PCR with Illumina technology creates a problem because of a lack of diversity at the beginning of the read. Such a problem can be avoided with an adapter primer with different length of variable region. You could have four times the same primer, but with 5N, 6N, 7N, and 8N between the adapter and the barcode. You could also increase PhiX concentration when running your sample.
42. NGS data analysis can be performed with pipelines such as DiMSum [28] or dms_tools [29].

Acknowledgments

We thank members of the Landry lab for helpful comments on the manuscript. Figures were created with BioRender.com. This work was supported by a CIHR Foundation grant 387697 and a HFSP grant RGP0034/2018 to CRL. CRL holds the Canada research chair in Cellular Systems and Synthetic Biology. RD holds a FRQ-S postdoctoral fellowship. SD was supported by a Mitacs Globalink scholarship. UD holds an Alexander-Graham-Bell Canada graduate scholarship and the Louis-Poirier scholarship of the Fondation du CHU de Québec-Université Laval. PCD holds a CIHR Vanier Canada Graduate Scholarship.

References

1. Rolland T, Taşan M, Charloteaux B et al (2014) A proteome-scale map of the human interactome network. Cell 159:1212–1226
2. Vo TV, Das J, Meyer MJ et al (2016) A proteome-wide fission yeast Interactome reveals network evolution principles from yeasts to human. Cell 164:310–323
3. Arabidopsis Interactome Mapping Consortium (2011) Evidence for network evolution in an Arabidopsis interactome map. Science 333:601–607
4. Tarassov K, Messier V, Landry CR et al (2008) An in vivo map of the yeast protein interactome. Science 320:1465–1470
5. Zhong Q, Simonis N, Li Q-R et al (2009) Edgetic perturbation models of human inherited disorders. Mol Syst Biol 5:321
6. Sahni N, Yi S, Zhong Q et al (2013) Edgotype: a fundamental link between genotype and phenotype. Curr Opin Genet Dev 23:649–657
7. Sahni N, Yi S, Taipale M et al (2015) Widespread macromolecular interaction perturbations in human genetic disorders. Cell 161: 647–660
8. Amberger J, Bocchini CA, Scott AF, Hamosh A (2008) McKusick's online Mendelian inheritance in man (OMIM®). Nucleic Acids Res 37: D793–D796
9. Podgornaia AI, Laub MT (2015) Pervasive degeneracy and epistasis in a protein-protein interface. Science 347:673–677
10. Dionne U, Bourgault É, Dubé AK et al (2021) Protein context shapes the specificity of SH3 domain-mediated interactions in vivo. Nat Commun 12:1597
11. Matreyek KA, Starita LM, Stephany JJ et al (2018) Multiplex assessment of protein variant abundance by massively parallel sequencing. Nat Genet 50:874–882
12. Diss G, Lehner B (2018) The genetic landscape of a physical interaction. Elife 7:e32472. https://doi.org/10.7554/eLife.32472
13. CampbelL-Valois F-X, Tarassov K, Michnick SW (2005) Massive sequence perturbation of a small protein. Proc Natl Acad Sci U S A 102: 14988–14993
14. Penn WD, McKee AG, Kuntz CP et al (2020) Probing biophysical sequence constraints within the transmembrane domains of rhodopsin by deep mutational scanning. Sci Adv 6: eaay7505
15. Karim AS, Curran KA, Alper HS (2013) Characterization of plasmid burden and copy number in Saccharomyces cerevisiae for optimization of metabolic engineering applications. FEMS Yeast Res 13:107–116
16. DiCarlo JE, Norville JE, Mali P et al (2013) Genome engineering in Saccharomyces cerevisiae using CRISPR-Cas systems. Nucleic Acids Res 41:4336–4343
17. Ryan OW, Skerker JM, Maurer MJ et al (2014) Selection of chromosomal DNA libraries using a multiplex CRISPR system. Elife 3:e03703. https://doi.org/10.7554/eLife.03703
18. Sadhu MJ, Bloom JS, Day L et al (2018) Highly parallel genome variant engineering with CRISPR–Cas9. Nat Genet 50:510–514
19. Biot-Pelletier D, Martin VJJ (2016) Seamless site-directed mutagenesis of the Saccharomyces cerevisiae genome using CRISPR-Cas9. J Biol Eng 10:6
20. Li A, Acevedo-Rocha CG, Reetz MT (2018) Boosting the efficiency of site-saturation mutagenesis for a difficult-to-randomize gene by a

two-step PCR strategy. Appl Microbiol Biotechnol 102:6095–6103

21. Freschi L, Torres-Quiroz F, Dubé AK, Landry CR (2013) qPCA: a scalable assay to measure the perturbation of protein-protein interactions in living cells. Mol Biosyst 9:36–43
22. Ryan OW, Poddar S, Cate JHD (2016) CRISPR–Cas9 genome engineering in Saccharomyces cerevisiae cells. Cold Spring Harb Protoc 2016:db.prot086827
23. Yachie N, Petsalaki E, Mellor JC et al (2016) Pooled-matrix protein interaction screens using barcode fusion genetics. Mol Syst Biol 12:863
24. Green R, Rogers EJ (2013) Transformation of chemically competent E. coli. Methods Enzymol 529:329–336
25. Woodman ME (2008) Direct PCR of intact bacteria (colony PCR). Curr Protoc Microbiol Appendix 3:Appendix 3D
26. Amberg DC, Burke DJ, Burke D, Strathern JN (2005) Methods in yeast genetics: a Cold Spring Harbor laboratory course manual. CSHL Press, Cold Spring Harbor, NY
27. Fowler DM, Stephany JJ, Fields S (2014) Measuring the activity of protein variants on a large scale using deep mutational scanning. Nat Protoc 9:2267–2284
28. Faure AJ, Schmiedel JM, Baeza-Centurion P, Lehner B (2020) DiMSum: an error model and pipeline for analyzing deep mutational scanning data and diagnosing common experimental pathologies. Genome Biol 21:207
29. Bloom JD (2015) Software for the analysis and visualization of deep mutational scanning data. BMC Bioinform 16:168

Chapter 15

Proteomic Mapping by APEX2-Catalyzed Proximity Labeling in *Saccharomyces cerevisiae* Semipermeabilized Cells

Birgit Singer-Krüger and Ralf-Peter Jansen

Abstract

Enzyme-catalyzed proximity labeling (PL) has proven to be a valuable resource for proteomic mapping of subcellular compartments and protein networks in living cells. We have used engineered ascorbate peroxidase (APEX2) to develop a PL approach for budding yeast. It is based on semipermeabilized cells to overcome poor cellular permeability of the APEX2 substrate biotin-phenol and difficulties in its delivery into the cell. The use of semipermeabilized cells has several advantages, in particular the avoidance of generating fragile spheroplasts and the opportunity of employing cells from a glucose-containing medium for APEX2 tagging. In this protocol we describe how to perform a ratiometric three-state stable isotope labeling by amino acids in cell culture (SILAC) approach that allows to map an open cellular compartment like the yeast nucleus. In particular, we focus on the proteomic sample preparation and provide instructions to achieve high-resolution mapping of a subcellular yeast proteome.

Key words Proximity labeling (PL), Engineered ascorbate peroxidase APEX2, Region of interest (ROI), Biotin-phenol (BP), True positive (TP), False positive (FP), Stable isotope labeling by amino acids in cell culture (SILAC), Quantitative mass spectrometry (MS)

1 Introduction

Enzyme-catalyzed proximity labeling (PL) has developed as a state-of-the-art approach to map subcellular compartments and protein networks in live cells [1]. Among the currently used proximity labeling enzymes, engineered ascorbic acid peroxidase (APEX2) is of particular advantage because of its fast labeling kinetics [1]. While its application in mammalian cells is well established since 2013 [2], the standard protocol of APEX2-mediated PL is not applicable in yeast due to poor cellular permeability of the substrate biotin-phenol (BP). Employing budding yeast spheroplasts in a glucose-free buffer [3] is not helpful as it suffers from several drawbacks, particularly those that arise from energy depri-

Frédéric Devaux (ed.), *Yeast Functional Genomics: Methods and Protocols*,
Methods in Molecular Biology, vol. 2477, https://doi.org/10.1007/978-1-0716-2257-5_15,

vation and protein biotinylation independent of exogenously added hydrogen peroxide, the essential cosubstrate of the peroxidase reaction [4].

Recently, we have developed a largely improved technique of APEX2 PL for budding yeast based on semipermeabilized cells obtained by gradual freezing [4]. A major benefit of this method is the avoidance of generating fragile spheroplasts, which is often time-consuming, and the opportunity of employing cells from a glucose-containing medium. Combined with stable isotope labeling by amino acids in cell culture (SILAC), we demonstrated the successful application of this approach by providing a quantitative high-quality mitochondrial matrix proteome [4] that rivals previous conventional proteomic approaches with purified mitochondria [5]. The concern that gradual freezing would generate lesions within subcellular membranes, through which reactive BP radicals might escape, was ruled out by demonstrating that tagging within the mitochondrial matrix yielded a proteome of very high specificity [4]. Another approach recently presented to adapt PL to the usage in microorganisms including budding yeast is to replace BP by alkyne-phenol [6]. After PL and cell lysis, the alkynyl group is derivatized with biotin-conjugated azide by click chemistry to allow visualization and streptavidin-mediated enrichment before proteomic mass spectrometry (MS) analysis [6]. This approach is also valid to follow spatially restricted RNA labeling [6]. Using semipermeabilized yeast cells generated with the approach described here, we could also demonstrate quantitative mapping of the nuclear chromatin, a more demanding semiopen compartment [4]. In addition to the capture of well-known chromatin components by APEX2-H2B, we identified proteins of unknown functions and an array of metabolic enzymes previously not connected to chromatin [4]. Finally, we challenged our new method of PL in combination with a lowly expressed protein of unknown function, Yer156C. To avoid overexpression of the Yer156C-APEX2 fusion, we used the endogenous *YER156C* promotor, thereby allowing a first proteomic profiling of the Yer156C neighborhood [4]. For all three applications, a ratiometric three-state SILAC protocol was applied, previously designed by the Ting lab to improve spatial specificity if semiopen compartments are mapped [7]. This method is based on cells derived from three cultures: one containing heavy (H) isotope of lysine expressing the APEX2 fusion of interest, a second containing medium (M)-isotope lysine expressing a construct that resides outside the region of interest (ROI), and a third culture with light (L)-isotope lysine expressing the APEX2 fusion of interest but biotinylation is subsequently not carried out due to the omission of H_2O_2 (Fig. 1). Here, we describe in detail the critical steps of this approach.

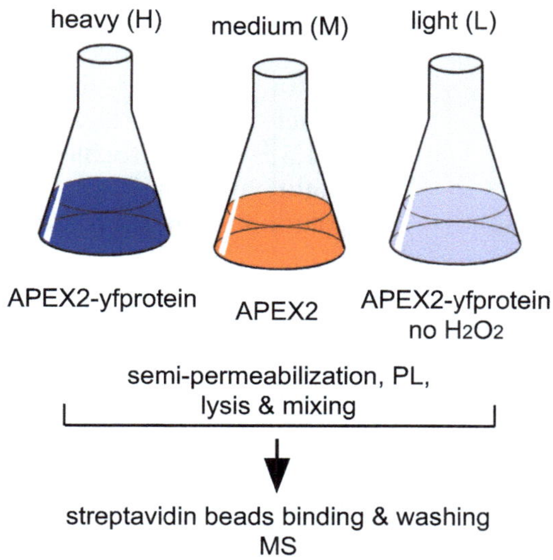

Fig. 1 Design and workflow of the three-state SILAC experiment. Yeast cells, expressing the experimental sample (APEX2 fusion of your favorite protein) are cultured in medium containing heavy (H) isotope lysine. Cells, expressing APEX2 alone or a different APEX2 fusion, residing outside the ROI, are cultured with medium (M) isotope lysine. Negative control cells expressing the experimental sample but in which later no PL is carried out, for example, by omitting H_2O_2, are grown in medium with light (L) isotope of lysine. Cells derived from the three cultures are semipermeabilized, and PL is conducted and analyzed as described in this protocol

2 Materials and Equipment

2.1 Yeast Media

Depending on the selectable marker of the plasmid encoding the APEX2-construct (*see* Subheading 2.2 and **Note 1**) and the amino acid(s) selected for SILAC, a suitable medium that matches the requirements needs to be chosen. Here, we describe the composition of the medium when using a *URA3*-based plasmid and the metabolic incorporation of isotope-labeled lysine.

1. SC-Ura medium: 0.67% w/v yeast nitrogen base without amino acids, 2% glucose, complete amino acid mixture minus uracil, autoclave for 20 min at 120 °C. For solid medium, add 2% agar before autoclaving and pour 25 mL in each plate.
2. Low biotin SC-Ura medium: 0.67% w/v yeast nitrogen base without amino acids and biotin, 2% glucose, complete amino acid mixture minus uracil, autoclave for 20 min at 120 °C. Supplement with biotin (125 ng/L final concentration) from a 1000-fold concentrated, filter-sterilized stock kept at −20 °C.
3. Low biotin SC-Ura-Lys medium for SILAC: 0.67% w/v yeast nitrogen base without amino acids and without biotin (see

above), 2% glucose, complete amino acid mixture minus uracil and lysine, autoclave for 20 min at 120 °C. Supplement with biotin (125 ng/L final concentration, see above) and with either L-lysine-$^{13}C_6/^{15}N_2$ hydrochloride (heavy), L-lysine-4,4,5,5-d4 hydrochloride (medium), or natural L-lysine (light), each at 30 mg/L (final) from a filter-sterilized stock solution (10 mg/mL) kept at −20 °C.

2.2 Transformed Yeast Cells, Carrying APEX2 Constructs

1. Yeast strains auxotrophic for lysine and/or arginine can be used to perform SILAC experiments. Another auxotrophy is required to allow transformation with a plasmid encoding the desired APEX2 fusion. We recommend centromeric plasmids like pRS315 or pRS316 carrying the selectable markers *LEU2* and *URA3*, respectively, and expression from the endogenous promotor or a promotor that avoids abnormally high expression levels of APEX2 fusions, which might interfere with correct subcellular localization and protein interactions (*see* **Note 1**). The yeast strain will be transformed with the plasmids using a standard transformation protocol. Purified transformants are stored in sterile 1 mL 15% glycerol at −80 °C.
2. To create a control strain, the same yeast strain will be transformed with a centromeric plasmid containing APEX2 alone. APEX2 alone should be expressed from the same promotor that was used for the APEX2 fusion with your favorite gene (*see* **Note 1**). Store transformants as described above.

2.3 Storage of Semipermeabilized Yeast Cells for Proximity Labeling

1. 1.5 mL polypropylene SafeSeal microtubes.
2. Freezing buffer: 15% (w/vol) glycerol, 150 mM KOAc, 2 mM MgOAc, 20 mM HEPES–NaOH pH 7.2, 0.5% glucose.
3. Cardboard or plastic box for −80 °C freezer.

2.4 Proximity Labeling, Detection by Western Blotting, Affinity Purification for Mass Spectrometry Analysis

1. Biotin-phenol (BP), Iris-Biotech GmbH, 50 mM stock in DMSO, store in 10–15 μL aliquots at −80 °C.
2. 1 M sodium ascorbate, prepare freshly on the day of the experiment.
3. 1 M sodium azide, prepare freshly on the day of the experiment.
4. 0.5 M Trolox in DMSO, prepare freshly on the day of the experiment.
5. 30% H_2O_2 (w/w), this reagent is approximately 10 M.
6. 10× PBS.
7. Quenching buffer: 1× PBS with 10 mM sodium ascorbate, 10 mM sodium azide, and 5 mM Trolox, diluted from the stock solutions above. This solution is stable for 30 min.

8. Mixed cellulose ester filters (0.45 μm pore size, 25 mm diameter).
9. SDS lysis buffer: 5% (w/v) SDS, 50 mM Tris–HCl, pH 9, 100 mM DTT, with protease inhibitor cocktail (e.g., cOmplete, Roche). SDS lysis buffer complemented with 10 mM sodium ascorbate, 10 mM sodium azide, and 5 mM Trolox, this buffer is stable for 30 min.
10. RIPA lysis buffer: 50 mM Tris–HCl, pH 8, 150 mM NaCl, 0.1% SDS, 0.5% Na-deoxycholate, 1% NP40. This buffer is stable at 4 °C. RIPA lysis buffer with freshly added 10 mM sodium ascorbate, 10 mM sodium azide, and 5 mM Trolox, this buffer is stable for 30 min.
11. 4× Laemmli sample buffer for SDS-PAGE.
12. Blotting buffer: 1× PBS, 2.5% (w/v) BSA, 0.2% Tween 20.
13. Phosphatase-labeled Streptavidin.
14. 660 nm protein assay kit (Pierce).
15. Streptavidin-coated magnetic Sepharose.
16. 1 M KCl.
17. 0.1 M Na_2CO_3.
18. 2 M Urea in 10 mM Tris–HCl, pH 8.
19. 3× Laemmli sample buffer for SDS-PAGE: supplemented with 20 mM DTT and 2 mM biotin.
20. Precast 4–20% SDS-PAGE gels.
21. Coomassie stain for mass spectrometry (MS) sample [8].
22. Vacuum filtration apparatus for filtration and collection of yeast cells: Millipore glass microanalysis filter holder kit with Millipore vacuum filtering side-arm flask (250 mL capacity).

3 Methods

3.1 Preparing Semipermeabilized Yeast Samples for PL and Western Blot Analysis

1. Streak the APEX2 transformants that you would like to analyze from the −80 °C frozen glycerol stocks on SC-Ura plates and let them grow for 48 h at 30 °C.
2. Pick a single colony using a sterile inoculation loop and start a 5 mL culture with SC-Ura medium. Incubate at 30 °C for 48 h, 160 rpm.
3. Dilute saturated cultures 1:1000 in 100 mL of low biotin SC-Ura medium and incubate overnight at 30 °C, 160 rpm, until cell density reaches 2–3 × 10^7/mL (*see* **Note 2**).
4. Harvest cells: wash once with H_2O, once with freezing buffer. This and subsequent steps are performed at room temperature.

5. Determine wet weight of cells on a fine balance. Add twice the volume of freezing buffer and resuspend cells by vortexing.
6. Transfer 60 μL portions of cell suspension into the tip of 1.5 mL polypropylene SafeSeal tubes (one 60 μL aliquot will be used for one biotinylation reaction). From a 100 mL culture, we typically obtain 7–9 aliquots.
7. Place tubes with cell aliquots into cardboard or plastic box, previously placed to a − 80 °C freezer.
8. Freezing of cells occurs at least for 1 day and frozen cells are never thawed unless used for biotinylation experiments. Storage of frozen cells is possible for at least 6 months without affecting biotinylation.

3.2 Small Scale Proximity Labeling for Western Blotting

1. Freshly prepare stock solutions of sodium ascorbate, sodium azide, and Trolox. A total volume of 1 mL for each solution is sufficient for 5 reactions. Keep the stocks at room temperature.
2. Prepare 1× PBS at room temperature. A volume of 15 mL for each PL reaction is required.
3. Prepare 5% SDS lysis buffer, calculate 40 μL/reaction. Upon addition of sodium ascorbate, sodium azide, and Trolox the solution is stable for 30 min.
4. Place mixed cellulose ester filters with pointed forceps into a small beaker filled with H_2O to allow filter-wetting, calculate 1 filter/reaction.
5. Make sure that a centrifuge for 1.5 mL polypropylene tubes is in close proximity. Put the settings to 2300 × *g*. Balancing tubes with different volumes will be needed.
6. Assemble vacuum filtration device and connect tubing to a vacuum system. The vacuum will be applied later.
7. Prepare 15 mL quenching solution. This solution is stable for 30 min. Pour 6 mL of this solution into a second 15 mL tube. One mL of this aliquot will be used to quench the biotinylation reaction (*see* **step 15**). The rest, 5 mL, and at least another 5 mL, will be required during the filtration step (*see* **step 17**).
8. Thaw one aliquot of BP. A total volume of 10 μL is sufficient. Dilute 1:3 with DMSO.
9. Prepare 1 mL 25 mM H_2O_2 from the 30% stock. Keep at room temperature.
10. Have a timer ready. Have an empty 1.5 mL polypropylene tube available and label it. Set a heating block for 1.5 mL polypropylene tubes to 55 °C.
11. Set a micropipette with tip to 1 mL to be ready for use.
12. Place a mixed cellulose ester filter onto the filtration device and fix the filtration assembly with the clamp. Apply the vacuum.

13. Pick one aliquot of 60 μL semipermeabilized yeast cells from the −80 °C freezer (*see* Subheading 3.1, **step 8**). The solution will thaw within 1 min at room temperature.
14. Add to the aliquot in successive order: 1.8 μL BP (from **step 8** of this Subheading) and 2.4 μL H_2O_2 (from **step 9** of this Subheading). Mix gently on a vortex (*see* **Note 3**). Start the timer for 60 s. Importantly, for negative control reactions, omit either one or the other reagent.
15. After exactly 60 s, add 1 mL quenching solution with the preassembled micropipette.
16. Transfer the quenched cells in two pipetting steps into the 4 mL quenching solution (*see* **step 7** of this Subheading).
17. Pour this solution into the funnel of the filtration device to immediately start the filtration process. Pour approximately 5 mL of the remaining 10 mL quenching solution into the original tube to harvest and filter residual cells. This step allows quantitative collection of cells.
18. Remove vacuum and disassemble the filtration unit. Carefully remove the filter with the captured cells on top. Roll it up (face with cells should be inside) so that it fits into the opening of the 1.5 mL tube (prepared in **step 10** of this Subheading).
19. Carefully rinse off cells from filter with 1 mL quenching solution. This usually requires two 0.5 mL steps. Discard the filter (*see* **Note 4**).
20. Collect cells by centrifugation at 2300 × *g* for 1 min.
21. Carefully remove the entire supernatant. This usually requires a second 30-s centrifugation step.
22. Resuspend cells with 40 μL of 5% SDS lysis buffer (*see* **step 3** of this Subheading).
23. Transfer the tube to a 55 °C heating block and incubate for 30 min.
24. Add 80 μL RIPA lysis buffer with freshly added sodium azide, Trolox, and sodium ascorbate.
25. Spin for 10 min in a microcentrifuge at maximal speed. Remove 70 μL of supernatant. It can be stored at −20 °C. For SDS-PAGE, mix with equal volume of 2× Laemmli sample buffer. A volume of 5 μL of this sample will be resolved by SDS-PAGE and transferred to a nitrocellulose membrane for Western blotting using standard techniques.
26. To detect biotinylation in the sample versus negative controls lacking APEX2, BP or H_2O_2 (*see* **step 14** of this Subheading), apply a standard Western blotting method and detect biotinylated proteins using phosphatase-labeled streptavidin. The expected result is a collection of bands appearing within the

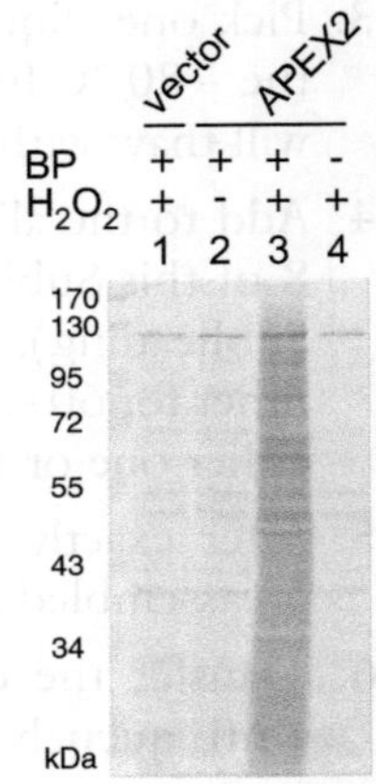

Fig. 2 Cell lysates obtained after PL using semipermeabilized yeast transformants containing an empty vector (lane 1), or expressing APEX2 (lanes 2–4) were analyzed by standard Western blotting using phosphatase-labeled streptavidin (*see* Subheading 3.2, **step 26**). As shown, the biotinylation is strictly dependent on APEX2, BP and H_2O_2

entire molecular-weight range that is dependent on the presence of APEX2, BP, and H_2O_2 (Fig. 2). In the negative control lanes, only bands corresponding to endogenously biotinylated proteins such as pyruvate carboxylase (Pyc1 and Pyc2) at 130 kDa should be detected (Fig. 2). With the APEX2 constructs we have generated, distinct banding patterns were not observed although the fusions are localized to different subcellular compartments [4]. Thus, the resolution and sensitivity of streptavidin blotting is likely insufficient to identify characteristic biotinylation "fingerprints" in Western blots.

3.3 Preparing Semipermeabilized SILAC Samples for PL and Mass Spectrometry Proteomics

1. Streak the two type of APEX2 transformants (APEX2 fusion of your favorite gene and APEX2 alone) from the −80 °C frozen glycerol stocks on SC-Ura plates and let them grow for 48 h at 30 °C.
2. For each, pick a single colony using a sterile inoculation loop and start a 5 mL preculture with low biotin SC-Ura-Lys medium, supplemented with biotin. Add sterile L-lysine-$^{13}C_6/^{15}N_2$ hydrochloride (heavy) to the culture with transformants expressing the APEX2 fusion of your favorite gene, L-lysine-4,4,5,5-d4 hydrochloride (medium) to the culture expressing APEX2 alone, and natural L-lysine (light) to the culture with cells expressing the APEX2 fusion of your favorite gene (Fig. 1). Incubate for at least 48 h at 30 °C, 160 rpm, to saturation.
3. Dilute the precultures 1:1000 into 250–400 mL main cultures with the corresponding medium for SILAC (same as described

in **step 2** of this Subheading). Incubate approximately 24 h at 30 °C, 160 rpm until cell density reaches 2×10^7 to 3×10^7/mL (*see* **Note 2**).

4. Harvest cells: wash once with H_2O, once with freezing buffer. This and subsequent steps are performed at room temperature.
5. Determine the wet weight of cells on a fine balance. Add twice the volume of freezing buffer and resuspend cells by vortexing.
6. Transfer 300 μL aliquots of the cell suspension into the tip of 1.5 mL polypropylene SafeSeal tubes (one 300 μL aliquot will be used for one biotinylation reaction). From a 250 mL culture, we typically obtain 3–4 aliquots.
7. Place tubes with cell aliquots into cardboard or plastic box, previously placed to a −80 °C freezer.
8. Freezing of cells occurs at least for 1 day and frozen cells are never thawed unless used for biotinylation experiments. Storage of frozen cells is possible for at least 6 months without affecting biotin tagging.

3.4 Three-State Proximity Labeling for MS Proteomics

1. Freshly prepare stock solutions of sodium ascorbate, sodium azide and Trolox. A total volume of 1.5 mL for each stock is sufficient for the three reactions. Keep at room temperature.
2. Prepare 1× PBS at room temperature. Three times 35 mL are required for the whole experiment with three PL reactions.
3. Prepare at least 1 mL 5% SDS lysis buffer. Prepare 3 × 0.3 mL portions. To one of them, freshly add 10 mM sodium ascorbate, 10 mM sodium azide and 5 mM Trolox. This solution is stable for 30 min.
4. Place mixed cellulose ester filters with pointed tweezers into a small dish filled with H_2O to allow filter-wetting. Three filters are needed for the complete experiment.
5. Make sure that a centrifuge for 1.5 mL polypropylene tubes is in close proximity. Put the settings to 2300 × *g*. Balancing tubes will be needed.
6. Assemble vacuum filtration device and connect tubing to a vacuum system. The vacuum will be applied later.
7. Prepare 35 mL quenching solution for the first PL. This solution is stable for 30 min. Pour approximately 10 mL of this solution into a second 15 mL tube. The rest (ca. 2 × 10 mL and 1.5 mL) will be required during filtration and rinsing steps.
8. Thaw one aliquot of BP. A total volume of 15 μL is sufficient for the entire experiment.
9. Prepare 1 mL of 50 mM H_2O_2 from the 30% stock. Keep at room temperature.

10. Have a timer ready. Have an empty 1.5 mL polypropylene tube available and label it. Set a heating block for 1.5 mL microtubes to 55 °C.
11. Set a micropipette with tip to 0.3 mL to be ready for use.
12. Place one mixed cellulose ester filter onto the filtration device and fix the filtration assembly with the clamp. Apply the vacuum.
13. Pick an aliquot of 300 μL semipermeabilized cells labeled with L-lysine-$^{13}C_6/^{15}N_2$ hydrochloride (heavy) from the −80 °C freezer (*see* Subheading 3.3, **step 8**). This is the experimental sample (Fig. 1). The suspension will thaw within 2–3 min at room temperature.
14. Add to the aliquot in successive order: 3 μL BP (**step 8** of this Subheading) and 6 μL H_2O_2 (**step 9** of this Subheading). Mix gently on a vortex (*see* **Note 3**). Start the timer for 60 s.
15. After exactly 60 s, transfer the entire PL reaction volume (slightly more than 0.3 mL) into the 10 mL aliquot with quenching solution with the preassembled micropipette.
16. Pour this solution into the funnel of the filtration device to immediately start the filtration process. Wash the cells on the filter for another two times, each with approximately 10 mL of quenching solution.
17. Remove vacuum and disassemble the filtration unit. Carefully remove the filter with the captured cells on top. Roll it up (face with cells inside) so that it fits into the opening of a 1.5 mL microtube (prepared in **step 10** of this Subheading).
18. Carefully rinse off cells with two times 0.75 mL quenching solution. Discard the filter (*see* **Note 4**).
19. Collect cells by centrifugation at 2300 × *g* for 30 s.
20. Carefully remove the entire supernatant. This usually requires a second 30-s centrifugation step.
21. Resuspend cells with 200 μL of 5% SDS lysis buffer (*see* **step 3** of this Subheading).
22. Transfer the tube to a 55 °C heating block and incubate for 30 min.
23. In the meantime, prepare 0.45 mL RIPA lysis buffer with freshly added sodium azide, Trolox, and sodium ascorbate (*see* Subheading 2.4, **step 10**).
24. After 30 min of incubation at 55 °C (**step 22** of this Subheading), add 0.4 mL RIPA lysis buffer with freshly added sodium azide, Trolox, and sodium ascorbate (**step 23** of this Subheading) and mix well by pipetting a few times up and down. Spin the sample in a microcentrifuge tube for 10 min at maximal

speed. Remove 580 μL of the supernatant and store it at 4 °C. Collect a small aliquot (30–50 μL) for control assays (*see*, e.g., **step 28** below).

25. At this point, the PL reaction and cell lysis for the first sample is completed.
26. In order to perform the second PL and lysis, **steps 9–24** of this Subheading are repeated, now using an aliquot of 300 μL semipermeabilized cells labeled with L-lysine-4,4,5,5-d4 hydrochloride (medium) (*see* Subheading 3.3, **step 8**). This is the control sample in which APEX2 alone will tag proteins outside the ROI (Fig. 1).
27. In order to perform the third PL reaction, **steps 9–24** are repeated, using the aliquot of 300 μL semipermeabilized cells labeled with natural L-lysine (light) (*see* Subheading 3.3, **step 8**). These are the cells expressing the APEX2 fusion of your favorite gene. Important: in **step 14** of this Subheading, omit the addition of H_2O_2 so that biotinylation via the APEX2 fusion will not occur.
28. For protein quantification of the three clarified lysates, each obtained in **step 24** of this Subheading, use the Pierce 660 nm protein assay according to the instructions of the manufacturer. If necessary, dilute samples so that the concentrations fall in the linear range of the assay.
29. Equilibrate 3 × 150 μL streptavidin-coated magnetic beads by washing each three times with 1 mL RIPA lysis buffer. Pellet the beads using a magnetic rack. Finally add 1 mL RIPA lysis buffer.
30. Mix the three whole-cell lysates from the heavy, medium and light samples in a 1:1:1 ratio by protein mass (determined in **step 28** of this Subheading). Dilute with 30 mL RIPA lysis buffer. Add the equilibrated streptavidin magnetic beads and rotate the sample at 4 °C overnight (*see* **Note 5**).
31. Pellet the beads on a magnetic rack. Part of the supernatant (unbound fraction) can be saved for further control analysis at −20 °C.
32. Wash the beads twice with RIPA lysis buffer, once with 1 M KCl, once with 0.1 M Na_2CO_3, once with 2 M urea in 10 mM Tris–HCl, pH 8, and twice with RIPA lysis buffer. Use 10 mL volumes for each wash. For each wash, resuspend the magnetic beads and rotate during a 5-min period at 4 °C. Transfer the beads during these washes first into one of the 15-mL tubes, finally into one 1.5 mL microtube (*see* **Note 5**). Cleanly remove all of the liquid, in particular before elution.
33. Eluate the biotinylated proteins by boiling in 60 μL 3× Laemmli sample buffer, supplemented with 20 mM DTT and

2 mM biotin for 10 min. After briefly vortexing beads and cooling the tube, centrifuge the sample for 2 min at max rpm to bring down the condensate and collect the eluate. It can be stored at −80 °C for several weeks if necessary.

34. Run 50 μL of the sample into a precast SDS-PAGE gel and process the gel for subsequent staining with colloidal Coomassie, for example, as described [8] (*see* **Note 6**).

3.5 MS Proteomics and MS Data Analysis

An excellent and recommended guideline for this part of the analysis is provided by the Ting lab [7]. We have adopted the approach to several of our yeast SILAC samples and provide here only a brief outline of critical steps.

1. After in-gel digestion of the proteins with trypsin or other proteases such as LysC and subsequent peptide extraction and desalting, measure the peptide mixture by LC-MS/MS.
2. Process the MS data using for example the established MaxQuant software and submit the extracted MS/MS peak lists to database search using the Andromeda search engine to query a target-decoy database of the *Saccharomyces cerevisiae* proteome.
3. For protein group identification and quantification, select a false discovery rate of 1% and a minimum of two quantified peptides.
4. For each detected protein, a SILAC ratio is calculated using its peptide SILAC ratios. Start with the nonnormalized SILAC ratios provided by MaxQuant.
5. For data normalization and filtering, among the identified proteins, generate a precise list of true positives (TPs) that are known to reside within the ROI. Make this list as comprehensive as possible, as it is utilized to determine cutoffs for SILAC ratios and to estimate the depth of coverage of the final proteome.
6. Among the identified proteins generate two comprehensive lists of false positives (FPs), FP-A and FP-B. The FP-A list represents proteins that should not be biotinylated to any degree, for example, due to their presence in a compartment that is separated from the APEX2 fusion by a nonpermeable membrane. This would, for example, apply for cytoplasmic proteins, if the APEX2 fusion is targeted to the mitochondrial matrix. The FP-B list contains proteins that are not localized to the ROI but could be biotinylated, if tagging occurs in a compartment with a porous membrane. This would happen with cytosolic proteins, if an APEX2 fusion is active in the nucleus.

7. To normalize H/L SILAC ratios (which reflect the extent of biotinylation) first calculate the median of distribution of ratios for the subset of FP-As and divide the ratio for each detected protein by this median. Take the $\log_2$ of those values. This normalization will center the distribution of nonspecific binders around $\log_2(H/L)$ to zero.
8. Analogously, normalize the H/M SILAC ratios (that reflect enrichment of proteins to the ROI, thus spatial specificity) provided by MaxQuant by calculating the median of the ratios for the group of FP-As. Divide the ratio for each detected protein by this median and take the $\log_2$ of those values.
9. Generate histograms of $\log_2(H/L)$ and $\log_2(H/M)$ values that depict the distribution of TPs versus FP-As or FP-Bs, respectively. They both should reveal a right-shifted distribution of the TP population compared to both FP populations thereby indicating a good enrichment of biotinylated proteins in the ROI.
10. To define the final proteome, confirm the enrichment of TPs over FPs by plotting receiver operating characteristic (ROC) curves for each SILAC ratio and then determine filter cutoffs for $\log_2(H/L)$ and $\log_2(H/M)$ at the maximum TPR-FPR value. Retain only the proteins with $\log_2 H/M$ and $\log_2 H/L$ ratios higher than these specifically determined cutoffs.

4 Notes

1. For the experimental design of a new APEX2 fusion, it is advisable to generate and test different constructs at different expression levels (e.g., by using different promotors) to select the one that permits proper subcellular targeting to the ROI and avoids mis-localization or aggregation due to overexpression. To test for functionality of the construct, complementation of a phenotype of the deletion mutant of your favorite gene is the method of choice, if applicable. An additional epitope tag present in the APEX2 construct such as HA, Myc, or FLAG, that also enables detection of the APEX2 fusion by immunoblotting and indirect immunofluorescence, might also be useful to confirm the interaction with a known binding partner by immunoprecipitation.
2. Higher cell densities of the main culture might affect cell lysis as cell wall stability usually increases in late growth phases.
3. To avoid uncontrolled spreading of the reaction volume in the tube during vortexing, select a suitable setting of your vortex. Aim for gentle mixing and a compact sample volume.

4. When handling the filter, gloves must always be worn. After rinsing off cells from the filter with quenching solution, squeeze the filter between thumb and forefinger to collect and harvest the majority of cells.
5. After mixing the three whole-cell lysates in a 1:1:1 ratio, a 10- to 20-fold dilution with RIPA lysis buffer is required to allow for the interaction between biotin and streptavidin. As a consequence, the increased reaction volume makes it necessary to use either one 50 mL or three 15 mL tubes during the overnight rotation. Afterward, make sure to quantitatively recover the magnetic beads by carefully rinsing tubes during the washing and concentration steps.
6. To estimate how much protein was isolated for MS, apply a known amount of protein like BSA or IgG in the range of 1–5 μg per lane. This sample(s) also serves as a positive control during the gel staining step.

References

1. Bosch JA, Chen C-L, Perrimon N (2020) Proximity-dependent labeling methods for proteomic profiling in living cells: an update. WIREs. Dev Biol 10:e392
2. Rhee HW, Zou P, Udeshi ND et al (2013) Proteomic mapping of mitochondria in living cells via spatially restricted enzymatic tagging. Science 339:1328–1331
3. Hwang J, Espenshade PJ (2016) Proximity-dependent biotin labelling in yeast using the engineered ascorbate peroxidase APEX2. Biochem J 473:2463–2469
4. Singer-Krüger B, Fröhlich T, Franz-Wachtel M et al (2020) APEX2-mediated proximity labeling resolves protein networks in Saccharomyces cerevisiae cells. FEBS J 287:325–344
5. Vögtle FN, Burkhart JM, Gonczarowska-Jorge H et al (2017) Landscape of submitochondrial protein distribution. Nat Commun 8:29
6. Li Y, Tian C, Liu K et al (2020) A clickable APEX probe for proximity-dependent proteomic profiling in yeast. Cell Chem Biol 27: 858–865
7. Hung V, Udeshi ND, Lam SS et al (2016) Spatially resolved proteomic mapping in living cells with the engineered peroxidase APEX2. Nat Protoc 11:456–475
8. Dyballa N, Metzger S (2009) Fast and sensitive colloidal Coomassie G-250 staining for proteins in polyacrylamide gels. J Vis Exp 30:1431

Chapter 16

Quantitative Proteomics in Yeast: From bSLIM and Proteome Discoverer Outputs to Graphical Assessment of the Significance of Protein Quantification Scores

Nicolas Sénécaut, Pierre Poulain, Laurent Lignières, Samuel Terrier, Véronique Legros, Guillaume Chevreux, Gaëlle Lelandais, and Jean-Michel Camadro

Abstract

Simple light isotope metabolic labeling (bSLIM) is an innovative method to accurately quantify differences in protein abundance at the proteome level in standard bottom-up experiments. The quantification process requires computation of the ratio of intensity of several isotopologs in the isotopic cluster of every identified peptide. Thus, appropriate bioinformatic workflows are required to extract the signals from the instrument files and calculate the required ratio to infer peptide/protein abundance. In a previous study (Sénécaut et al., J Proteome Res 20:1476–1487, 2021), we developed original open-source workflows based on OpenMS nodes implemented in a KNIME working environment. Here, we extend the use of the bSLIM labeling strategy in quantitative proteomics by presenting an alternative procedure to extract isotopolog intensities and process them by taking advantage of new functionalities integrated into the Minora node of Proteome Discoverer 2.4 software. We also present a graphical strategy to evaluate the statistical robustness of protein quantification scores and calculate the associated false discovery rates (FDR). We validated these approaches in a case study in which we compared the differences between the proteomes of two closely related yeast strains.

Key words Mass spectrometry, Proteomics, Quantification, Isotopic labeling, Yeast, Metabolism, bSLIM, Proteome Discoverer, Minora, KNIME

1 Introduction

The yeast *Saccharomyces cerevisiae* is a biological model that is widely used for the development and validation of global analytical methods in functional genomics and genetics. Yeast has been extensively studied for many years, resulting in a solid understanding of its physiology and metabolism. Yeast is the first eukaryotic

Gaëlle Lelandais and Jean-Michel Camadro are co-senior authors.

Frédéric Devaux (ed.), *Yeast Functional Genomics: Methods and Protocols*,
Methods in Molecular Biology, vol. 2477, https://doi.org/10.1007/978-1-0716-2257-5_16, © The Author(s) 2022

organism for which the genome was fully sequenced [1]. This has opened up new avenues for the exploration of living organisms, notably through the analysis of gene and protein expression using the amazing recent technical developments in transcriptomics and proteomics. In-depth knowledge of the yeast genome has enabled the construction of complete collections of haploid or diploid strains carrying modified alleles, for example, disruptions, deletions, and ORF or promoter fusions with a large number of reporter genes for use as probes to assess gene function and associated regulatory networks, laying the foundation for systems biology.

One specific aspect of yeast is its ability to grow in the presence (aerobiosis) or absence (anaerobiosis) of molecular oxygen. This is made possible by a metabolic switch that allows passage from respiratory to fermentative metabolism, provided that the carbon source available to the yeast can be metabolized by fermentation. This requires processes in which mitochondrial functions are essential, making yeast a critical organism for deciphering the genetics, biochemistry, and physiology of this energy producing organelle.

Yeast cells are capable of growing on synthetic media in which the vitamins and essential trace elements are provided or on complex media containing yeast hydrolysates or peptones. In wild-type yeast grown on synthetic media, cell metabolism is based on the assimilation of organic nitrogen (usually ammonium sulphate or chloride) and the catabolism of a single carbon source, such as glucose, glycerol, or acetate. The genetics of yeast has been brought to light through the selection and use of mutants affected in the synthesis of certain nucleotides (e.g., Ura3) or amino acids (e.g., Lys, His, Leu, Met, Arg, Trp). Such auxotrophic mutants require the addition of the defective bases and/or amino acids in the synthetic growth media.

Beyond the identification of proteins in complex extracts, mass spectrometry-based proteomic analysis allows the quantification of differences in the proteome between several biological states. Several bottom-up quantitative proteomics approaches have been reported [2], providing critical information in yeast biology. They are based either on in vitro labeling of peptides by isobaric chemical probes, releasing fragments in MS/MS, of which the measured intensity reflects the abundance of the protein in the initial extract (e.g., ICAT or TMT labeling), or on differential metabolic labeling in vivo, in which the cells are cultured in the presence of "light" (unlabeled) or "heavy" (labeled) amino acids that will be incorporated into the proteins, allowing their quantification after tryptic digestion and the measurement of a heavy–light ratio for each peptide/protein (e.g., SILAC and derived methods) (for a review, *see* [3]). Despite the fact that TMT- and SILAC-based quantitative proteomics allow multiplexing multiple samples in a single run, one of the most widely used proteomics approaches is

Table 1
Relative abundance of the stable isotopes of the elements found in proteins

$^{12}C = 98.93\%$	$^{13}C = 1.07\%$		
$^{1}H = 99.9885\%$	$^{2}D = 0.0115\%$		
$^{16}O = 99.757\%$	$^{17}O = 0.0373\%$	$^{18}O = 0.2057\%$	
$^{14}N = 99.632\%$	$^{15}N = 0.368\%$		
$^{32}S = 94.93\%$	$^{33}S = 0.76\%$	$^{34}S = 4.29\%$	$^{36}S = 0.02\%$
$^{31}P = 100\%$			

still the label-free approach, in which individual LC-MS/MS runs are compared and the intensity of the peptide ions in MS1 are measured [4] to determine differences in protein abundance.

Recently we presented an innovative quantification method, called simple light-isotope metabolic (SLIM) labeling [5]. The SLIM-labeling strategy uses the fundamental property of the living matter in which all the biomolecules are basically composed of carbon, nitrogen, hydrogen, oxygen, and sulfur, with several additional elements, such as phosphorus, selenium, or iodine. Most of these elements, except phosphorus, are naturally present in the form of several stable isotopes for which the abundance is fixed (Table 1). It is thus possible to infer their isotopic abundance in biomolecules, such as amino acids by solely taking into account their average elemental composition: $C_{4.9384}$ $H_{7.7583}$ $N_{1.3577}$ $O_{1.4773}$ $S_{0.0417}$ [6].

This has important consequences in terms of high-resolution MS-based peptide/protein analysis. Every peptide is measured as a series of ions (m/z) in an isotope cluster of similar charge (z) but with the mass ranging from the monoisotopic mass m_0, containing only the lightest isotopes of each element, to higher masses resulting from the statistical distribution of additional neutrons present in the stable isotopes (isotopologs). The intensities of the various isotopologs within an isotope cluster therefore depend on the elemental composition of the peptide and follows a Poisson distribution that can be accurately modeled using dedicated software, such as MIDAS [7]. The basic principle of SLIM labeling is to manipulate the elemental composition of proteins in vivo from the natural abundance of the isotopes of the atoms present in proteins (C, H, N, O, S, P), defining the "NC" (natural carbon) condition as the condition named "12C" in which the proteins are enriched in the light isotope of carbon (^{12}C) and, eventually, nitrogen (^{14}N) (referred as to "12C14N" condition). Considering the main routes for amino-acid biosynthesis in yeast (Fig. 1) [8], we hypothesized that providing yeast cells with U-[^{12}C]-glucose as the sole carbon source would result in the rapid synthesis of U-[^{12}C]-

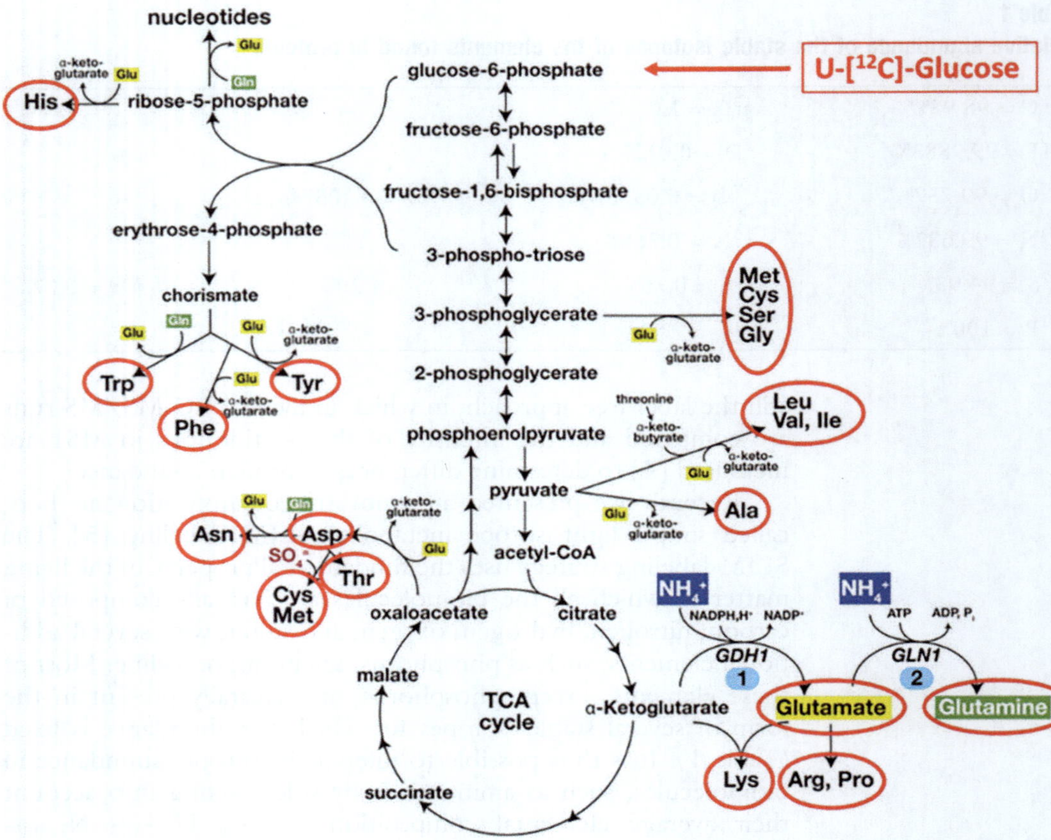

Fig. 1 Yeast Metabolism (adapted from Ljungdahl PO et al. YeastBook 2012 [5]

amino acids and their incorporation into newly synthesized proteins. Applying this labeling method allowed us to experimentally evaluate the half-life of the proteome in *Candida albicans*, and measure the effect of the proteasome inhibitor MG132 and a broad-specificity serine-protease inhibitor, PMSF, on the dynamics of the proteome in this organism [5].

Increasing the amino-acid content in ^{12}C (and to a lesser extent in ^{14}N) results in a different and simpler isotopic cluster that always remains within the boundaries of that observed with a natural isotopic composition, but with the intensity of the monoisotopic ion greatly enhanced. This has significant impact on downstream analyses, that is, allowing better signal-to-noise discrimination, more precise mass determination, and better MS/MS fragmentation spectra. As a result, higher scores for peptide identification and protein sequence coverage are obtained (*see* characteristic mass spectra in Fig. 2a–c). We took advantage of these characteristics to develop a new quantitative proteomics method in which peptides originating from the NC condition are mixed in equimolar amounts with peptides from the 12C condition. The intensity of

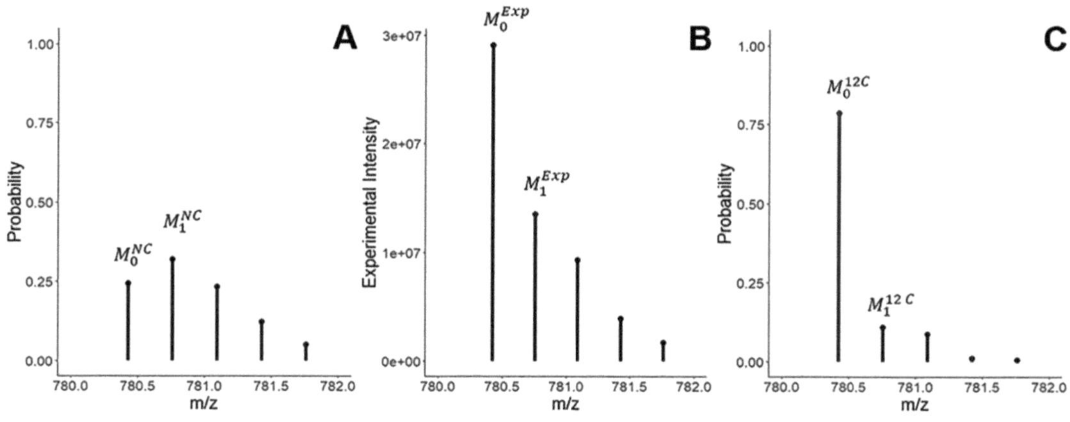

Fig. 2 Theoretical spectra of the peptide DQPILFWGGATAVGQMLIQLAK from the protein YNL134c under NC (**a**) and 12C (**c**) conditions and the corresponding experimental spectrum (**b**) of the same peptide from a 1:1 mixture of total extract from the 12C and NC conditions

every isotopolog in any isotope cluster is thus the sum of the intensity of the isotopologs from each condition (Fig. 2b). Therefore, measuring the ratio between the experimental values of the monoisotopic ion (M_0) and the next ion containing one more neutron (M_1), *modulo* the values of their theoretical intensity, expressed as the probability of occurrence, in each condition allows calculation of the molar fraction of the peptide originating from the NC- and 12C conditions. We recently described the full formalism for the quantification of ^{12}C incorporation into proteins/peptides and its use in quantitative proteomics, and we developed the data processing tools required to smoothly run SLIM labeling experiments [9].

One critical step in the SLIM-labeling quantification procedure is the accurate extraction of the intensities for all isotopologs in each isotope cluster from the experimental spectra. In our initial study, we used commercial software, Progenesis QI for metabolomics, but it does not provide the possibility to automatically link the quantification files with the identification files [5]. This prompted us to develop another workflow, referred to as bSLIM [9], in which only the intensities of the identified peptides are used to extract the data using an OpenMS node, FeatureFinderIdentification [9], which was modified to fit every mass trace in the isotope cluster. This approach only required us to install and run the KNIME (Konstanz Information Miner) environment for computation [9], together with the latest versions of OpenMS [10–12], and hence is fully independent of any commercial software.

Here, we present an alternative integrated procedure that takes advantage of the tools available in the proprietary software suite Thermo Scientific Proteome Discoverer. Proteome Discoverer (PD v2.4) is a popular program for the analysis of peptide-centric proteomics data, with a high level of integration with Thermo Fisher Scientific high-resolution, high-sensitivity Orbitrap instruments. This analytical platform includes many algorithms developed by Thermo or third parties, such as the IMP Protein Chemistry facility (https://pd-nodes.org/) [13, 14] and others. Proteome Discoverer is therefore widely used on a routine basis in many MS-based proteomics laboratories. It associates and integrates both raw spectra processing and filtering, peptide identification through database searches in data-dependent analyses, diverse quantification routines, and convenient spectra viewers. The output of the Proteome Discoverer analyses is written in ".msf" or ". pdresult" files. A key feature of ".pdresult" files is that they are SQLite relational databases (https://www.sqlite.org) that can be queried using SQL. The possibility to visualize individual annotated spectra for peptides up to the level of their isotope clusters with their associated intensity prompted us to develop the appropriate tools to extract this data and use it as input for our bSLIM labeling quantitative proteomics strategy. We accessed the individual mass trace intensities by taking advantage of the capabilities of the newly developed node, Minora, initially designed for label-free quantification.

We also present a solution to assess the robustness of the protein quantification scores calculated using bSLIM which was missing from our previous data analyses workflow. Derived from the SAM (Significance Analysis of Microarrays [15]) method, the general idea is to randomize the original bSLIM output data sets multiple times and calculate the associated "random scores." These scores are graphically compared to the "real scores" obtained from the original bSLIM data. Proteins for which the real scores vary the most from the random scores are thus easily detected and worth considering for further analyses. In this chapter, we present a case study to illustrate the different outputs from the various workflows that we developed. We compared differences between the proteomes of two "wild-type" *Saccharomyces cerevisiae* strains with the same genetic background, but with one strain (BY4742) harboring the deletion of four genes (Ura3, His3, Leu2, and Lys2) relative to the reference strain S288c (*see* **Note 1** for data availability). The proteomes of the two strains are expected to be very similar and therefore represent a challenging test to assess the sensitivity and specificity of our quantification methods.

Overall, we expect that these alternative solutions implemented in the bSLIM data analysis workflow will be useful for proteomics laboratories running Orbitrap-based mass spectrometers, which are very familiar with Proteome Discoverer. This is an original way to combine the completeness and reproducibility of routine proprietary software with the power of open-source tools.

2 Materials

1. Reagents for yeast synthetic growth media and appropriate supplements.
2. SLIM-labeling specific reagent: U-[^{12}C]-glucose (e.g., Cambridge Isotope Laboratories).
3. Lysis buffer: 40 mM HEPES–KOH, pH 7.5, 350 mM NaCl, 10% glycerol, 0.1% Tween-20.
4. Acid-washed silica beads (0.45–0.5 mm Ø).
5. 200 μg/mL trypsin solution, prepared by dissolving 20 μg trypsin (Proteomic grade) in 100 μL of 1 mM HCl.
6. Cold Acetone.
7. 50 mM ammonium carbonate (NH_4HCO_3).
8. 0.1% formic acid (MS grade).
9. Dry incubator at 37 °C.
10. Vacuum dryer (Speed Vac).
11. Low-binding microcentrifuge tubes.
12. 4–12% polyacrylamide gradient gels.
13. Coomassie blue (MS friendly, such as SimplyBlue SafeStain, Invitrogen).
14. Bradford protein assay reagent.
15. An instrument setup for LC-MS/MSMS data acquisition (*see* **Note 2**).
16. Appropriate software suites for quantification and identification of the peptide/protein content of the samples analyzed (Fig. 3).

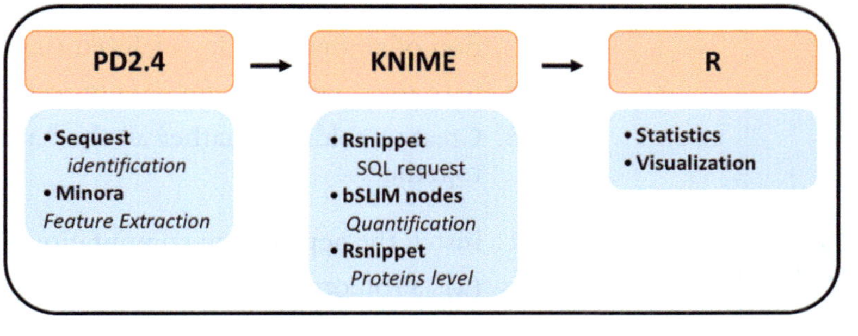

Fig. 3 General organization of the data processing workflows

3 Methods

3.1 Cell Growth and Preparation of Protein Extracts

1. Grow the cells to be compared in a synthetic medium with either regular glucose (NC-condition), or U-[^{12}C]-glucose (12C-condition) as the sole carbon source (*see* **Note 3**).
2. At the appropriate cell density (mid-exponential phase of growth), collect the cells by centrifugation for 10 min at 4000 × *g* at 4 °C.
3. Wash the cell pellet with cold water, resuspend the cells in lysis buffer at a cell density of 0.6 g/mL, and lyse the cells by adding 0.32 mL acid-washed, heat-sterilized silica glass beads (0.45–0.5 mm∅) to 0.6 mL cell suspension and vortexing the resulting suspension three times for 5 min, leaving the tubes on ice for 5 min between each vortexing.
4. Centrifuge the lysed cells for 5 min at 3000 × *g* and collect the supernatant, referred to as the cell homogenate.
5. Carefully measure the protein concentration using the Bradford Protein microassay and validate the protein measurement by running an aliquot on an SDS-PAGE gel and staining with Coomassie Blue.
6. Precipitate a 50-μg protein aliquot using 6 vol. cold acetone for 2.5 h at −20 °C.
7. Resuspend the dry protein pellet in 50 mM ammonium carbonate buffer by heating for 15 min at 95 °C.
8. Add 5 μL of 200 μg/mL trypsin stock solution and incubate for 12 h at 37 °C in a dry incubator.
9. Remove all solvents by vacuum drying.
10. Resuspend the peptides in 0.1% formic acid.
11. Carefully mix an equal amount of peptides from the NC- and 12C-conditions.
12. Inject the samples, typically 5 μg in ≤5 μL, into the LC-MSMS instrumental setup (*see* **Note 4**).
13. Ensure that your instrumental setup allows the isotopic resolution of all the peptides analyzed (*see* **Note 5**).
14. Save the ".raw" files for data processing and signal extraction.
15. Create a folder to gather all the ".raw" files from one project together.

3.2 Data Processing Workflows

1. Install the appropriate computational resources.
   (a) Proteome Discoverer 2.4 or higher with a valid activation key.

(b) KNIME (v4.2.3) with all available extensions (https://www.knime.com/downloads): the OpenMS nodes (v2.6.0) are part of the "community nodes."

(c) R (v4.0.2), including the dplyr, dbplyr, RSQLite, sqldf, readr, raster, RMySQL packages and libraries (https://cran.r-project.org/bin/windows/).

3.2.1 Proteome Discoverer Analysis

1. Open Proteome Discoverer 2.4.
2. Create a new study and add your Thermo Fisher Scientific mass spectrometry *.raw* files.
3. Select the appropriate "processing." The basic processing workflow is composed of the following nodes: spectrum files, spectrum selector, sequestHT (1.1.0.189), Percolator (3.02.1), and IMP-ptmRS. To this Processing workflow, add the "Minora Feature Detector" node linked to the "Spectrum Files" node and set the correct advanced parameters) (*see* **Note 6**).
4. Select the appropriate "consensus" workflow composed of the following nodes: MSF Files, PSM Grouper, Peptide Validator, Peptide and Protein Filter with a link to Protein annotation, Protein Scorer with a link to Protein FDR Validator, and Protein Grouping (*see* **Note 7**).
5. Enable the postprocessing node "Display Settings."
6. Run the analysis, **one file at the time**, by giving nonambiguous names to the output files. The produced results files from PD2.4 have the extension .pdresult and are used as input in our KNIME workflow.

3.2.2 Isotopolog Intensity Extraction and Peptide/Protein Quantification Using a Dedicated bSLIM KNIME Workflow (Fig. 4)

1. Open KNIME 4.2.3.
2. Import from https://zenodo.org (DOI 10.5281/zenodo.4467829), the bSLIM quantification workflow "*File > Import KNIME workflow*" (file extension is ".knwf"). The workflow is a modification of the original workflow presented in our previous study [9]. The adaption is very simple (disconnection between metanode 1.2 "FFiDData filtering" and connection with the Row filter "erase ModifiedPeptides").

   This workflow contains three main parts.

   (a) The ".pdresult" file, which extracts all data concerning every peptide, including their identification and the intensity of the isotopologues in the isotope clusters. This procedure uses an R script written specifically for this study. It is embedded in a dedicated RSnippet as a part of the KNIME quantification workflow (*see* **Note 8** on SQL request formalism).

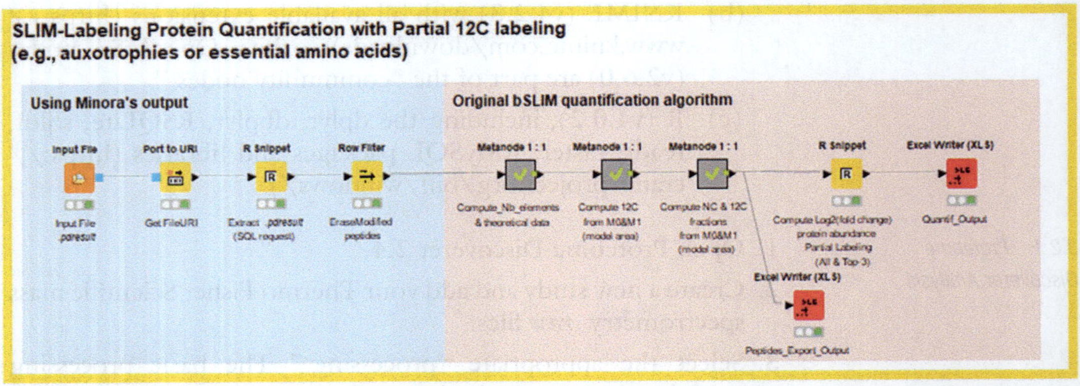

Fig. 4 KNIME workflow of the integrated ".pdresult" processing node connected to the quantification nodes

(b) The peptide/protein quantification workflow based on our previous study [9]. As previously described, two biological conditions are considered: complete labeling of the proteins (true wild-type strain) or partial labeling of the proteins (strains that are auxotrophic for specific amino acids).

(c) The procedure to compute the statistics on the identified peptides/proteins. These two procedures use R scripts embedded in specific RSnippets.

3. Check that your installation of Rserver allows all RSnippets to run smoothly.

4. There are two possible cases:

   (a) The samples come from an autotrophic yeast strain and the SLIM labeling is complete: follow "*cases of total labeling experiment.*"

   (b) The samples come from auxotrophic cells for which certain amino acids are not labeled. The SLIM labeling is thus incomplete: follow "*case of incomplete labeling experiment.*" In the latter case, the essential amino acids are defined as containing "Carbon B," corresponding to carbon atoms of natural isotopic abundance. It is therefore necessary to open the meta-node "Compute_Nb_elements& theoretical data" and then the meta-node "Compute Elemental Composition" and, finally, the carbon B calculus node (Compute Nb_Carbon-B (Ex: HKL)) to add the correct total number of carbons to each exogenous amino acid by typing the regular expression in the form "(Nb_H*6)+ (Nb_K*6)+ (Nb_L*6)", as exemplified for the BY4742 strain auxotrophic for histidine (H), lysine (K), and leucine (L). The number of carbon atoms for each amino acid is shown in Table 2.

Table 2
Number of carbon atoms per amino acid

Amino acid	A	C	D	E	F	G	H	I	K	L
Nb carbon atoms	3	3	4	5	9	2	6	6	6	6
Amino acid	M	N	P	Q	R	S	T	V	W	Y
Nb carbon atoms	5	4	5	5	6	3	4	5	11	9

5. Set the Excel exporter nodes with an appropriate name for file output (scores for quantifications at proteins or peptide levels) (*see* **Note 9**).

3.2.3 Statistics and Graphical Assessment of Score Significance

For statistical analysis of differential expression, we reproduced the SAM methodology, adapting it to the specifics of the quantitative measurements of protein abundance that are obtained at the end of the bSLIM workflow (Fig. 5) (*see* **Note 10**). Workflows available at https://zenodo.org (DOI 10.5281/zenodo.4467882).

1. Aggregate the protein quantification results from individual experiments (replicates) into a single ".tsv" table: Accession/ "name of column-2"/ "name of column-3"/ . . . / "name of column-n." Typically, column-2 to –n represents the log2(Fold change) of protein abundance per experimental condition.
2. Load the R scripts developed in this study to compute the scoring functions, and save the analysis.
3. Use the graphical package ggplot2, within the script, to produce the figures showing differentially expressed proteins.
4. Retrieve the table produced by the script containing the proteins for which the over- or underexpression is statistically significant between the different experimental conditions.

3.3 Case Study: BY4742 Vs S288c Proteome Comparison

To test and illustrate the different outputs from the various workflows, we compared the differences between the proteomes of two "wild-type" *Saccharomyces cerevisiae* strains with the same genetic background but with one strain (BY4742) harboring the deletion of four genes (Ura3, His3, Leu2, and Lys2) versus the reference strain S288c. The proteomes of the two strains are expected to be very similar (*see* **Note 3** for experimental details).

As shown in Fig. 6, the graphical representation of the distribution of quantification quality scores shows the efficiency of the workflows to identify proteins that are underexpressed (yellow) or overexpressed (blue) in the laboratory wild-type strain BY4742 relative to the reference strain S288c. All the proteins encoded by the genes that are deleted in BY4742 appear as the most significantly diminished (indeed absent) in BY4742, showing the sensitivity and specificity of the proposed quantification methods.

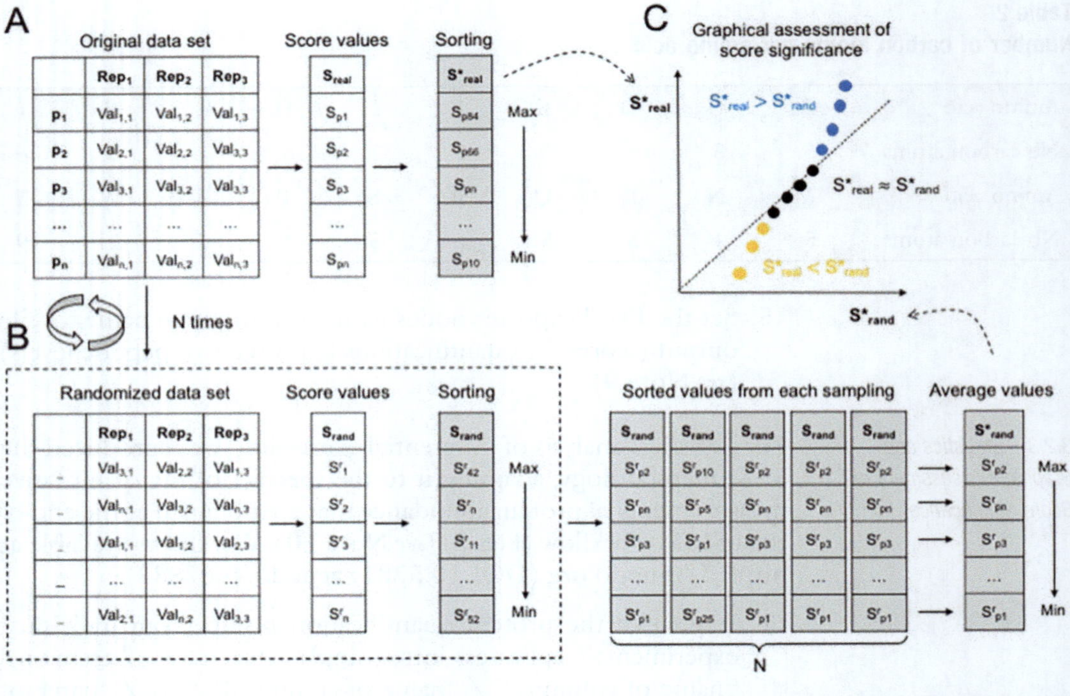

Fig. 5 Schematic representation of the methodology used to assess the relevance of the bSLIM results exported for each protein. (**a**) The original bSLIM output data set is organized as a table in which the proteins are presented in rows and the repetitions of the experiments in columns. For all proteins, score values are calculated and are next sorted from the highest to the lowest. (**b**) Random data set is created from the original dataset, randomly sampling the values in each column. New scores are next calculated and sorted from the highest to the lowest. This process is repeated N times, resulting in a final table with N columns, comprised of the sorted values from each sampling. The average values of all sorted scores are finally calculated. Note that the maximal average value is derived from calculation of the mean between the maximal score values obtained in each sampling. (**c**) The significance of the scores obtained from the original data set are graphically assessed by plotting the S^*real values (**a**) against the S^*rand values (**b**). Significant scores are those that are higher (colored in blue) or lower (colored in yellow) than the random scores. False discovery rates are finally calculated, comparing for each protein score, the average number of other scores in the table of sorted values (from each sampling, *see* (**b**)) that are higher (respectively lower) than the number of scores that are higher (respectively lower) in the original dataset. This is the same method as detailed in [15]

4 Notes

1. The original data sets are publicly available in the ProteomeXchange platform under the Pride submission number PXD021329.
2. The instrumental setup in our laboratory consists of Orbitrap Fusion Tribrid ETD and Orbitrap Q-Exactive Plus mass spectrometers, equipped with Easy-Spray nanoelectrospray ion sources. The LC setup consists of Easy nano-LC Proxeon 1000 or 1200 systems equipped with an Acclaim PepMap100

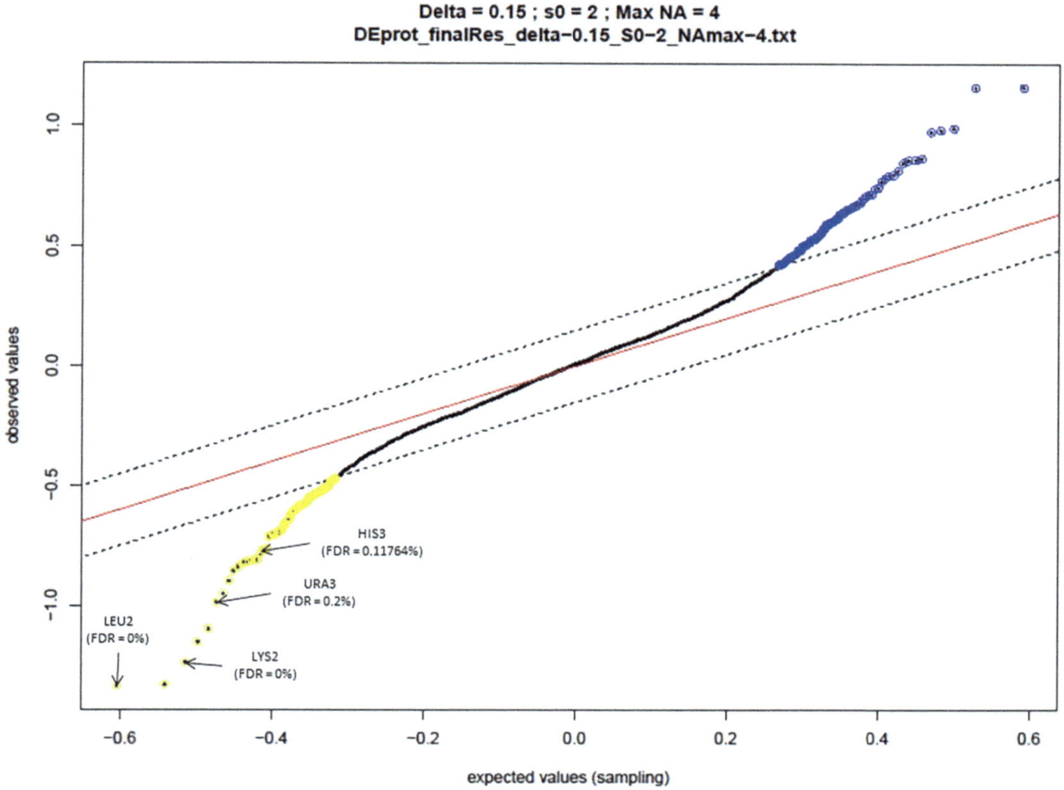

Fig. 6 Experimental distribution of the statistical distribution of the protein quantification quality scores in the characterization of the differences between the BY4742 and S288c proteomes

C18 precolumn and a Pepmap-RSLC Proxeon C18 column. These devices are all from Thermo Fisher Scientific (Bremen, Germany and San Jose, CA, USA).

3. In the case study presented here, the *S. cerevisiae* strains S288c (MATα SUC2 gal2 mal2 mel flo1 flo8-1 hap1 ho bio1 bio6) [16] and its isogenic derivative BY4742 (MATα his3Δ1 leu2Δ0 lys2Δ0 ura3Δ0) are grown on a synthetic medium made of 6.7 g/L Yeast Nitrogen Base (YNB) with ammonium sulfate, without amino acids, with 0.5% glucose as the sole carbon source. The auxotrophy of the BY4742 strain is complemented with uracil (20 mg/L), histidine (20 mg/L), and leucine and lysine (both 30 mg/L). The carbon source was either regular D(+)-glucose anhydrous, defining the normal condition (NC condition), or U-[^{12}C]-glucose (Cambridge Isotope Laboratories), defining the 12C condition. The 10% glucose stock solutions were filter-sterilized.
4. Liquid chromatography coupled to mass spectrometry data acquisition:

In the case study presented here, the chromatographic separation of peptides was performed using the following parameters: Acclaim PepMap100 C18 precolumn (2 cm, 75 μm i.d., 3 μm, 100 Å), Pepmap-RSLC Proxeon C18 column (75 cm, 75 μm i.d., 2 μm, 100 Å), and 300 nL/min flow. The chromatographic separation of peptides was obtained with a gradient consisting of 95% solvent A (water, 0.1% formic acid) to 35% solvent B (99.9% acetonitrile, 0.1% formic acid) in 90 min, followed by column regeneration for 15 min, giving a total run time of 1 h and 45 min.

5. Peptides masses were analyzed in the Orbitrap cell in full ion scan mode at a resolution of 70,000 with a mass range of *m/z* 375–1500 and an AGC target of 3.10^6. MS/MS were performed in a Top 20 DDA mode. Peptides were selected for fragmentation by Higher-energy C-trap Dissociation (HCD) with a Normalized Collisional Energy of 27%, and a dynamic exclusion of 30 s. Fragment masses were measured in the Orbitrap cell at a resolution of 17,500, with an AGC target of 2.10^5. Monocharged peptides and unassigned charge states were excluded from the MS/MS acquisition. The maximum ion accumulation times were set to 50 ms for MS and 45 ms for MS/MS acquisitions respectively.
6. All MS/MS data are processed using the SequestHT (v1.1.0.189) node. The mass tolerance is set to 6 ppm for precursor ions and 0.02 Da for fragments when using an Orbitrap Q-Exactive Plus mass spectrometer. The following alterations are used for various modifications: carbamidomethylation (C), if the sample is reduced and alkylated, and oxidation (M). Phosphorylation (STY) and acetylation (K, N-term) are generally added for additional analyses of trypsin digests. The maximum number of missed cleavages by trypsin is limited to two. MS/MS data are searched against the Uniprot *Saccharomyces cerevisiae* reference proteome UP000002311 (https://www.uniprot.org/proteomes/UP000002311, 6049 protein counts).
7. The Consensus workflow is very basic, because using the Minora node, as presented here, strictly requires that only one results file is processed per run (Fig. 7: Proteome Discoverer 2.4 consensus workflow for presentation).
   (a) The R snippet uses SQL query to link the table of identification with the isotopic intensities. The data are then incorporated in the KNIME workflow.
   (b) The computation is rapid and can be performed as a side analysis during the bSLIM experiment. In cases of auxotrophy, only the amino acids synthesized by the yeast are labeled, whereas the exogenous amino acids that need to

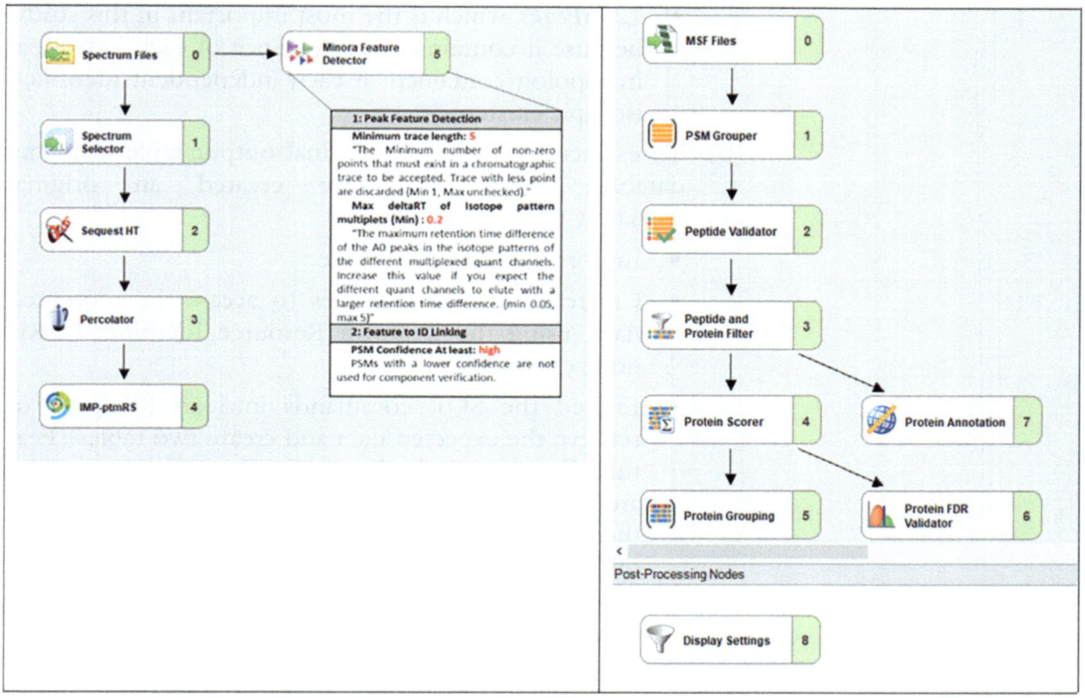

Fig. 7 Details of Proteome Discoverer 2.4 processing (left panel) and consensus (right panel), including the specific Minora node parameters to extract the isotopolog intensities for every isotope cluster

be added to the media are not, resulting in mixed labeling. Quantification is possible with the introduction of a new calculation to accommodate this type of analysis. Experimental data were analyzed using BY4742 auxotrophic yeast.

8. For each identified peptide, we extracted the following items, which are combined in a single output table: FeatureId / MassOverCharge / ParentProteinAccessions / ParentProteinDescriptions / MasterScanNumbers / RetentionTime / Charge / Sequence / Modifications / MonoisotopicMassOverCharge / Area / Intensity / NumberOfIsotopes / MassOverChargeIsotope / PeakHeight (M_0) / PeakArea (M_0) / PeakHeight (M_1) / PeakArea (M_1) / ... / PeakHeight (M_4) / PeakArea (M_4).

   (a) The ".pdresult" file is an SQLite relational database and the information can be accessed using SQL queries. The database contains all PD search parameters, variables, and results. In our quantification workflow, we only need to access three tables that contain the relevant information:

      - *TargetPsms*, containing the peptide IDs.
      - *LcmsFeatures*, with the description of the MS1 cluster used for the identification.

- *LcmsPeaks*, which is the most important in this study, because it contains the abundance of each individual isotopolog contained in each independent identified isotopic cluster.

(b) To extract and produce the final output table from the database (".pdresult"), we created an original workflow to:

- Import the ".pdresults" file.
- Create explicit path names to access the requested data, using the Uniform Resource Identifier (URI) node of KNIME.
- Embed the SQL commands into an RSnippet to retrieve the expected data and create two tables (Feature_Data and Peak_Data in the R code). These tables are further joined using a connection link between them as an intermediate table used as a "dictionary" of ID equivalents.
- Define the ordered rank of the isotopologs extracted by sequential numbering of the lines related to each PSM (Protein Spectrum Match).

(c) Within the R script, the complete SQL request for generating Table Feature_Data is as follows:

```
select TargetPsms.MassOverCharge, TargetPsms.ParentProtei-
nAccessions, TargetPsms.ParentProteinDescriptions, Tar-
getPsms.MasterScanNumbers, TargetPsms.RetentionTime,
TargetPsms.Charge, TargetPsms.Sequence, TargetPsms.Modifica-
tions, LcmsFeatures.MonoisotopicMassOverCharge, LcmsFeatures.
Area, LcmsFeatures.Intensity, LcmsFeatures.NumberOfIsotopes,
LcmsFeatures.Id as FeatureId
 from TargetPsms, TargetPsmsLcmsFeatures, LcmsFeatures
 where TargetPsms.PeptideID = TargetPsmsLcmsFeatures.Tar-
getPsmsPeptideID
 and TargetPsmsLcmsFeatures.LcmsFeaturesId = LcmsFeatures.Id
```

(d) Within the R script, the complete SQL request for generating Table Peak_Data is as follows:

```
select LcmsFeatures.Id as FeatureId, LcmsPeaks.MassOverCh-
arge, LcmsPeaks.PeakHeight, LcmsPeaks.PeakArea
 from LcmsFeatures, LcmsFeaturesLcmsPeaks, LcmsPeaks
 where LcmsFeaturesLcmsPeaks.LcmsFeaturesId = LcmsFeatures.Id
 and LcmsFeaturesLcmsPeaks.LcmsPeaksId = LcmsPeaks.Id
```

(e) The intensity of the isotopolog ions is defined by the peak area.

(f) We restrict the number of isotopologs during extraction in the final dataset to five, as we only require M_0 and M_1 for further quantification calculations.

9. The produced table is used in the bSLIM workflow for complete or incomplete labeling with the correct exogenous (non-labeled) amino acids given in parameters. The code proceeds to the ratio of M_1 over M_0 to quantify the molar fraction, the key variable for the quantification. The ratio of the molar fraction/(1-molar fraction) is calculated. For protein levels, all top N peptides log2(Ratio) are grouped together to obtain classical fold changes for biological interpretations.
10. A key question in proteomics data analysis is the distinction between noteworthy (or significant) results from other observations, which are false positives, that is, acquired by random chance. Indeed, the large amount of data arising from proteomics technologies is associated with an increase in the possibility to observe atypical "by chance" values in the dataset. In this context, statistical methodologies generally assume that all variations in the data are due to random fluctuations and, accordingly, derive a probability to observe variations that are greater than those present in the data. Random fluctuations can be modelled in two different ways. In the first, a mathematical function is chosen (often normal or student laws) and a statistical hypothesis is used to discriminate "significant" from "nonsignificant" observations, based on a predefined error rate (generally 5%). In the second, random permutations of the original dataset are performed to define empirical distributions, which will be used to assess potential random fluctuations. It is a remarkably interesting approach, especially when the theoretical probability distributions of the studied parameters are not demonstrated, as is the case with the bSLIM output dataset.

Acknowledgments

We would like to thank Dr. Bernard Delanghe (Thermo Fisher Scientific) for providing us with the table structure of the ".pdresult" file that allowed us to develop the present workflow. NS received a thesis grant from CRI-Paris. This work was supported by the ARN, grant ANR-18-CE44-0014. The English text was edited by Alex Edelman & Associates (http://www.alexedelman.com/).

References

1. Goffeau A, Barrell BG, Bussey H, Davis RW, Dujon B, Feldmann H, Galibert F, Hoheisel JD, Jacq C, Johnston M et al (1996) Life with 6000 genes. Science 274:546. 563–567
2. Wilm M (2009) Quantitative proteomics in biological research. Proteomics 9:4590–4605
3. Chahrour O, Cobice D, Malone J (2015) Stable isotope labelling methods in mass spectrometry-based quantitative proteomics. J Pharm Biomed Anal 113:2–20
4. Leger T, Garcia C, Videlier M, Camadro JM (2016) Label-free quantitative proteomics in yeast. Methods Mol Biol 1361:289–307
5. Leger T, Garcia C, Collomb L, Camadro JM (2017) A simple light isotope metabolic labeling (SLIM-labeling) strategy: a powerful tool to address the dynamics of proteome variations in vivo. Mol Cell Proteomics 16:2017–2031
6. Senko MW, Beu SC, McLafferty FW (1995) Automated assignment of charge states from resolved isotopic peaks for multiply charged ions. J Am Soc Mass Spectrom 6:52–56
7. Alves G, Ogurtsov AY, Yu YK (2014) Molecular isotopic distribution analysis (MIDAs) with adjustable mass accuracy. J Am Soc Mass Spectrom 25:57–70
8. Ljungdahl PO, Daignan-Fornier B (2012) Regulation of amino acid, nucleotide, and phosphate metabolism in Saccharomyces cerevisiae. Genetics 190:885–929
9. Sénécaut N, Alves G, Weisser H, Lignieres L, Terrier S, Yang-Crosson L, Poulain P, Lelandais G, Yu YK, Camadro JM (2021) Novel insights into quantitative proteomics from an innovative bottom-up simple light isotope metabolic (bSLIM) Labeling data processing strategy. J Proteome Res 20:1476–1487
10. Sturm M, Bertsch A, Gropl C, Hildebrandt A, Hussong R, Lange E, Pfeifer N, Schulz-Trieglaff O, Zerck A, Reinert K et al (2008) OpenMS - an open-source software framework for mass spectrometry. BMC Bioinform 9:163
11. Rost HL, Sachsenberg T, Aiche S, Bielow C, Weisser H, Aicheler F, Andreotti S, Ehrlich HC, Gutenbrunner P, Kenar E et al (2016) OpenMS: a flexible open-source software platform for mass spectrometry data analysis. Nat Methods 13:741–748
12. Pfeuffer J, Sachsenberg T, Alka O, Walzer M, Fillbrunn A, Nilse L, Schilling O, Reinert K, Kohlbacher O (2017) OpenMS—a platform for reproducible analysis of mass spectrometry data. J Biotechnol 261:142–148
13. Doblmann J, Dusberger F, Imre R, Hudecz O, Stanek F, Mechtler K, Durnberger G (2019) apQuant: accurate label-free quantification by quality filtering. J Proteome Res 18:535–541
14. Griss J, Stanek F, Hudecz O, Durnberger G, Perez-Riverol Y, Vizcaino JA, Mechtler K (2019) Spectral clustering improves label-free quantification of low-abundant proteins. J Proteome Res 18:1477–1485
15. Tusher VG, Tibshirani R, Chu G (2001) Significance analysis of microarrays applied to the ionizing radiation response. Proc Natl Acad Sci U S A 98:5116–5121
16. Mortimer RK, Johnston JR (1986) Genealogy of principal strains of the yeast genetic stock center. Genetics 113:35–43

Chapter 17

A Strong Cation Exchange Chromatography Protocol for Examining N-Terminal Proteoforms

Esperanza Fernández, Annelies Bogaert, Evy Timmerman, An Staes, Francis Impens, and Kris Gevaert

Abstract

Especially in eukaryotes, the N-terminal acetylation status of a protein reveals translation initiation sites and substrate specificities and activities of N-terminal acetyltransferases (NATs). Here, we discuss a bottom-up proteomics protocol for the enrichment of N-terminal peptides via strong cation exchange chromatography. This protocol is based on depleting internal tryptic peptides from proteome digests through their retention on strong cation exchangers, leaving N-terminally acetylated/blocked peptides enriched among the nonretained peptides. As such, one can identify novel N-terminal proteoforms and quantify the degree of N-terminal protein acetylation.

Key words N-terminomics, N-terminal acetylation, N-terminal acetyltransferases

1 Introduction

The N-terminus of a protein may be part of a functional motif or domain that dictates protein interactions [1–3], influences protein function and stability [4–6], and governs protein localization [7]. A high number of N-terminal proteoforms, these being proteins stemming from the same gene but with different N-termini (e.g., due to several ways of transcript splicing or different types and degrees of modifications of the protein's alpha-amino group), exist and some have already been linked to human diseases [8]. Of note, about 80–90% of the cytosolic proteins of (higher) eukaryotes are estimated to be N-terminally acetylated, a typical cotranslational modification catalyzed by N-acetyltransferases (NATs) that transfer an acetyl group from acetyl-coA to the free α-amino group of a nascent protein [9]. As this modification occurs during translation, identifying it serves as a proxy for translation initiation events [10].

Frédéric Devaux (ed.), *Yeast Functional Genomics: Methods and Protocols*,
Methods in Molecular Biology, vol. 2477, https://doi.org/10.1007/978-1-0716-2257-5_17,

N-terminomics is a branch of proteomics that aims at specifically analyzing such N-terminal parts of proteins to provide functional insights on protein N-termini [11, 12]. N-terminomics approaches have been applied to identify N-terminal modifications [13–18], to characterize protease specificity and protease substrates [19–25], and to identify N-terminal proteoforms [4, 26, 27]. Several such techniques have been developed, and these are typically classified as positive selection techniques, directly enriching N-terminal peptides, or as negative selection techniques, indirectly enriching N-terminal peptides by depleting internal peptides [11, 12]. Our lab was the first to develop an N-terminomics approach, a negative selection technique called N-terminal COFRADIC (COmbined FRActional DIagonal Chromatography) [20]. Since then, COFRADIC has been tailored and regularly adapted to specific applications. In a previous book series, we elaborated on the application of COFRADIC to profile protease substrates [28]. Here, we describe a method that focuses on detecting the degree of in vivo N-terminal protein acetylation, also allowing to identify new translation initiation sites besides monitoring the substrate specificity and activity of N-terminal acetyltransferases [29]. In this updated protocol, we eliminate the series of consecutive reverse-phase (RP)-HPLC fractionation steps in the COFRADIC protocol, originally called the primary and secondary runs [28, 30]. RP-HPLC fractionations tend to be lengthy and were considered the main hurdle of the overall COFRADIC workflow. Due to the very high scan speed and increased sensitivity of current-day mass spectrometers [31], the strong cation exchange (SCX) chromatography step that preceded the RP-HPLC fractionations, coupled to mass tagging of amines (for identifying the degree of N-terminal modification), was found to suffice for in-depth N-terminomics profiling [29]. Indeed, such an approach gives an enrichment of N-terminal peptides of over 90%, similar to the original COFRADIC workflow [30, 32]. Furthermore, in this updated version, we also include the possibility of changing the required protein desalting steps in the COFRADIC protocol for Amphipols (APols) A8-35, a desalting polymer. APols are a class of mild detergents that can be used for protein extraction [33]. In acidic conditions, APols-protein micelles co-precipitate even at low protein concentration ratios, providing a fast method for concentrating proteins and also for desalting and buffer-exchange through the resuspension of the APols-proteins micelles into a neutral pH buffer [34].

The steps of this SCX procedure are outlined in Fig. 1. In summary, protein cysteines are reduced and blocked by alkylation, upon which alpha-amino groups (from N-terminal amino acids) and epsilon-amino groups (from lysine side-chains) are acetylated with an N-hydroxysuccinimide (NHS) ester of a $^{13}C_1D_3$-form of

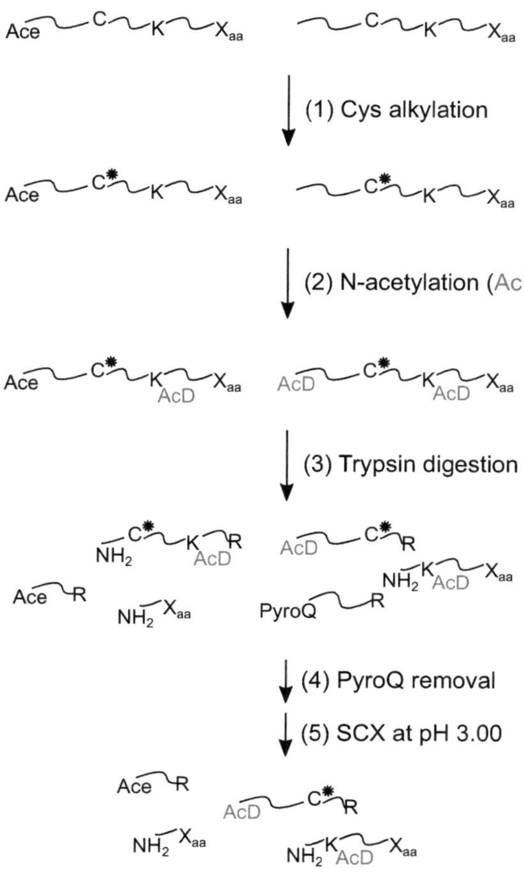

Fig. 1 Schematic workflow of the N-terminomics workflow. *Ace* in vivo acetylated N-terminal amino acid, *AcD* in vitro acetylated N-terminal amino acid and Lys, *C** alkylated Cys, X_{aa} C-terminal amino acid, NH_2 tryptic peptide, *PyroQ* pyro glutamate, *SCX* strong cation exchange

acetic acid (see further). Following trypsin digestion, subsequent steps include the removal of cyclized N-terminal glutamines (pyroglutamate) and lowering of the pH to 3.00. The peptide mixture then undergoes N-terminal peptide enrichment through SCX chromatography at pH 3.00 in which peptides with a net charge of zero or with a negative net charge (mainly N-terminally acetylated peptides, C-terminal peptides and phoshopeptides at this pH) (Fig. 2) will not be retained on the SCX-resin, thus flowing through the column for collection and subsequent analysis by liquid chromatography–tandem mass spectrometry (LC-MS/MS).

Important for quantitative N-terminal acetylome studies is the use of the NHS ester of $^{13}C_1D_3$-acetic acid (AcD), 4 Da heavier than the natural acetyl group (Ace). AcD therefore allows to distinguish in vitro from in vivo acetylated N-termini based on this mass

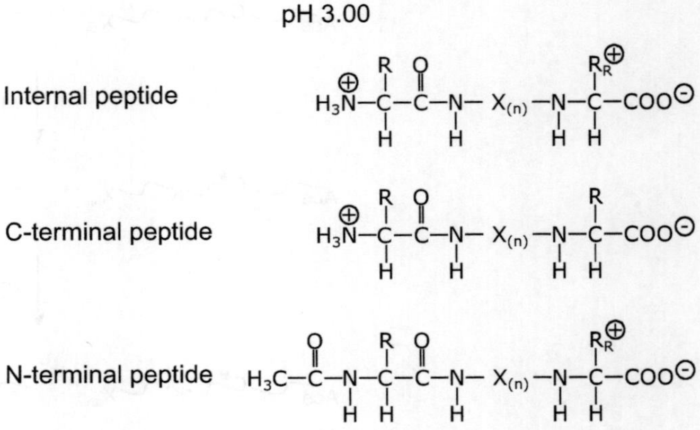

Fig. 2 Charges of an internal tryptic peptide, C-terminal tryptic peptide and N-terminally acetylated tryptic peptide at a pH of 3.00. *R* any amino acid side chain, R_R arginine side chain, $X_{(n)}$ any amino acid (*see* **Note 19**)

difference observed in MS-spectra. Further, the ratio of the signal intensities from both forms of N-terminal peptides is a proxy for the degree of in vivo acetylation for studying the substrate specificities and repertoires of N-terminal acetyltransferases (NATs) [35] or of any other protein, molecule, molecular pathway, or cellular condition that may have an impact on NAT activity.

2 Materials

2.1 pGAPase Purification

1. TAGZyme Qcyclase/pGAPase enzymes (Qiagen).
2. His SpinTrap (GE Healthcare).
3. Binding Buffer: 20 mM imidazole in 20 mM ammonium bicarbonate.
4. Elution Buffer: 500 mM imidazole in 20 mM ammonium bicarbonate.

2.2 Cell Lysis and Protein Extraction

1. cOmplete™ EDTA-free protease inhibitor cocktail, tablets.
2. Lysis buffer: 50 mM Tris–HCl, pH 7.6, 12 mM EDTA, 250 mM NaCl, 140 mM Na_2HPO_4, cOmplete™ protease cocktail tablet (*see* **Note 1**).
3. PBS: 135 mM NaCl, 2.7 mM KCl, 7 mM Na_2HPO_4.
4. Glass beads (425–600 μm size).
5. Amphipols A8-35 (APols).
6. Guanidinium hydrochloride (Gu.HCl).
7. Bio-Rad DC Protein Assay Kit.

2.3 Sample Preparation for N-Terminal Peptide Enrichment

1. Tris(2-carboxyethyl)phosphine hydrochloride (TCEP–HCl).
2. Iodoacetamide.
3. Sodium hydroxide.
4. Formic acid.
5. Disposable desalting columns packed with Sephadex TM G-25 (GE Healthcare).
6. Guanidinium hydrochloride (Gu.HCl).
7. N-hydroxy-succinimidyl $^{13}C_1D_3$-acetate [28, 30].
8. Acetylation buffer: 50 mM sodium phosphate, pH 8.0.
9. Glycine.
10. Hydroxylamine.
11. Ammonium bicarbonate.
12. Triethyl ammonium bicarbonate (TEAB).
13. Bio-Rad DC Protein Assay Kit.
14. Sequencing grade modified trypsin.

2.4 Pyroglutamate Removal

1. Purified TAGZyme Qcyclase/pGAPase enzymes (Qiagen, Manchester, UK) (*see* Subheading 3.2).
2. Pyro-Glu buffer: 10 mM NaCl, 0.5 mM EDTA, pH 8.0, 3 mM cysteamine, and 50 μM aprotinin.

2.5 Strong Cation Exchange (SCX) Enrichment of N-Terminal Peptides

1. Trifluoroacetic acid 1% (v/v) in water.
2. Clean-Up PCX2 SPE Cartridges 50 mg, 200 mg, or 500 mg (United Chemical Technologies) (*see* **Note 2**).
3. 12-port vacuum manifold for SPE cartridges (Agilent).
4. SS needles (Agilent).
5. Manifold Luer plugs (Agilent).
6. Vacuum pump.
7. SCX solution A1: 60 mg sodium phosphate monobasic + 50 mL H_2O.
8. SCX solution A2: 57.6 mg H_3PO_4 + 50 mL H_2O.
9. SCX solution B1: 60 mg sodium phosphate monobasic + 15 mL H_2O + 35 mL acetonitrile, HPLC grade.
10. SCX solution B2: 57.6 mg H_3PO_4 + 15 mL H_2O + 35 mL acetonitrile, HPLC grade.
11. SCX buffer A: combine SCX solutions A1 and A2 until the pH reaches 3.0 (approximately A1/A2 3/1 (v/v)).
12. SCX buffer B: combine SCX solutions B1 and B2 until the pH reaches 3.0 (approximately B1/B2 1/2.5 (v/v)).

13. 5 M sodium hydroxide.
14. 5 mM sodium chloride.

2.6 Oxidation and Cleanup of Purified N-Terminal Peptides

1. Hydrogen peroxide, 30% (w/w).
2. Bond-Elut OMIX C18 tips (Agilent).
3. Preequilibration solvent: 80% acetonitrile, HPLC grade, 0.1% trifluoroacetic acid.
4. Solvent A: 0.1% trifluoroacetic acid.
5. Solvent B: 60% acetonitrile, HPLC grade, 0.1% trifluoroacetic acid.

3 Methods

3.1 pGAPase Purification

The pGAPase (pyroglutamyl aminopeptidase) enzyme in the TAG-Zyme Qcyclase/pGAPase enzymes kit contains a polymer that interferes with LC-MS analysis through peptide ionization suppression (*see* Fig. 2 in [28]). This section describes the removal of this polymer prior to the use of the enzyme. As pGAPase contains a poly His-tag, it can be purified by immobilized metal ion affinity chromatography (IMAC) [28].

1. Invert and shake the His SpinTrap column to resuspend the medium.
2. Twist off the bottom closure and loosen the cap a little.
3. Place the column in a 2 mL Eppendorf tube and spin for 30 s at 100 × *g* to remove the storage solution.
4. Equilibrate the spin column by loading 600 μL binding buffer and let it spin for 30 s at 100 × *g*.
5. Add 400 μL (10 U) of enzyme (each vial of pGAPase contains 400 μL) (*see* **Note 3**).
6. Spin the column for 30 s at 100 × *g*.
7. Wash the spin column by applying 600 μL of binding buffer and spin for 30 s at 100 × *g*.
8. Put the spin column in a fresh Eppendorf tube.
9. Elute the enzyme by adding 200 μL elution buffer and spin for 30 s at 100 × *g*.
10. Repeat **step 9**.
11. Aliquot the purified pGAPase (25 μL per Eppendorf tube (approximately 625 mU)) and store at −20 °C.
12. To assess removal of the polymer, analyze an aliquot of the purified enzyme by LC-MS after diluting the sample 100 times (*see* Fig. 2 in [28]).

3.2 Cell Lysis and Protein Extraction

The enrichment of N-terminal peptides can be performed on a wide range of samples. Here, we provide two methods compatible with N-terminal enrichment of peptides through SCX. The methods differ in the type of the protein denaturing compounds used and must be chosen according to the protocol application (*see* **Note 4**).

3.2.1 Yeast Cell Wall Disruption

1. Grow 300 mL of yeast and harvest the cells in the exponential growth phase ($OD_{600\ nm} = 3.0$).
2. Pellet the cells and wash them twice with 50 mL PBS.
3. Resuspend the cells in 1 mL of yeast lysis buffer.
4. Add 0.6 g of acid-washed glass beads and vortex 10 times for 30 s (place the beads on ice for 30 s in between).
5. Centrifuge the cell lysates for 10 min at 7500 × *g* at 4 °C.
6. Collect the supernatant and determine the protein concentration using the Bio-Rad DC Protein Assay Kit.
7. Continue with the sample preparation. The yeast protein extract can be solubilized either with 4 M Gu.HCl (Subheading 3.2.2) or APols at 1 mg/mL (Subheading 3.2.3).

3.2.2 Sample Preparation Using Guanidinium Hydrochloride

1. Take a volume of supernatant corresponding to 2 mg of protein material and add solid G. HCl (764 mg) to a final concentration of 4 M (*see* **Notes 5** and **6**).
2. Adjust the volume to 2 mL with 50 mM sodium phosphate, pH 7.9.
3. Take 1 mg of sample to perform the rest of the protocol (*see* **Note 7**).

3.2.3 Sample Preparation Using APols

1. Take the supernatant and adjust the protein concentration to 1 mg/mL with 50 mM sodium phosphate, pH 7.9.
2. Take 1 mg of protein and add Apols (1 mg/mL) (*see* **Notes 5** and 7).
3. Leave the proteins on ice for 30 min.

3.3 Sample Preparation for N-Terminal Peptides Enrichment

We recommend to start with at least 1 mg of protein material at an ideal concentration of 1 mg/mL.

1. Reduce and alkylate the protein extract by adding a final concentration of 15 mM TCEP-HCl and 30 mM iodoacetamide. Adjust the pH of the sample to 7.60–7.90 with 1 M sodium hydroxide. Let the reaction proceed for 30 min at room temperature in the dark (*see* **Notes 8** and **9**). The excess of reduction and alkylation reagents are removed through different methods according to the sample preparation procedure used.
2. For the protein mixture that was obtained as described in Subheading 3.2.2, desalt on a NAP-10 column (*see* **Note 10**)

in freshly prepared 1.33 M Gu.HCl in acetylation buffer. Reduce the volume of the resulting 1.5 mL eluate to 1 mL by vacuum drying to obtain a final concentration of Gu.HCl of 2 M.

3. For the protein mixture that was obtained as described in Subheading 3.2.3, acidify by adding 10% formic acid until the sample reaches a pH of 3.5–3.7 (*see* **Note 11**). Incubate the sample for 5 min at room temperature. Precipitate the micelles formed by APols and proteins by centrifuging for 10 min at 16,000 × *g* at room temperature and discard the supernatant. Wash the pellet by adding 500 μL of 0.1% (vol/vol) formic acid without disturbing the pellet and centrifuging for 5 min at 16,000 × *g* at room temperature. Discard the supernatant (*see* **Note 12**). Resuspend the pellet containing the proteins and APols in 1 mL of acetylation buffer.
4. Prepare a stock solution of 500 mM N-hydroxy-succinimidyl $^{13}C_1D_3$-acetate reagent in acetonitrile/water, 1/1 (v/v).
5. Add the reagent to the sample to a final concentration of 10 mM (*see* **Note 13**). Incubate the reaction for 60 min at 30 °C while shaking at 700 rpm.
6. Repeat **step 5**, resulting in a final concentration of 20 mM of the NHS-ester.
7. To quench unreacted NHS-ester, add 40 μL of 1 M glycine (2× molar excess of glycine over the NHS-ester) and incubate for 10 min at 30 °C (*see* **Note 14**).
8. Reverse potential O-acylation of Ser, Thr or Tyr side-chains by adding 4 μL of hydroxylamine (4× molar excess over the NHS-ester) and incubate for 10 min at 30 °C (*see* **Note 15**). Remove the chemical reagents through different methods according to the protein extraction procedure.
9. For proteins that were obtained as described in Subheading 3.2.2, desalt on a NAP-10 column in 10 mM freshly prepared ammonium bicarbonate, pH 8.0.
10. For proteins that were obtained as described in Subheading 3.2.3, acidify by adding 10% formic acid until the sample reach a pH of 3.5–3.7. Incubate the sample for 5 min at room temperature. Precipitate the micelles formed by APols and proteins by centrifuging for 10 min at 16,000 × *g* at room temperature and discard the supernatant. Wash the pellet with 500 μL of 0.1% (vol/vol) formic acid and centrifuging for 5 min at 16,000 × *g* at room temperature. Discard the supernatant. Resuspend the pellet containing the APols and proteins in 1 mL of 50 mM TEAB buffer, pH 8.0.

11. Determine the protein concentration using the Bio-Rad DC Protein Assay Kit when the proteins were desalted on a NAP-10 column (*see* **Note 16**).
12. Heat the sample for 5 min at 95 °C and place directly on ice for 10 min.
13. Add trypsin in a trypsin/substrate ratio of 1/50 (w/w) and incubate overnight at 37 °C. The trypsin reaction is continued according to the sample preparation procedure.
14. Vacuum-dry the peptide mixture from proteins that were obtained as described in Subheading 3.2.2.
15. Acidify the peptide mixture from proteins that were obtained as described in Subheading 3.2.3 by adding 10% formic acid until the sample reach a pH of 3.5–3.7. Incubate the sample for 5 min at room temperature. Precipitate the micelles formed by APols and undigested proteins by centrifuging for 10 min at 16,000 × *g* at room temperature and discard the pellet. Collect the supernatant containing the peptides and dry completely under vacuum (*see* **Note 17**).

3.4 Pyroglutamate Removal

Peptides with an N-terminal glutamine may cyclize into pyroglutamate (pyro-Glu) catalyzed by the presence of weak acid at elevated temperatures and an acidic pH [36, 37]. Since pyro-Glu-starting peptides are less basic than peptides starting with a primary α-amine, they will be poorly retained by the SCX resin and coelute with N-terminal peptides, by which they somehow "contaminate" the samples. Therefore, to remove these peptides, a new primary α-amine is introduced upon enzymatically removing the pyro-Glu residue.

1. Add 1 μL of 800 mM NaCl, 1 μL of 50 mM EDTA, pH 8.0 and 11 μL of freshly prepared 50 mM cysteamine to a 25 μL aliquot of purified pGAPase (Subheading 3.1) and activate the pGAPase by incubation at 37 °C for 10 min.
2. Redissolve the peptide sample in 212 μL of freshly prepared pyro-Glu buffer (*see* **Note 18**).
3. Add the activated pGAPase (625 mU) to the sample together with 25 μL Qcyclase (1250 mU) and incubate for 1 h at 37 °C to remove all N-terminal pyroglutamate residues.

3.5 Strong Cation Exchange (SCX) Enrichment of N-Terminal Peptides

As a result of the protein modification steps, N-terminal tryptic peptides carry a net charge of zero at pH 3.00. Hence, at this pH, such peptides will not be retained on the SCX resin. In contrast, internal tryptic peptides which are positively charged at pH 3.00 remain bound to the resin, allowing efficient enrichment of N-terminal peptides (*see* **Note 19**).

1. Add 700 μL of SCX solution A2 to the pyroglutamate reaction and monitor the pH.
2. Acidify the sample to pH 3.00 by dropwise addition of 1% (vol/vol) trifluoroacetic acid (*see* **Note 20**).
3. Centrifuge the sample for 15 min at 16,000 × *g* at room temperature (*see* **Note 21**).
4. Take the supernatant and measure the peptide concentration.
5. Remeasure the pH and adjust to 3.00 if necessary.
6. Put a Clean-Up PCX2 SPE cartridge on a vacuum manifold and equilibrate the cartridge subsequently with 1 mL acetonitrile, 1 mL H_2O, and 2 mL of SCX buffer A (*see* **Notes 22** and **23**).
7. Load the sample onto the equilibrated cartridge.
8. Add 1 mL SCX solvent A to the cartridge (*see* **Note 24**).
9. Disrupt hydrophobic interactions by adding 5 mL of SCX buffer B to the cartridge and collect the eluted N-terminal peptides in a 15 mL tube.
10. Add 1 mL of SCX buffer B containing 5 mM NaCl to the cartridge and collect the eluate in the 15 mL tube of **step 9** (*see* **Note 25**). For the 500 mg cartridge, add 1 mL of SCX buffer B to the cartridge and collect the eluate in the 15 mL tube of **step 9** (*see* **Note 26**).
11. Increase the pH of the eluate to about 6 by dropwise addition of 5 M sodium hydroxide (*see* **Note 27**).
12. Dry the collected peptides under vacuum until the volume reaches approximately 500 μL.
13. Collect the 500 μL and transfer them to a LoBind 1.5 mL Eppendorf tube.
14. Add 750 μL of 80% (vol/vol) acetonitrile to the 15 mL tube, vortex thoroughly and transfer the solution to the LoBind 1.5 mL Eppendorf tube containing the peptides (*see* **Note 28**).
15. Dry the collected peptides into the LoBind 1.5 mL Eppendorf tube completely under vacuum.

3.6 Oxidation and Clean-up of Purified N-Terminal Peptides

The sulfur atom of methionine is susceptible to oxidize to a sulfoxide derivative when exposed to air. Therefore, we suggest to oxidize all methionines to methionine-sulfoxide to create uniform and more oxidation-stable peptide mixtures [32].

1. Redissolve the peptide sample in 120 μL of solvent A.
2. Make a 3% (w/w) hydrogen peroxide solution in water, add 30 μL of this solution to the sample, and incubate for 30 min at 30 °C to oxidize methionine to methionine-sulfoxides (*see* **Note 29**).

3. Acidify with 8 μL of trifluoroacetic acid and immediately proceed to desalt the peptides.
4. Preequilibrate the Bond-Elut OMIX C18 tips with the preequilibration solvent by pipetting fresh solvent for 5 times.
5. Remove the preequilibration solvent by pipetting 150 μL of solvent A for 5 times (*see* **Note 30**).
6. Pipet the acidified peptide solution for 20 times in order for the peptides to bind to the C18 beads.
7. Wash the peptides by pipetting 150 μL of solvent A for 3 times.
8. Elute the peptides by pipetting 70 μL of solvent B for 10 times.
9. Repeat the previous step and combine both eluates.
10. Transfer the peptides to an HPLC vial and dry completely under vacuum.
11. Send samples to LC-MS/MS analysis. Any LC-MS/MS method and mass spectrometer instrument can be applied.

3.7 Data Analysis

Data analysis includes the creation of peak list files from the recorded mass spectra and their search against a protein sequence database. We typically use the Mascot search engine to perform these tasks. We also recommend to use Mascot Daemon as controller software and Mascot Distiller to create peak lists. These allow to quantify MS and MS/MS data and calculate the labeling ratios for each peptide identified. Any other search engine and data analysis package able to perform these tasks can also be used. A simultaneous search for N-terminal acetylated peptides and tryptic peptides other than N-terminal acetylated is not possible with Mascot. Therefore, two independent searches must be done: one specific search for N-terminal acetylated peptides, to analyze the degree of labeling (acetylation), and second and independent search to identify tryptic peptides other than N-terminally acetylated peptides that have also been eluted through the SCX column (*see* **Note 31**).

3.7.1 Search Settings for N-Terminally Acetylated Peptides

1. Set fixed modifications to methionine oxidation (sulfoxide), $^{13}C_1D_3$-acetate (AcD) of lysine side-chains and S-carbamidomethylation of cysteine residues.
2. Set the quantitation to acetylation (Ace) versus $^{13}C_1D_3$-acetate (AcD) of peptide alpha-N-termini.

3.7.2 Search Settings for Peptides Other Than N-Terminally Acetylated Peptides

1. Set fixed modifications to methionine oxidation (sulfoxide), $^{13}C_1D_3$-acetylation (AcD) of lysine side-chains and S-carbamidomethylation of cysteine residues.
2. Set variable modifications to pyroglutamate formation of Gln-starting peptides (*see* **Note 32**). Pyro-carbamidomethyl formation of Cys-starting peptides and oxidation of S-carbamidomethyl cysteine residues can optionally be set as additional variable modifications.

The following search settings are common for Subheadings 3.7.1 and 3.7.2:

3. Set enzyme setting to semi-ArgC/P. Although trypsin has been used, only cleavage C-terminal to arginine residues can occur as trypsin cleavage after lysine residues is inhibited by acetylation of the epsilon amino group (*see* **Note 33**).
4. Semi-ArgC/P cleavage allows to identify Arg cleavage followed by a Pro, that otherwise will be missed.
5. Missed cleavages are set to zero as peptides with one or more missed cleavages will be likely retained in the SCX resin.
6. Other settings such as the mass tolerance of the precursor and fragment ions are specific for the mass spectrometer used.

For Mascot identifications, we suggest only peptides scoring above the threshold of a 99% confidence level are retained (*see* **Note 34**).

4 Notes

1. Dissolve 1 tablet in 100 mL of buffer instead 50 mL as described by the manufacturer's instructions. High concentrations of protease inhibitors can cause interfering signals during mass spectrometric analysis.
2. These Clean-Up PCX2 SPE cartridges substitute the Bond-Elut cartridges (Agilent) previously used to perform SCX chromatography. The latter have been modified by the supplier and were found to contain polymers that affect LC-MS/MS analysis.
3. The maximum capacity of this spin column is 600 μL and 750 μg of His-tagged protein.
4. APols is a mild detergent that solubilizes the plasma membrane, the membranes of the nucleus and other organelles, thus solubilizing membrane proteins. Due to its property of precipitating proteins in acidic conditions, it can be used as a fast desalting approach. However, this property is lost when used in combination with most detergents—except for NP-40—and chaotropes. Furthermore, APols is compatible with trypsin at concentrations that range up to 10 mg/mL, and also compatible with mass spectrometry analysis. The addition of Gu.HCl helps to denature the proteins and to abolish any remaining protease activity.
5. The minimal protein starting amount for N-terminal peptide enrichment following cell lysis is 1 mg. As a rule of thumb, we suggest to start with 300 mL of yeast culture at $OD_{600} = 3.0$ that will yield approximately 3 mg of protein following cell disruption. Samples should be kept on ice as much as possible.

6. Addition of this amount of Gu.HCl increases the sample volume with about 33%.
7. The rest of the lysate can be frozen at −80 °C.
8. The reduction/alkylation step will also inhibit cysteine protease activity and prevents protein refolding.
9. Always prepare a fresh iodoacetamide stock solution. Adjust the TCEP-HCl solution to pH 8 with 5 M sodium hydroxide by adding about 60 μL of 5 M sodium hydroxide to 200 μL of 750 mM TCEP-HCl (check the pH with indicator strips), further lowering the TCEP-HCl stock concentration to 570 mM. Adjusting this pH, and if necessary also the pH of the reaction mixture, is critical to ensure efficient and complete alkylation of cysteine residues. Note that longer incubation times should be avoided as these will promote alkylation at other groups, for instance of the epsilon-amines of lysine side-chains [27, 36]. Also, the reaction should proceed in the dark as iodoacetamide is a light-sensitive reagent.
10. Depending on the sample volume, different disposable desalting columns packed with Sephadex TM G-25 can be used. NAP-10 columns can load a maximum of 1 mL of sample, while 0.5 mL and 2.5 mL can be loaded on NAP-5 and PD-10 columns, respectively.
11. Proteins and APols form aggregates at acidic pH, which can be observed by the formation of a cloudy solution.
12. If the lysis buffer contains NP-40, repeat 3 times the washing step with 0.1% (vol/vol) formic acid to remove any residual NP-40.
13. The concentration of acetonitrile in the reaction must be below 5%.
14. Addition of glycine ensures, together with the subsequent desalting step, that no NHS-reagent is left during trypsin digestion as this could lead to unwanted acetylation of the alpha-amino groups of newly generated tryptic peptides.
15. Hydroxylamine should be incubated at least for 10 min to ensure complete reversion of O-acylation of Thr, Ser, and Tyr residues. Incubations longer than 30 min should be avoided to prevent chemical cleavage of proteins.
16. Protein concentration cannot be determined in APols lysates by using acidic protein assays as the Bio-Rad DC Protein Assay Kit. Protein concentration can be determined by assays based on the reaction of Cu^{1+} with bicinchoninic acid.
17. Peptides are not precipitated by APols and remain in the supernatant [26, 37].

18. The pyro-Glu buffer contains the protease inhibitor aprotinin, which will prevent trypsin from cleaving pGAPase and Qcyclase (glutamine cyclotransferase).
19. Peptides containing a histidine residue carry an extra positive charge at a pH of 3.0. Therefore, N-terminal tryptic peptides carrying a histidine will be retained by the SCX resin.
20. The pH has to be between 2.95 and 3.05, so if possible use a pH probe rather than a pH indicator paper.
21. A small pellet containing undigested proteins and trypsin could be observed.
22. The size of the PCX2 SPE Cartridges is chosen according to the amount of peptide material. 50 mg cartridges are used for <50 μg of peptides; 200 mg cartridges for 50–150 μg of peptides; and 500 mg cartridges for 150–500 μg of peptides.
23. These volumes are used for cartridges containing 50 mg and 200 mg of resin. The volumes have to be increased three times when using 500 mg cartridges.
24. The PCX2 SPE cartridges are mixed-bed mode cartridges containing both C8 and SCX resins. The C8 resin retains peptides based on hydrophobic interactions, while the SCX resin retains peptides based on ionic interactions. In this way, N-terminal peptides with a net charge of zero or a negative charge at pH 3.00 will be retained on the C8 resin through hydrophobic interaction. Compared to a SCX resin-only system, the mixed bed resin renders the preenrichment step less sensitive to the presence of salts. Even in case of a moderate salt concentration in the samples, N-terminal peptides are initially retained in SCX buffer A by hydrophobic interactions with the C8 resin. Once the salt is washed away, hydrophobic interactions are disrupted by acetonitrile present in SCX buffer B, and retention is solely based on ion exchange.
25. Adding a small concentration of sodium chloride helps to elute N-terminal peptides.
26. The bed volume of 500 mg cartridges is about 1.5 mL. We add 1 mL more of SCX buffer B to allow the SXC buffer B containing 5 mM NaCl elutes from the cartridge.
27. During vacuum drying extreme pH conditions should be avoided. Therefore, we increase the pH after the SCX preenrichment step to 6.
28. We perform a washing step with 80% acetonitrile to collect eluted peptides that might adhere to the tube inner walls following vacuum drying.
29. The oxidation reaction uses 0.6% H_2O_2. The reaction should be stopped after 30 min to prevent further, less controlled oxidation of methionine sulfoxides to their sulfone form.

30. All solvents are discarded once pipetted except for the sample that will be continuously pipetted for 20 times.
31. The ratio of the peptides holding a net charge at pH 3 over the total number of identified peptides illustrates the enrichment rate of the SCX procedure.
32. Pyroglutamates followed by a proline residue cannot be removed by pyroglutamyl peptidase type I [38–42]. Therefore, pyro-Glu-Pro starting peptides will remain present after the pyro-Glu removal step.
33. If Arg-C is set as enzyme setting, only peptides cleaved after Arg residues can be identified. In this way, many neo-N-terminal peptides, that is, those not generated by cleavage after Arg residues, would not be identified. By using the "semi" enzyme setting, we allow the search engine to also identify peptides that are cleaved after an Arg residue only at one side of the peptide. Hence, cleavage events, for example, following important of nuclear-encoded proteins in mitochondria, can be identified.
34. Optimally, only one or few N-terminal peptides are expected per protein.

Acknowledgments

K.G. acknowledges support from The Research Foundation—Flanders (FWO), project number G008018N.

References

1. Calligaris R, Bottardi S, Cogoi S et al (1995) Alternative translation initiation site usage results in two functionally distinct forms of the GATA-1 transcription factor. Proc Natl Acad Sci U S A 92:11598–11602
2. Claus P, Doring F, Gringel S et al (2003) Differential intranuclear localization of fibroblast growth factor-2 isoforms and specific interaction with the survival of motoneuron protein. J Biol Chem 278:479–485
3. Thomas D, Plant LD, Wilkens CM et al (2008) Alternative translation initiation in rat brain yields K2P2.1 potassium channels permeable to sodium. Neuron 58:859–870
4. Gawron D, Ndah E, Gevaert K et al (2016) Positional proteomics reveals differences in N-terminal proteoform stability. Mol Syst Biol 12:858
5. Hartmann EM, Armengaud J (2014) N-terminomics and proteogenomics, getting off to a good start. Proteomics 14:2637–2646
6. Varshavsky A (2019) N-degron and C-degron pathways of protein degradation. Proc Natl Acad Sci U S A 116:358–366
7. Kazak L, Reyes A, Duncan AL et al (2013) Alternative translation initiation augments the human mitochondrial proteome. Nucleic Acids Res 41:2354–2369
8. Bogaert A, Fernandez E, Gevaert K (2020) N-terminal proteoforms in human disease. Trends Biochem Sci 45(4):308–320
9. Ree R, Varland S, Arnesen T (2018) Spotlight on protein N-terminal acetylation. Exp Mol Med 50:1–13
10. Van Damme, P., T. Arnesen and K. Gevaert (2011) Protein alpha-N-acetylation studied by N-terminomics. FEBS J 278(20): 3822–3834
11. Bogaert A, Gevaert K (2020) Protein amino-termini and how to identify them. Expert Rev Proteomics 17:581–594
12. Perrar A, Dissmeyer N, Huesgen PF (2019) New beginnings and new ends: methods for

large-scale characterization of protein termini and their use in plant biology. J Exp Bot 70: 2021–2038

13. Arnesen T, van Damme P, Polevoda B et al (2009) Proteomics analyses reveal the evolutionary conservation and divergence of N-terminal acetyltransferases from yeast and humans. Proc Natl Acad Sci U S A 106: 8157–8162
14. Goetze S, Qeli E, Mosimann C et al (2009) Identification and functional characterization of N-terminally acetylated proteins in Drosophila melanogaster. PLoS Biol 7:e1000236
15. Nevitt C, Tooley JG, Schaner Tooley CE (2018) N-terminal acetylation and methylation differentially affect the function of MYL9. Biochem J 475:3201–3219
16. Thompson CR, Champion MM, Champion PA (2018) Quantitative N-terminal footprinting of pathogenic mycobacteria reveals differential protein acetylation. J Proteome Res 17: 3246–3258
17. van Damme P, Stove SI, Glomnes N et al (2014) A Saccharomyces cerevisiae model reveals in vivo functional impairment of the Ogden syndrome N-terminal acetyltransferase NAA10 Ser37Pro mutant. Mol Cell Proteomics 13:2031–2041
18. van Damme P, van Damme J, Demol H et al (2009) A review of COFRADIC techniques targeting protein N-terminal acetylation. BMC Proc 3(Suppl 6):S6
19. Bhagwat SR, Hajela K, Bhutada S et al (2020) Identification of unexplored substrates of the serine protease, thrombin, using N-terminomics strategy. Int J Biol Macromol 144:449–459
20. Gevaert K, Goethals M, Martens L et al (2003) Exploring proteomes and analyzing protein processing by mass spectrometric identification of sorted N-terminal peptides. Nat Biotechnol 21:566–569
21. Impens F, Colaert N, Helsens K et al (2010) MS-driven protease substrate degradomics. Proteomics 10:1284–1296
22. Mahrus S, Trinidad JC, Barkan DT et al (2008) Global sequencing of proteolytic cleavage sites in apoptosis by specific labeling of protein N termini. Cell 134:866–876
23. Tam LX, Aigner H, Timmerman E et al (2015) Proteomic approaches to identify substrates of the three Deg/HtrA proteases of the cyanobacterium Synechocystis sp. PCC 6803. Biochem J 468:373–384
24. Tsiatsiani L, Timmerman E, de Bock PJ et al (2013) The Arabidopsis metacaspase9 degradome. Plant Cell 25:2831–2847
25. Xu G, Shin SB, Jaffrey SR (2009) Global profiling of protease cleavage sites by chemoselective labeling of protein N-termini. Proc Natl Acad Sci U S A 106:19310–19315
26. van Damme P, Gawron D, van Criekinge W et al (2014) N-terminal proteomics and ribosome profiling provide a comprehensive view of the alternative translation initiation landscape in mice and men. Mol Cell Proteomics 13: 1245–1261
27. van der Meijden E, Kazem S, Dargel CA et al (2015) Characterization of T antigens, including middle T and alternative T, expressed by the human polyomavirus associated with Trichodysplasia spinulosa. J Virol 89:9427–9439
28. Staes A, van Damme P, Timmerman E et al (2017) Protease substrate profiling by N-terminal COFRADIC. Methods Mol Biol 1574:51–76
29. Drazic A, Aksnes H, Marie M et al (2018) NAA80 is actin's N-terminal acetyltransferase and regulates cytoskeleton assembly and cell motility. Proc Natl Acad Sci U S A 115: 4399–4404
30. Staes A, Impens F, van Damme P et al (2011) Selecting protein N-terminal peptides by combined fractional diagonal chromatography. Nat Protoc 6:1130–1141
31. Meier F, Brunner AD, Koch S et al (2018) Online parallel accumulation-serial fragmentation (PASEF) with a novel trapped ion mobility mass spectrometer. Mol Cell Proteomics 17: 2534–2545
32. Staes A, van Damme P, Helsens K et al (2008) Improved recovery of proteome-informative, protein N-terminal peptides by combined fractional diagonal chromatography (COFRADIC). Proteomics 8:1362–1370
33. Ning Z, Seebun D, Hawley B et al (2013) From cells to peptides: "one-stop" integrated proteomic processing using amphipols. J Proteome Res 12:1512–1519
34. Ning Z, Hawley B, Seebun D et al. (2014) APols-aided protein precipitation: a rapid method for concentrating proteins for proteomic analysis. J Membr Biol 247:941–947
35. Ning Z, Seebun D, Hawley B et al. (2013) From cells to peptides: "one-stop" integrated proteomic processing using amphipols. J Proteome Res 12:1512–1519
36. Aksnes H, van Damme P, Goris M et al (2015) An organellar nalpha-acetyltransferase, naa60, acetylates cytosolic N termini of transmembrane proteins and maintains Golgi integrity. Cell Rep 10:1362–1374
37. Dimarchi RD, Tam JP, Kent SB et al (1982) Weak acid-catalyzed pyrrolidone carboxylic

acid formation from glutamine during solid phase peptide synthesis. Minimization by rapid coupling. Int J Pept Protein Res 19: 88–93

38. Rehder DS, Dillon TM, Pipes GD et al (2006) Reversed-phase liquid chromatography/mass spectrometry analysis of reduced monoclonal antibodies in pharmaceutics. J Chromatogr A 1102:164–175
39. Bovine pitu itary pyrrolidonecarboxylyl peptidase. Endocrinology 93:1428–1434
40. Browne P, O'Cuinn G (1983) An evaluation of the role of a pyroglutamyl peptidase, a post proline cleaving enzyme and a post-proline dipeptidyl amino peptidase, each purified from the soluble fraction of guinea-pig brain, in the degradation of thyroliberin in vitro. Eur J Biochem 137:75–87
41. Ning Z, Hawley B, Seebun D et al (2014) APols-aided protein precipitation: a rapid method for concentrating proteins for proteomic analysis. J Membr Biol 247:941–947
42. van Damme P, Arnesen T, Gevaert K (2011) Protein alpha-N-acetylation studied by N-terminomics. FEBS J 278:3822–3834

Part IV

Genotypic Screens and Phenotypic Profiling

Chapter 18

RNA Interference (RNAi) as a Tool for High-Resolution Phenotypic Screening of the Pathogenic Yeast *Candida glabrata*

Andreas Tsouris, Joseph Schacherer, and Olena P. Ishchuk

Abstract

After its discovery RNA interference (RNAi) has become a powerful tool to study gene functions in different organisms. RNAi has been applied at genome-wide scale and can be nowadays performed using high-throughput automated systems (robotics). The simplest RNAi process requires the expression of two genes (Dicer and Argonaute) to function. To initiate the silencing, constructs generating either double-strand RNA or antisense RNA are required. Recently, RNAi was reconstituted by expressing *Saccharomyces castellii* genes in the human pathogenic yeast *Candida glabrata* and was used to identify new genes related to the virulence of this pathogen.

In this chapter, we describe a method to make the *C. glabrata* pathogenic yeast competent for RNAi and to use RNA silencing as a tool for low- or high-resolution phenotypic screening in this species.

Key words RNAi, Yeast, Pathogenic yeast, *Candida glabrata*, Functional genomics

1 Introduction

1.1 RNA Interference, Gene Silencing Pathway of Eukaryotes

An RNA interference (RNAi) is the process of the targeted inhibition of gene expression (or gene silencing) at posttranscriptional level with the help of RNA molecules. RNAi is triggered by the formation of double-strand RNA (dsRNA) and results in the degradation of an mRNA target, which is complementary to the RNA used. To function, RNAi uses the following components, which are encoded by single or multiple genes: RNA-dependent RNA polymerase (RdRP), Dicer and Argonaute. RdRP converts single-strand RNA into dsRNA. The dsRNA is further cleaved by Dicer (RNase III family protein) into small interfering RNAs (siRNAs), which are 21–25 nt long RNA duplexes (depending on the species). The resulting siRNAs are then bound and delivered to mRNA target by Argonaute proteins, which are a component of an

Frédéric Devaux (ed.), *Yeast Functional Genomics: Methods and Protocols*,
Methods in Molecular Biology, vol. 2477, https://doi.org/10.1007/978-1-0716-2257-5_18,

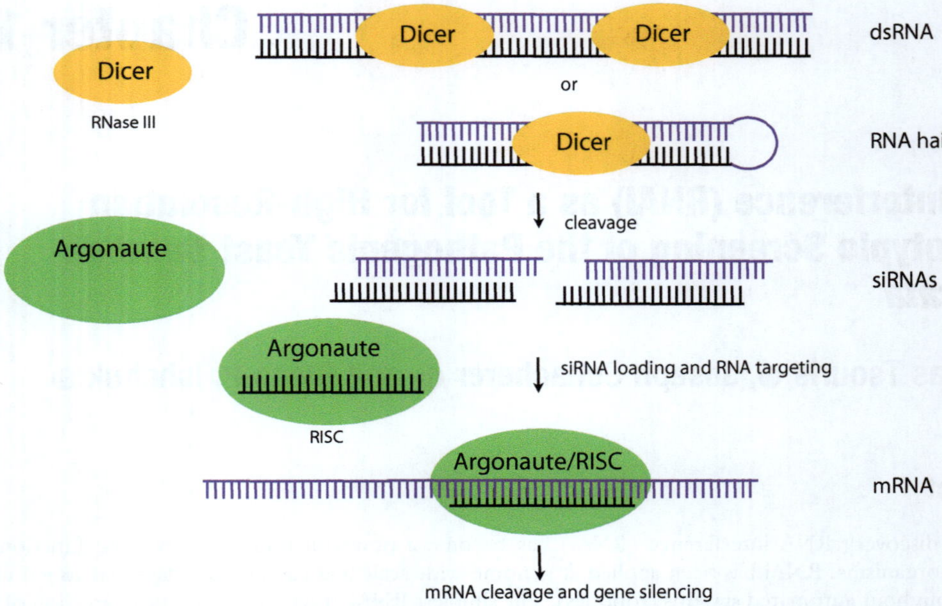

Fig. 1 Schematic overview of RNAi tool. dsRNA or RNA hairpin (dsRNA with secondary structure) are substrates for RNase III Dicer that cleaves dsRNAs into siRNAs. The generated siRNAs are then loaded into the Argonaute protein to be delivered by sequence homology to mRNA-target. Argonaute is a part of a bigger protein complex (RISC), which slices and inactivates the mRNA-target and thus silences gene expression

RNA-induced silencing complex (RISC) and are involved in mRNA cleavage. The mRNA cleavage silences the gene expression (Fig. 1).

The process was discovered and named "RNAi" in 1998 when the injection of dsRNAs into nematode *Caenorhabditis elegans* silenced gene expression [1]. In this experiment, the targets of RNAi were coding sequences of native genes (myofilament, myosin and myoD-family proteins) and heterologous model genes (GFP reporters). On the contrary, the dsRNA targeting introns and promoter regions did not result in any phenotypic changes. In *C. elegans* it was initially shown that the dsRNAs are more potent in gene silencing than sense or antisense single-strand RNAs [1].

Unless the process is initiated with exogenous dsRNA, the RNAi requires RNA-dependent RNA polymerase (RdRP) [2]. RdRP uses the single-strand RNA as a template for its polymerase activity to generate dsRNA (with or without siRNAs as primers) [3, 4].

The endoribonuclease Dicer (RNase III protein) is generating most small regulatory RNAs, including RNAi component siRNAs [5–8]. Depending on the organism, the genome can carry one or multiple genes encoding Dicer. To fulfill its activity, Dicer possesses the following domains: PAZ (responsible for the recognition of and binding to the 3′-end of siRNA precursors (dsRNAs)), dsRBD (facilitates dsRNAs binding and cleavage), two RNase III domains (two domains form one single dsRNA cleavage center), and N terminal helicase domain (as reviewed in ref. [9]). The siRNAs size is determined by the physical distance between the PAZ domain and the active site of Dicer [10].

Products of Dicer and the mediator component of RNAi, siRNAs, were detected in plants [11]. The siRNAs are short 21–25 nt RNA duplexes with 2 nucleotide 3′-overhangs, and they are homologous to the gene being suppressed. The siRNAs size differs by few ribonucleotides depending on the species. For example, siRNAs have a length of 23 nt in yeast (*S. castellii*), 22 nt in fly (*Drosophila melanogaster*), and 21–25 nt in plants [11–14].

Argonaute proteins are a part of the multicomponent nuclease RISC, and they are responsible for its mRNA slicing activity. The Argonaute proteins are highly conserved across different species with ~85% sequence similarity and four conserved domains: PIWI (similar to RNase H domain and is responsible for slicing activity), PAZ (also found in Dicer proteins, anchors the 3′-end of siRNAs), MID (binds the 5′-end of siRNAs), and N (required for full catalytic activity) [15].

To prevent constitutive RNAi silencing, the process is regulated at different stages. Recently, two noncanonical nucleotidyltransferases Cid14 and Cid16 were found to modify the 3′-end of the siRNAs bound to Argonaute and to target them to the exosomes for degradation [16]. The adenylation of siRNAs by Cid14 recruits either Rrp6 or Triman, while Rrp6 is involved in degradation; Triman deadenylates the siRNAs and thus helps to reuse the siRNAs [16]. Argonaute protein expression was found to be regulated at the transcriptional and posttranslational level [15].

RNAi is evolutionarily conserved and is naturally found in fungi, plants, animals, and humans. RNAi serves different purposes in the cell, for example, posttranscriptional gene silencing, heterochromatin formation, viral defense, transposon silencing, and others.

RNAi functions as the cell's "immune system" for silencing repetitive sequences, like transposons; thus, RNAi plays a role in genome stability [17–24]. For example, in fission yeast, plants, and animals, transposons are clustered at centromeres. By targeting these repeats the RNAi machinery silences their transcription and forms heterochromatin at these loci. With the help of RNAi, the heterochromatin is assembled at centromeric regions during cell division [25–27]. RNAi recruits methyltransferase complexes to produce the repressive methylated histone H3K9me and further recruits proteins of the Heterochromatin Protein 1 (HP1) family [21, 28]. RNAi also affects the transposase expression and transposon excision, and with loss of RNAi and DNA methylation the recombination rates at transposons by centromeres are elevated [24, 29].

RNAi also plays a role in the defense against infections by pathogenic viruses and mutations affecting RNAi pathway result in increased susceptibility to infection and/or virus replication [12, 18, 30–37]. To counteract RNAi, some viruses outcompete Dicer and the RISC complex by binding siRNAs and thus they suppress the silencing or slicing of their RNAs [38].

Fungal comparative genomics studies showed that the RNAi and heterochromatin related genes were lost in many and likely most of Saccharomycotina species while some species carry the remnants of the pathway [14, 39]. It is possible, that it was lost to maintain the RNA virus encoding Killer Toxin (e.g., in *Saccharomyces cerevisiae* it is a beneficial trait for the competition with sensitive strains) [40], or RNAi was not needed any more due to limited virus infections in other species. On the contrary, all RNAi pathway components, for example, genes encoding Dicer, Argonaute, and RNA-directed RNA polymerase can be found in the genome of other classes of fungal species, which also carry heterochromatin proteins [39, 41]. Fission yeast, *Schizosaccharomyces pombe,* for example, carries three genes encoding RdRP, Dicer and Argonaute, respectively [42]. There is a difference in the heterochromatin organization of *S. cerevisiae* and *S. pombe*, particularly at centromeric regions where RNAi takes part [39].

A decade ago, it was found that some Saccharomycotina species carry functional RNAi with a Dicer protein (Dcr1) of noncanonical structure [14]. In contrast to canonical Dicer proteins that carry tandem RNase III domains (e.g., from Human, *Arabidopsis thaliana* and *S. pombe*), Dicer of *S. castellii,* for example, has only one RNase III domain, and no helicase domain, thus it resembles the Rnt1 protein's domain organization [14]. Due to this unusual structure, noncanonical Dicer was not found before 2009. It was hypothesized that Dicer of *S. castellii* must acts as a homodimer for cleavage on both strands of RNA duplex [14]. In addition, *S. castellii* possesses the 5′-to-3′ exoribonuclease Xrn1p, which acts as a cofactor of RNAi in this yeast and aids in loading siRNAs into Argonaute, affecting the siRNAs composition, and degradation of sliced RNAs [43]. Because RNAi was found to take part in several cellular processes, it is unclear what is the primary evolutionary function of RNAi in fungi. It is likely, that in evolutionary history the RNAi in budding yeast was mainly involved in transposon silencing. While in fission yeast (*S. pombe*) RNAi is involved in transcriptional silencing with chromatin remodeling, the RNAi of filamentous fungus *Neurospora crassa* does not maintain chromatin [41]. In some species, RNAi aids in the adaptation to pathogenic lifestyle. For instance, in human fungal pathogen *Mucor circinelloides,* RNAi is involved in the development of antibiotic resistance. It inhibits the expression of *fkbA* in *M. circinelloides* resulting in resistance to tacrolimus [44–46].

1.2 *Candida glabrata* Opportunistic Human Pathogen

Although a part of healthy human microflora, *Candida glabrata* is an opportunistic human pathogenic yeast. This yeast normally does not cause any diseases unless the host immune system is compromised or under condition of treatment by broad-spectrum antibiotics, and after several surgeries [47]. *C. glabrata* is closely related to the *S. cerevisiae* baker's yeast [48]. Both *S. cerevisiae* and *C. glabrata* are Whole-Genome Duplication yeast, which underwent substantial gene loss after the genome duplication event in their common ancestor. However, *C. glabrata* lost more genes under strong selection of adaptation to the pathogenic lifestyle. For example, *C. glabrata* lost genes of nicotinic acid biosynthesis, and the availability of nicotinic acid is crucial for its survival in the host and regulates the adhesion to host tissues [49]. Nowadays, *C. glabrata* is considered the most common cause of human candidiasis infections as often as *Candida albicans* (as reviewed in refs. [50, 51]). While occupying the same ecological niches (human host), *C. glabrata* is only distant relative to *C. albicans* [48], and both species differ in their strategies to colonize and invade host [51]. Compared to *C. albicans*, *C. glabrata* pathogenicity is also less studied. While gene knockouts is a powerful tool to study gene function, the gene knockouts are less efficient in *C. glabrata* compared to *S. cerevisiae*. Indeed, the nonhomologous recombination is higher in *C. glabrata* than that in *S. cerevisiae* [52, 53]. For this reason, the introduction of new tools such as CRISPR-Cas9 and RNAi is an important addition to the *C. glabrata* genetics toolbox to study *C. glabrata* gene functions [54, 55]. Creating and analyzing the genome-wide gene knockout and/or gene knockdown collections of *C. glabrata* strains will contribute greatly to understanding the biology of this human pathogen.

1.3 RNAi as a Tool, Applications in Yeast

As RNAi silences or downregulates gene expression with high specificity, it has become an important tool to study gene functions and is being used for the development of diseases treatment, including human diseases. While absent in most budding yeast, the functional RNAi pathway was found in *S. castellii* in 2009 by the D. Bartel lab [14]. It was also successfully introduced to other budding yeast as a tool. By expressing *S. castellii* genes encoding Dicer and Argonaute under strong constitutive promoters in other Saccharomycotina species such as *S. cerevisiae* and *C. glabrata*, which lost the RNAi during their evolution, it was possible to achieve targeted gene silencing [14, 55]. In *S. cerevisiae*, RNAi tool was used in different strains to identify new genes and construct improved strains for the production of para-hydroxybenzoic acid (PHBA), isobutanol, cellulase, increased tolerance to isobutanol and lactic acid, improved α-amylase secretion, and so on (as reviewed in ref. [56]).

Recently introduced to *C. glabrata*, the RNAi toolbox was used to identify new genes involved in virulence of this human pathogen [55]. The new *C. glabrata* RNAi strain was transformed

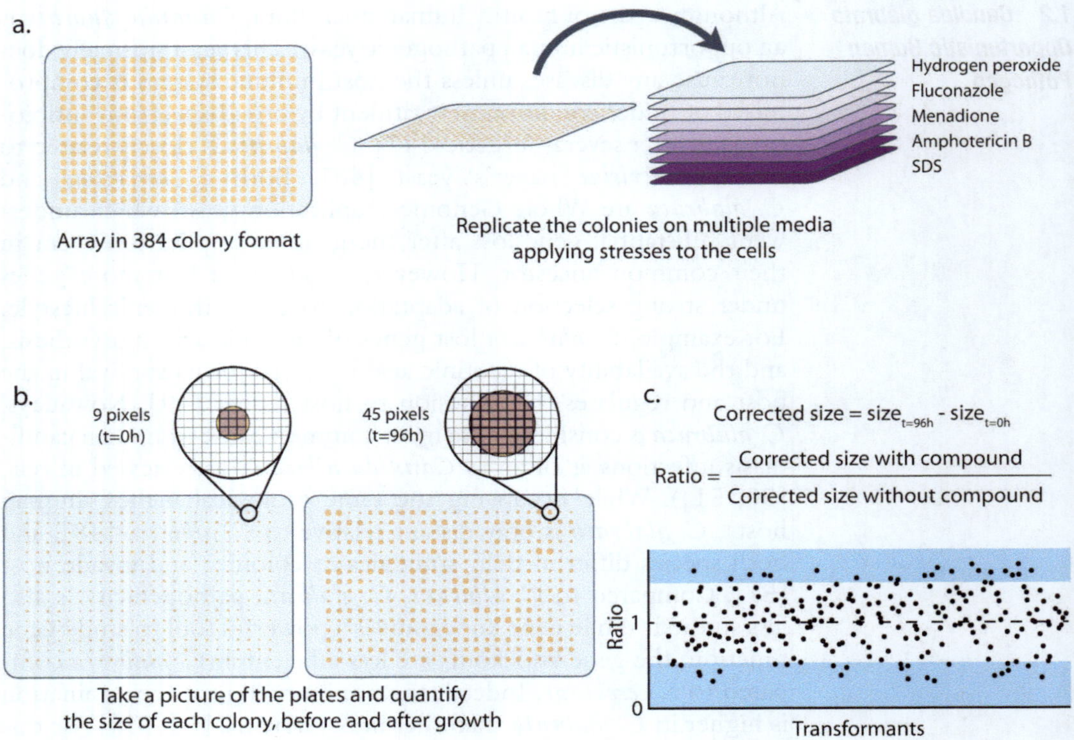

Fig. 2 High-throughput screening of RNAi transformants under various stresses. (**a**) Strains transformed with the RNAi plasmid and a control strain were arrayed on a plate in a 384-colony format. The transformants were replicated on solid media containing various compounds that apply stress on the cells and on a control medium without any compounds added. (**b**) The transformants were photographed on the various media before growth and every 12 h during their 96 h growth period. The size of each colony at each time point was quantified as the number of pixels that it occupied. (**c**) The transformants with the most extreme phenotypes were selected for further analysis in macrophages

with a gene library. To assess the effect of each gene on virulence, we screened the mitotic growth of thousands of transformants under various stresses using a high-throughput phenotypic strategy (Fig. 2) [55, 57, 58]. The selected stresses were similar to those the cells would face while infecting another organism (ROS, antifungal drugs and cell membrane stressors). To apply these stresses to the cells, we used solid media (YNB glucose 2%) containing menadione, hydrogen peroxide (ROS); fluconazole, amphotericin B (antifungal drugs) or sodium dodecyl sulfate (SDS, cell membrane stressor). The cells were also grown in duplicate on solid media without any additional stress to estimate the influence of the different stresses on the growth of each transformant. Photographs of the colony of each transformant were taken before growth and at different time points (every 12 h). The number of pixels occupied by each colony were counted to estimate the colony size. The size of the colony before growth (t = 0 h) was subtracted from the colony size at different time points in order to calculate the absolute growth as the increase in colony size. We then calculated the ratio between the average absolute growth of a transformant upon each stress to its average absolute growth in the absence of stress. Finally, the transformants with the most extreme growth ratios were selected for further analysis in human macrophages [55]. Using this RNAi tool and a robotic platform several genes were identified behind specific phenotypes, for example, genes affecting *C. glabrata* plasma membrane, protein desumoylation, ribosomal RNAs, and others [55].

1.4 Outlook

By expressing the genes encoding Dicer and Argonaute from *S. castellii* and different silencing constructs, we reconstructed a functional RNAi pathway in the pathogenic yeast *C. glabrata* (*see* Subheading 3.1). This tool proved useful in the high-resolution phenotypic screening of new virulence-genes with a random gene library that generated RNA antisense constructs (*see* Subheadings 3.2 and 3.3). Our next steps shall be the construction of barcoded gene libraries, which upon transformation shall generate double-strand RNA constructs like hairpins, which shall be applicable for every *C. glabrata* gene. Upon strictly regulated inducible promoter these constructs could be expressed in *C. glabrata* RNAi strains and their impact on virulence could be studied in vivo.

2 Materials

2.1 Expression of RNAi and Silencing Constructs in *C. glabrata*

- P1062 expression plasmid vector (pUC19 backbone carrying the *Streptoalloteichus hindustanus ble* gene, *C. glabrata* promoter *TEF1* fused to *S. castellii DCR1* ORF and *C. glabrata* terminator *CYC1*, *C. glabrata* promoter *TDH3* fused to *S. castellii AGO1* ORF and *C. glabrata* terminator *TDH3*) [55].
- P1028 expression plasmid vector (pUC19 backbone carrying the *C. glabrata LYS2* gene) [55].
- *Sph*I restriction enzyme.
- GeneJET PCR purification kit (Thermo Fisher Scientific).
- Zeocin™.
- $CuSO_4 \times 5H_2O$.
- YNB with ammonium sulfate.
- D-Glucose.
- Agar.
- Yeast extract.
- Peptone.
- Primers for amplification of the expression cassettes for *DCR1* (Primer1 TTTGGATCCCGCTGATCTCTGTGCGTGTCC

and Primer2 CCGGGATCCCTTAGGAGTTTTGTAGTATATGTACATG), *AGO1* (Primer3 GCCCCAAGCTTCTGTGCCACCAATAAATGTATG and Primer4 CCGCCGCAAGCTTGATTCTTCTTATAAAATGATATTTTAAATAC) and silencing constructs (Primer5 CGGAGCACCTGGCTGATTTACCATATGATGTG and Primer2 CCGGGATCCCTTAGGAGTTTTGTAGTATATGTACATG) are used.

2.2 Gene Library Construction

1. P1226 expression plasmid vector (pUC19 backbone carrying *C. glabrata LYS2* gene, ARS-like sequence (the *S. cerevisiae COXII* gene from pMIR4 [59], *C. glabrata* promoter *TEF1* and *C. glabrata* terminator *CYC1*).
2. Lyticase.
3. *Sau*3AI restriction enzyme.
4. GeneJET PCR purification kit (Thermo Fisher Scientific).
5. GeneJET Plasmid Maxiprep kit (Thermo Fisher Scientific).
6. Ampicillin.
7. T4 DNA ligase.
8. UltraPure™ 0.5 M EDTA, pH 8.0.

2.3 Screening Using Robotics

1. Rotor HDA high-throughput screening robot (Singer Instruments).
2. D-Glucose.
3. YNB with ammonium sulfate.
4. Hydrogen peroxide.
5. Fluconazole.
6. 250 μg/mL Amphotericin B, solution.
7. Menadione.
8. SDS, 20% solution.
9. PlusPlates (Singer Instruments).
10. Nunc 96-well flat bottom plates.
11. Agar.

2.4 Macrophages Differentiation

1. FCS, fetal bovine serum.
2. Triton X-100.
3. RPMI-1640 media with L-glutamine.
4. rh M-CSF, recombinant human Macrophage Colony Stimulating Factor, (Immunotools).
5. CD14 magnetic beads (Miltenyi Biotech).
6. PBS buffer.

3 Methods

3.1 RNAi, Hairpin, and Antisense Constructs for Gene Silencing in C. glabrata

The first step of the process is to construct a RNAi-competent *C. glabrata* strain by expressing the genes encoding Dicer and Argonaute of *S. castellii* under control of strong constitutive promoters, for example, using *C. glabrata* promoters *TEF1* and *TDH3* (*see* **Note 1**).

Then, to knock down specific genes, two types of constructs that silence the genes of interest can be used: hairpin and antisense (Fig. 3) (*see* **Note 2**).

1. Linearize plasmid vector P1062 carrying both Dicer and Argonaute of *S. castellii* expression cassettes by the restriction enzyme *Sph*I. Purify the digested plasmid using GeneJET PCR purification kit (*see* **Note 3**).

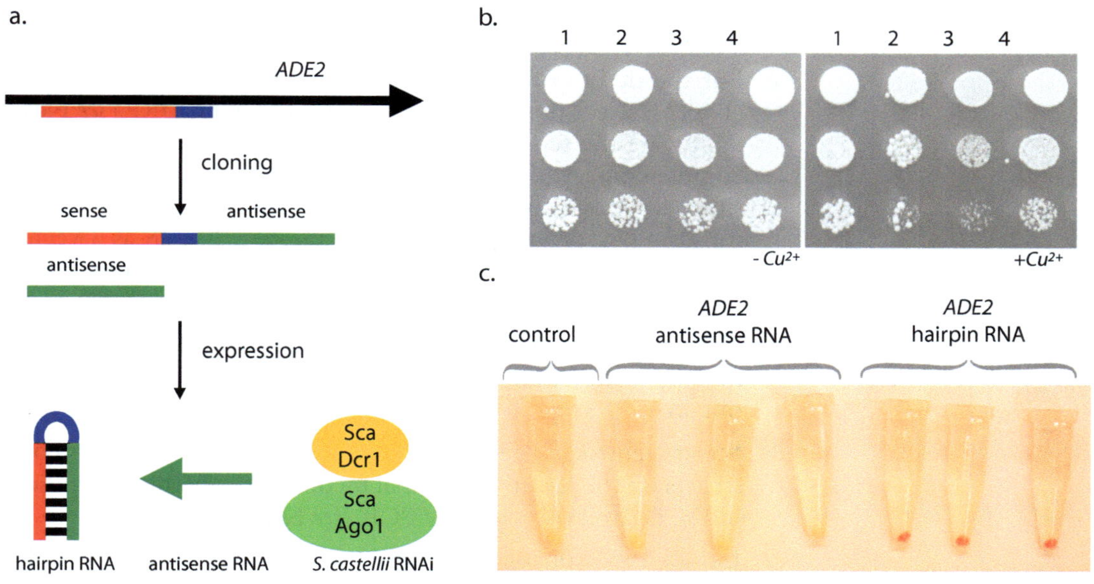

Fig. 3 RNAi components and *ADE2* gene silencing constructs in *C. glabrata*. (**a**) To downregulate gene expression (*ADE2* as an example) silencing constructs are cloned. For RNA hairpin expression, the sense region of the gene ORF is selected and is of 418 nt long (1–339 nt region (red), 340–418 (blue)). It is cloned into the expression vector together with DNA fragment complementary to the red region (antisense part (green)); the 3′-end of the sense region of 79 nt (blue) will form a loop structure of transcribed RNA. The antisense fragment (green) could also be expressed separately. Two heterologous genes of *S. castellii* (*DCR1* and *AGO1* encoding Dicer and Argonaute, correspondingly) needs to be coexpressed with silencing constructs; (**b**) Growth of *C. glabrata* transformants carrying the *ADE2* antisense construct expressed from *CUP1* promoter on media with and without addition of copper. The growth inhibition is observed under promoter induction (with copper). Strains: 1— control (RNAi strain with empty vector); 2 and 3—*ADE2* antisense construct; 4—control (strain with *ADE2* antisense but without RNAi); (**c**) Production of P-ribosylamino imidazole observed only in *C. glabrata* transformants carrying *ADE2* hairpin, which suggests a strong inhibition of the model gene

2. Use *Sph*I-linearized plasmid for lithium-acetate transformation [60] of strain of interest (*see* **Note 4**). For expression of *S. hindustanus ble* gene, which is a selective marker on the plasmid, 24 h incubation in liquid YPD on the shaker is required before plating onto selective medium (YPD medium supplemented with zeocin at 200 μg/mL).
3. Select transformants carrying integrated P1062 plasmid on the YPD medium supplemented with zeocin at 200 μg/mL. Analyze the presence of both *DCR1* and *AGO1* expression cassettes in the transformants by PCR (*see* **Note 5**).
4. For the design of the antisense construct for a gene of interest (Fig. 3), select the antisense 339 nt region of gene ORF (*see* **Note 6**). Clone this fragment under control of strong inducible promoter of *C. glabrata*, for example, promoter *CUP1* and plasmid P1028 as a vector (*see* **Note 7**).
5. Transform the obtained antisense construct as linearized plasmid into the strain carrying both *DCR1* and *AGO1* genes obtained by **steps 1–3** (*see* **Note 8**).
6. For the design of the hairpin construct for a gene of interest (Fig. 3), select the sense 418 nt region of gene ORF (*see* **Note 9**). Ligate the sense region with the antisense region, which is complementary to 1–339 nt sense part (Fig. 3), and the expression vector carrying strong inducible promoter of *C. glabrata*, for example, promoter *CUP1* and plasmid P1028 as a vector (*see* **Note 10**). The 79 nt part with no complementary counterpart will form the loop of the RNA hairpin structure.
7. Transform the obtained hairpin construct as a linearized plasmid into the strain carrying both *DCR1* and *AGO1* genes obtained by **steps 1–3** (*see* **Note 8**).
8. Select transformants carrying either antisense or hairpin constructs on the selective minimal medium. Verify the integration of the desired construct by PCR (*see* **Note 11**).
9. Induce the expression of antisense and hairpin constructs by culturing the transformants in the minimal medium with 0.05 mM $CuSO_4$ for promoter *CUP1* induction (*see* **Note 12**).
10. Assess the gene expression downregulation by RT-Q-PCR (*see* **Note 2**).

3.2 Gene Library Construction

To assess the impact of knocking down genes on the fitness of the cells on a genome-wide scale, a library of random genome fragments inserted in an expression vector is constructed, assuming that a significant portion of clones in the library will carry the antisense regions of genes to induce RNAi (*see* **Note 13**).

1. Isolate the genomic DNA of the strain of interest using the modified protocol published previously, which results in high-

molecular weight DNA [61, 62]. First, prepare yeast spheroplasts by the incubation with 0.1 mg/mL of lyticase in 0.5 mM EDTA at 37 °C for 1 h. Lyse the spheroplasts in 300 μL of lysis solution (0.2% SDS, 50 mM EDTA, pH 8.0) at 70 °C for 15 min. Place the cell lysate on ice. Then add 100 μL of 5 M potassium acetate pH 5.0 to precipitate proteins. Collect the supernatant and mix it with an equal volume of 2-propanol to precipitate the DNA. Further, dissolve the precipitated DNA in TE buffer (10 mM Tris–HCl, 1 mM EDTA, pH 8.0) and treat it with RNase A to remove RNA. At the final step, to remove RNase A and remaining protein treat the samples with phenol–chloroform–isoamyl alcohol mixture (25:24:1). Precipitate the pure DNA with ethanol and resuspend the pellet in TE buffer. Assess the quality and size of isolated DNA by loading in on 1% agarose gel. High–molecular weight and undamaged DNA is required for the next step (*see* **Note 14**).

2. Perform a partial digestion of the isolated genomic DNA with *Sau*3AI restriction enzyme. Digest 20 μg of genomic DNA with different amounts of *Sau*3AI (2–20 units) in a total volume of 100 μL (*see* **Note 15**). To obtain the desirable fragment size, add EDTA at the final concentration of 0.01 M to stop the reaction (*see* **Note 16**).
3. After digestion, purify the digested DNA with the GeneJET PCR purification kit following manufacturer's recommendations.
4. Clone the purified digested DNA into the *Bam*HI site of the multi-copy plasmid P1226 under control of promoter *TEF1* (*see* **Note 17**).
5. Transform the ligation mix into bacteria and select transformants on the LB medium with 100 μg/mL ampicillin (*see* **Note 18**).
6. Assess the library composition in terms how many clones carry the insert and estimate the insert size by sequencing of obtained plasmids. This is performed by isolating the plasmids from bacterial transformants and analyzing the insert size by PCR using primers for gene library expression cassette: promoter *TEF1* and terminator *CYC1* of *C. glabrata* (Primer1 and Primer2).
7. Wash out and collect ~400,000 colonies of obtained transformants. To resuspend the colonies from the surface of plates, add a small amount of LB medium (~2–3 mL) on each plate and use a sterile spatula to move the medium around the plate by applying slight pressure to the surface of the plate. Collect the medium with 400,000 colonies, add 100 μg/mL of ampicillin and cultivate at 37 °C for 2 h.

8. After 2 h isolate plasmid DNA from bacterial culture of collected colonies using GeneJET Plasmid Maxiprep kit (*see* **Note 19**).
9. Transform the obtained gene library into the RNAi strain of *C. glabrata* using the lithium-acetate protocol (*see* **Note 4**) with 500 ng of gene library DNA per sample.

3.3 Phenotypic Profiling Using Robotics

1. Array the transformant and control strain (empty vector) in 96 well plates in 100 μL of selective minimal medium YNB, 2% glucose and incubate overnight at 37 °C.
2. Transfer the transformants and control strain on PlusPlates with selective minimal medium YNB, 2% glucose, 2% agar in a 384-colony format. In every PlusPlate replicate two copies of two different plates (Fig. 2).
3. Incubate the cells overnight at 30 °C.
4. Replicate the cells in YNB PlusPlates with 2% glucose and the growth conditions to be tested, one at a time. In our case we tested the following.
   (a) 3 mM hydrogen peroxide.
   (b) 12 mM hydrogen peroxide.
   (c) 0.0025% SDS.
   (d) 0.01% SDS.
   (e) 0.05 μg/mL Amphotericin B.
   (f) 0.25 μg/mL Amphotericin B.
   (g) 0.02 mM Menadione.
   (h) 0.06 mM Menadione.
   (i) 50 μg/mL Fluconazole.
   (j) 400 μg/mL Fluconazole.
   (k) No additional compound (control plate).
5. Take a picture of every plate ($t = 0$ h).
6. Incubate the plates at 37 °C for 96 h and take a photograph of each plate every 12 h.
7. Quantify the size of every colony at each time point using the R package gitter [63] (*see* **Note 20**).
8. For every colony, subtract the initial colony size ($t = 0$ h) from the colony size of the other time points to calculate the absolute growth for each time point (*see* **Note 21**).
9. For each time point calculate the mean value of the absolute growth of the replicates of each transformant or control strain (*see* **Note 21**).

10. For each time point calculate the relative growth ratio of each transformant by dividing its mean absolute growth value by the mean absolute growth value of the control strain (*see* **Note 21**).
11. Identify the transformants with the most extreme relative growth ratio in each condition, those with the highest or lowest relative growth values (*see* **Note 22**).

3.4 Macrophages

As *C. glabrata* evolved to survive and replicate after being engulfed by host macrophages during infection [51, 64–66]. The corresponding virulence related traits can be studied in gene library transformants using a macrophage model. The protocol of preparation of human macrophages from peripheral blood monocytes is based on the method described before [64, 66].

1. Dilute blood with PBS buffer and centrifuge at 500 × *g* for 30 min.
2. After centrifugation, from three cell layers obtained collect the middle layer of cells and resuspend them in fresh PBS buffer (*see* **Note 23**).
3. Isolate monocytes using CD14 magnetic beads [64, 66].
4. Incubate CD14-positive monocytes in RPMI-1640 media with L-glutamine, 10% heat-inactivated FCS, and 50 ng/mL recombinant human macrophage colony stimulating factor (rh M-CSF) for 7 days in CO_2 incubator at 37 °C.
5. During 7 days of incubation, the monocytes are differentiated into M2 macrophages.
6. After 7 days, detach M2 macrophages for the plate surface with 10 mM EDTA in PBS.
7. Transfer macrophages into 96-well plates with 4×10^4 cells/well) in RPMI with 50 ng/mL rh M-CSF and 10% FCS. Incubate plates overnight in a CO_2 incubator at 37 °C.
8. Inoculate the yeast strain and cultivate it for 24 h in YNB medium prior the macrophage infection.
9. The next day, prepare the yeast for macrophage infection. For this purpose, wash the stationary yeast YNB cultures three times with PBS buffer.
10. Add yeast cells to M2 macrophages at multiplicity of infection of one (MOI 1) in RPMI w/o FCS. Incubate for 3 h in a CO_2 incubator at 37 °C.
11. Remove unattached yeast cells by washing the macrophages two times with 60 μL of PBS buffer.
12. Add 20 μL of 0.5% Triton X-100 to lysate the macrophages, incubate for 10 min under gentle shaking.

13. After the incubation, dilute cells and plate on YPD plates. Incubate plates at 37 °C for 1 day for yeast colonies to appear.
14. To determine yeast survivor rate count CFU of yeast.

4 Notes

1. Other promoter activities can be assessed using a reporter gene, for example, encoding deoxyribonucleoside kinase of *Drosophila melanogaster*, as done for *C. glabrata* [55].
2. Depending on the construct, ~30% (antisense) and 70% (hairpin) of gene knockdowns have been achieved in *C. glabrata* [55].
3. For the restriction reaction, 10 μg of plasmid DNA and 30 units of *Sph*I in 100 μL of the total volume is used to linearize the plasmid at 37 °C.
4. The lithium-acetate transformation is performed by the protocol by Boretsky et al. [60]. After the lithium-acetate treatment, cells are pelleted and resuspended in a fresh lithium-acetate solution to obtain the concentration of 5×10^9 cells/mL. Then 2–5 μg of the linearized plasmid is dissolved in 10 μL of MQ, which is then added to 100 μL of cell suspension.
5. For the amplification of the expression cassette carrying the promoter *TEF1* with *DCR1* ORF and the terminator *CYC1*, Primer1 and Primer2 (of Subheading 2.1) are used in a PCR reaction with the annealing temperature of 61 °C. For the amplification of the expression cassette carrying the promoter *TDH3* with *AGO1* ORF and the terminator *TDH3*, Primer3 and Primer4 (*see* Subheading 2.1) are used in the PCR reaction with the annealing temperature of 58 °C.
6. We selected the size of 339 nt, which works for *S. cerevisiae* [14]. However, from our gene library, we isolated the plasmids with the antisense fragments of size greater than 339 nt, which are effective in gene expression inhibition in *C. glabrata* [55].
7. The plasmid P1028 is a vector for gene expression in *C. glabrata* (*see* Subheading 2.1), and P1028 is used for integration into the genome by nonhomologous recombination. The *Sac*I-restriction site is used for gene cloning into P1028: For example, the expression cassettes (carrying the promoter *CUP1*, antisense/hairpin construct, and the terminator *CYC1*) are cloned into this site. The plasmid P1028 is linearized by either *Sal*I or *Sma*I prior to the transformation procedure.
8. The transformation method is the same as specified in **step 2**.

9. For this purpose, the exon regions of the gene are generally selected (because the RNAi method works on the level of mRNA).
10. Perform the ligation reaction of three DNA fragments using the T4 DNA ligase according to the chosen manufacturer's recommended protocol.
11. To verify the integration, use the Primer5 and Primer2 pair (*see* Subheading 2.1) to amplify one fragment carrying the promoter *CUP1* fused to antisense or hairpin with terminator *CYC1,* in a PCR reaction with annealing temperature of 61 °C.
12. When the cells, which have been grown in the minimal medium with glucose, attain their stationary phase, collect them. Then dilute these cells to final OD_{600} of 0.2 and cultivate the cell cultures with 0.05 mM $CuSO_4$ for 24 h. (The *CUP1* promoter is leaky, because without copper induction it is expressed at a very low level; for example, its leakiness has been confirmed by a promoter analysis with a reporter gene in *C. glabrata* [55].)
13. In our model, the antisense constructs inhibit the gene expression less than the hairpins. Consequently, the antisense constructs require a high expression level, and so the gene library was constructed using a strong constitutive promoter and a multicopy vector.
14. When the molecular weight is greater than 10 kb, undamaged DNA is detected as a single band. Additional bands or smears below 10 kb indicate that DNA is damaged and should not be used for the later steps of this protocol.
15. The reaction is performed at room temperature (22 °C) to slow down the speed of the reaction. At 37 °C the enzyme fully digests DNA, producing (almost exclusively) short fragments of DNA (roughly 1 kb).
16. To obtain DNA from 1 kb to 5 kb (the average size range of yeast ORFs), perform the reaction in time intervals from 1 to 10 min.
17. Mix the vector DNA with inserts (partially digested DNA) in the ratio of 1:1. For example, 5 μg of the vector DNA is mixed with 5 μg of the inserts in 300 μL of the ligation mixture.
18. Perform the transformation protocol, which provides high transformation for *E. coli*, for example, using commercially available chemically competent cells.
19. Cultivation stops at 2 h, so that the plasmids with smaller inserts do not overpopulate the plasmids with larger inserts.
20. We used the default parameters, and the colony positions were identified independently for each time point.

21. We continued the analysis in R, but the colony size data produced by gitter can be exported and analyzed with your software of choice.

22. Transformants with relative growth ratios close to 1 are resistant to the stress applied. Transformants with relative growth ratios significantly lower than the control strain have increased susceptibility to the stress applied.

23. The middle layer is composed of a mixture of white blood cells (lymphocytes, monocytes, and platelets).

References

1. Fire A, Xu S, Montgomery MK et al (1998) Potent and specific genetic interference by double-stranded RNA in *Caenorhabditis elegans*. Nature 391(6669):806–811
2. Cogoni C, Macino G (1999) Gene silencing in *neurospora crassa* requires a protein homologous to RNA-dependent RNA polymerase. Nature 399(6732):166–169
3. Sijen T, Fleenor J, Simmer F et al (2001) On the role of RNA amplification in dsRNA-triggered gene silencing. Cell 107(4):465–476
4. Makeyev EV, Bamford DH (2002) Cellular RNA-dependent RNA polymerase involved in posttranscriptional gene silencing has two distinct activity modes. Mol Cell 10(6): 1417–1427
5. Bernstein E, Caudy AA, Hammond SM et al (2001) Role for a bidentate ribonuclease in the initiation step of RNA interference. Nature 409(6818):363–366
6. Nicholson RH, Nicholson AW (2002) Molecular characterization of a mouse cDNA encoding dicer, a ribonuclease III ortholog involved in RNA interference. Mamm Genome 13(2): 67–73
7. Carmell MA, Hannon GJ (2004) RNase III enzymes and the initiation of gene silencing. Nat Struct Mol Biol 11(3):214–218
8. Song MS, Rossi JJ (2017) Molecular mechanisms of dicer: endonuclease and enzymatic activity. Biochem J 474(10):1603–1618
9. Kurzynska-Kokorniak A, Koralewska N, Pokornowska M et al (2015) The many faces of dicer: the complexity of the mechanisms regulating dicer gene expression and enzyme activities. Nucleic Acids Res 43(9):4365–4380
10. Hammond SM (2005) Dicing and slicing: the core machinery of the RNA interference pathway. FEBS Lett 579(26):5822–5829
11. Hamilton AJ, Baulcombe DC (1999) A species of small antisense RNA in posttranscriptional gene silencing in plants. Science 286(5441): 950–952
12. Lee YS, Nakahara K, Pham JW et al (2004) Distinct roles for drosophila Dicer-1 and Dicer-2 in the siRNA/miRNA silencing pathways. Cell 117(1):69–81
13. Zamore PD (2004) Plant RNAi: how a viral silencing suppressor inactivates siRNA. Curr Biol 14(5):R198–R200
14. Drinnenberg IA, Weinberg DE, Xie KT et al (2009) RNAi in budding yeast. Science 326(5952):544–550
15. Müller M, Fazi F, Ciaudo C (2020) Argonaute proteins: from structure to function in development and pathological cell fate determination. Front Cell Dev Biol 7:360
16. Pisacane P, Halic M (2017) Tailing and degradation of Argonaute-bound small RNAs protect the genome from uncontrolled RNAi. Nat Commun 8:15332
17. Fischer SE, Pan Q, Breen PC et al (2013) Multiple small RNA pathways regulate the silencing of repeated and foreign genes in *C. elegans*. Genes Dev 27(24):2678–2695
18. Hu Y, Stenlid J, Elfstrand M et al (2013) Evolution of RNA interference proteins dicer and argonaute in Basidiomycota. Mycologia 105(6):1489–1498
19. Ito H, Kakutani T (2014) Control of transposable elements in *Arabidopsis thaliana*. Chromosome Res 2:217–223
20. Gebert D, Rosenkranz D (2015) RNA-based regulation of transposon expression. Wiley Interdiscip Rev RNA 6(6):687–708
21. Ugolini I, Halic M (2018) Fidelity in RNA-based recognition of transposable elements. Philos Trans R Soc Lond B Biol Sci 373(1762):20180168
22. Fischer SEJ, Ruvkun G (2020) *Caenorhabditis elegans* ADAR editing and the ERI-6/7/MOV10 RNAi pathway silence endogenous

viral elements and LTR retrotransposons. Proc Natl Acad Sci U S A 117(11):5987–5996
23. Pérez-Arques C, Navarro-Mendoza MI, Murcia L et al (2020) A non-canonical RNAi pathway controls virulence and genome stability in Mucorales. PLoS Genet 16(7):e1008611
24. Talbert PB, Henikoff S (2020) What makes a centromere? Exp Cell Res 389(2):111895
25. Holmes RJ, Cohen PE (2007) Small RNAs and RNAi pathways in meiotic prophase I. Chromosome Res 15(5):653–665
26. Folco HD, Pidoux AL, Urano T et al (2008) Heterochromatin and RNAi are required to establish CENP-A chromatin at centromeres. Science 319(5859):94–97
27. Lejeune E, Bayne EH, Allshire RC (2010) On the connection between RNAi and heterochromatin at centromeres. Cold Spring Harb Symp Quant Biol 75:275–283
28. Jih G, Iglesias N, Currie MA et al (2017) Unique roles for histone H3K9me states in RNAi and heritable silencing of transcription. Nature 547(7664):463–467
29. Wallis DC, Nguyen DAH, Uebel CJ et al (2019) Visualization and quantification of transposon activity in *Caenorhabditis elegans* RNAi pathway mutants. G3 (Bethesda) 9(11): 3825–3832
30. Xie Z, Fan B, Chen C et al (2001) An important role of an inducible RNA-dependent RNA polymerase in plant antiviral defense. Proc Natl Acad Sci U S A 98(11):6516–6521
31. Tabara H, Yigit E, Siomi H et al (2002) The dsRNA binding protein RDE-4 interacts with RDE-1, DCR-1, and a DExH-box helicase to direct RNAi in *C. elegans*. Cell 109(7): 861–871
32. Lu R, Maduro M, Li F et al (2005) Animal virus replication and RNAi-mediated antiviral silencing in *Caenorhabditis elegans*. Nature 436(7053):1040–1043
33. Jeang KT (2012) RNAi in the regulation of mammalian viral infections. BMC Biol 10:58
34. Zhou R, Rana TM (2013) RNA-based mechanisms regulating host-virus interactions. Immunol Rev 253(1):97–111
35. Sabin LR, Zheng Q, Thekkat P et al (2013) Dicer-2 processes diverse viral RNA species. PLoS One 8(2):e55458
36. Palmer WH, Hadfield JD, Obbard DJ (2018) RNA-interference pathways display high rates of adaptive protein evolution in multiple invertebrates. Genetics 208(4):1585–1599
37. Andika IB, Kondo H, Suzuki N (2019) Dicer functions transcriptionally and posttranscriptionally in a multilayer antiviral defense. Proc Natl Acad Sci U S A 16(6): 2274–2281
38. Rawlings RA, Krishnan V, Walter NG (2011) Viral RNAi suppressor reversibly binds siRNA to outcompete dicer and RISC via multiple turnover. J Mol Biol 408(2):262–276
39. Axelson-Fisk M, Sunnerhagen P (2006) Comparative genomics and gene finding in fungi. In: Sunnerhagen P, Piskur J (eds) Comparative genomics using fungi as models. Topics in current genetics, vol 15. Springer-Verlag, Berlin, Heidelberg
40. Drinnenberg IA, Fink GR, Bartel DP (2011) Compatibility with killer explains the rise of RNAi-deficient fungi. Science 333(6049): 1592
41. Nakayashiki H, Kadotani N, Mayama S (2006) Evolution and diversification of RNA silencing proteins in fungi. J Mol Evol 63(1):127–135
42. Wood V, Gwilliam R, Rajandream M et al (2002) The genome sequence of *Schizosaccharomyces pombe*. Nature 415(6874):871–880
43. Getz MA, Weinberg DE, Drinnenberg IA et al (2020) Xrn1p acts at multiple steps in the budding-yeast RNAi pathway to enhance the efficiency of silencing. Nucleic Acids Res 48(13):7404–7420
44. Calo S, Shertz-Wall C, Lee SC et al (2014) Antifungal drug resistance evoked via RNAi-dependent epimutations. Nature 513(7519): 555–558
45. Villalobos-Escobedo JM, Herrera-Estrella A, Carreras-Villaseñor N (2016) The interaction of fungi with the environment orchestrated by RNAi. Mycologia 108(3):556–571
46. Chang Z, Heitman J (2019) Drug-resistant Epimutants exhibit organ-specific stability and induction during murine infections caused by the human fungal pathogen. Mucor circinelloides. MBio 10(6):e02579–e02619
47. Pfaller MA, Diekema DJ (2007) Epidemiology of invasive candidiasis: a persistent public health problem. Clin Microbiol Rev 20(1): 133–163
48. Dujon B, Sherman D, Fischer G et al (2004) Genome evolution in yeasts. Nature 430(6995):35–44
49. Scannell DR, Butler G, Wolfe KH (2007) Yeast genome evolution-the origin of the species. Yeast 24(11):929–942
50. Fidel PL Jr, Vazquez JA, Sobel JD (1999) *Candida glabrata*: review of epidemiology, pathogenesis, and clinical disease with comparison to *C. albicans*. Clin Microbiol Rev 12(1): 80–96

51. Brunke S, Hube B (2013) Two unlike cousins: *Candida albicans* and *C. glabrata* infection strategies. Cell Microbiol 15(5):701–708
52. Cormack BP, Falkow S (1999) Efficient homologous and illegitimate recombination in the opportunistic yeast pathogen *Candida glabrata*. Genetics 151(3):979–987
53. Corrigan MW, Kerwin-Iosue CL, Kuczmarski AS et al (2013) The fate of linear DNA in *Saccharomyces cerevisiae* and *Candida glabrata*: the role of homologous and non-homologous end joining. PLoS One 8(7):e69628
54. Enkler L, Richer D, Al M et al (2016) Genome engineering in the yeast pathogen *Candida glabrata* using the CRISPR-Cas9 system. Sci Rep 6:35766
55. Ishchuk OP, Ahmad KM, Koruza K et al (2019) RNAi as a tool to study virulence in the pathogenic yeast *Candida glabrata*. Front Microbiol 10:1679
56. Chen Y, Guo E, Zhang J et al (2020) Advances in RNAi-assisted strain engineering in *Saccharomyces cerevisiae*. Front Bioeng Biotechnol 8: 731
57. Hou J, Schacherer J (2017) Fitness trade-offs Lead to suppressor tolerance in yeast. Mol Biol Evol 34(1):110–118
58. Fournier T, Abou Saada O, Hou J et al (2019) Extensive impact of low-frequency variants on the phenotypic landscape at population-scale. Elife 8:e49258
59. Hanic-Joyce PJ, Joyce PB (1998) A high-copy-number *ADE2*-bearing plasmid for transformation of *Candida glabrata*. Gene 211(2): 395–400
60. Boretsky YR, Pynyaha YV, Boretsky VY et al (2007) Development of a transformation system for gene knock-out in the flavinogenic yeast *Pichia guilliermondii*. J Microbiol Methods 70(1):13–19
61. Mauersberger S, Ohkuma M, Schunck W-H et al (1996) Candida maltosa. In: Wolf K (ed) Nonconventional yeasts in biotechnology. Springer, Berlin, pp 411–580
62. Wach A, Pick H, Philipsen P (1994) Procedures for isolating yeast DNA for different purposes. In: Johnston JR (ed) Molecular genetics of yeast. A practical approach. IRL Press, Oxford, pp 1–16
63. Wagih O, Parts L (2014) Gitter: a robust and accurate method for quantification of colony sizes from plate images. G3 (Bethesda) 4(3): 547–552
64. Seider K, Brunke S, Schild L et al (2011) The facultative intracellular pathogen *Candida glabrata* subverts macrophage cytokine production and phagolysosome maturation. J Immunol 187(6):3072–3086
65. Kasper L, Seider K, Gerwien F et al (2014) Identification of *Candida glabrata* genes involved in pH modulation and modification of the phagosomal environment in macrophages. PLoS One 9(5):e96015
66. Seider K, Gerwien F, Kasper L et al (2014) Immune evasion, stress resistance, and efficient nutrient acquisition are crucial for intracellular survival of *Candida glabrata* within macrophages. Eukaryot Cell 13(1):170–183

Chapter 19

High-Throughput Gene Mutagenesis Screening Using Base Editing

Philippe C. Després, Alexandre K. Dubé, Nozomu Yachie, and Christian R. Landry

Abstract

Base editing is a CRISPR-Cas9 genome engineering tool that allows programmable mutagenesis without the creation of double-stranded breaks. Here, we describe the design and execution of large-scale base editing screens using the Target-AID base editor in yeast. Using this approach, thousands of sites can be mutated simultaneously. The effects of these mutations on fitness can be measured using a pooled growth competition assay followed by DNA sequencing of gRNAs as barcodes.

Key words Base editing, Yeast, Target-AID, CRISPR-Cas9, Systems biology, Mutagenesis, High-throughput, Functional genomics

1 Introduction

Base editors combine the programmable aspects of CRISPR-Cas9 genome editing tools with the activity of DNA-modifying enzymes to perform precise mutagenesis [1, 2]. The first generation of base editors combined partially inactivated Cas9 enzymes with cytidine deaminases to selectively deaminate cytosines to uracil. This DNA lesion can be repaired through two different pathways, leading to either a C-to-T or a C-to-G mutation. An uracil glycosylase inhibitor (UGI) was then fused to the editor to channel mutation outcomes to C to T and increase efficiency in mammalian cells, leading to the creation of Target-AID and BE3. Both of these first generation editors can be used in yeast, as shown either in the original manuscript for Target-AID or by later work in the case of BE3 [3]. In yeast, removing the UGI does not drastically reduce editing rates and allows us to explore more genotypes during the course of the experiment. In this chapter, we focus on the use of the pRS315e_pGal-nCas9(D10A)-PmCDA1 vector, which encodes for Target-AID without the UGI fusion as a way to generate

Frédéric Devaux (ed.), *Yeast Functional Genomics: Methods and Protocols*,
Methods in Molecular Biology, vol. 2477, https://doi.org/10.1007/978-1-0716-2257-5_19,

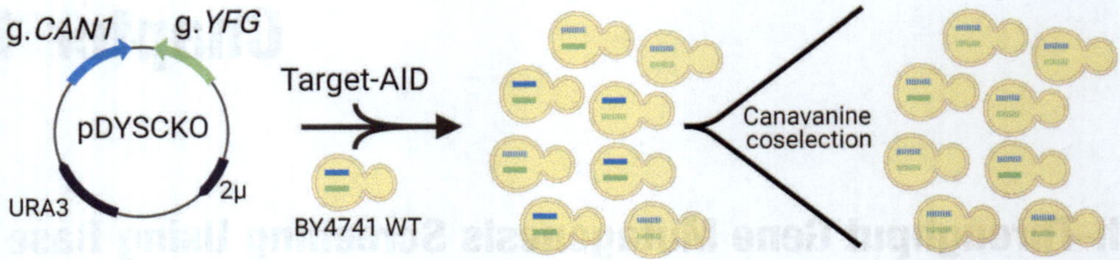

Fig. 1 The pDSYCKO vector for base editing with enrichment by coselection. The vector contains two gRNA expression cassettes, as well as the *URA3* marker, a 2μ origin and an ampicillin resistance gene for selection in bacteria (not shown). After cotransformation with a base editor, both *CAN1* and *YFG* (*Your Favorite Gene*) are mutated. Even if the two loci are not edited at the same rate overall, successful mutagenesis at one locus is correlated with successful mutagenesis at the other, thus increasing edition rate. Selecting for cells with a *CAN1* loss of function thus increases the fraction of cells where *YFG* has been mutated, increasing the effective efficiency of base editing

diversity when performing high-throughput screening. To make the text easier to read, here we will refer to this construct as Target-AID.

To improve the mutagenesis rates in our edited cell population, we perform a coselection step where only cells in which a negative selection marker was edited can survive and move on to the competition screen [4–6]. In yeast, we found this could increase two to threefold the editing efficiency [7]. We target the amino acid permease *CAN1*, which imports the toxic arginine analog canavanine. *CAN1* loss of function blocks the entry of canavanine in the cell, leading to drug resistance [8]. As such, the pDYSCKO vector used for large-scale screening contains two gRNA in a tail to tail configuration: the constant *CAN1*-targeting coselection gRNA, and a variable gRNA targeting *Your Favorite Gene* (*YFG*) to edit the locus of interest (Fig. 1). Before performing the large-scale assay, we recommend performing preliminary assays using gRNAs with easily identifiable outcomes. For example, a *ADE1* loss of function leads to the accumulation of a red pigment that changes the colony color of edited cells. This allows for measurement of editing rates without having to use high-throughput sequencing. A pDYSCKO plasmid bearing a gRNA creating a *ADE1* loss of function is available from Addgene (#120264). The *CAN1* inactivating gRNA on pDYSCKO depends on a C-to-G mutation to generate a loss of function. As such, using the protocol we present here with an editor that has the UGI fusion will require a different coselection gRNA.

The first step in performing a large-scale base editing screen is the selection of gRNAs to be included in the library. The total number of gRNAs will dictate many aspects of the experiment during the library construction in bacteria and yeast transformation, as well as the depth of high-throughput sequencing required. From there, the gRNA library is synthesized as a double stranded DNA (dsDNA) pool and cloned into the pDYSCKO coselection

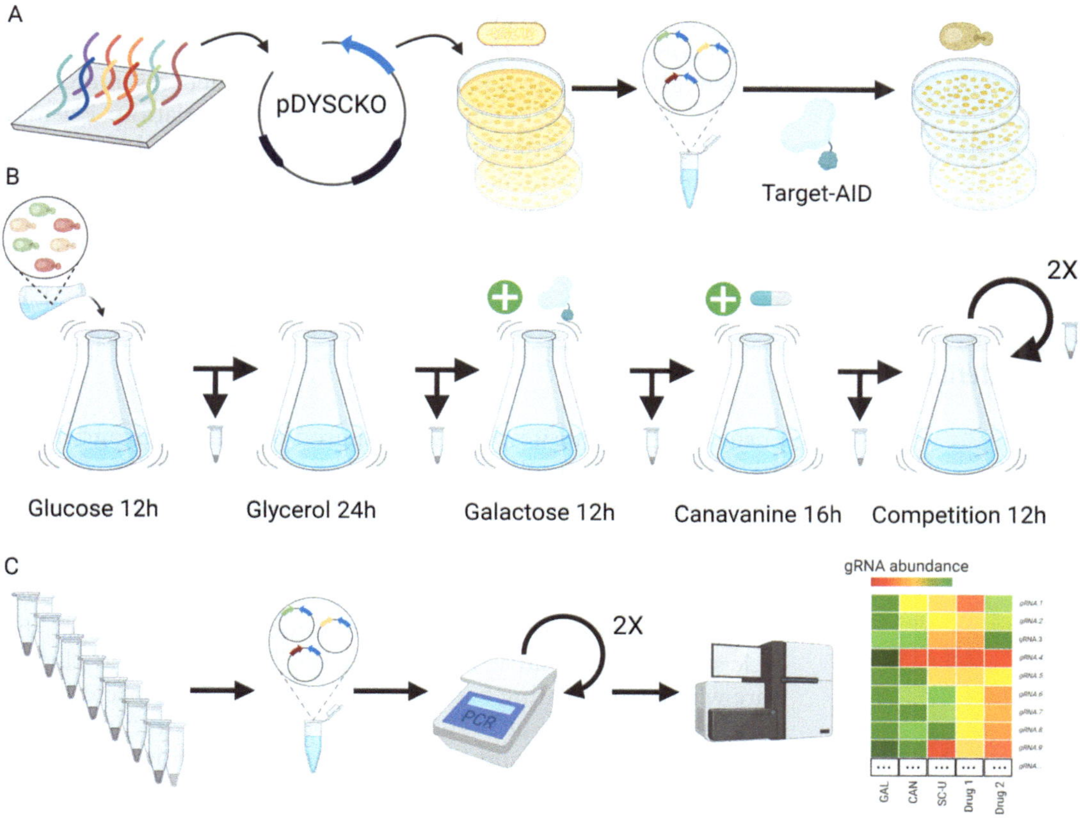

Fig. 2 Workflow for the base editing screen. (**a**) After the gRNA library is designed and synthesized as dsDNA, it is cloned into the pDYSCKO coselection vector using Golden Gate assembly. Multiple cloning reactions are pooled to increase coverage, and plasmid DNA is extracted. The gRNA library is then transformed into a yeast strain already bearing a galactose inducible base editor. Multiple transformations are pooled to ensure homogeneous coverage. (**b**) Once the plasmid library has been transformed in yeast, the screen can begin. At each step, cell pellets are sampled for plasmid DNA extraction. After transformation, the yeast cell pool is amplified during a glucose culture round. Cells are then transferred to glycerol media to prime them for galactose induction. When they are switched to galactose, the expression of the base editor is induced, leading to mutagenesis of the target genes. The cells are then diluted in canavanine containing media for a coselection round. Finally, two rounds of competition are performed. (**c**) Plasmid DNA is extracted from cell pellets at the time points of interest. Using two PCR steps, the gRNA region is amplified and fused with adapters for high-throughput sequencing. Changes in frequency of gRNAs in the amplicon pools provide information on the fitness effects of the mutations they induced

vector using Golden Gate assembly (Fig. 2a). The cloned library is transformed into a yeast strain already containing Target-AID under the control of a galactose inducible promoter. The yeast population goes through two preediting rounds: one to amplify the library from the cells recovered from transformation plates, and another using glycerol as a carbon source to lower cell to cell variation during galactose induction [9]. After Target-AID expression induction, the cells are diluted in coselection media to enrich edited cells. The cells then compete for two rounds, as in a standard pooled competition assay (Fig. 2b). Because the editing outcomes of each gRNA are predictable, we can use them as molecular barcodes to measure mutant abundance in a population. Plasmid DNA is extracted at the different timepoints, and used to prepare high-throughput sequencing amplicon libraries. This data is used to measure the variation in relative abundance of the different gRNAs and infer the effect of the mutations they induce on cell growth (Fig. 2c).

2 Materials

Ensure all prerequisite packages required to run the programs in the analysis are installed and function properly. Make sure to backup all scripts and raw data files securely, and consider uploading them to a public archive (ex.: github for code and the NCBI SRA for sequencing).

All solutions should be prepared using distilled, deionised water unless stated otherwise. When autoclaving is required, the cycle should be 15 min at 121 °C. Use sterile glassware and plasticware when manipulating bacteria and yeast cells. Store bacteria and yeast glycerol stocks at −80 °C. Please use best practices when disposing of chemical and biological waste, and follow the manufacturer's instruction when using kits unless stated otherwise.

2.1 gRNA Library Creation (Computational)

1. A computer with a linux distribution, such as ubuntu, or a virtual machine that can emulate such an environment (*see* **Note 1**).
2. Python version ≥3.6.1.
3. Base editing gRNA selection software: beditor (https://github.com/rraadd88/beditor).
4. A list of mutations to be processed by beditor (*see* **Note 2**).

2.2 gRNA Library Cloning

1. Synthesized gRNA library as dsDNA fragments: Resuspend and aliquot library according to the manufacturers' instructions (*see* **Note 3**).
2. PCR grade water.
3. Golden Gate Assembly Mix (New England Biolabs).
4. pDYSCKO plasmid (Addgene #120263) [7] plasmid DNA prep, diluted to 75 ng/μL.
5. Competent *E. coli*: Chemically competent (*see* **Note 4**) MC1061 [10]. We usually prepare competent cells in-house (*see* **Note 5**).
6. An ice bucket.

7. Bacteria recovery media 2YT: 10 g/L yeast extract, 16 g/L tryptone, 2 g/L glucose, 5 g/L NaCl. Mix in ~80% of the desired final volume of water. Adjust volume and autoclave. Store at room temperature.
8. A 42 °C water bath.
9. Sterile glass beads.
10. Assembly selection media 2YT + Ampicillin plates: 10 g/L yeast extract, 16 g/L tryptone, 2 g/L glucose, 5 g/L NaCl, 100 μg/mL ampicillin, 20 g/L agar. Mix the four first compounds in 75% of the desired final volume of water. Adjust volume and add agar. Autoclave. When the media has cooled down to ~60 °C, add 1 mL per liter of 100 mg/mL ampicillin. Pour plates and let the media set before flipping the plates over to reduce condensation. Store at 4 °C, but take out to dry overnight before use.
11. Petri dish scraping: sterile water, 1 L, autoclaved. Store at room temperature.
12. 80% Glycerol: Mix 80 mL of 100% glycerol and 20 mL water, then autoclave. Store at room temperature.
13. Screw-cap tubes, sterile.
14. Midi-prep kit.
15. NanoDrop spectrophotometer.

2.3 Prepare Yeast Competent Cells with the Target-AID Base Editor

1. Yeast strain: We use BY4741 (*MATa his3Δ1 leu2Δ0 ura3Δ0 met15Δ0*) as a standard lab strain [11].
2. Yeast competent cells: Competent cells are prepared using the lithium acetate method [12] (*see* **Note 6**).
3. pGAL-Target-AID (Addgene) #79617 [1], plasmid DNA prep, diluted to 200 ng/μL.
4. Plate mixture: For 100 mL, mix 90 mL 40% PEG 3350, 10 mL 1 M LiOAc, 1 mL 1 M Tris–HCl, pH 7.5, 200 μL 0.5 M EDTA. All solutions should be sterile. Store at room temperature.
5. YPD: 10 g/L yeast extract, 20 g/L tryptone and 20 g/L glucose. Dissolve in 80% of the desired final volume of water. Adjust volume, then autoclave. Store at room temperature.
6. Transformant selection plates (Synthetic complete media minus leucine, solid (SC-L, *see* **Note 7**): 1.74 g/L yeast nitrogen base without ammonium sulfate, without amino acids, 0.97 g/L -L drop-out mix, 5 g/L Ammonium sulfate, 20 g/ L glucose, 20 g/L agar. Dissolve the three first compounds in 20% of the desired final volume water, and adjust volume to 25% of the final volume. Separately (*see* **Note 8**), dissolve glucose in 60% of the final volume of water, and adjust to

75% of the final volume. Add agar to the dissolved glucose. Autoclave both solutions. After autoclaving, combine the two solutions and let cool to ~60 °C. Pour plates and let the media set before flipping the plates over to reduce condensation. Store at 4 °C, but take out to dry overnight before use.

7. A plate reader (*see* **Note 9**).
8. Competent cells preparation media (Synthetic complete media minus leucine, liquid (SC-L)): 1.74 g/L yeast nitrogen base without ammonium sulfate, without amino acids, 0.97 g/L -L drop-out mix, 5 g/L Ammonium sulfate, 20 g/L glucose. Dissolve in 75% of the desired final volume of water. Adjust to final volume. Autoclave and store at room temperature.
9. SORB: 1 M Sorbitol, 100 mM LiOAc, 10 mM Tris–HCl, pH 8.0, 1 mM EDTA. Filter-sterilize, and store at room temperature.
10. Carrier DNA: 10 mg/mL Salmon sperm DNA, 10 mM Tris–HCl, 1 mM EDTA, pH 8.0. Pass the solution in a 23G needle 20 times (10 in and 10 out). Aliquot in 1.5 mL tubes, and boil for 10 min. Store at −20 °C. Boil for 5 min and then cool on ice before use.
11. DMSO.

2.4 Library Transformation in Yeast

1. Transformant selection plates (Synthetic complete media lacking uracil and leucine, solid (SC-UL)): 1.74 g/L yeast nitrogen base without ammonium sulfate, without amino acids, 0.90 g/L -UL drop-out mix, 5 g/L ammonium sulfate, 20 g/L glucose, 20 g/L agar. Dissolve the three first compounds in 20% of the desired final volume of water. Adjust volume to 25% of the final volume. Separately, dissolve glucose in 60% of the desired final volume of water, and adjust to 75% of the final volume. Add 20 g agar to the dissolved glucose. Autoclave both solutions. After autoclaving, combine the two solutions and let cool to ~60 °C. Pour plates and let the media set before flipping the plates over to reduce condensation. Store at 4 °C, but take out to dry overnight before use.
2. Petri dish scraping: sterile water, 1 L, autoclaved. Store at room temperature.
3. 80% Glycerol: Mix 80 mL 100% glycerol and 20 mL water, then sterilize using autoclave. Store at room temperature.

2.5 Large-Scale Screen

1. Library outgrowth media (SC-UL + 2% glucose, liquid): 1.74 g/L yeast nitrogen base without ammonium sulfate, without amino acids, 0.90 g/L -UL drop-out mix, 5 g/L ammonium sulfate, 20 g/L glucose. Dissolve compounds in 75% of the desired final volume of water. Adjust volume. Autoclave and store at room temperature.

2. Preinduction media (SC-UL + 5% glycerol, liquid): For 1 L, dissolve 1.74 g/L yeast nitrogen base without ammonium sulfate, without amino acids, 0.90 g/L -UL drop-out mix, 5 g/L ammonium sulfate, 5% glycerol. Dissolve all compounds except glycerol in 75% of the desired final volume of water. Adjust volume to 93.7% of the final volume. Autoclave, cool down, and then add 6.25% of the final volume of sterile 80% glycerol and mix well. Store at room temperature.
3. 40% Galactose (*see* **Note 10**). Filter-sterilize. Store at room temperature.
4. Induction media (SC-UL + 5% galactose, liquid): 1.74 g/L yeast nitrogen base without ammonium sulfate, without amino acids, 0.90 g/L -UL drop-out mix, 5 g/L Ammonium Sulfate, 5% galactose. Dissolve all compounds except galactose in 75% of the desired final volume of water. Adjust volume to 87.5% of the final volume. Autoclave, cool down, and then add 12.5% of the final volume of 40% galactose and mix well. Store at room temperature.
5. 50 mg/mL Canavanine stock solution. Filter-sterilize, and split in 1 mL aliquots. Store at −20 °C.
6. Coselection media (SC-ULR + 2% glucose + canavanine 50 μg/mL, liquid) (*see* **Note 11**). For 1 L, dissolve 1.74 g/L yeast nitrogen base without ammonium sulfate, without amino acids, 0.83 g/L -ULR drop-out mix, 5 g/L ammonium sulfate, 20 g/L glucose, 50 μg/mL canavanine. Dissolve all compounds except canavanine in 75% of the desired final volume of water. Adjust volume. Autoclave and store at room temperature. Before use, add 0.1% of final volume of canavanine stock solution and mix well.
7. Competition media (SC-U + 2% glucose, liquid): 1.74 g/L yeast nitrogen base without ammonium sulfate, without amino acids, 1.27 g/L -U drop-out mix, 5 g/L Ammonium Sulfate, 20 g/L glucose. Dissolve in 75% of the desired final volume of water. Adjust volume. Autoclave and store at room temperature.

2.6 Sequencing Library Preparation

1. ChargeSwitch Plasmid Yeast Mini Kit (Invitrogen).
2. 4000 U/mL Zymolyase.
3. PCR thermocycler.
4. PCR 1 oligonucleotides: These oligonucleotides amplify the gRNA region of the pDYSCKO plasmid for deep sequencing and add an overhang that has homology to the primers used to barcode samples and add the Illumina sequencing adapters; the forward oligonucleotide sequence is

TAACTTACGGAGTCGCTCTACGATGTTTCGGCGTTC GAA , and the reverse sequence is GGATGGGATTCTT TAGGTCCTGCGAAGGCTTTAATTTGC.

5. A high fidelity DNA polymerase. We have used either Phusion polymerase or the KAPA HiFi PCR Kit. In this protocol, the reaction parameters for Phusion are used.
6. Agarose gel electrophoresis system.
7. Electrophoresis buffer (50× TAE): 2 M Tris base, 1 M acetic acid, 0.05 M EDTA, pH 8.0. Store at room temperature. Dilute to 1× in water before use.
8. Agarose gel, 1.5%: 15 g/L agarose, 1× TAE. Dissolve the agarose using heat, and pour gel after cooling to ~60 °C. Add the appropriate comb to run the desired number of samples.
9. A DNA ladder covering 50–1500 bp.
10. A dsDNA dye.
11. Agarose gel imaging device compatible with the dsDNA dye.
12. Gel extraction kit OR magnetic beads. In this protocol, we use a gel extraction kit, but magnetic beads are also suitable and might provide better yield and quality.
13. Phusion Mastermix-HF (NEB), or the KAPA HiFi PCR Kit.
14. PCR 2 oligonucleotides: These oligonucleotides are used to add the Illumina sequencing adapters to the purified amplicon and to add a unique sample barcode (*see* **Note 12**, shown here in lowercase). We provide here an example pair, but *see* [13] for a detailed description of the approach or more additional barcode sequences.

    Forward: ATGATACGGCGACCACCGAGATCTA CACTCTTTCCCTACACGACGCTCTTCC GATCTNNNNN cactaatgg TAACTTACGGAGTCGCTCTACG

    Reverse:

    CAAGCAGAAGACGGCATACGAGATCGGTCTCGG-CATTCCTGCTGAACCGCTCTTCCGATCTNNNNNagga-cattcGGATGGGATTCTTTAGGTCCTG.
15. KAPA Library Quantification Kit for Illumina.
16. Illumina HiSeq sequencer and SBS kit (*see* **Note 13**).

3 Methods

3.1 gRNA Library Creation (Computational)

1. Use the beditor software with your list of mutations to be created, or select gRNAs from the list of gRNAs provided in **Note 2**. *See* **Note 14** for some considerations when designing gRNA libraries for base editing screens.

2. Filter out gRNAs containing a *BsaI* restriction site (forward: GGTGTC, reverse: GAGACC).
3. For use with pDYSCKO, add the following flanking sequences to the gRNA sequence; 5′: CATAAGGAGGGTCTCCTGATC, 3′:GTTTAGAGACCATGAGCGA. The final sequence should look like this: CATAAGGAGGGTCTCCTGATC-gRNA-GTTTAGAGACCATGAGCGA.
4. Have the oligonucleotide library synthesized commercially (or in-house if you have the facilities).
5. Determine the number of Golden Gate assembly transformants and yeast transformants that will be required for good coverage of the library by multiplying the number of gRNAs in the library by the coverage. For example, a 1000 gRNA library covered 50× would require at least 50,000 (1000 × 50) independent clones. We recommend at least 100× coverage at both steps, and going above this threshold will reduce variance in the initial abundance of the different gRNAs in the library (*see* **Note 15**).

3.2 gRNA Library Cloning

1. Thaw a dsDNA fragment library aliquot and a pDYSCKO plasmid aliquot. Dilute the dsDNA library to 2 ng/μL.
2. Prepare the following reaction in a PCR tube: 2 μL 10× NEB Golden Gate buffer, 1 μL dsDNA pool, 1 μL pDYSCKO plasmid, 1 μL NEB Golden Gate mix and 15 μL PCR grade water. Prepare a negative control reaction by replacing the DNA insert pool with PCR grade water. Incubate in a thermocycler using the following cycle:
   (a) 1 h at 37 °C.
   (b) 5 min at 55 °C.

   As soon as the assembly reaction is completed, move to the transformation step.
3. Use the ligation mix to perform four transformations/reaction, and include one negative control. First, thaw the appropriate amount of competent cells (we use 100 μL per transformation) on ice. While the cells thaw, cool down round bottom plastic tubes on ice.
4. When the cells are thawed, split them (100 μL per tube) and add 5 μL of ligation mix to the cells. Mix by lightly tapping the tube with your finger, and incubate on ice for 15 min.
5. Perform a heat shock in a 42 °C water bath for 1 min.
6. Put the cells back on ice for 5 min.
7. Add 900 μL 2YT to the tubes, and incubate for 1 h at 37 °C with agitation.

8. Spin down the tubes, remove the supernatant, and resuspend the cells in 100 μL 2YT. Plate the cells on 2YT + Amp using sterile glass beads. Perform a series of tenfold dilutions (1/100, 1/1000, 1/10,000) platings in at least three replicates to estimate the total number of clones generated (*see* **Note 16**). Incubate overnight at 37 °C.
9. Add 5 mL sterile water to each plate and incubate 5 min at room temperature. Use a sterile glass rake to scrape and resuspend colonies in water. Pour the resulting cell suspension in a 15 mL or 50 mL conical tube. Vortex to make sure the suspension is homogeneous.
10. For each tube, make a glycerol stock by combining 810 μL cells and 190 μL 80% glycerol in a screw-cap tube. Store at −80 °C.
11. Depending on the observed cloning efficiency and desired library coverage, combine tubes and perform midi-prep extractions following manufacturers instruction (*see* **Note 17**).
12. Measure the DNA concentration of each plasmid pool using a NanoDrop Spectrophotometer. Make a master library stock by mixing equal quantities of each pool to a concentration of 250 ng/μL. Store at −20 °C until used for yeast transformation (*see* **Note 18**).

3.3 Yeast Competent Cell Preparation with the Target-AID Base Editor

1. Prepare yeast competent cells using the BY4741 strain.
2. Thaw a 20 μL BY4741 competent cell aliquot and a pGAL-Target-AID plasmid aliquot on ice.
3. Add 5 μL pGAL-Target-AID and 100 μL Plate mixture to the competent cell tube. Mix by vortexing. Incubate for 30 min at room temperature.
4. Add 15 μL DMSO, and incubate for 20 min at 42 °C.
5. Pellet the cells by centrifugation at 400 × *g* for 3 min. Remove the supernatant, and resuspend the cells in 100 μL YPD.
6. Incubate at 30 °C without agitation for 4 h.
7. Plate the cells on SC-L selective media using sterile glass beads, and incubate for 48 h at 30 °C.
8. From the transformation plates, start 2 precultures by picking a few colonies and resuspending them in 6 mL SC-L in culture tubes. Incubate overnight at 30 °C with agitation.
9. The following day, dilute the cells to 0.15 OD in SC-L liquid media in flasks. The total culture volume should be 200 mL.
10. Let cells reach exponential phase (0.6–0.7 OD). Split the culture into four different 50 mL tubes.
11. Pellet cells by centrifuging at 500 × *g* for 5 min. Remove the supernatant.

12. Wash the cells by adding 5 mL sterile water to the tubes and centrifuging at 500 × *g* for 5 min again.
13. Wash the cells by adding 5 mL SORB solution and centrifuge at 500 × *g* for 5 min again (*see* **Note 19**).
14. Using aspiration, remove as much SORB as possible from the tube. Resuspend the cell pellet in 360 μL and 40 μL boiled carrier DNA for each 50 mL tube.
15. Make 200 μL aliquots and store at −80 °C until use.

3.4 Library Transformation in in Yeast

1. Thaw 8 competent cells aliquots and 250 μL pDYSCKO master library on ice (*see* **Note 20**).
2. Split the competent cells in 40 tubes containing 40 μL competent cells each. Add 5 μL pDYSCKO master library to each tube.
3. Add 200 μL plate mixture and mix by vortexing. Incubate for 30 min at room temperature.
4. Add 30 μL DMSO and incubate 20 min at 42 °C.
5. Pellet cells by centrifuging at 400 × *g* for 3 min. Remove supernatant, and resuspend cells in 200 μL YPD.
6. Incubate at 30 °C without agitation for 4 h.
7. Combine the different transformations in a single tube, and plate 100 μL aliquots on SC-UL solid media using sterile glass beads. Perform a series of tenfold dilutions (1/100, 1/1000, 1/10,000) in at least three replicates to estimate the number of transformants obtained. Incubate at 30 °C for 48 h.
8. Add 5 mL sterile water to the transformation plates, and incubate for 5 min at room temperature. Recover the cells using a sterile glass rake and combine the cells suspensions in a sterile flask.
9. Prepare library glycerol stocks by mixing 450 μL 80% glycerol and 950 μL cell suspension in a screw-cap tube. Store at −80 °C.
10. If ready to perform screen, use a spectrophotometer to measure cell suspension OD, and make one 60 OD units pellet per screen replicate by centrifuging at 500 × *g* for 5 min. Perform all the following steps in parallel for the different replicates.

3.5 Large-Scale Screen

1. Wash the starter pellets by adding 5 mL water and centrifuge at 500 × *g* for 5 min. Repeat for a second wash.
2. Inoculate 100 mL SC-UL + 2% glucose with the cell pellet (final concentration 0.6 OD/mL). Incubate the culture with shaking at 30 °C for 8 h.
3. Pellet 100 OD units by centrifuging at 500 × *g* for 5 min. Remove supernatant and wash with 5 mL sterile water. Use the

pellet to inoculate a 100 mL SC-UL + 5% glycerol culture (final concentration 1 OD/mL). Incubate with shaking at 30 °C for 24 h. Take 2 OD units pellets of each sample for downstream sequencing (*see* **Note 21**).

4. Pellet 50 OD units by centrifuging at 500 × *g* for 5 min. Remove supernatant and wash with 5 mL sterile water. Use the pellet to inoculate a 100 mL SC-UL + 5% galactose culture (final concentration 0.5 OD/mL). Incubate with shaking at 30 °C for 12 h. Take 2 OD units pellets of each sample for downstream sequencing.
5. Pellet 10 OD units by centrifuging at 500 × *g* for 5 min. Remove supernatant and wash with 5 mL sterile water. Use the pellet to inoculate a 100 mL SC-UL + 2% glucose + canavanine culture (final concentration 0.1 OD/mL). Incubate with shaking at 30 °C for 16 h. Take 2 OD units pellets of each sample for downstream sequencing.
6. Pellet 5 OD units by centrifuging at 500 × *g* for 5 min. Remove supernatant and wash with 5 mL sterile water. Use the pellet to inoculate a 100 mL SC-U + 2% glucose (final concentration 0.05 OD/mL) to start the first competition round (*see* **Note 22**). Incubate with shaking at 30 °C for 12 h. Take 2 OD units pellets of each sample for downstream sequencing.
7. Perform another competition round by repeating **step 6** (*see* **Note 23**).
8. Take 2 OD units pellets in duplicate from the last competition round. Make glycerol stocks as described in **step 9** of Subheading 3.4.

3.6 Sequencing Library Preparation

1. Thaw cell pellets from the time points of interest on ice.
2. Extract plasmid DNA using the ChargeSwitch Plasmid Yeast Mini Kit, with the following modifications to the suggested protocol.
   (a) Use 4000 U/mL Zymolyase instead of lyticase.
   (b) At **step 4**, incubate for 1 h at room temperature, switch to −80 °C for 1 min, and then incubate for another 15 min at room temperature. This replaces the single 30 min incubation at room temperature in the kit instructions.
   (c) Elute in 70 μL elution buffer (10 mM Tris–HCl, pH 8.5).
3. Store at −20 °C until ready for the next step of library preparation.
4. Prepare the following reaction (total volume 25 μL, 1 per sample): 5 μL 5× Phusion HF buffer, 0.5 μL 10 mM dNTPs, 1.25 μL 10 μM PCR1 forward, 1.25 μL 10 μM PCR 2 reverse,

0.5 μL 2000 U/mL Phusion polymerase, 5 μL Plasmid DNA (*see* **Note 24**), 11.7 μL PCR grade water.

5. Perform the first PCR amplification round using the following cycle:
   (a) 98 °C for 30 s.
   (b) 16 cycles of 98 °C for 10 s, 58 °C for 15 s, 72 °C for 5 s.
   (c) 72 °C for 5 s.
6. Verify products on a 1.5% agarose gel. The expected product is 230 bp long.
7. Purify products using a gel extraction kit (*see* **Note 25**).
8. Prepare the following reaction (20 μL): 10 μL Phusion Mastermix-HF, 3.75 μL 1.333 μM PCR 2 forward barcoded oligo, 3.75 μL 1.333 μM PCR 2 reverse barcoded oligo, 2.5 μL 2 ng/μL purified PCR 1 product.
9. Perform the second PCR amplification round using the following cycle:
   (a) 98 °C for 30 s.
   (b) 15 cycles of 98 °C for 10 s, 60 °C for 10 s, 72 °C for 1 min.
   (c) 72 °C for 5 min.
10. Verify products on a 1.5% agarose gel. The expected product is 377 bp long.
11. Purify products using a gel extraction kit (*see* **step 2** of this section).
12. Perform quality control and determine sample quality using the KAPA Library Quantification Kit for Illumina.
13. Pool samples in equimolar amounts and sequence using Illumina SBS technology, following manufacturer's instructions (*see* **Notes 26** and **27**).

4 Notes

1. While a linux distribution is best, the software should be able to run on macOS. For windows users, the windows subsystem for linux is a good option as it allows the installation and use of linux programs and gives access to a fully functional bash shell. A drawback is that GUIs are not supported, meaning that the beditor interface will not be available.
2. Alternatively, we provide here a list of curated gRNA suitable for base editing across yeast annotated coding sequences https://github.com/PhilDesp/MiMB_2021.
3. Many dsDNA oligonucleotide pool manufacturers recommend performing a PCR round on the library before using it for

cloning. A ssDNA oligonucleotide library can also be used in this protocol, provided it has been amplified by PCR into dsDNA. Follow the recommendations of the manufacturer regarding PCR parameters and reagents.

4. Our in-house competent cells routinely reach transformation efficiencies above 2×10^7 CFU/μg. As transformation efficiency is crucial to obtain good coverage of the library, make sure the competent cell and transformation protocol you use can reliably attain this threshold.
5. As bacteria competent cell preparation is not the focus of this chapter, please see [14] for the details of our competent cell preparation protocol.
6. The yeast competent cell protocol we use is detailed in Subheading 2.3, **steps 8–15**. The BY4741 competent cells can be prepared using the same procedure, substituting SC-L for YPD.
7. The Synthetic complete amino acid drop-out mix compositions are shown in Table 1. We usually prepare drop-out by mixing all powders with a mortar and pestle and then aliquot the mixture in 50 mL tubes.

Table 1
Composition of the amino acid drop-out mixes

Nutrient	DO-L	DO-UL	DO-ULR	DO-U
Adenine	0.5 g	0.5 g	0.5 g	0.5 g
Aspartate	2.0 g	2.0 g	2.0 g	2.0 g
Arginine	2.0 g	2.0 g	NA	2.0 g
Glutamate	2.0 g	2.0 g	2.0 g	2.0 g
Histidine	2.0 g	2.0 g	2.0 g	2.0 g
Leucine	NA	NA	NA	10.0 g
Lysine	2.0 g	2.0 g	2.0 g	2.0 g
Methionine	2.0 g	2.0 g	2.0 g	2.0 g
Phenylalanine	2.0 g	2.0 g	2.0 g	2.0 g
Serine	2.0 g	2.0 g	2.0 g	2.0 g
Threonine	2.0 g	2.0 g	2.0 g	2.0 g
Tryptophan	2.0 g	2.0 g	2.0 g	2.0 g
Tyrosine	2.0 g	2.0 g	2.0 g	2.0 g
Uracil	2.0 g	NA	NA	NA
Valine	2.0 g	2.0 g	2.0 g	2.0 g

8. We find that autoclaving the agar separately is preferable for synthetic media, as it allows it to solidify better. Adding the glucose as well reduces bubbling during autoclaving and ensures it does not spill outside the flask.

9. For yeast, we use a plate reader (TECAN infinite F200) and a 96 well plate (Greiner) instead of a spectrophotometer to measure cell density. We find that extrapolating the OD measurement of 1 mL from 200 μL provides a good approximation for OD as measured by a spectrophotometer, and is much easier to perform at high throughput. The formula for OD when a measurement is made with 200 μL culture is OD = (measured well – blank media well) × 5. Make tests with your plate reader to make sure this approximation also holds for you.

10. Some heating might be required to facilitate dissolution. Wait until the solution has cooled before filtering.

11. The presence of arginine in the media will drastically reduce the efficiency of canavanine. Make sure all media containing canavanine is made with arginine-free amino acid drop-out.

12. To be able to differentiate samples, each of them should be amplified using a different combination of barcodes (BC). For example, a set of 4 forward oligonucleotides (BC 1, 2, 3, 4) and 4 reverse oligonucleotides (BC 1, 2, 3, 4) can barcode 16 (4 × 4) samples. Please note that these barcodes are not Illumina indexes. As such, your libraries will not be automatically demultiplexed by the sequencer proprietary software and will be output in the unassigned directory. Inform the person in charge of sequencing your samples of this particularity to avoid confusion.

13. As sequencing kits are constantly being added and removed from the market, we cannot provide a specific reagent here. We have successfully sequenced amplicons generated by this two-step protocol on the following machines: MiSeq, HiSeq, and iSeq. Sequencing with paired ends is necessary to read both barcodes, and the SBS kit should be at least 150 cycles long so that the gRNA region is adequately covered by the forward read.

14. Since the signal of a gRNA that generates too many genotypes will be hard to interpret, we apply filters to limit the number of editable nucleotides in the editing window. Using this approach, a simple model covering the four most probable mutagenesis outcomes can reliably predict the genotype of a majority of edited cells [15]. You should also choose the number of gRNAs that best fit the resources you want to invest in the screen. More sequencing depth allows better detection of fitness effects, but this can lead to higher costs if the library has

tens of thousands of gRNA. The size of your library should take into account these trade-offs to make sure it can fulfill your research objectives.

15. The abundance of each clone in the library can be simulated with a multinomial distribution to help inform the target number of transformants. For example, in a 10,000 gRNA library with a 10× coverage, the standard deviation from the targeted 10 occurrences is on average 3.16 (>30% of the target). Increasing the coverage to 100× lowers the standard deviation relative to target down to 10% and 500× to 4.4%. As there are many other biases that can influence the distribution of gRNA counts, we usually try to play it safe and aim for as high a coverage as possible without making the experiments unmanageable. For example, we assume that every colony observed is an independent clone, while there might be multiple colonies that derive from the same plasmid acquisition event, lowering the effective library coverage.
16. The target number of clones that is desirable will depend on the size of the gRNA library and the observed cloning efficiency. We recommended genotyping ~10 colonies on the serial dilution plates by Sanger sequencing to make sure that most colonies bear a correctly constructed plasmid. If the success rate is below 90%, some further optimization of the assembly reaction might be required.
17. This assumes relatively even efficiency across the different transformations. If this is not the case, normalize the volume of cells in the prep so the number of colonies are roughly equivalent.
18. This is a good time to perform quality control on the library using high-throughput sequencing. Use the protocol described in **steps 4–13** of Subheading 3.6, but using plasmid from your master library diluted to 1 ng/μL as template.
19. We usually start boiling the Carrier DNA at the start of this centrifugation as this also takes 5 min. Put the boiled Carrier DNA on ice before getting the competent cells out of the centrifuge to make sure it has time to cool down.
20. Adjust quantities based on the scale of your experiment.
21. We usually take multiple pellets at every time point as a contingency for failed extractions.
22. In this protocol, we do not apply any additional selection on the edited cells and only measure the difference in cell growth rate. It is possible to change the environment during the competition round to apply additional constraints (e.g., growth at 37 °C instead of 30 °C, or an antibiotic such as rapamycin). Take care to maintain the selection on the pDYSCKO plasmid (the *URA3* marker) in your media by not supplying uracil.

23. In this protocol, we only perform two rounds of competition. Adding more rounds will increase the differences in abundance due to lower growth rates, at the cost of potentially introducing noise due to the apparition of spontaneous mutants with higher fitness. To limit this bias, we usually limit our competition experiments to a maximum of 25 generations. Using the widely-used 2 h/division estimate for yeast growth in the exponential phase, we consider that each 12 h competition round adds 6 generations. As the cells were also competing during base editing (12 h, 6 generations) and coselection (16 h, 8 generations), stopping at two competition rounds stops the experiment just a bit over the 25 generation mark $(6 + 8 + 6 + 6 = 26)$.
24. The plasmid DNA yield might be too low for accurate Nanodrop quantification and as such we recommend using 5 μL to reduce the bottleneck effect on sequencing library preparation.
25. Magnetic beads are also suitable if product purity is good.
26. Much like the number of clones required in the library preparation steps, the total sequencing coverage per sample will depend on the number of gRNAs. Additionally, sequencing with more depth will increase the confidence that can be put in the observed increases and decreases in gRNA abundance. A target of 100 reads/gRNA is usually enough to detect most important fitness effects. For example, a 2000 gRNA library would require 200,000 reads/sample to cover at 100×. More data and experimental considerations on the design of bar-seq experiments can be found in this helpful manuscript [16]. In short, biological replicates (different yeast cell populations) add much more statistical power than technical replicates (different libraries for the same populations), and having three to four biological replicates with lower coverage can be preferable than two with a higher coverage.
27. Analyzing the NGS libraries to find enriched or depleted gRNAs is in effect very similar to RNA-Seq differential expression analysis. As such, many existing packages (e.g., *DESeq* [17]) can be used to measure the relative abundance of gRNAs in different conditions and perform statistical testing once the libraries have been demultiplexed.

Acknowledgments

We thank the members of the Yachie and Landry laboratories for helpful discussions on this method. This work was supported by the Canadian Institutes of Health Research Foundation grant 387697 to C.R.L. and by a Vanier graduate scholarship to P.C.D., and the Japan Society for the Promotion of Science grant numbers S15734 and S17161 to C.R.L. and N.Y.

References

1. Nishida K, Arazoe T, Yachie N, Banno S, Kakimoto M, Tabata M et al (2016) Targeted nucleotide editing using hybrid prokaryotic and vertebrate adaptive immune systems. Science 353:aaf8729. https://doi.org/10.1126/science.aaf8729
2. Komor AC, Kim YB, Packer MS, Zuris JA, Liu DR (2016) Programmable editing of a target base in genomic DNA without double-stranded DNA cleavage. Nature 533:420–424
3. Tan J, Zhang F, Karcher D, Bock R (2019) Engineering of high-precision base editors for site-specific single nucleotide replacement. Nat Commun 10:439
4. Agudelo D, Duringer A, Bozoyan L, Huard CC, Carter S, Loehr J et al (2017) Marker-free coselection for CRISPR-driven genome editing in human cells. Nat Methods 14:615–620
5. Billon P, Bryant EE, Joseph SA, Nambiar TS, Hayward SB, Rothstein R et al (2017) CRISPR-mediated base editing enables efficient disruption of eukaryotic genes through induction of STOP codons. Mol Cell 67:1068–1079.e4
6. Kane NS, Vora M, Varre KJ, Padgett RW (2017) Efficient screening of CRISPR/Cas9-induced events in drosophila using a co-CRISPR strategy. G3 7:87–93
7. Després PC, Dubé AK, Nielly-Thibault L, Yachie N, Landry CR (2018) Double selection enhances the efficiency of target-AID and Cas9-based genome editing in yeast. G3 8:3163–3171
8. Ahmad M, Bussey H (1986) Yeast arginine permease: nucleotide sequence of the CAN1 gene. Curr Genet 10:587–592
9. Stockwell SR, Landry CR, Rifkin SA (2015) The yeast galactose network as a quantitative model for cellular memory. Mol Biosyst 11:28–37
10. Casadaban MJ, Cohen SN (1980) Analysis of gene control signals by DNA fusion and cloning in Escherichia coli. J Mol Biol 138:179–207
11. Brachmann CB, Davies A, Cost GJ, Caputo E, Li J, Hieter P et al (1998) Designer deletion strains derived from Saccharomyces cerevisiae S288C: a useful set of strains and plasmids for PCR-mediated gene disruption and other applications. Yeast 14:115–132
12. Gietz RD, Woods RA (2002) Transformation of yeast by lithium acetate/single-stranded carrier DNA/polyethylene glycol method. Methods Enzymol 350:87–96
13. Yachie N, Petsalaki E, Mellor JC, Weile J, Jacob Y, Verby M et al (2016) Pooled-matrix protein interaction screens using barcode fusion genetics. Mol Syst Biol 12:863
14. Green R, Rogers EJ (2013) Twenty eight—transformation of chemically competent E. coli. In: Lorsch J (ed) Laboratory methods in enzymology: DNA. Academic Press, Waltham, MA, pp 329–336
15. Després PC, Dubé AK, Seki M, Yachie N, Landry CR (2020) Perturbing proteomes at single residue resolution using base editing. Nat Commun 11:1871
16. Robinson DG, Chen W, Storey JD, Gresham D (2014) Design and analysis of Bar-seq experiments. G3 4:11–18
17. Anders S, Huber W (2010) Differential expression analysis for sequence count data. Genome Biol 11:R106

Chapter 20

SAturated Transposon Analysis in Yeast (SATAY) for Deep Functional Mapping of Yeast Genomes

Agnès H. Michel and Benoît Kornmann

Abstract

Genome-wide transposon mutagenesis followed by deep sequencing allows the genome-wide mapping of growth-affecting loci in a straightforward and time-efficient way.

SAturated Transposon Analysis in Yeast (SATAY) takes advantage of a modified maize transposon that is highly mobilizable in *S. cerevisiae*. SATAY allows not only the genome-wide mapping of genes required for growth in select conditions (such as genetic interactions or drug sensitivity/resistance), but also of protein sub-domains, as well as the creation of gain- and separation-of-function alleles. From strain preparation to the mapping of sequencing reads, we detail all the steps for the making and analysis of SATAY libraries in any *S. cerevisiae* lab, requiring only ordinary equipment and access to a Next-Gen sequencing platform.

Key words Transposon, Ac/Ds, Genome-wide insertional mutagenesis, Next-generation sequencing

1 Introduction

Genome-wide saturated mutagenesis can be achieved in many microbes by generating dense libraries of transposon insertion mutants and identifying the insertion sites by deep sequencing. The number of sequencing reads mapping to a defined genomic location can then serve as a proxy for the fitness of the strain bearing this particular mutation, in the conditions used [1]. Increased and decreased fitness translate as increased and decreased number of sequencing reads for these transposons. Therefore, in haploid genomes, essential genes are typically devoid of sequencing reads.

A screen involves (1) inducing transposition in a chosen organism and strain, (2) growing and selecting the transposon mutants in defined conditions, (3) harvesting the mutant library and extracting genomic DNA, (4) amplifying the transposon–genome

Supplementary Information The online version contains supplementary material available at [https://doi.org/10.1007/978-1-0716-2257-5_20].

Frédéric Devaux (ed.), *Yeast Functional Genomics: Methods and Protocols*,
Methods in Molecular Biology, vol. 2477, https://doi.org/10.1007/978-1-0716-2257-5_20, © The Author(s) 2022

junctions selectively by PCR, (5) identifying the transposon insertion sites by deep sequencing and mapping reads onto the reference genome.

Such an approach has been developed and successfully used in a variety of bacteria [1–4] and eukaryotic microbes (including *S. pombe* [5], *S. cerevisiae* [6–8], *S. uvarum* [9], *C. albicans* [10, 11], *C. glabrata* [12], *R. toruloides* [13], and *P. pastoris* [14]), using various combinations of transposons/transposase systems, transposon–genome junction amplification methods, and sequencing platforms (recently reviewed in ref. [15]).

Here we describe a protocol for SAturated Transposon Analysis in Yeast (SATAY [6, 7]), which allows the rapid and effortless generation and sequencing of high-density transposon libraries in the budding yeast *S. cerevisiae* (Fig. 1). In SATAY, a modified non-autonomous Maize transposon (Mini*Ds*) interrupts the *ADE2* gene [16]. A galactose-induced hyperactive transposase (Ac) mobilizes the transposon. Homology-directed repair of

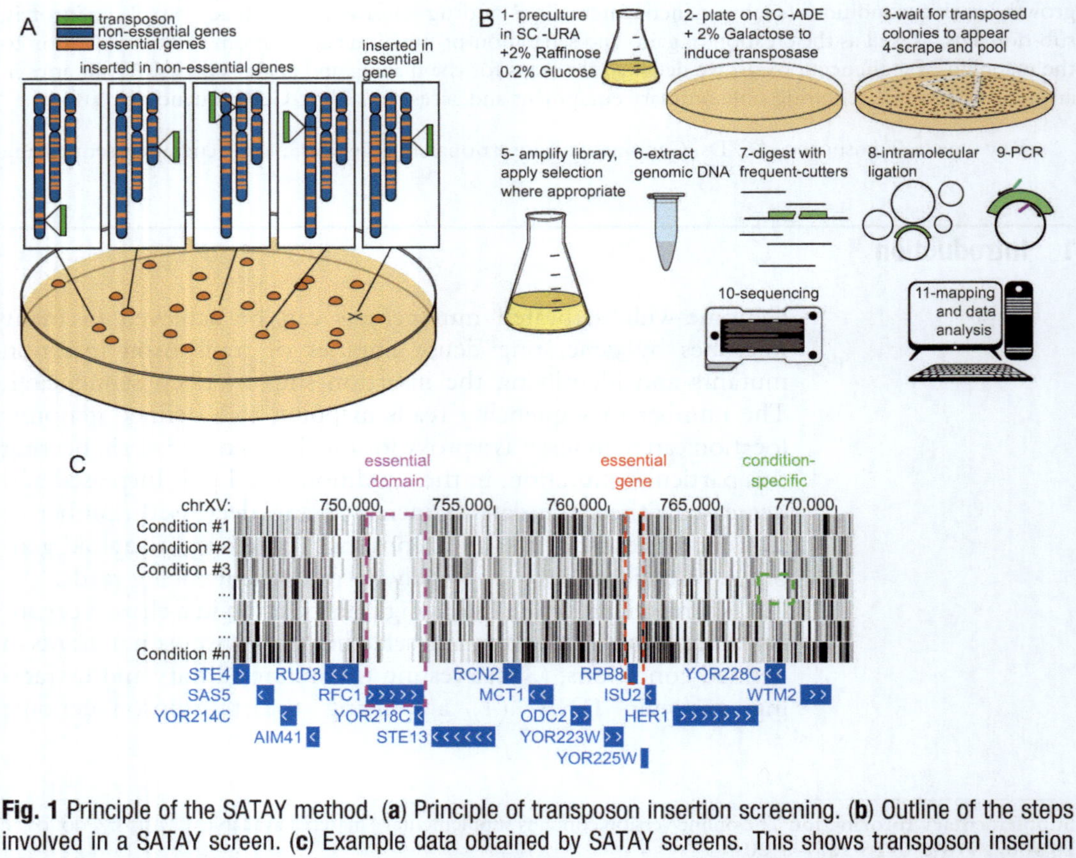

Fig. 1 Principle of the SATAY method. (**a**) Principle of transposon insertion screening. (**b**) Outline of the steps involved in a SATAY screen. (**c**) Example data obtained by SATAY screens. This shows transposon insertion sites in a section of *S. cerevisiae* genetic map. Screens were performed in different conditions/genetic backgrounds. Data reveal essential genes (red dashed box), essential protein domains (magenta dashed box), and genes necessary for growth in defined conditions (green dashed box)

ADE2 then confers adenine prototrophy to the transposed cells. A screen begins by transforming a plasmid (pBK549) encoding the galactose-induced transposase, the *ADE2* gene interrupted by the Mini*Ds* transposon and a *URA3* selection marker, into an *ade2- ura3-* strain [7]. After some sanity checks, transformed clones are amplified and densely plated onto +galactose-adenine plates, allowing for (1) the induction of the transposition and (2) the growth of transposed Ade+ cells into colonies. These colonies are then scraped off the plate, re-grown in +glucose -adenine medium (plus any chosen environmental perturbation), before being processed for DNA extraction, amplification, sequencing, and mapping. Specific amplification of transposon–genome junction is achieved by circular PCR. Simultaneous sequencing of multiple libraries is achieved using the Illumina® system.

Optimized plating conditions yield 35,000–100,000 transposed colonies per 9.0 cm Ø petri dish, meaning that 30–70 petri dishes are sufficient to yield two million colonies. In our hands, this number allows mapping of transposons at ~250,000–400,000 independent genomic locations, averaging a resolution of one transposon every 50 bp [7].

Compared to synthetic genetic arrays and other high-throughput mutagenesis methods, SATAY possesses significant advantages. First, MiniDs targets the genome with little sequence specificity, minimizing insertion biases [16]. Second, its insertion within a gene does not preclude the expression of either the upstream or downstream sequences, allowing not only the mapping of genes required for growth in all or select conditions, but also of protein sub-domains, as well as the creation of gain- and separation-of-function alleles [6, 7, 17].

SATAY has been used to map genome-wide genetic interactions, as well as to identify genes involved in resistance and sensitivity to various chemicals, yielding insight into processes as diverse as signaling, organelle biology, lipid transport and metabolism, DNA repair, and transcriptional regulation [6, 7, 17–21].

2 Materials

All solutions are made with MilliQ (MQ) water.

2.1 Equipment

1. UV-Vis Spectrophotometer.
2. Stereomicroscope with transillumination.
3. Bright field microscope with >10× long-distance objective.
4. Shaking incubator (needs to accommodate up to 5 L Erlenmeyer flasks at 30 °C for yeast work, and culture tubes at 37 °C for bacteria work).

5. 30 °C static incubator (needs to accommodate at least 50 petri plates).
6. 37 °C static incubator for bacteria work.
7. Horizontal gel electrophoresis apparatus.
8. Repeater pipette, 5-mL sterile tips.
9. Erlenmeyer flasks, sterile.
10. Bunsen burner.
11. Mini-Beadbeater (e.g., Disruptor Genie, Scientific Industrial).
12. Thermomixer.
13. Refrigerated microcentrifuge.
14. Centrifuge (needs to accommodate 50-mL centrifuge tubes at 600 × *g* and 1600 × *g*, and 500-mL centrifuge tubes at 1600 × *g*).
15. PCR thermocycler.
16. Optional: Hemocytometer (Neubauer Improved Hemocytometer Counting Chamber).

2.2 Growth Media

1. 10× Amino Acid mix: 0.3 g/L L-Isoleucine, 3 g/L L-Valine, 0.4 g/L Adenine Hemisulfate salt, 0.2 g/L L-Arginine monohydrochloride, 0.2 g/L L-Histidine monohydrochloride monohydrate, 1 g/L L-Leucine, 0.3 g/L L-Lysine monohydrochloride, 0.2 g/L L-Methionine, 0.5 g/L L-Phenylalanine, 2 g/L L-Threonine, 0.4 g/L L-Tryptophan, 0.3 g/L L-Tyrosine, 0.2 g/L Uracil, 1 g/L L-Glutamic Acid monosodium salt hydrate, 1 g/L L-Aspartic Acid sodium salt monohydrate. Dissolve under constant agitation and gentle heat (~1 h). Autoclave for 15 min at 121 °C. Store at 4 °C.
2. 10× Drop Out mix: 10× Amino Acid mix in which the chosen component has been omitted.
3. 10× carbohydrate solution: 200 g/L of either D-(+)-glucose (Dextrose), D-(+)-Galactose, or D(+)-Raffinose. Aid dissolution by stirring on a heated (~40 °C) stirrer unit. Autoclave for 15 min at 121 °C. Store at room temperature.
4. 4× S solution: 6.8 g/L Yeast Nitrogen base (without Amino Acid and Ammonium Sulfate), 20 g/L Ammonium Sulfate. Autoclave for 15 min at 121 °C. Store at room temperature.
5. SC plates: dissolve 1.7 g Yeast Nitrogen base (without Amino Acid and ammonium sulfate) with 5 g ammonium sulfate in 320 mL MQ water. In a separate flask, suspend 20 g granulated agar in 500 mL MQ water. Autoclave both the solutions for 15 min at 121 °C. Combine under constant stirring, allowing the mix to cool down to ~60 °C. Add 100 mL 10× carbohydrate solution and 100 mL 10× Drop Out mix (for instance,

SC-D -ADE is obtained by adding 10× Dextrose solution and 10× -ADE Drop Out mix). Pour 25 mL in 9.0-cm Ø petri dishes.

6. SC-Rd -URA liquid medium: 1× S solution, 1× -URA Drop Out mix, 1× carbohydrate solution (Raffinose), 0.1× carbohydrate solution (Dextrose). Store at 4 °C.
7. SC-D -ADE liquid medium: 1× S solution, 1× -ADE Drop Out mix, 1× carbohydrate solution (Dextrose). Store at 4 °C.
8. YPD: mix and autoclave 20 g BactoPeptone, 10 g Yeast Extract, 55 mg Adenine Hemisulfate salt in 900 mL. Add 100 mL 10× Dextrose solution. Store at room temperature.

2.3 DNA Work

1. Filter pipette tips.
2. Nonstick microfuge tubes (e.g., Ambion).
3. Safe-Lock 1.5-mL microtubes.
4. Glass beads, 0.5 mm, unwashed.
5. PCR clean-up kit.
6. PvuII.
7. EcoRV.
8. DpnII.
9. NlaIII.
10. 5 Weiss Units/μL T4 ligase.
11. 5 U/μL Taq DNA Polymerase with ThermoPol Buffer (NEB).
12. 5× TBE: 0.45 M Tris–Base, 0.45 M Boric Acid, 0.01 M EDTA, pH 8.0.
13. Cell Breaking Buffer: 2% Triton X-100, 1% SDS, 100 mM NaCl, 100 mM Tris–HCl, pH 8.0, 1 mM EDTA. All stock solutions autoclaved, except 100% Triton X-100.
14. Ultrapure Phenol:Chloroform:Isoamylalcohol 25:24:1.
15. 25 mM dNTPs Mix. Store at −20 °C as 50 μL aliquots.
16. TE: 10 mM Tris–HCl, pH 8.0, 1 mM EDTA.
17. Ethanol 100% and 70%.
18. 3 M Sodium Acetate, pH 5.2.
19. RNAse A:10 mg/mL RNase A, 10 mM Sodium Acetate, 0.1 M Tris–HCl, pH 7.4. Vortex RNAse in Sodium Acetate. Boil for 15 min. Adjust pH to 7.4 by adding 1 mL 1 M Tris–HCl pH 7.4. Aliquot in microtubes. Store at −20 °C.
20. 5 μg/μL Linear acrylamide.
21. Agarose for electrophoresis.

2.4 Yeast Work

1. 9.0 cm Ø petri dishes.
2. 50-mL centrifuge tubes.
3. Corning® 500-mL PP Centrifuge Tubes with Plug Seal Cap.
4. Corning® 500-mL Polyetherimide Centrifuge Tube Cushions.
5. Wood toothpicks, sterile.
6. Glass pipettes, sterile.
7. Glass beads 3 mm.
8. Pasteur pipettes plain glass 230 mm.
9. 50% PEG 3350. Autoclave for 15 min at 121 °C. Store at room temperature. Use within a month.
10. 1 M LiOAc. Autoclave for 15 min at 121 °C. Store at room temperature.
11. 2 mg/mL salmon sperm DNA sodium salt. Dispense 1 mL aliquots in sterile 1.5-mL microtubes. Incubate at 95 °C for 5 min to denature the double stranded DNA. Immediately store at −20 °C. Thaw when needed, do not refreeze.
12. Transformation mix: 40% PEG3350, 0.1 M LiOAc, 0.3 mg/mL salmon sperm DNA. Dispense 162 μL aliquots in 0.5-mL sterile microtubes. Store at −20 °C. Use one aliquot per transformation reaction. Thaw when needed, do not refreeze.

2.5 Sequencing

1. MiSeq Reagent Kit v3 (150 Cycles) (Illumina®) or NextSeq 500/550 Mid Output Kit v2.5 (150 Cycles) (Illumina®) or NextSeq 500/550 High Output Kit v2.5 (75 Cycles) (Illumina®).
2. PhiX Control v3 (Illumina®).
3. 1–200 μL Gel loading Tips, round (Exacta Cruz).

3 Methods

3.1 Generate Starting Yeast Clones

3.1.1 Verify pBK549 by Restriction Digest

1. Miniprep 4–10 colonies of pBK549 using your favorite miniprep procedure.
2. Digest 150 ng DNA with 20 units PvuII in a total volume of 20 μL.
3. Digest 150 ng DNA with 20 units EcoRV in a total volume of 20 μL.
4. Load 4 μL of each restriction digest on a 0.6% Agarose 1× TBE electrophoresis gel. Run the gel for 1.5 h at 80 V then image. Optional: run the gel for a further 30 min at 40 V.
5. Identify and proceed with the DNA preparations that contain little or no recombined product (Fig. 2a–c, *see* **Note 1**).

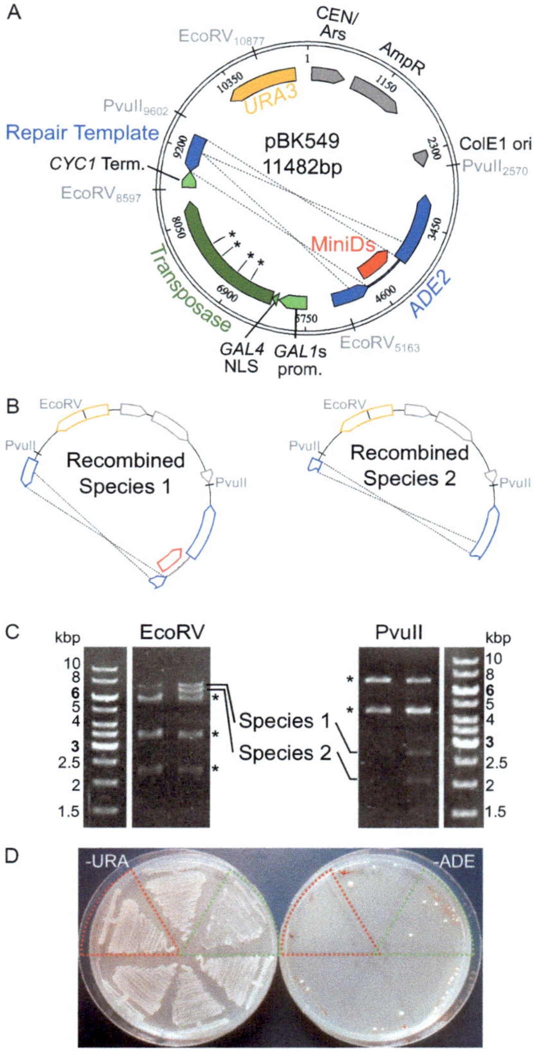

Fig. 2 The plasmid pBK549. (**a**) Map of the plasmid. Dashed lines represent homologies for repair of the plasmid after transposon (MiniDs) excision. Stars represent mutations that make the Ac transposase hyperactive. (**b**) Maps of recombinant species that appear spuriously in a transposase-independent manner, both upon bacterial plasmid expansion and yeast growth. (**c**) Diagnostic restriction digests of two pBK549 preparations, one with little recombinant species and perfectly usable (left), the other with more recombinant species (right). Fragment sizes for the diagnostic digest of the correct plasmid are, for EcoRV 5768, 3434 and 2284 bp, and for PvuII, 7036 and 4450 bp (marked with stars). Recombinant species 1 yields an EcoRV fragment of 6498 bp, and PvuII fragments of 4450 and 2048 bp. Recombinant species 2 yields an EcoRV fragment of 7140 bp, and PvuII fragments of 4450 and 2690 bp. (**d**) Streak of six independent pBK549-transformed yeast clones. The colonies appearing rarely on -ADE plates indicate that the clone bears an intact pBK549 (e.g., green sector) capable of generating Ade + progeny by spontaneous recombination. Clones that fail to produce any colonies on -ADE (red sector) might bear an already recombined species that is transposition-deficient. The clone in the green sector was then scraped off the -URA plate and processed further (*see* Subheading 3.2.1)

3.1.2 Transform pBK549 in Strains of Interest

1. Inoculate the strain(s) of interest in 5 mL YPD. Grow at 30 °C, 140–200 rpm until saturation (*see* **Notes 2** and **3**).
2. Measure the OD_{600} of each saturated culture. Dilute to OD_{600} 0.2 in 20 mL fresh YPD. Incubate at 30 °C, 140–200 rpm until the culture has reached OD_{600} 0.8.
3. Spin each culture at 600 × *g* for 5 min at room temperature.
4. Discard the supernatants. Resuspend the cells in 20 mL sterile MQ water and pellet the cells as above.
5. Discard the supernatants. Resuspend the cells in 1 mL sterile MQ water, dispense 180 μL of cell suspension into a fresh 1.5-mL microtube and pellet the cells as above.
6. Discard the supernatants. Add 16 μL pBK549 (7 ng/μL, 112 ng). Add 162 μL of transformation mix then mix well using the same pipette.
7. Incubate at 30 °C for 30 min, no agitation.
8. Incubate at 42 °C for 20 min, no agitation.
9. Centrifuge at 600 × *g* for 5 min, room temperature.
10. Discard the supernatants. Resuspend each cell pellet in 1 mL sterile MQ water. For each strain transformed, spread one SC-D -URA plate with 200 μL of the resuspended cells.
11. Incubate at 30 °C for 2 days or until the colonies have reached the desired size. Store the plates at 4 °C (*see* **Note 4**).

3.1.3 Selection of Yeast Transformants

1. Streak 12–18 independent transformants on 2–3 SC-D-URA and 2–3 SC-D -ADE plates (*see* **Note 5**).
2. Pursue with clones that produce a low but detectable background of Ade+ colonies (Fig. 2d) (*see* **Note 6**).

3.2 Transposition

3.2.1 Day 0

1. Scrape cells off selected sectors from the SC-D-URA plate and inoculate 40–60 mL SC-Rd -URA at OD_{600} 0.2 (*see* **Notes 7–9**).
2. Inoculate 3–5 pre-cultures from 3 to 5 independent sectors (*see* **Notes 3** and **10**).
3. Grow at 30 °C for 16–20 h, with shaking at 160–200 rpm.
4. Take the SC-Gal -ADE (transposition plates) and SC-D -ADE (control plates) plates from the cold room. Store these plates at room temperature overnight (*see* **Notes 11** and **12**).

3.2.2 Day 1

1. Measure the OD_{600} of each pre-culture (*see* **Note 13**).
2. Optional. Take a small aliquot of each overnight culture, dilute it 40,000× in sterile MQ water and plate 200 μL on one SC-D and one SC-D -URA plate (*see* **Note 14**).

3. Transfer the cells to a 50-mL centrifuge tube and centrifuge for 5 min at 600 × *g* at room temperature.
4. Calculate the concentration factor required to resuspend the cells at OD_{600} 40 and remove the appropriate volume of supernatant using a pipette (*see* **Notes 15** and **16**).
5. Resuspend the cell pellet using the same pipette.
6. Arrange the SC-Gal -ADE plates in stacks of 10 plates. Dispense 200 μL of cell suspension on each plate of the first stack. Spread the cells with glass beads, keeping the plates horizontal and using a gentle circular motion. Ensure that no area of the agar surface has been missed in any of the plates (*see* **Note 17**).
7. Repeat with the next set of 10 plates (*see* **Note 18**).
8. The liquid spread in point 6 above should now be fully absorbed and the cells should form a uniform opaque film covering the agar surface entirely (*see* **Note 19**). Gently shake the stack of plates, this time using straight motions, two to three times in four directions, keeping the plates horizontal. Patterns on the agar surface become visible (*see* **Notes 20** and **21**).
9. Remove the beads immediately. Do not allow the beads to remain on the agar surface.
10. Proceed with the next stack. Repeat until all plates have been spread with the cells.
11. As a negative control, dispense 200 μL of the same cell suspension onto one SC-D -ADE plate. Spread as above (*see* **Note 22**).
12. Store the plates in plastic sleeves loosely sealed with tape, ensuring air circulation but preventing plates from drying out during the long incubation period. Incubate at 30 °C until ready to harvest (*see* **Note 23**).

3.2.3 Day 4

1. Observe the plates after 3 days of incubation. Draw a square centimeter on the dish and count the number of clones contained within it using a stereomicroscope (Fig. 3a).

3.2.4 Days 5–20

1. Observe the plates every 3–4 days for the appearance of new colonies, for as long as 20 days (*see* **Notes 24** and **25**).
2. When no new colonies appear, estimate the final number of transposed colonies and proceed to harvest (Fig. 3) (*see* **Note 26**).

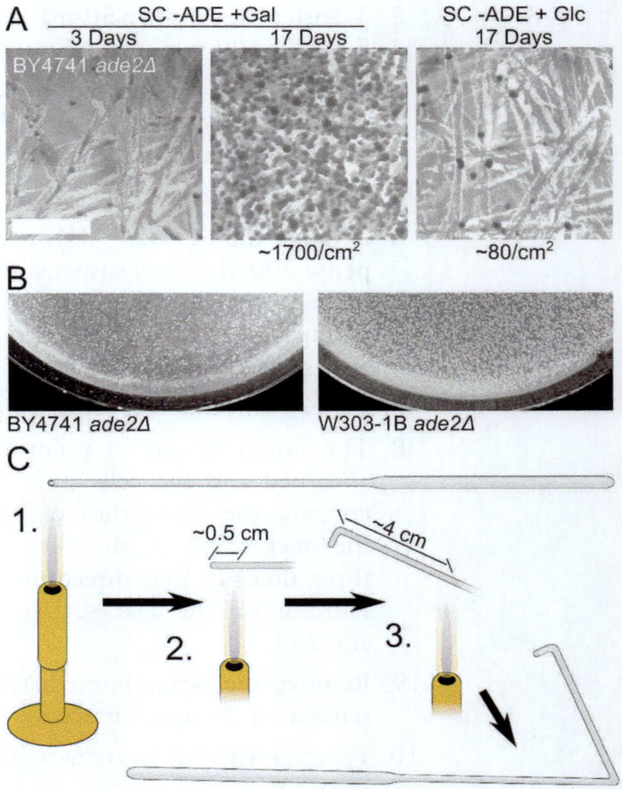

Fig. 3 Growth and harvest of transposed clones. (**a**) Appearance of Ade+ colonies in galactose (inducing transposase) and glucose (repressing transposase) over time. Images are taken with a transilluminated stereomicroscope. Scale bar, 1 mm. (**b**) General aspect of transposed SC -ADE +Galactose plates after 17 days. Two genetic backgrounds (BY4741 and W303-1B) yield colonies of different sizes and number. (**c**) Manufacturing of harvesting rakes from long glass Pasteur pipettes. The opening of the pipette is first closed (1), then a first ~90° bend is generated ~0.5 cm from the tip (2). A second 45° bend is then generated ~4 cm from the tip in the same plane as the first bend (3)

3.3 Plate Harvest, Library Expansion, and Collection

3.3.1 Harvest the Plates

1. Proceeding five plates at a time, dispense 2 mL of SC-D -ADE medium per plate using a glass pipette (*see* **Note 27**).
2. Scrape the cells off the surface of the plate using a rake (Fig. 3c, *see* **Note 28**). Leave the plates tilted for the medium to collect on one side.
3. Collect the cell suspensions using a pipette. Dispense into a sterile container (*see* **Note 29**).
4. Proceed with the next five plates until all plates have been harvested (*see* **Note 30**).

3.3.2 Reseed

1. Mix the harvested cell suspensions well.
2. Measure the OD_{600} of each harvested cell suspension (*see* **Note 31**).
3. Reseed the harvests in a total of 2 L SC-D -ADE at a density of OD_{600} 0.2 (~2E+06 cells/mL) (*see* **Note 32**).
4. Dilute an aliquot of the reseeded culture 1000×. Spread 200 μL on one SC-D and one SC-D -ADE plate using glass beads. Incubate for 3 days at 30 °C. Count the colonies to estimate the total number of cells, number of Ade + cells and number of cells per transposition event reseeded (*see* **Note 33**).
5. Incubate at 30 °C in a shaking incubator for 16–24 h, 140–200 rpm.
6. Optional. Determine the number of cells per transposition event after re-growth of the library (*see* **Note 34**).
7. Optional. Dilute and re-grow the cultures to apply further treatments (e.g., exposure to drugs [6, 7, 17–19] (*see* **Note 35**).
8. Optional. Freeze the excess harvest for further use (*see* **Note 36**).

3.3.3 Final Harvest

1. Collect at least 500 mL of saturated culture (OD_{600} ~ 4), 5 min, 1600 × *g*, room temperature. Harvest the entire culture if centrifuge space is not limited (*see* **Note 37**).
2. Resuspend each pellet in 150 mL MQ water. Centrifuge at room temperature for 5 min, 1600 × *g*.
3. Weigh a 50-mL conical centrifuge tube.
4. Resuspend each pellet in 40 mL MQ water and transfer the cells into the 50-mL tube.
5. Repeat the centrifugation, remove all the supernatant and weigh the pellet by subtracting the tube's weight from the total weight.
6. Resuspend the pellet in an equal amount (weight/volume) of MQ water. Distribute 1 mL in 1.5-mL microtubes, so that each tube contains ~500 mg of cells in a dense suspension.
7. Centrifuge at room temperature for 5 min at 1600 × *g* or for a few seconds at 16,000 × *g*. Remove all the supernatant using a P1000 followed by a P200 pipette, store the 500 mg cell pellets at −20 °C until further use.

3.4 DNA Template Preparation

The procedure below works with high efficiency and is easy, but generates phenol:chloroform waste. The method can be replaced by commercial DNA isolation kits.

3.4.1 Genomic DNA Extraction

1. Resuspend 500 mg cell pellet in 500 μL of Cell Breaking Buffer.

2. Distribute the cell suspension into 280 μL aliquots in 1.5-mL Safe-Lock microtubes.
3. For each aliquot: add ~300 μL 0.5 mm unwashed glass beads, then 200 μL Phenol:Chloroform:Isoamylalcohol 25:24:1.
4. Ensure that no glass bead is stuck between the wall of the tube and the cap, as this could prevent tight sealing and cause leakage.
5. Vortex for 10 min at 4 °C using a Mini-Beadbeater.
6. Add 200 μL 1× TE to the lysate. Vortex briefly then centrifuge for 5 min at 16,100 × *g*, 4 °C.
7. Transfer the clear upper layer (~400 μL) to a fresh 1.5-mL tube, without disturbing the white interphase.
8. Add 2.5 vol (1 mL) of 100% Ethanol and mix by inversion.
9. Pellet the nucleic acids for 5 min at 16,100 × *g*, 20 °C (*see* **Note 38**).
10. Pipette out as much supernatant as possible.
11. Pulse centrifuge to ensure that all the liquid collects at the bottom of the tube.
12. Remove the remaining supernatant by pipetting.
13. Dilute 10 mg/mL RNase A to 250 μg/mL in MQ water. Prepare 220 μL of 250 μg/mL RNase A per nucleic acid pellet.
14. Add 200 μL of 250 μg/mL RNase A on top of the pellet.
15. Incubate for 15 min at 55 °C, 1000 rpm. Vortex briefly to completely resuspend the DNA pellet.
16. Pulse centrifuge to ensure that all the liquid collects at the bottom of the tube.
17. Add 2.5 vol (500 μL) 100% Ethanol and 0.1 vol (20 μL) 3 M Sodium Acetate, pH 5.2. Mix well by inversion or by pipetting.
18. Centrifuge for 5 min at 16,100 × *g*, 20 °C to pellet the DNA.
19. Pipette out as much supernatant as possible. Centrifuge briefly to ensure that all the liquid collects at the bottom of the tube. Pipette out the remaining supernatant.
20. Add 1 mL 70% Ethanol. Place the tube in the opposite orientation to the previous centrifugation so that the pellet washes through the 70% Ethanol and collects on the opposite side of the tube. Centrifuge for 5 min at 16,100 × *g*, 20 °C.
21. Remove most of the supernatant using a P1000 pipette. Centrifuge briefly to collect all the remaining 70% Ethanol to the bottom of the tube, then remove it entirely using a P20 pipette.
22. With the microtube cap open, dry the pellet for 10 min at 37 °C.

23. Distribute a total of 200 μL MQ water equally to all samples. Resuspend the pellets completely for 20–40 min at 55 °C on a thermomixer shaking at 700 rpm. Pool the resuspended genomic DNA aliquots.

3.4.2 Genomic DNA Quantification on Gel (See **Note 39**)

1. Cast a 0.6% Agarose 1× TBE electrophoresis gel.
2. Mix one volume of 6× home-made or commercial loading dye buffer with 3 volumes MQ water to make a working loading dye solution.
3. Dilute 2 μL gDNA prep with 18 μL of loading buffer prepared above.
4. Load various amounts of 1 kb GeneRuler™ and 5 μL of each diluted gDNA sample (Fig. 4a, *see* **Note 40**).
5. Run the gel for 1.5 h at 80 V and image the gel avoiding saturation (*see* **Note 41**).
6. Quantify using Fiji/ImageJ [22] as shown in Fig. 4a, b.

3.4.3 Genomic DNA Digestion with NlaIII and DpnII

1. In parallel, digest 2 μg of genomic DNA with 50 units of DpnII, and 2 μg with 50 units of NlaIII. Both digestions take place overnight in 50 μL at 37 °C in Nonstick microfuge tubes.
2. Heat-inactivate the reactions for 20 min at 65 °C.
3. Run 2 μL (80 ng) of each reaction on a 1% Agarose 1× TBE electrophoresis gel (Fig. 5a, b).
4. The digested DNA can be stored at −20 °C until further use.

3.4.4 Digested gDNA Circularization

1. To the tube containing 48 μL digested DNA, add in the following order (*see* **Note 42**).
   (a) 307 μL MQ water.
   (b) 40 μL 10× T4 ligase buffer, mix well.
   (c) 5 μL T4 ligase, mix gently.
2. Incubate for 6 h at 22 °C.
3. Add 40 μL 3 M Sodium Acetate pH 5.2, 1 μL (5 μg) linear acrylamide and mix well (*see* **Note 43**).
4. Add 1 mL 100% Ethanol, mix well. Precipitate overnight or longer at −80 °C.
5. Centrifuge for 20 min at 16,100 × *g*, 4 °C.
6. Remove as much supernatant as possible without touching the pellet.
7. Add 1 mL 70% Ethanol. Break the pellet using a pipette.
8. Centrifuge for 20 min at 16,100 × *g*, 20 °C.

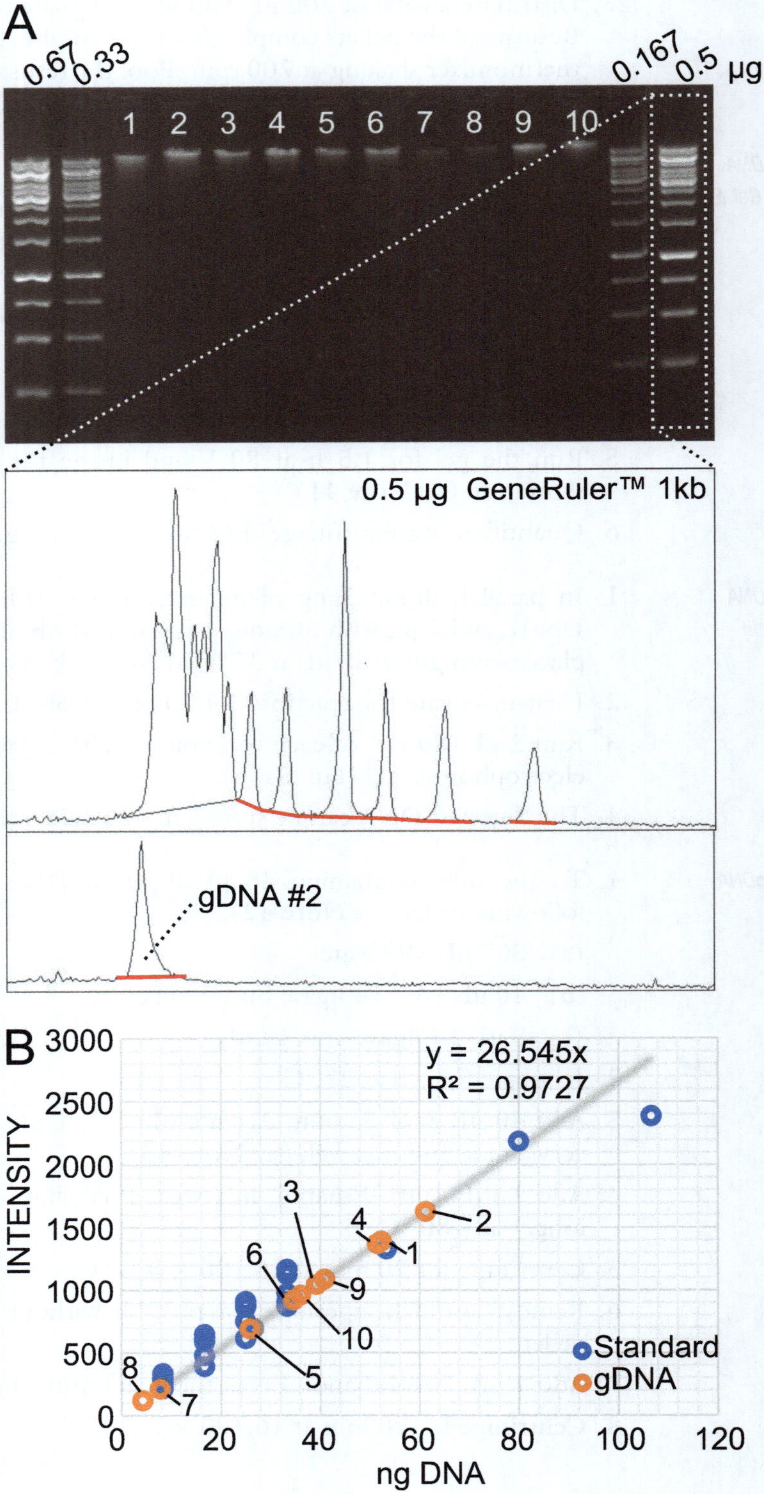

Fig. 4 DNA library generation. (**a**) Gel quantification of genomic DNA preps. Outer lanes contain indicated amounts of DNA ladder. Lower panel shows the quantification of indicated parts of the gel using Fiji/ImageJ. (**b**) Linear regression of the quantified standard DNA ladder bands and quantification of genomic preps 1–10

9. Remove most of the supernatant using a P1000 pipette. Pulse centrifuge and remove the remaining supernatant completely using a P20 pipette.
10. With the microtube cap open, dry the pellet for 10 min at 37 °C.
11. Add 105 μL MQ water, incubate at 55 °C for 15–30 min with optional shaking at 1000 rpm.
12. The ligated DNA can be stored at −20 °C until further use.

3.5 Library Amplification

3.5.1 Primer Design

1. Each library is amplified using primer P5_MiniDs (common to all libraries) and two unique indexed primers (P7_(index)_-MiniDs), one for the NlaIII sub-library and the other for the DpnII sub-library. The indexed primers are designed to specifically amplify the transposon-genome junctions and to generate products adapted to the Illumina® sequencing platforms (Fig. 5a, Table 1).
2. Choose the best-suited sequencing platform according to availability, number and complexity (number of clones) of libraries, and budget (*see* **Note 44**).
3. Choose combinations of indexes for the sequencing of pooled libraries according to Illumina®'s recommendations [23–25] (*see* **Note 45**).
4. Resuspend all primers as 100 μM stock solutions in sterile MQ water on a Thermomixer at 55 °C shaking at 1000 rpm for 10 min. Store aliquots at −20 °C.

3.5.2 PCR and Purification

1. Prepare one large reaction per sample by mixing the following reagents in the indicated order: 847.55 μL MQ water, 115 μL 10× ThermoPol Buffer, 9.2 μL 25 mM dNTPs, 11.5 μL Primer P5_MiniDs 100 μM, 11.5 μL Indexed Primer P7_(index)_MiniDs 100 μM, 40.25 μL (17.5 units) Taq DNA Polymerase. Omit the template at this stage.
2. Dispense 90 μL of the PCR mix in a fresh PCR tube. Add 10 μL MQ water to make a no-template negative control.
3. Add the remaining master mix into the thawed 105 μL aliquot of template DNA. Mix well using a P1000 pipette.
4. Aliquot 100 μL of reaction into 10 PCR tubes (or 125 μL into a strip of 8 PCR tubes).
5. Run the following PCR cycle: 95 °C 1 min [95 °C 30 s, 55 °C 30 s, 68 °C 3 min] × 35 cycles, 68 °C 10 min, 15 °C final hold.
6. Pool the reactions corresponding to the NlaIII-digestion.
7. Pool the reactions corresponding to the DpnII-digestion. Keep the NlaIII and DpnII pools separate.

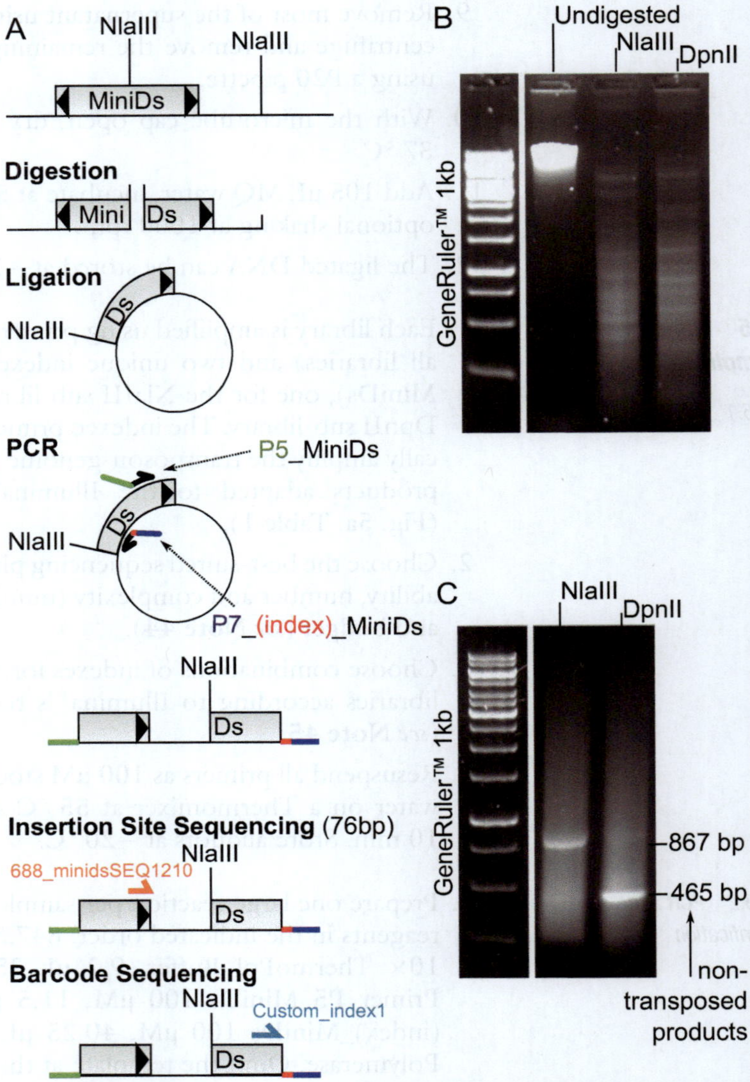

Fig. 5 Library sequencing. (**a**) Outline of the molecular biology steps. (**b**) Gel electrophoresis of genomic DNA before and after restriction digestion with indicated enzymes. (**c**) Gel electrophoresis of purified PCR products. The bands at 867 and 465 bp correspond to intact (untransposed) pBK549

8. Run 2 μL of each pool and 2 μL of corresponding negative controls on a 1% agarose gel, 1× TBE (*see* **Note 46**).
9. Purify 100 μL of each PCR pool using a PCR purification kit to generate the NlaIII and the DpnII sub-libraries (*see* **Note 47**).
10. Run 2 μL of each sub-library on a 1% agarose gel, 1× TBE (Fig. 5c).

Table 1
Primers

Name	Sequence	Use
Ade2PriFwd	GTATAAATTGGTGCGTAAAATCGTTGGATCTCTCTTCTAA cggatccccgggttaattaa	Delete *ADE2*
Ade2PriRev	TATGTATGAAGTCCACATTTGATGTAATCATAACAAAGCC gaattcgagctcgtttaaac	Delete *ADE2*
P5_MiniDs	AATGATACGGCGACCACCGAGATCTACtccgtcccgcaagttaaata	Amplify library
P7_D701_MiniDs	caagcagaagacggcatacgagatCGAGTAATacgaaaacgaacgggataaa[a]	Amplify and barcode library
P7_D702_MiniDs	caagcagaagacggcatacgagatTCTCCGGAacgaaaacgaacgggataaa[a]	Amplify and barcode library
P7_D703_MiniDs	caagcagaagacggcatacgagatAATGAGCGacgaaaacgaacgggataaa[a]	Amplify and barcode library
P7_D704_MiniDs	caagcagaagacggcatacgagatGGAATCTCacgaaaacgaacgggataaa[a]	Amplify and barcode library
P7_D705_MiniDs	caagcagaagacggcatacgagatTTCTGAATacgaaaacgaacgggataaa[a]	Amplify and barcode library
P7_D706_MiniDs	caagcagaagacggcatacgagatACGAATTCacgaaaacgaacgggataaa[a]	Amplify and barcode library
P7_D707_MiniDs	caagcagaagacggcatacgagatAGCTTCAGacgaaaacgaacgggataaa[a]	Amplify and barcode library
P7_D708_MiniDs	caagcagaagacggcatacgagatGCGCATTAacgaaaacgaacgggataaa[a]	Amplify and barcode library
P7_D709_MiniDs	caagcagaagacggcatacgagatCATAGCCGacgaaaacgaacgggataaa[a]	Amplify and barcode library
P7_D710_MiniDs	caagcagaagacggcatacgagatTTCGCGGAacgaaaacgaacgggataaa[a]	Amplify and barcode library
P7_D711_MiniDs	caagcagaagacggcatacgagatGCGCGAGAacgaaaacgaacgggataaa[a]	Amplify and barcode library
P7_D712_MiniDs	caagcagaagacggcatacgagatCTATCGCTacgaaaacgaacgggataaa[a]	Amplify and barcode library
P7_N701_MiniDs	caagcagaagacggcatacgagatTCGCCTTAacgaaaacgaacgggataaa[a]	Amplify and barcode library

(continued)

Table 1
(continued)

Name	Sequence	Use
P7_N702_MiniDs	caagcagaagacggcatacgagatCTAGTACGacgaaaacgaacgggataaa[a]	Amplify and barcode library
P7_N703_MiniDs	caagcagaagacggcatacgagatTTCTGCCTacgaaaacgaacgggataaa[a]	Amplify and barcode library
P7_N704_MiniDs	caagcagaagacggcatacgagatGCTCAGGAacgaaaacgaacgggataaa[a]	Amplify and barcode library
P7_N705_MiniDs	caagcagaagacggcatacgagatAGGAGTCCacgaaaacgaacgggataaa[a]	Amplify and barcode library
P7_N706_MiniDs	caagcagaagacggcatacgagatCATGCCTAacgaaaacgaacgggataaa[a]	Amplify and barcode library
P7_N707_MiniDs	caagcagaagacggcatacgagatGTAGAGAGacgaaaacgaacgggataaa[a]	Amplify and barcode library
P7_N710_MiniDs	caagcagaagacggcatacgagatCAGCCTCGacgaaaacgaacgggataaa[a]	Amplify and barcode library
P7_N711_MiniDs	caagcagaagacggcatacgagatTGCCTCTTacgaaaacgaacgggataaa[a]	Amplify and barcode library
P7_N712_MiniDs	caagcagaagacggcatacgagatTCCTCTACacgaaaacgaacgggataaa[a]	Amplify and barcode library
P7_N714_MiniDs	caagcagaagacggcatacgagatTCATGAGCacgaaaacgaacgggataaa[a]	Amplify and barcode library
P7_N715_MiniDs	caagcagaagacggcatacgagatCCTGAGATacgaaaacgaacgggataaa[a]	Amplify and barcode library
P7_N716_MiniDs	caagcagaagacggcatacgagatTAGCGAGTacgaaaacgaacgggataaa[a]	Amplify and barcode library
P7_N718_MiniDs	caagcagaagacggcatacgagatGTAGCTCCacgaaaacgaacgggataaa[a]	Amplify and barcode library
P7_N719_MiniDs	caagcagaagacggcatacgagatTACTACGCacgaaaacgaacgggataaa[a]	Amplify and barcode library

(continued)

Table 1
(continued)

Name	Sequence	Use
P7_N720_MiniDs	caagcagaagacggcatacgagatAGGCTCCGacgaaaacgaacgggataaa[a]	Amplify and barcode library
P7_N721_MiniDs	caagcagaagacggcatacgagatGCAGCGTAacgaaaacgaacgggataaa[a]	Amplify and barcode library
P7_N722_MiniDs	caagcagaagacggcatacgagatCTGCGCATacgaaaacgaacgggataaa[a]	Amplify and barcode library
P7_N723_MiniDs	caagcagaagacggcatacgagatGAGCGCTAacgaaaacgaacgggataaa[a]	Amplify and barcode library
P7_N724_MiniDs	caagcagaagacggcatacgagatCGCTCAGTacgaaaacgaacgggataaa[a]	Amplify and barcode library
P7_N726_MiniDs	caagcagaagacggcatacgagatGTCTTAGGacgaaaacgaacgggataaa[a]	Amplify and barcode library
P7_N727_MiniDs	caagcagaagacggcatacgagatACTGATCGacgaaaacgaacgggataaa[a]	Amplify and barcode library
P7_N728_MiniDs	caagcagaagacggcatacgagatTAGCTGCAacgaaaacgaacgggataaa[a]	Amplify and barcode library
P7_N729_MiniDs	caagcagaagacggcatacgagatGACGTCGAacgaaaacgaacgggataaa[a]	Amplify and barcode library
688_minidsSEQ1210	TTTACCGACCGTTACCGACCGTTTTCATCCCTA	Sequence transposon-genome junction
Custom_index1	GGTTTTCGATTACCGTATTTATCCCGTTCGTTTTCGT	Sequence barcode in barcoded library

All primers stored at −20 °C as 100 μM in MQ water. All primers ordered desalted, except P5_MiniDs (Desalted, PAGE, or HPLC purified), 688_minidsSEQ1210 (HPLC), Custom_index1 (HPLC). All primers ordered lyophilized

11. Quantify the purified products using a UV-spectrophotometer (*see* **Note 48**).

3.6 Sequencing

Refer to Illumina® system guides for handling of reagents and run set up (Chap. 3 Sequencing in refs. [26, 27]).

SATAY-specific instructions are indicated below.

3.6.1 Combine Libraries to Be Sequenced Together

1. Mix equal concentrations of the NlaIII and DpnII sub-libraries to generate a library (*see* **Note 49**).
2. If mixing multiple libraries, adjust their amount in the pool so that the number of molecules per transposition event is kept constant (*see* **Note 50**).
3. Quantify the pooled sample using a spectrophotometer (*see* **Note 48**). Confirm that the concentration is as expected.
4. Estimate the molarity of the pooled sample using the following estimate for the average molecular weight of PCR products: 500 bp × 650 g/(mole × bp) = 325,000 g/mole.

3.6.2 Library Preparation

1. Denature and dilute SATAY library and PhiX control to 20 pM in HT1 buffer according to manufacturer recommendations (Protocol A [27, 28]) (*see* **Note 51**).
2. Combine the denatured SATAY library with 1% denatured PhiX control v3 to the following final volumes and concentrations in pre-chilled HT1 buffer (*see* **Note 52**). The library is ready to load.
   - MiSeq Kit v3 (150 cycles). 20 pM denatured DNA in 650 μL.
   - NextSeq 500/550 High Output Kit v2.5 (75 cycles). 2.6–3 pM denatured DNA in 1350 μL.
   - NextSeq 500/550 Mid Output Kit v2.5 (150 cycles). 2.5 pM denatured DNA in 1350 μL.

3.6.3 Loading of Library

1. Miseq Kit v3 (150 cycles). Load 600 μL into Load Samples well [28].
2. NextSeq 500/550 High Output Kit v2.5 (75 cycles). Load 1.3 mL into well 10 [29].
3. NextSeq 500/550 Mid Output Kit v2.5 (150 cycles). Load 1.3 mL into well 10 [29].

3.6.4 Loading of Sequencing Primer (Sequencing of the Transposon–Genome Junctions)

1. MiSeq Kit v3 (150 cycles). Load 3.4 μL of 100 μM Primer 688_minidsSEQ1210 into well 12 [30] (*see* **Note 53**).
2. NextSeq 500/550 High Output Kit v2.5 (75 cycles). Load 5.16 μL of 100 μM Primer 688_minidsSEQ1210 into well 20 [30].
3. NextSeq 500/550 Mid Output Kit v2.5 (150 cycles). Load 4.02 μL of 100 μM Primer 688_minidsSEQ1210 into well 20 [30].

3.6.5 Loading of Custom Index Primer (Sequencing of Indexes)

1. MiSeq v3 (150 cycles). Load 600 μL of 0.5 μM Primer Custom_index1 diluted in pre-chilled HT1 buffer, into well 19 (section Preparing and Loading Custom Primers in ref. [31]).
2. NextSeq 500/550 High Output Kit v2.5 (75 cycles). Load 2 mL of 0.3 μM Primer Custom_index1 diluted in pre-chilled

HT1 buffer, into well 9 (section Prepare and Add Custom Primers in ref. [32]).

3. NextSeq 500/550 Mid Output Kit v2.5 (150 cycles). Load 2 mL of 0.3 μM Primer Custom_index1 diluted in pre-chilled HT1 buffer, into well 9 (section Prepare and Add Custom Primers in ref. [32]).

3.6.6 Instrument Setup

1. Generate a sample sheet following standard procedures (for instance, using Illumina® Experiment Manager, IEM), with the following parameters:

   Miseq:

   (a) Index Reads: 1.

   (b) Read Type: Single Read.

   (c) Cycles Read 1: 76 (*see* **Note 54**).

   (d) Workflow: FastQ only.

   (e) FastQ Only Workflow-Specific Settings: √ Custom Primer for Index.

   NextSeq:

   (a) Index Reads: 1.

   (b) Read Type: Single Read.

   (c) Read Length: Read 1 = 76, Read 2 = 0, Index 1 = 8, Index 2 = 0.

   (d) Custom Primers: √ Index 1.

   (e) Workflow: FastQ only.

2. Upload the Sample Sheet to the instrument.

3. Run the sequencing reaction.

3.7 Analysis

The analysis is performed using R and Bioconductor packages ShortRead [33], Rsamtools [34], and Rbowtie [35, 36].

The sequencing run generates one or more FASTQ file [37] per indexed sub-library (depending on the instrument). The analysis is conducted on raw FASTQ files directly.

You can either run an R script (Subheading 3.7.1, *see* **Note 55**) or use a web server (Subheading 3.7.2).

3.7.1 Reads Mapping Using R Script

1. Install a working copy of Rstudio (https://www.rstudio.com/) or any other R working environment.

2. Download the Fastq.gz files from the sequencing machine or service and store them into a dedicated folder locally (fastqFolder) (*see* **Note 56**).

3. Download the analysis script (Supplement File 1), or if you want to be guided through the analysis, download the corresponding R notebook (Supplement File 2).

4. Download a copy of the genome annotation (saccharomyces_-cerevisiae_R64–2-1_20150113.gff) from the https://yeastgenome.org website and save it in a dedicated folder (genomepath). The indexed genome sequence used by the script will also be stored in this folder.
5. Edit the script under the section "Declare variables" so as to point to your fastqFolder and genomepath.
6. Run the script.

3.7.2 Reads Mapping Using Web Server

1. Download the Fastq.gz files and store them into a folder locally.
2. Generate *.zip or *.tar.gz archive files, one file per sample.
3. Upload each archive individually on the web server at https://kornmann.bioch.ox.ac.uk/satay/submission.html.
4. Download the resulting *.bed, *.wig, and *pergene.txt files. The *pergene.txt contains the number of transposon and reads mapping to all ORFs (*see* **Note 57**).

3.7.3 Visualizing Transposon Maps

1. Open a browser and head to http://genome-euro.ucsc.edu/cgi-bin/hgTracks?hgS_doOtherUser=submit&hgS_otherUserName=benjou&hgS_otherUserSessionName=23bDePuYrk [38].
2. Click "Manage Custom Tracks" the "add custom track."
3. Click "Browse" and select the *.bed or *.wig file from the previous analysis, then click submit.
4. Click on "go" and adjust zoom level in the genome browser.
5. Adjust displayed tracks and settings as per genome-browser's documentation.

4 Notes

1. pBK549 recombines occasionally in a transposase-independent manner, both in bacteria (including *recA*-strains) and in yeast. The resulting species are transposition-deficient. The presence of some amount of recombined species in a pBK549 DNA prep is not an issue. Yeast clones that inherit them by transformation will be eliminated before starting the screen (Subheading 3.1.3). The highest the amount of recombined species relative to intact pBK549, the more yeast transformants you will need to check. Plasmid preps containing a mixture of species can be purified by retransformation in *E. coli*. To keep the amount of recombined pBK549 species as low as possible, we recommend to:
   (a) Systematically verify new plasmid DNA preps.
   (b) Extract plasmid DNA from isolated bacteria colonies only.

(c) Inoculate the bacteria culture from a colony, not a pre-culture (even for MidiPrep).

(d) Freeze an aliquot of bacteria culture in 30% glycerol, immediately after taking it out of the incubator. Use stocks corresponding to DNA preps that satisfy quality control to produce more DNA preps.

2. The strains must be *ura- ade-*. Strains such as W303 are *ade2-* and can be used as is. Deletion of *ADE2* (56 nt before ATG to 62 nt after the STOP) can be achieved using primers Ade2-PriFwd and Ade2PriRev (Table 1) on any template from ref. [39].
3. SATAY will pick up genetic interactions with secondary mutations spuriously present in a transposed strain. We recommend generating independent strains of the genotype of interest, if possible. Thanks to barcoding, each library can be divided into sub-libraries, providing biological replicates. The important and most labor-intensive step is to obtain a total of 2–4 million independent clones per condition/genotype (*see* **Notes 16** and **26**).
4. Transformation plates can be used for up to 1 month, after which we recommend freshly re-transforming the vector.
5. Here the goal is not to obtain isolated colonies but to propagate the transformants and check for spontaneous recombination. A small percentage of transformed cells will receive recombined species 1 or 2 and grow into clones that are transposition-deficient (Fig. 2b). These clones can be distinguished from clones harboring intact plasmids by streaking on -ADE medium. Clones bearing intact plasmids will yield a few (1–20) Ade+ colonies by spontaneous (transposase-independent) homologous recombination, while those bearing recombined species 1 and 2 will fail to yield any Ade + progeny. The ability to produce a few Ade+ colonies is a means to recognize transposition-proficient clones than can be used further. Use a flat toothpick to pick an isolated colony from the transformation plate, make a thick streak on SC-D-ADE first, then on SC-D-URA. Proceed to spread on the SC-D-URA plate with the same toothpick. Use a fresh toothpick to spread on the SC-D-ADE plate.
6. To minimize any further appearance of Ade + spontaneous recombinants, store the streaked SC-D-URA plates at 4 °C and use within a week. Beyond a week, repeat the verification procedure directly from the transformation plate.
7. Scrape the cells using a tip mounted on a P1000 pipette. Inoculate in a 1.5-mL microtube pre-filled with 1 mL sterile MQ water. Pipette the cells up and down to mix them well.

Measure the OD_{600} of a 50× dilution of this cell suspension. Calculate the volume of cell suspension required to inoculate 40–60 mL SC-Rd-URA medium at OD_{600} 0.2.

8. One OD_{600} unit for cells scraped from a plate corresponds to ~1.6E+07 cells. One 1/6-plate sector is enough to inoculate 40–60 mL at OD_{600} 0.2.

9. The 0.2% glucose allows for a smooth diauxic transition from glucose to raffinose. The pre-culture in raffinose primes cells for subsequent galactose induction by minimizing glucose repression.

10. 2 mL of saturated (OD_{600} ~ 4) pre-culture per transposition plate or 100 mL for 50 plates.

11. The goal is to ensure that the plates are dry before plating. Plating on moist plates will result in a long drying time and inhomogeneous plating.

12. The number of SC-Gal -ADE (transposition) plates required depends on the transposition efficiency, which may vary with the genetic background. Aim to obtain 2–4 million transposed clones per condition/genotype. One SC-D -ADE (control) plate is required per pre-culture.

13. The pre-cultures should have reached saturation (OD_{600} ~ 4 in our hands). Before measuring the OD_{600}, dilute the culture to ensure the readout value is within the linear range of the spectrophotometer (usually 0.1–1).

14. This step checks that most cells are viable and contain the plasmid. As a guideline, we find ~450 (By4741 background, from 2.5E+07 cells/OD as measured with a hemocytometer) and ~250 (W303 background, from 1E+07 cells/OD as measured with a hemocytometer) colonies on SC-D plates. Expect that up to 20% of the cells have become *ura*-, which is not an issue.

15. For example, if the overnight pre-culture is at OD_{600} 4, then you need to concentrate the cells 40/4 = 10-fold, therefore remove 90% of the initial volume. It might be necessary to perform two consecutive rounds of centrifugation to harvest the required amount of cells. In this case, remove the supernatant from the first centrifugation entirely, add the remaining culture and repeat the centrifugation. Remove the appropriate volume of the supernatant to attain the desired cell concentration.

16. The procedures described here were optimized for our laboratory conditions (spectrophotometer, growth media, strains). We recommend testing several plating concentrations at small scale to determine which one leads to the highest number of transposed colonies. Plating at OD_{600} 35–40 was optimal in our hands. Plating at OD_{600} 70 decreased the yield, firstly

because it became difficult to plate homogeneously and secondly because the cell density on the plate was so high that it prevented the growth of transposed clones. Plating below OD_{600} 35 decreased the yield of transposed clones proportionally.

17. Use a repeater pipette. Use a different tip for each pre-culture. Keep the tip in a sterile beaker while not in use.
18. The number of plates to process at once depends on how fast the cell suspension is absorbed on the surface of the plate. The dryer the plates, the faster the liquid will be absorbed, therefore the fewer the plates that should be processed at once.
19. Do not proceed further if the plates are still wet. Set the wet plates aside until they are dry while processing the remaining plates.
20. Homogeneous plating is key. We determined that spreading 200 μL using 10–15 glass beads per plate was optimal. Fewer beads will result in inhomogeneous plating. More beads decrease the number of cells seeded on the plate. Dispense the beads on all the plates before plating. Store the plates upside down such that the beads remain in the lid until needed. Leaving the beads on the agar will produce dents, resulting in inhomogeneous plating and inhomogeneous size and distribution of the transposed colonies.
21. Check the plating using a microscope equipped with a long-distance objective (e.g., a tetrad dissection microscope). Aim for a monolayer of cells covering 90% of the plate. Interspersed mounds of cells are perfectly normal.
22. Perform one control plating on SC-D -ADE for each independent pre-culture. This control is necessary to estimate the number of Ade + cells present prior to transposition induction. Ade + cells can result from spontaneous transposase-independent recombination of pBK549 (*see* **Note 5**). Cells containing such recombination products will form colonies on both the SC-D -ADE and SC-Gal -ADE plates within the first 2–3 days after plating, while genuinely transposed clones will develop over the course of 10–15 days.
23. In very dry incubators, drying of the plates can be prevented by incubating the sleeves in a humid chamber, for example, a plastic box containing moist towel paper.
24. The incubation time might vary with the strain and media composition. We recommend performing a small-scale time course experiment to monitor the appearance of transposed clones. The number of clones as a function of the incubation time is reproducible and generally plateaus after 10 days. SC-Gal -ADE plates can be kept at 4 °C for at least 3 months before plating, without alteration of transposition efficiency.

25. The clones might be difficult to differentiate from the background lawn of cells after only 3 days of incubation but will become more distinct rapidly. Unlike mounds of cells resulting from plating, colonies are characterized by their round shape, defined edges, and a distinctive refraction. Mark the colonies using permanent pens of different colors or take photos of the same area to follow the appearance of new ones. Colonies overlap above a density of ~500 per cm^2, therefore estimate the number of overlapping colonies within a mound.
26. ByK352 (By4741 *ade2::HIS3*) consistently produces 1200–1700 colonies/cm^2 (66,000–93,500 colonies/plate) and ByK832 (W303-1B *ade2::NatMX6*), 350–500 colonies/cm^2 (19,250–27,500 colonies/plate). Transposition-independent Ade + background colonies represent 2–10% of the final number of Ade + clones.
27. Process a few plates at a time to avoid the liquid being absorbed into the agar before the cells have been harvested.
28. The best rakes are made by bending a long Pasteur pipette with a Bunsen burner (Fig. 3c). Commercial glass rakes used for spreading are sometimes too thick to scrape the cells efficiently.
29. To reduce the volume of medium needed, it is also possible to reuse the collected cell suspension to harvest the next set of plates.
30. Because liquid absorbs in the agar, about 50 mL are collected from 50 harvested plates. The harvested material is thick and composed of ~70% Ade + cells. The red coloration is due to the presence of untransposed *ade-* cells that accumulated a red pigment on the SC-Gal -ADE transposition plates.
31. The OD_{600} of the harvested material is ~50. Dilute to read within the linear range of the spectrophotometer.
32. This step serves to expand the library and to dilute the *ade-* cells. The expanded library is then either processed for sequencing or exposed to various treatments and conditions prior to sequencing. It is possible to skip the expansion phase and to apply treatments in -ADE medium directly.
33. 2 L of reseed medium for a library of 2–4 million harvested colonies corresponds to reseeding ~700–1400 cells per transposition event. It is important that the number of cells per transposition event remains high to avoid losing complexity. This number is estimated as follows:

    If the Ade + cells represent 70% of the total number of cells and OD_{600} 0.2 represents 2E+06 cells/mL, then

    [2000 mL × 2E+06 cells/mL × 70%] / 2–4E+06 clones = 700 – 1400 cells/transposition event.

    This calculation can be refined once the SC-D and SC-D -ADE plates have been counted.

34. Spread 200 μL of a 40,000-fold dilution of the saturated culture on one SC-D and one SC-D -ADE plate using glass beads. Incubate the plates for 3 days at 30 °C. Count the colonies to determine the number of cells per mL.
35. To avoid decreasing the complexity of the library, it is important that each transposition event remains represented by a fair number of cells (*see* **Note 33**). We recommend more than 100 cells per event as a safe guideline.
36. The harvest is in vast excess and can be frozen in 30% glycerol. Since viability upon freezing thawing is decreased, we recommend freezing small aliquots to measure cell viability post freezing, in addition to larger aliquots (to re-seed the library accordingly).
37. 1 L of culture at OD_{600} 4 produces ~5 g of pellet.
38. The dark red coloring of the nucleic acid pellet is due to the pigment produced by the *ade-* cells. This pigment is insoluble in ethanol but soluble in water. It does not alter the subsequent steps of library preparation and sequencing.
39. The genomic DNA preparation contains RNAse and therefore cannot be quantified accurately using a spectrophotometer. Moreover, we surmise that the presence of the red pigment extracted from the *ade-* cells might alter DNA quantification depending on the method. Gel electrophoresis combined with staining with a DNA intercalating agent is insensitive to either of these issues.
40. Any other quantified ladder as well as previously quantified genomic DNA preparations can serve as standards.
41. Homogeneous staining is best achieved by staining the gel post running, especially when using charged intercalating agents like ethidium bromide.
42. To ensure no genomic material is lost, all the following steps, including the assembly of the PCRs, are performed in the same tube as the digestion reaction.
43. Linear acrylamide serves as a carrier during precipitation and is inert in subsequent reactions.
44. We recommend sequencing a maximum of:
    (a) 2 libraries per MiSeq (Reagent Kit v3, 150 cycles) run.
    (b) 5 libraries per NextSeq (Mid Output Kit v2.5, 150 cycles) run.
    (c) 20 libraries per NextSeq (High Output Kit v2.5, 75 cycles) run.
45. It is important that the indexes differ by at least three consecutive nucleotides and are chosen with an appropriate balance of dyes for sequencing. The MiSeq platform uses four channels,

while two are used by the NextSeq platform. During the first three cycles of index sequencing, no channels must be omitted. Information on selecting indexes with optimized color balance can be found in the Illumina® documentation (section Color Balance in ref. [23]).

46. The NlaIII and DpnII sub-library products appear as smears centered around 500 bp, running from 250 bp up to 10 kb or beyond. Both the sub-libraries form discrete characteristic bands that correspond to the amplification of untransposed pBK549 (NlaIII, 867 bp; DpnII, 465 bp, Fig. 5c). The intensity of these untransposed bands relative to the rest of the smear diminishes with the length of the regrowth in SC-D -ADE (see Subheading 3.3.2). The negative control lanes should be clear except for possible primer dimers around 70 bp.

47. Making 1 mL PCR and purifying only 100 μL appear to be a waste but the rationale is the following. To avoid losing complexity in the library, it is important that each transposon is represented by a fair number of molecules in the PCR. 2 μg DNA contains ~10^8 genomes. Therefore, each of the 2–4 million clones of a typical library is present in ~20–40 copies. Thus 2 μg DNA appears like a minimal amount, and 1 mL a minimal PCR volume for this amount of DNA. After amplification, the PCR product consists exclusively of transposon–genome junctions; therefore, much less is necessary downstream.

48. The following instructions are based on quantification of the PCR products using a NanoDrop spectrophotometer. Other methods of quantification might lead to different values. It might be necessary to adjust the concentration of library on the sequencing cartridge.

49. For example: if the concentration of the NlaIII sub-library is 300 ng/μL and that of the DpnII sub-library is 150 ng/μL, then mix 1 and 2 volumes of both the sub-libraries, respectively. The final concentration will be $(300 + 2 \times 150) / 3 = 200$ ng/μL.

50. For example: if the complexity of library A is twice that of library B, double the amount (in ng) of library A over that of B.

51. The PhiX standard DNA is optional but serves as quality control and may allow Illumina® technical support to diagnose problematic runs.

52. These are optimal DNA concentrations for SATAY libraries on Illumina® platforms in our hands.

53. Use a thin gel loading tip to ensure the primer is loaded in the solution contained in well 12. Mix by moving the tip sideways.

54. Set the "read length" on the instrument to 76 bp to actually read 75 bp.
55. Other options are available from third parties (e.g., https://github.com/leilaicruz/LaanLab-SATAY-DataAnalysis).
56. Depending on whether using the MiSeq or the NextSeq platform, the machine will output one or four *fastq.gz files per sample, respectively. This is because the MiSeq v3 flowcell harbors one sequencing lane versus four for the NexSeq v2.5 flowcell. Make sure that the files corresponding to the NlaIII- and DpnII-digested libraries contain "NlaIII" and "DpnII" in their file name, respectively, as this pattern will be used by the script to identify them.
57. If submitting multiple samples, the output for each sample will be concatenated in the *.bed, *.wig, and *pergene.txt files.

Acknowledgments

This work was supported by Wellcome Trust Investigator Award grant 214291/Z/18/Z.

References

1. Cain AK, Barquist L, Goodman AL et al (2020) A decade of advances in transposon-insertion sequencing. Nat Rev Genet 21: 526–540. https://doi.org/10.1038/s41576-020-0244-x
2. van Opijnen T, Bodi KL, Camilli A (2009) Tn-seq: high-throughput parallel sequencing for fitness and genetic interaction studies in microorganisms. Nat Methods 6:767–772. https://doi.org/10.1038/nmeth.1377
3. Christen B, Abeliuk E, Collier JM et al (2014) The essential genome of a bacterium. Mol Syst Biol 7:528–528. https://doi.org/10.1038/msb.2011.58
4. Girgis HS, Liu Y, Ryu WS, Tavazoie S (2007) A comprehensive genetic characterization of bacterial motility. PLoS Genet 3:e154. https://doi.org/10.1371/journal.pgen.0030154
5. Guo Y, Park JM, Cui B et al (2013) Integration profiling of gene function with dense maps of transposon integration. Genetics 195: 599–609. https://doi.org/10.1534/genetics.113.152744
6. Michel AH, Hatakeyama R, Kimmig P et al (2017) Functional mapping of yeast genomes by saturated transposition. eLife 6:e23570. https://doi.org/10.7554/eLife.23570
7. Michel AH, van Schie S, Mosbach A et al (2019) Exploiting homologous recombination increases SATAY efficiency for loss- and gain-of-function screening. In: bioRxiv. p 866483. https://doi.org/10.1101/866483
8. Edskes HK, Mukhamedova M, Edskes BK, Wickner RB (2018) Hermes transposon mutagenesis shows [URE3] prion pathology prevented by a ubiquitin-targeting protein: evidence for carbon/nitrogen assimilation cross talk and a second function for Ure2p in Saccharomyces cerevisiae. Genetics 209: 789–800. https://doi.org/10.1534/genetics.118.300981
9. Sanchez MR, Payen C, Cheong F et al (2019) Transposon insertional mutagenesis in Saccharomyces uvarum reveals trans-acting effects influencing species-dependent essential genes. Genome Res 29:396–406. https://doi.org/10.1101/gr.232330.117
10. Segal ES, Gritsenko V, Levitan A et al (2018) Gene essentiality analyzed by in vivo transposon mutagenesis and machine learning in a stable haploid isolate of Candida albicans. mBio 9:e02048–e02118. https://doi.org/10.1128/mBio.02048-18
11. Gao J, Wang H, Li Z et al (2018) Candida albicans gains azole resistance by altering

sphingolipid composition. Nat Commun 9: 4495. https://doi.org/10.1038/s41467-018-06944-1

12. Gale AN, Sakhawala RM, Levitan A et al (2020) Identification of essential genes and fluconazole susceptibility genes in Candida glabrata by profiling Hermes transposon insertions. G3 (Bethesda) 10:3859–3870. https://doi.org/10.1534/g3.120.401595
13. Coradetti ST, Pinel D, Geiselman GM et al (2018) Functional genomics of lipid metabolism in the oleaginous yeast Rhodosporidium toruloides. eLife 7:e32110. https://doi.org/10.7554/eLife.32110
14. Zhu J, Gong R, Zhu Q et al (2018) Genome-wide determination of gene essentiality by transposon insertion sequencing in yeast Pichia pastoris. Sci Rep 8:10223. https://doi.org/10.1038/s41598-018-28217-z
15. Levitan A, Gale AN, Dallon EK et al (2020) Comparing the utility of in vivo transposon mutagenesis approaches in yeast species to infer gene essentiality. Curr Genet 66: 1117–1134. https://doi.org/10.1007/s00294-020-01096-6
16. Lazarow K, Du M-L, Weimer R, Kunze R (2012) A hyperactive transposase of the maize transposable element *activator* (*ac*). Genetics 191:747–756. https://doi.org/10.1534/genetics.112.139642
17. Serbyn N, Noireterre A, Bagdiul I et al (2020) The aspartic protease Ddi1 contributes to DNA-protein crosslink repair in yeast. Mol Cell 77(5):1066–1079. https://doi.org/10.1016/j.molcel.2019.12.007
18. Haribowo AG, Hannich JT, Michel AH et al (2019) Cytotoxicity of 1-deoxysphingolipid unraveled by genome-wide genetic screens and lipidomics in Saccharomyces cerevisiae. MBoC 30:2814–2826. https://doi.org/10.1091/mbc.E19-07-0364
19. Nicastro R, Raucci S, Michel AH et al (2021) Indole-3-acetic acid is a physiological inhibitor of TORC1 in yeast. PLoS Genet 17:e1009414. https://doi.org/10.1371/journal.pgen.1009414
20. Serbyn N, Bagdiul I, Michel AH et al (2020) SUMO orchestrates multiple alternative DNA-protein crosslink repair pathways. Cell Rep 37(8):110034. SSRN: https://ssrn.com/abstract=3625817. https://doi.org/10.2139/ssrn.3625817
21. Jorgensen JR, Tei R, Baskin JM et al (2020) ESCRT-III and ER–PM contacts maintain lipid homeostasis. MBoC 31:1302–1313. https://doi.org/10.1091/mbc.E20-01-0061
22. Schindelin J, Arganda-Carreras I, Frise E et al (2012) Fiji: an open-source platform for biological-image analysis. Nat Methods 9: 676–682. https://doi.org/10.1038/nmeth.2019
23. Illumina® (2018) Index Adapters Pooling Guide (1000000041074). https://support.illumina.com/content/dam/illumina-support/documents/documentation/chemistry_documentation/experiment-design/index-adapters-pooling-guide-1000000041074-05.pdf
24. Illumina® (2021) Indexed sequencing on Illumina Systems (15057455). https://support-docs.illumina.com/SHARE/IndexedSeq/indexed-sequencing.pdf
25. Illumina® (2016) Library pooling guidelines for the NextSeq 500/550 and MiniSeq systems. https://emea.support.illumina.com/bulletins/2016/10/library-pooling-guidelines-for-the-nextseq-and-miniseq-systems.html
26. Illumina® (2018) MiSeq System Guide (1000000061014). Available at https://support.illumina.com/content/dam/illumina-support/documents/documentation/system_documentation/miseq/miseq-system-guide-for-miseq-reporter-1000000061014-00.pdf
27. Illumina® (2019) NextSeq 550 System Guide (15069765). https://support.illumina.com/content/dam/illumina-support/documents/documentation/system_documentation/nextseq/nextseq-550-system-guide-15069765-06.pdf
28. Illumina® (2019) MiSeq System Denature and Dilute Libraries Guide (15039740). Available at https://support.illumina.com/content/dam/illumina-support/documents/documentation/system_documentation/miseq/miseq-denature-dilute-libraries-guide-15039740-10.pdf
29. Illumina® (2018) NextSeq System Denature and Dilute Libraries Guide (15048776). Available at https://support.illumina.com/content/dam/illumina-support/documents/documentation/system_documentation/nextseq/nextseq-denature-dilute-libraries-guide-15048776-09.pdf
30. Illumina® (2020) Spiking Custom Primers into the Illumina Sequencing Primers. Available at https://emea.support.illumina.com/bulletins/2016/04/spiking-custom-primers-into-the-illumina-sequencing-primers-.html
31. Illumina® (2016) MiSeq System Custom Primers Guide (15041638). Available at https://support.illumina.com/content/dam/

illumina-support/documents/documentation/system_documentation/miseq/miseq-system-custom-primers-guide-15041638-01.pdf

32. Illumina® (2020) NextSeq 500 and NextSeq 550 Sequencing Systems Custom Primers Guide (15057456). Available at https://emea.support.illumina.com/content/dam/illumina-support/documents/documentation/system_documentation/nextseq/nextseq-500-550-custom-primers-guide-15057456-03.pdf
33. Morgan M, Anders S, Lawrence M et al (2009) ShortRead: a bioconductor package for input, quality assessment and exploration of high-throughput sequence data. Bioinformatics 25:2607–2608. https://doi.org/10.1093/bioinformatics/btp450
34. Li H, Handsaker B, Wysoker A et al (2009) The sequence alignment/map format and SAMtools. Bioinformatics 25:2078–2079. https://doi.org/10.1093/bioinformatics/btp352
35. Wei Z, Zhang W, Fang H et al (2018) esATAC: an easy-to-use systematic pipeline for ATAC-seq data analysis. Bioinformatics 34:2664–2665. https://doi.org/10.1093/bioinformatics/bty141
36. Langmead B, Salzberg SL (2012) Fast gapped-read alignment with bowtie 2. Nat Methods 9:357–359. https://doi.org/10.1038/nmeth.1923
37. Cock PJA, Fields CJ, Goto N et al (2010) The sanger FASTQ file format for sequences with quality scores, and the Solexa/Illumina FASTQ variants. Nucleic Acids Res 38:1767–1771. https://doi.org/10.1093/nar/gkp1137
38. Kent WJ, Sugnet CW, Furey TS et al (2002) The human genome browser at UCSC. Genome Res 12:996–1006. https://doi.org/10.1101/gr.229102
39. Longtine MS, McKenzie A, Demarini DJ et al (1998) Additional modules for versatile and economical PCR-based gene deletion and modification in Saccharomyces cerevisiae. Yeast (Chichester, England) 14:953–961. https://doi.org/10.1002/(SICI)1097-0061(199807)14:10<953::AID-YEA293>3.0.CO;2-U

Chapter 21

High-Throughput, High-Precision Colony Phenotyping with *Pyphe*

Stephan Kamrad, Jürg Bähler, and Markus Ralser

Abstract

Colony fitness screens are powerful approaches for functional genomics and genetics. This protocol describes experimental and computational procedures for assaying the fitness of thousands of microbial strains in numerous conditions in parallel. Data analysis is based on *pyphe*, an all-in-one bioinformatics toolbox for scanning, image analysis, data normalization, and interpretation. We describe a standard protocol where endpoint colony areas are used as fitness proxy and two variations on this, one using colony growth curves and one using colony viability staining with phloxine B. Different strategies for experimental design, normalization and quality control are discussed. Using these approaches, it is possible to collect hundreds of thousands of data points, with low technical noise levels around 5%, in an experiment typically lasting 2 weeks or less.

Key words Screen, Colony, Fitness, Functional genomics, Phenomics, Microbiology, Python software, Large-scale phenotyping, Cell viability, Growth curve

1 Introduction

Measuring properties of densely arrayed microbial colonies is a powerful and efficient technique for determining the fitness of large panels of microbial strains. Today, numerous knockout or overexpression mutant libraries, wild strain collections, segregant panels, and synthetic genetic array construction methods are available to underpin systematic, data-driven investigations. Owing to specialized robotics, colony screens are largely automated and can be done at huge scales, e.g., with millions of double knockout mutants [1]. Colony screens have been employed to measure the fitness of large segregant libraries for the dissection of complex traits [2, 3] and the characterization of wild strain libraries [4–6]. All such experiments require scalable computational approaches for data acquisition, data management, batch correction, and sta-

Frédéric Devaux (ed.), *Yeast Functional Genomics: Methods and Protocols*,
Methods in Molecular Biology, vol. 2477, https://doi.org/10.1007/978-1-0716-2257-5_21,

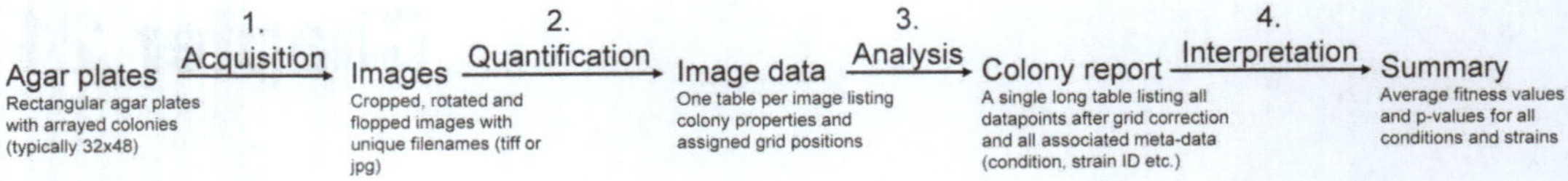

Fig. 1 Overview of data analysis workflow

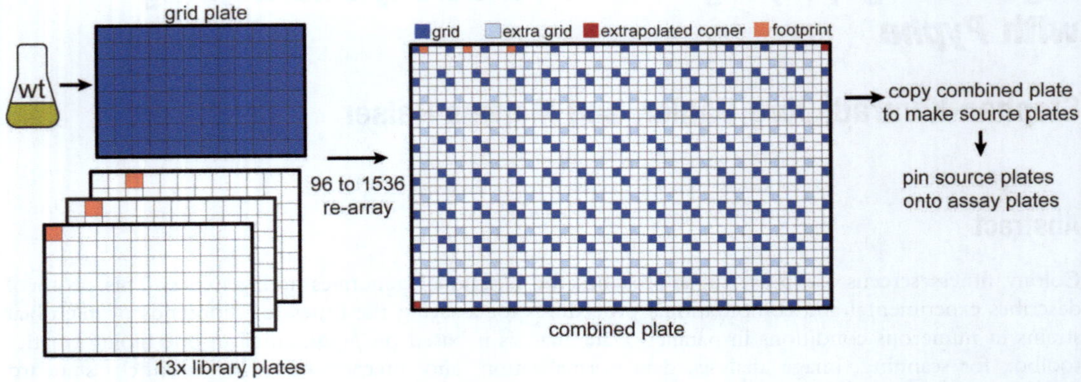

Fig. 2 Assembly of assay plates

tistical analysis. To this end, we have recently published a new toolbox named *pyphe* which enables fast, reproducible, and flexible pipelines for colony screen analysis (Fig. 1) [7].

One key challenge of colony screens is that the observed colony phenotype is affected by factors other than the strain's fitness. Batch effects, e.g., originating from media preparation or incubation time, make colony sizes not directly comparable between different plates. Even within plates, nutrient, moisture, and temperature gradients as well as differences in inoculation mass due to uneven pinning can result in variation between colonies located in different areas of the plate. A well-known example is the "edge effect", where the outermost colonies grow larger due to a lower degree of nutrient competition. Normalization is therefore key for obtaining reliable fitness estimates. Edge effects, and other effects that affect rows and columns evenly, can in some cases be rectified by dividing by the row/column median, and other, regional biases can sometimes be rectified by comparing colonies to their neighbors (e.g., as implemented in SGAtools [8]). Both these normalization methods only work if the null effect can be reliably estimated, i.e., if the majority of colonies show no growth effect. An alternative approach is to include a grid of control strains on every plate (Fig. 2) [9]. Each colony is then compared to essentially a weighted average of the neighboring controls, resulting in a relative fitness that is standardized within and across plates (as long as the same control strain is used). *Pyphe* implements row/column median normalization as well as a grid normalization, giving the user full control over the normalization strategy. Using grid normalization, it is possible to achieve technical noise levels of around 5%, enabling the detection of subtle differences in fitness.

Classically, plates are imaged once and colony footprint areas are used as fitness proxy. Recent methods have expanded on this in two areas. ScanLag [10], Quantitative Fitness Analysis [11], Colony-live [12], and Scan-o-matic [9] use repeated imaging of colonies and extract parameters from the resulting growth curve. Additionally, Colony-live and Scan-o-matic use transmission scanning where the gray-value of a pixel can be used as a proxy for colony thickness. These methods achieve lower noise values than endpoint measurements but require substantially higher investments in scanners (which need to be housed in a temperature-controlled environment), data storage, and analysis. We recently found that, at least of the parameters tested, the growth rate parameters extracted from timeseries correlate tightly with endpoint sizes [7]. This is compatible with a model where colonies grow until the media is exhausted by the ensemble of colonies, giving each colony a fixed time-window for growth. This is an important difference to screens conducted in multi-well plates with liquid media, where each culture is given a fixed amount of resources. *Pyphe* supports the acquisition and analysis of colony growth curves as well as traditional endpoint measurements, allowing the user to choose the method most suitable for their experiment and laboratory setup.

Another variation on classical screens is the use of phloxine B as a dead cell stain [13]. If this is included in the media, the color of a colony reflects the proportion of dead cells in it [7]. This provides an orthogonal readout that is often uncorrelated to colony sizes after both readouts are standardized to the control strain. As phloxine B in the media generally does not affect cell growth, it can be included in all plates, whereby a second fitness dimension is added to the experiment at little extra cost and effort. *Pyphe* can acquire and analyze color images and extract quantitative redness scores which then integrate seamlessly into the rest of the analysis pipeline.

We here present a detailed protocol for colony fitness screens, split into an experimental and computational part. The basic protocol describes an endpoint experiment using colony areas as fitness proxies. In our opinion, this should be the default for most investigations and the starting point for new users. Two variations on this are described: How to record and analyze colony growth curves and how to analyze colony viability by phloxine B staining. These, together with the detailed accompanying notes, are meant as a starting point for users to implement their own workflows. By providing a framework of data formats and analysis steps, much of it will be applicable to researchers not using *pyphe* but looking to implement their own computational or experimental procedures. Experimental details will vary between microbial species, assay conditions, and laboratory settings. *Pyphe* is set up as an open-source, collaborative software project, and we welcome user feedback and contributions for its continuous improvement.

2 Materials

2.1 Experimental

1. Sterile microbial growth media (liquid and with agar).
2. Rectangular agar plate dishes, e.g., Singer PlusPlates.
3. Microwave (or autoclave) for melting agar media.
4. 50-mL serological pipettes and pipette pump.
5. Conical flask (e.g., 250 mL), shaking incubator.
6. Pinning robot and consumables (e.g., Singer RoToR with 96 long RePads, 96 short RePads, and 1536 short RePads).
7. Cling film.
8. Plate incubator.
9. Strain library cryostocks in 96 or 384 format, ideally with unique footprints.
10. Cryostock of a suitable control strain.

Optional

11. Incubator to house scanners (for time-course imaging).
12. Phloxine B (for colony viability measurements).
13. Reagents required for testing specific growth conditions of interest (e.g., drugs, chemicals or nutrients).

2.2 Scanning and Data Processing

1. Scanner (Epson V800 or V850 recommended) with a fixture to hold plates into place on the scanner surface (laser cutting guide available on https://github.com/Bahler-Lab/pyphe/tree/master/Documentation/Scan). Please see Fig. 3 for more information on how to set up the scanner.
2. A computer with Ubuntu (or another Linux operating system) and Python3 (e.g., Anaconda3) for scanning.
3. Any computer with any operating system with Python3 installed for the rest of the analyses.

3 Methods

3.1 General Method: Endpoints

3.1.1 Experimental Design and Plating

1. Pick an appropriate control strain for your library. For knock-out, overexpression, and similarly constructed libraries, this should usually be the library background strain ("wild type"). For collections of wild isolates or segregant libraries, the choice of control strain is more flexible, but it should not fall on either extreme side of the distribution of fitness phenotypes.

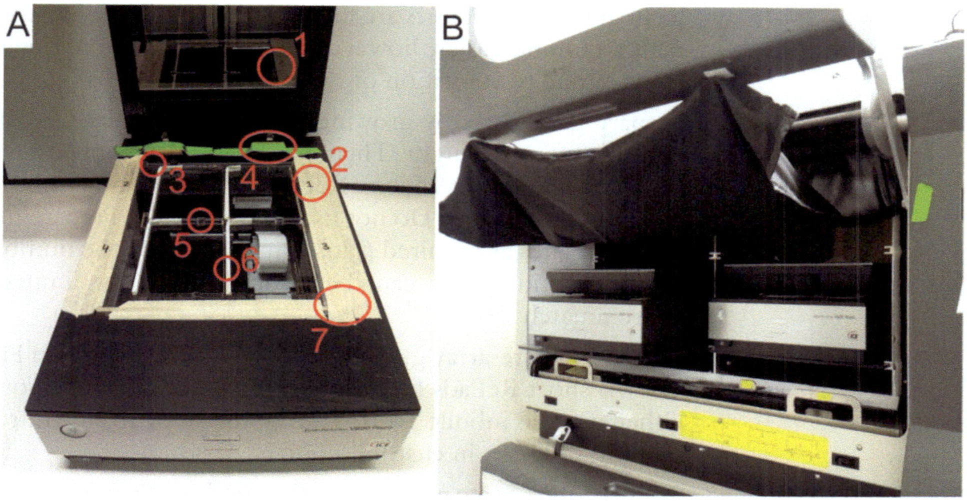

Fig. 3 Physical setup of scanners. (**a**) Epson V800 scanner fitted with a fixture. Note that the white background insert on the scanner lid has been removed in preparation for transmission scanning (1), the plate positions are clearly numbered to avoid mixing up plates (2), the uppermost area of the scanning bed is free and not covered by the fixture as this area is used for sensor calibration (3), small "bumper pads" of cardboard have been attached next to the hinges to ensure an optimal and consistent distance between the lid and the scanner bed once the lid is closed (4), the fixture contains small notches that fit into pockets of Singer PlusPlates which ensure plates can only be inserted in the correct orientation (5), additional tape has been applied to further stabilize the position of plates within the fixture (6) and that the fixture has been securely attached to the scanner (7). For best results, place plates into the scanners without their lids. (**b**) Scanners are positioned in an incubator for time-course imaging. Scanners are placed on a custom-built shelf for easy access and fully covered with black fabric. The cover prevents light interference and air currents from the incubator fan hitting the plates directly, which would lead to faster and uneven drying

2. Plan the arrangement of library and control strains on the final assay plates. Overall, five different types of plates are required throughout the workflow: library plates, grid plates, combined plates, source plates, and assay plates (Fig. 2). Consider how many replicates are required to answer your biological question (*see* **Note 1**). Plan how to rearrange the library plates around the control grid and prepare the layout file of your assay plates (*see* **Notes 2** and **3**).
4. Prepare sterile growth media and pour agar plates for waking up the library and for the preparation of the grid plate (*see* **Note 4**). We usually use rich media for waking up strains and preparing the grid plate, even if the screen is in the end carried out on minimal media.
5. Wake up the control strain and your library. This protocol is applicable to a wide range of microbial organisms and growth media, so we cannot specify incubation times for plates. The general advice for the preparatory phase is that plates should be

incubated until colonies are fully (or almost fully) grown as this makes pinning easier, however, not so long that colonies get stressed due to starvation or desiccation.

6. Prepare grid plates by growing a 40 mL liquid culture of the control strain overnight. Then, pour the culture into an empty PlusPlate and, using 96 long RePads, pin from liquid to solid to make the grid plates. Do not use target mixing. Prepare multiple grid plates as required (plus some spares) as the number of times a grid plate can be pinned from is limited to approximately 10.
7. Once grown, re-array grid and library plates as shown in Fig. 2. Use 96 short RePads for this. On the Singer RoToR, you can either use the inbuilt 16-to-1 array mode or make your own pinning program in custom mode.
8. To make the source plates, copy the combined plates onto fresh agar plates, using 1536 short pins. This step is important because the combined plates are a patchwork and should therefore be copied again to make consistent, evenly arrayed source plates. Besides, this step can serve as an amplification step for larger screens with many assay plates. Simply prepare several copies of each source plate, approximately one for every 10 assay plates.
9. Prepare your assay plates with drugs, inhibitors, or particular nutrient composition as required (*see* **Note 4**). Make one or two spares per condition in case something goes wrong during pinning.
10. Copy your source plates onto the assay plates using 1536 short pins. Consistent, small inoculums are obtained with low pinning pressure (5–10%). Clearly label each plate with a unique ID (*see* **Note 5**). Monitor the quality of pinned plates (*see* **Note 6**).
11. Wrap plates in cling film and place in an incubator. Plates should always rest upside-down to prevent condensate dropping onto the colonies. Ideally, there should be no strong air currents or temperature gradients in the incubator which can lead to plates drying out and colonies growing unevenly.
12. Let the colonies grow until they stop or barely grow anymore (usually 2–3 days). See ref. [7] for detailed reasoning for this choice of time point.

3.1.2 Acquisition

1. To install *pyphe*, run the following command in your terminal (*see* **Note 7**).

    $ pip install pyphe.

    ImageMagick needs to be installed, the scanner needs to be accessible with SANE (Scanner Access Now Easy), and the TPU8x10 scanning mode must be available. This is the case by default for recent versions of SANE, which is included in the standard Ubuntu installation.

2. Image your plates once they have grown. Turn on the scanner and run

   $ pyphe-scan --fixture som3 --postfix POSTFIX --nplates N

   where POSTFIX is the name of your experiment and N the total number of plates you have to scan. The fixture argument specifies how to crop individual plate images from the original scan. Parameters for our fixture model are pre-programmed (use som3_edge and som3 for images containing or not containing the edge, respectively), but it is possible to configure your own fixtures in the *pyphe-scan* script. Plates are always placed in the scanner with the agar on the bottom (i.e., not upside-down) and with the A1 colony position in the back-right corner (Fig. 3). Remove the plate lids, lower the scanner top, and make sure it is even. Press "y" when prompted to start scanning. Swap plates as instructed. *Pyphe-scan* will create a folder with the current date and the POSTFIX in the current directory in which images will be saved.

3. In parallel, prepare a table mapping the scan sequence to the plate ID. Later, expand this to include information about the layout, experimental condition, and any additional meta-data (date, batch, comments, etc.) you want to keep track of. This will be your experimental design table (EDT, *see* **Note 5**).

3.1.3 Quantification

1. Extract colony sizes from images by running

   $ pyphe-quantify batch --grid auto_1536 --pattern "images/*.jpg"

   The batch parameter indicates the mode in which to run *pyphe-quantify*. In batch mode, images are analyzed one at a time and colony areas are extracted. The --grid option specifies how to assign identified colonies to grid positions. In automatic mode, as used here, the grid is detected based on pixel row and column intensity peaks. There are other ways to manually define grid positions when this is required, see the documentation. The --pattern argument can be used to specify which images should be analyzed; in this case, it is assumed that all your images are located in a folder "images" in the current working directory and are in jpg format.

2. Carefully check whether the results are as expected. A QC image is produced for every single analyzed image. Investigate if colonies were identified correctly and that colonies are assigned the correct row and column (*see* **Note 8**). *Pyphe-quantify* has numerous parameters that can be adjusted, check the documentation for details.

3.1.4 Analysis

1. Edit your EDT table to include a column "Data_path" which contains relative file paths to each image's data file (these are located in the "pyphe_quant" directory by default). If you have to exclude individual images for QC reasons, simply delete the corresponding lines from the EDT table.
2. Aggregate data across plates and perform grid and subsequent row/column median normalization by running

   $ pyphe-analyse --edt EDT.csv --format pyphe-quantify-batch --gridnorm standard1536 --rcmedian --qc_plots analyse-qc

   The --edt option specifies the relative path to your experimental design table and should be adjusted if required. The --format option describes the type of input data, in this case colony areas produced by *pyphe-quantify*. The --gridnorm argument enables grid correction, and the following parameter specifies where the grid colonies are located on the plate. "standard1536" refers to the recommended position of two 96 grids in the top left and bottom right corners. In some cases, when your grid strain grows differently than your library, it might not be appropriate to use row/column normalization (*see* **Note 9**) after grid correction. In those cases, simply leave out the --rcmedian parameter. The --qc_plots option tells *pyphe-analyse* to produce a QC report for each plate in the experiment, which will be saved in the specified folder. There are additional options which are explained in the documentation.
3. Once the command has finished, join your layout table onto the produced colony report (in R or python or the programming language of your choice). Alternatively, you can use the --load-layouts option to load strain IDs directly into your colony report.
4. Perform quality control suited to your experiment (*see* **Note 10**). For each plate, check that the footprints are empty and check the coefficient of variation of the additional grid colonies. Discard plates from the analysis as required.

3.1.5 Interpretation

1. Compute summary statistics and p-values comparing each strain to the control strain (repeated across all conditions).

   $ pyphe-interpret --ld pyphe-analyse_data_report.csv --circularity 0.85 --control CONTROL-STRAIN --grouping_column Condition --axis_column Strain

   The --ld option specifies the path to the colony report generated in the previous step. The --circularity option triggers a filter that excludes all colonies with a circularity below the specified value. This is useful in order to exclude image analysis artefacts from further analysis. The --grouping_column argument specifies a column in the colony report for subsetting your data. In each subset, i.e., for each unique entry in the grouping column, a different set of *t*-tests will be performed. The --axis_column argument specifies the explanatory variable of your *t*-test. The --control option specifies your control strain, and this needs to be the string identifier also found in the axis column of your colony report.

2. Compute summary statistics and p-values comparing each condition to the control condition (repeated across all strains). This tests for condition-specific differences in fitness, correcting for basal growth of the strain.

   $ pyphe-interpret --ld pyphe-analyse_data_report.csv --circularity 0.85 --control CONTROL-CONDITION --grouping_column Strain --axis_column Condition

   For example, in a scenario where a slow-growing mutant (relative to the wild-type grid strain) has been measured in a large number of conditions and shows a consistent growth defect (e.g., a fitness of 0.8) in all conditions (including a control condition without stress), the first command would result in a significant p-value for this strain in all conditions as each time it is compared only to the wild type (which has a fitness of 1) in the same condition. This second approach would show no significant difference in fitness as the fitness in each condition is the same as in the control condition.

3. Proceed to analyze your data according to your question. For calling hits, it is usually useful to define an effect size threshold additional to your p-value cut-off.

3.2 Variation 1: Time-Course Analysis

Pyphe implements methods to analyze growth curves of colonies, obtained by automated, regular imaging of plates. Characteristics of the growth curve, usually the maximum slope, are extracted and corrected for plate position effects, similarly as endpoint colony sizes. This method typically achieves lower noise values, but we nonetheless only recommend it for specialized applications as the maximum slopes extracted from growth curves usually correlate tightly with much more easily obtained endpoints [7]. Also *see* **Note 11** for a list of other software resources, some dedicated specifically to colony growth curve analysis.

To record and analyze growth curves, prepare your assay plates as described in Subheading 3.1.1. Then proceed as follows:

1. Immediately after pinning, place the assay plates into the scanner without lids. Take a note of the plate ID corresponding to each position. Close the lid so that it rests horizontally on the plates. If desired, place the scanner in an incubator or temperature-controlled environment.

2. On the scanner computer, run *pyphe-scan-timecourse*.

   $ pyphe-scan-timecourse --nscans NSCANS --interval INTERVAL --postfix POSTFIX --fixture som3

NSCANS is the number of scans in the time-course, INTERVAL is the time between the end of a scan and the start of the next (we recommend 20–30 minutes). After the first scan, check that everything is working as expected and that the first images look ok.

3. After scanning has finished, manually inspect the images. Often, it is necessary to remove some images at the end of the time course, e.g., because colonies started touching each other, contaminations or condensation appeared, or the plates dried out.
4. For every plate separately, extract growth curves using *pyphe-quantify* in time-course mode.

   $ pyphe-quantify timecourse --grid auto_1536

   This will produce a table of growth curves, where individual colonies are the columns and timepoints are the rows.
5. For every plate separately, extract growth curve parameters using *pyphe-growthcurves.*

   $ pyphe-growthcurves --input INPUT --plots

   Input is the path of the csv file produced by *pyphe-quantify.*
6. Proceed with analysis as above. Run *pyphe-analyse* using the --format pyphe-growthcurves option. The Data_path column of your EDT table should contain paths to the files produced by *pyphe-growthcurves. Pyphe-analyse* will perform corrections on the maximum slope of the growth curves but will retain and include other growth curve parameters in the colony report.

3.3 Variation 2: Colony Viability Analysis

Addition of the dead-cell stain phloxine B to the media results in colonies with varying redness which is correlated to the fraction of dead cells in the colony. This can be performed in parallel to the standard workflow, imaging each plate twice: once in grayscale transmission mode (for colony sizes) and once in color reflective mode (for viability analysis). In our hands, viability analysis with phloxine B works better in 384 format and in conditions which are not too stressful, which both result in larger colonies with a stronger signal. Note that phloxine B staining has only been used in a quantitative screening context in *S. pombe*, so some initial validation will be required when applied to other species.

1. Proceed to prepare your assay plates as in Subheading 3.1.1 but include 5 mg/L phloxine B in the media. A 1000× aqueous stock can be stored in the fridge protected from light for several weeks.
2. Grow plates as usual, but protected from light (phloxine B is light sensitive).
3. Scan plates with the white background attached to the scanner lid.

$ pyphe-scan --mode Color --fixture som3-color --postfix POSTFIX --nplates N.

4. Analyze the images using *pyphe-quantify* in redness mode
   $ pyphe-quantify redness --grid auto_1536 --pattern "images/*.jpg".
5. Proceed with the analysis in *pyphe-analyse.* We recommend using only row/column median correction if possible.
   $ pyphe-analyse --edt EDT.csv --format pyphe-quantify-redness --rcmedian --qc_plots analyse-qc.

4 Notes

1. **Power calculations.** At the beginning of the planning phase of an experiment, consider the resolving power required of it to answer your biological question. This will allow you to design cost- and labor-efficient experiments. In the simplest case, you might only be interested in the binary classification of growth/ no growth, e.g., when testing the essentiality of genes in specific contexts. In another setting, the goal might be to characterize growth differences down to a couple of percentage points, e.g., when identifying genetic variants of wild strains which have subtle effects on growth. The number of replicates required to address these two questions differs drastically. You can design your experiment appropriately by calculating the number of required replicates based on the minimal effect size you are trying to detect, the noise of the method (5–10% is a good, conservative estimate in general) and the desired statistical power (chance of correctly rejecting the null hypothesis). This can be done using the stats.power module of the python statsmodels package or the power.t.test function in R, or other online tools. For example, to have an 80% chance of detecting a 5% difference in growth at 5% noise (standardized effect size = 1), with a p-value cutoff of 0.05, you would require 17 samples in each group (two-sided Student's *t*-test). You also need to consider that correcting for multiple testing will in the end decrease the statistical power of your test. In the case of the Bonferroni correction (which is not recommended), you can easily calculate this loss of power by dividing your alpha by the number of tests. For preferred FDR methods, this is not straightforward and requires you to make an educated guess.
2. **Plate layouts**. T-tests generally require the samples to be independent. What this means in practice is often not entirely clear, and it will be up to you to decide the details of your experiment design. In general, we would not consider replicates located next to each other on the same plate as independent as they are subject to the same pinning errors as well as local nutrient,

moisture, and temperature regimes. It is therefore not recommended to generate replicates using 1-to-4 or 1-to-16 multiplexing pinning programs where a single 96 plate is replicated into 384 or 1536 respectively. Instead, replicates should be obtained by recombining library plates in different combinations into the assay plate, mixing up the extract location and neighbors. Colonies grow in competition, and lack of neighbors usually results in increased growth. This is commonly observed as an edge effect where plates on the borders grow bigger. For the same reason, colony screens are sensitive to empty areas of the plate, effectively creating internal edges. If library plates are only partly filled, we recommend filling empty spots with control strains or random strains. This is less important if empty positions are scattered and more so if entire corners, areas, rows, or columns are empty. Despite generally discouraging empty positions, we strongly recommend leaving one individual, unique position per library plate empty. Such footprints serve two important purposes. First, they are negative control positions which should always be empty, if they are not this indicates a source of contamination. Second, they aid identification of specific plates in the case of multiple assay plate layouts. Images will not contain the labeling information written on the side of the plates, so in case of a mix-up, the footprints can provide crucial clues to the identity of the plate.

3. **Layout files.** Preparing the layout of your assay plates is a key task that will require some programming/data processing, at least for larger experiments. This is best done with layouts in long format, i.e., using a table with at least three columns (Row, Column, Strain) where every line of the table describes a single colony position. For every library plate in 96 format, take a note of the position into which it is pinned onto the combined plates. When pinning from 96 to 1536 format, there are 16 of these positions (rows 1–4 and columns 1–4). Let us call the row position p_r and the column position p_c. The row and column positions of a colony on the assay plate ($1 \leq a_r \leq 32$ and $1 \leq a_c \leq 48$) are then related to the positions ($1 \leq s_r \leq 8$ and $1 \leq s_c \leq 12$) on the source plate by:

   $a_r = 4*(s_r - 1) + p_r$

   $a_c = 4*(s_c - 1) + p_c$.

   Using this formula, transform the row and column values for each 96 library plate depending on their target position. Create a layout for the grid plate and include it too. Then, concatenate the tables for all plates and sort by row and column.

4. **Agar plate preparation.** To achieve high data completeness and low technical noise, it is crucial that assay plates are flat,

without bubbles and of suitable dryness. For this, ensure the following:

- Always let the media cool down to approximately 60 °C before pouring plates. Otherwise, the contraction of the agar during cooling will result in unwanted ripples on the surface. When potentially temperature-sensitive drugs are to be included in the assay plates, add these when the media has cooled down and right before pouring to minimize the exposure to high temperatures.
- Always pour plates on a level and even surface. In our experience, this can be done on a standard lab bench without a sterile environment, as long as the plate lid is immediately replaced after pouring.
- Always add a consistent amount (we use 40 mL) of agar medium to each plate. This will result in plates with a consistent height and also avoid other artefacts. Thicker plates mean more nutrients are available to each colony which will change the colony size. Use a serological 50-mL pipette and take up 5 mL more than required. This will prevent bubbles. If you spot any bubbles, suck them back up with the pipette.
- Plates should be dried for a consistent time (we use 45 min without lids) before use. Alternatively, closed plates can be left unwrapped on the bench overnight which in our experience results in plates with a suitable dryness for immediate use. Plates can be stored in the fridge/cold room if not used immediately but will require extra drying before use.

5. **Preparation of the EDT table.** Preparing a correct and complete experimental design table (EDT) is a key requisite for obtaining results with *pyphe-analyse*. All plates in your experiment should have a unique ID, and this should be clearly written on the edge of the plate (not on the bottom where it would show up in the scanned images). For example, plate IDs could follow the format date_layout_condition_replicate. While scanning, take a note of which plate ID corresponds to which image in the scan sequence in table format. This table can then later easily be transformed into the EDT required by *pyphe* by adding extra columns with additional meta-data. The final table must be in csv format, and the first column must contain the unique plate IDs. There must also be a column called "Data_path" which points to the image data file produced by *pyphe-quantify*. Any additional meta-data, such as date, condition, library versions, comments, batch information, can be included and will be carried through to the colony report. See the Documentation folder in the *pyphe* github repository for an example file. Note that there is no need for

all the data files to be located in the same folder, which is convenient if you have large experiments containing several batches. Generally, file paths should not contain spaces or non-standard characters or those with special meanings in the terminal (%,>,?,/,",*,&, etc.). Use _ or . or - to separate words.

6. **Quality control during pinning.** Even with precise robots, the transfer of colonies by pinning can be prone to errors which, if unspotted, will result in missing or wrong data. A common problem that occurs when target plates are uneven is that entire areas or corners of the plate have no colony inoculums. Such pinning errors are dangerous as they could result in colonies which are absent for technical reasons to be interpreted as unviable phenotypes. *Pyphe* helps spot these errors by detecting missing control colonies and setting all neighboring colonies to NA, but this only occurs after the damage is done. We therefore strongly advise to check every plate for missing corners and correct pinning by eye. Have one or two spare plates at hand to repeat transfers that have failed. If the problem persists, it can help to increase the target plate pinning pressure. With the Singer RoToR, it helps to avoid pinning errors if plates are consistently placed and pushed into the same corner in their holder.
7. **Use of command line interface.** To use *pyphe*, you need to be familiar with some standard characteristics of command line programs, especially the concept of the working directory and relative file paths. In the terminal, navigate to the base directory of your experiment and run *pyphe* commands from there. File paths need to be defined in relation to that directory. Commands in this protocol are marked with $, indicating that this is a line of code to be run in the terminal (the $ is not part of the command). Help for each *pyphe* tool is available by running the tool with the --help parameter (e.g., $ pyphe-analyse --help).
8. **Quality control during image analysis.** *Pyphe-analyse* produces a qc image for every image analyzed. It is crucial to look at the images, even for huge datasets, to make sure colonies have been correctly identified and correctly matched to their grid positions. Issues with colony detection can usually be remedied by adjusting the threshold parameter (using the --t parameter), setting a hard threshold (using the --hardImageThreshold parameter) or using local thresholding (activated with the --localThresh parameter). The last is especially recommended for images with uneven brightness, e.g., those obtained with the Singer Phenobox. By default, *pyphe-analyse* excludes very small objects. If your colonies are very small, adjust this exclusion parameter using the --s or --hardSizeThreshold parameters. For gridding issues, it can be worth to switch from automatic grid placement to manual one. See the *pyphe-analyse* documentation.

9. **Normalization strategies.** *Pyphe-analyse* gives the user several options for normalization strategies. If neither the --rcmedian nor --gridnorm options are set, no normalization is performed and the raw data from the plates is simply aggregated and summarized. For screens without a reference grid, *pyphe* can still be used (with row/column normalization only or no normalization). However, lowest noise values are obtained if both options are set. In that case, grid normalization is performed followed by row/median column normalization. This second normalization can correct an artefact of the grid normalization method which slightly over-corrects colonies next to the edge. This is essentially because colonies just off the edge are compared to colonies on the edge and therefore appear relatively smaller (see Fig. 1, figure supplement 2B and Appendix 2 of ref. [7]).

   However, row/column median normalization can and should only be used if the majority of strains in each row and column show no effect (i.e., the null effect can be reliably estimated by taking the median). This is usually the case for library screens where most of the mutants behave like wild-type, and there is only a few outlying "hits." For wild strain libraries, this case is harder to argue, but it could still work if your strains are arranged randomly. It certainly will not work if your grid strain grows differently to the rest of your library (because only some rows/columns have a lot of replicates of the grid strain). In those cases, performing row/column normalization after grid normalization will do more damage than good. *Pyphe-analyse* produces qc plots for every plate analyzed, and you should inspect these carefully to check that the normalization is working as expected.

10. **Noise statistics for quality control.** Biological noise, technical noise, and experimental errors (incorrect plate preparation, mis-labeling, pinning errors) will impact your data and can result in wrong conclusions if they go undetected. We therefore highly recommend to perform extensive quality control to quantify the unexplained variation and spot experimental errors. The use of negative-control positions (footprints) is key and plates with contaminated footprints should be discarded from the analysis. Furthermore, we include a number of control strains on every assay plate. One easy way to achieve this is to include an additional 96 grid of control strains on each assay plate. These colonies are not used during reference grid correction but are expected to have fitness values of 1. Based on these internal controls, it is possible to calculate two key noise indicators: First, the coefficient of variation (CV), the ratio of the standard deviation of the corrected fitness of these control colonies to their mean is an excellent indicator of the level of

noise present in the assay. We usually exclude all plates which exceed a certain CV threshold (e.g., 10%). Second, the fraction of unexplained variance (FUV) is the ratio of the variance of the control strains and the variance of all strains. That is, an FUV of 1 indicates that the spread of values is equally broad in the control as it is across the library, which can indicate that the observed variation across the library strains is purely technical. A suitable cutoff for exclusion of individual plates and conditions will depend strongly on your library. Certainly, an FUV of greater than 1 would be highly unusual and deserved further investigation. It can also be of value to exclude plate and conditions which show very small uncorrected colony sizes. This would indicate that the stressor included was too strong or the nutrients did not support any growth. In these cases, grid correction can introduce artefacts as small colony sizes are extremely noisy. These QC steps differ greatly between experiments, so they need to be performed manually on the *pyphe-analyse* long data report, removing spurious lines or setting them to NA. Once completed, you should proceed with hit calling using *pyphe-interpret*.

11. **Alternative software solutions.** Scan-o-matic [9] is a sophisticated platform for scanning, image analysis, and spatial normalization. Scan-o-matic pioneered spatial correction with reference grids and additionally supports calibrating pixel intensities to cell numbers, enabling exact quantification of population sizes. Scan-o-matic uses time-course imaging and requires a specialized hardware setup, including pixel calibration strips, modifications to scanners, and a local area network.

    Gitter [14] is an R package for determining colony sizes from images. It can work with a range of input image types and has robust algorithms for thresholding and grid assignment. *Pyphe* can work with image quantification data from gitter. Unfortunately, gitter is currently archived by CRAN as it requires outdated packages. For experts, it is possible to install it from the archive but this requires manual installation of several dependencies.

    SGAtools [8] provides tools for spatial normalization (not including reference grid correction) and statistical analysis of colony screen data. Available as a web service without installation.

    Colonyzer [15] has been developed for the Quantitative Fitness analysis workflow [11]. Colonyzer has been designed to work on colonies which have been stamped on plates with manual replicators, resulting in a spot (i.e., a spread-out inoculum).

    IRIS [16] is an advanced image analysis tool for single or time-course images, specializing in detecting additional colony morphology features such as bacterial biofilm formation.

CellProfiler [17] is a powerful, multi-purpose image analysis tool with which one can assemble analysis pipelines to count and measure colony sizes as well as potentially other morphological parameters.

Acknowledgments

We would like to thank Natalie Barthel for critical reading of the manuscript. The study was supported by the Francis Crick Institute, which receives its core funding from Cancer Research UK (FC001134), the UK Medical Research Council (FC001134), and the Wellcome Trust (FC001134), and received specific funding from the Wellcome Trust (200829/Z/16/Z) and the ERANET COBIOTEC network "SyCoLim" under grant agreement BMBF-031B0931.

References

1. Costanzo M, VanderSluis B, Koch EN et al (2016) A global genetic interaction network maps a wiring diagram of cellular function. Science 353(6306):aaf1420
2. Bloom JS, Ehrenreich IM, Loo WT et al (2013) Finding the sources of missing heritability in a yeast cross. Nature 494:234–237
3. Märtens K, Hallin J, Warringer J et al (2016) Predicting quantitative traits from genome and phenome with near perfect accuracy. Nat Commun 7:11512
4. Jeffares DC, Rallis C, Rieux A et al (2015) The genomic and phenotypic diversity of Schizosaccharomyces pombe. Nat Genet 47:235–241
5. Peter J, De Chiara M, Friedrich A et al (2018) Genome evolution across 1,011 Saccharomyces cerevisiae isolates. Nature 556:339–344
6. Kamrad S, Grossbach J, Rodríguez-López M et al (2020) Pyruvate kinase variant of fission yeast tunes carbon metabolism, cell regulation, growth and stress resistance. Mol Syst Biol 16: e9270
7. Kamrad S, Rodríguez-López M, Cotobal C et al (2020) Pyphe, a python toolbox for assessing microbial growth and cell viability in high-throughput colony screens. elife 9:e55160
8. Wagih O, Usaj M, Baryshnikova A et al (2013) SGAtools: one-stop analysis and visualization of array-based genetic interaction screens. Nucleic Acids Res 41:W591–W596
9. Zackrisson M, Hallin J, Ottosson L-G, et al (2016) Scan-o-matic: high-resolution microbial phenomics at a massive scale. G3 6:3003–3014
10. Levin-Reisman I, Fridman O, Balaban NQ (2014) ScanLag: high-throughput quantification of colony growth and lag time. J Vis Exp 89:51456
11. Banks AP, Lawless C, Lydall DA (2012) A quantitative fitness analysis workflow. J Vis Exp 66:4018
12. Takeuchi R, Tamura T, Nakayashiki T et al (2014) Colony-live—a high-throughput method for measuring microbial colony growth kinetics—reveals diverse growth effects of gene knockouts in Escherichia coli. BMC Microbiol 14:171
13. Lie S, Banks P, Lawless C et al (2018) The contribution of non-essential Schizosaccharomyces pombe genes to fitness in response to altered nutrient supply and target of rapamycin activity. Open Biol 8(5):180015
14. Wagih O, Parts L (2014) Gitter: a robust and accurate method for quantification of colony sizes from plate images. G3 4:547–552
15. Lawless C, Wilkinson DJ, Young A et al (2010) Colonyzer: automated quantification of micro-organism growth characteristics on solid agar. BMC Bioinformatics 11:287
16. Kritikos G, Banzhaf M, Herrera-Dominguez L et al (2017) A tool named Iris for versatile high-throughput phenotyping in microorganisms. Nat Microbiol 2:17014
17. Lamprecht MR, Sabatini DM, Carpenter AE (2007) CellProfiler: free, versatile software for automated biological image analysis. BioTechniques 42:71–75

Chapter 22

Bulk-Fitness Measurements Using Barcode Sequencing Analysis in Yeast

Claire A. Chochinov and Alex N. Nguyen Ba

Abstract

The use of DNA barcodes for determining changes in genotype frequencies has been instrumental to increase the scale at which we can phenotype strain libraries by using next-generation sequencing technologies. Here, we describe the determination of strain fitness for thousands of yeast strains simultaneously in a single assay using recent innovations that increase the precision of these measurements, such as the inclusion of unique-molecular identifiers (UMIs) and purification by solid-phase reverse immobilization (SPRI) beads.

Key words DNA barcodes, Bulk fitness assays, Strain fitness determination, Unique molecular identifiers

1 Introduction

Characterizing strain fitness is a crucial aspect of genetic studies and evolution. Methods to quantify relative fitness have traditionally relied on crude measurements such as side-by-side spot dilution assays [1] and growth-curve fitting [2]. More recently, methods involving the mixing of the two strains of interest [3] (competitive fitness assays) have become popular for their ease and flexibility. These techniques rely on measuring the relative change in frequency of the strains over time by using benign markers such as fluorescence [4] which can accurately track the frequency of each strain by flow cytometry. Although these methods all produce similar results [5], the advantage of competitive fitness assays is that it is amenable to parallelization using DNA barcoding technologies [6, 7]. By transforming strains of interests with random nucleotide barcodes, a *bulk* competitive fitness assay can be performed by DNA sequencing (Fig. 1). This procedure is inexpensive

Supplementary Information The online version contains supplementary material available at [https://doi.org/10.1007/978-1-0716-2257-5_22].

Frédéric Devaux (ed.), *Yeast Functional Genomics: Methods and Protocols*,
Methods in Molecular Biology, vol. 2477, https://doi.org/10.1007/978-1-0716-2257-5_22,

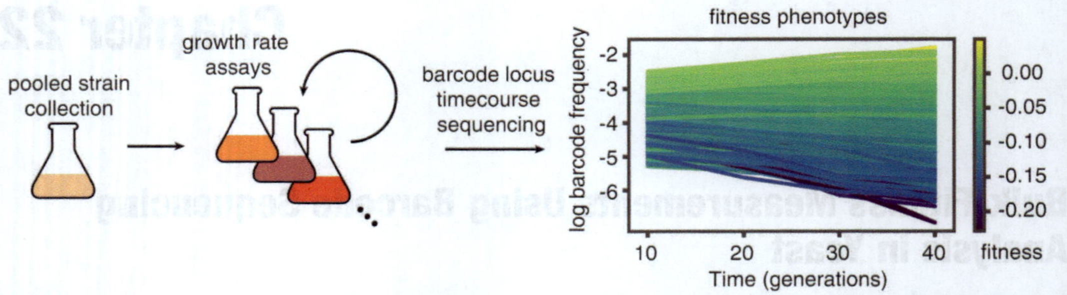

Fig. 1 Schematic of the experiment. Pooled barcoded strains are grown in bulk in a desired environment and passaged for a few generations by repeated serial dilution. Barcode locus sequencing at each passage is used to obtain strain frequencies and infer fitness

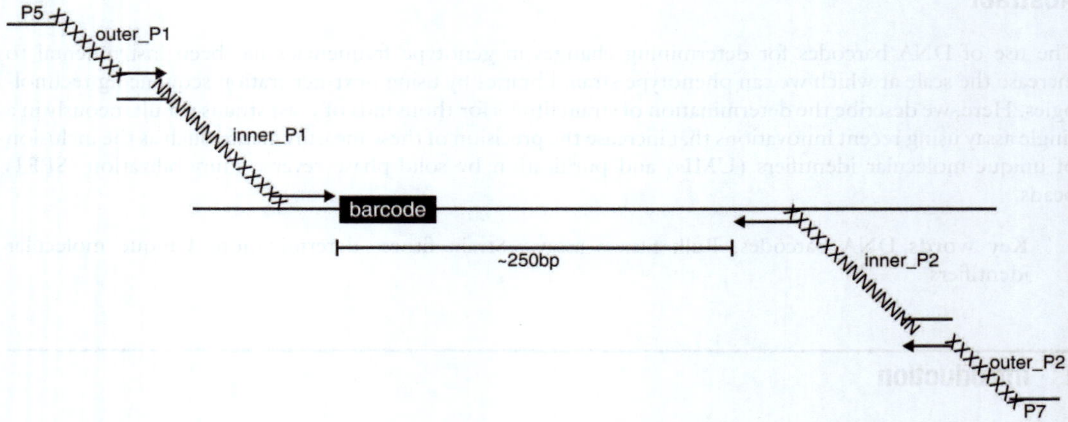

Fig. 2 Schematic of the amplification reaction. Two-stage PCR is used to amplify the barcodes from pooled strains. The first stage, using inner_P1 and inner_P2 primers is a limited cycling reaction, which adds unique-molecular identifiers (UMIs) to amplification products. Following purification by solid-phase reversible immobilization (SPRI beads), a second amplification using outer_P1 and outer_P2 is used to attach Illumina adapter sequences

(at the time of writing, the cost of these pooled competitive fitness measurements can be as low as 0.2 cent per strain, per environment) and efficient for large panels of strains: we routinely perform these assays for >100,000 strains simultaneously in multiple conditions with replicates, with each set of assays taking at most 1 h per day over a week. At this scale, we obtain measurements with errors in competitive fitness of approximately 0.1–1%, which is similar to what has been observed in pairwise assays using fluorescence.

We describe here an experimental protocol and analysis for these bulk competitive fitness assays using barcoded yeast, which we chose due to its long history in systematic genetics studies [8]. We incorporate several innovations to obtain more accurate measurements, including a refinement on the enzymes and reagents used for the procedure and the introduction of unique-molecular identifiers to prevent PCR exponential noise [9] (Fig. 2). This protocol can be easily adapted to different species, by changing the growth media and genomic DNA extraction protocol. The experimental scale can also be modified, for example, by performing these growth assays and genomic extractions in microtiter plates. Finally, any phenotype that can be measured by the relative change in frequency of the strains of interest can be assayed using this method.

2 Materials

Prepare all solutions using molecular biology grade water. All reagents can be stored at room temperature unless noted.

2.1 Yeast Genomic DNA Extraction Using Silica Mini-Preparative Columns

1. Yeast lysis buffer: 1 M Sorbitol, 100 mM Sodium Phosphate buffer, pH 7.4, 10 mM EDTA, 0.5% SB3-14 (3-(*N*,*N*-dimethylmyristylammonio)propanesulfonate), 200 μg/mL pancreatic Rnase A, and 20 mM DTT, and 5 mg/mL Zymolyase 20T. Made in 15-mL falcon tube aliquots by mixing the non-enzymatic components first, and stored in a −20 °C freezer after mixing. The reagent is thawed at room temperature and mixed by vortex before use (*see* **Notes 1–3**).
2. Binding buffer: 100 mM MES, pH 5, 4.125 M Guanidine thiocyanate, 25% isopropanol, 10 mM EDTA (*see* **Notes 4** and **5**).
3. Wash buffer 1: 0.85 M Guanidine thiocyanate, 25% isopropanol, 10 mM EDTA.
4. Wash buffer 2: 80% Ethanol, 10 mM Tris–HCl, pH 8 (*see* **Note 6**).
5. Elution buffer: 10 mM Tris–HCl, pH 8.5.
6. Silica mini-preparative columns and collection tubes (*see* **Note 7**).

2.2 Two-Step PCR

1. Q5 polymerase kit: containing 5× Q5 polymerase buffer, and Q5 polymerase enzyme. Stored at −20 °C, thawed on ice before use.
2. KAPA HiFi PCR kit: containing 5× KAPA HiFi polymerase buffer, 10 mM dNTP mix, and KAPA HiFi polymerase enzyme. Stored at −20 °C, thawed on ice before use (*see* **Note 8**).
3. 10 mM dNTP mixture. Stored at −20 °C, thawed on ice before use.
4. AMPure XP beads or equivalent. Stored at 4 °C, warmed to room temperature, and vortexed before use.

Table 1
List of primers

Primers for first-stage PCR (containing UMIs and optionally an index)	
Inner_P1	TCGTCGGCAGCGTCAGATGTGTATAAGAGACAG NNNNNNNN XXXXXX YYYYYYYYYYYYYYYYYYYY
Inner_P2	GTCTCGTGGGCTCGGAGATGTGTATAAGAGACAG NNNNNNNN XXXXXX YYYYYYYYYYYYYYYYYYYY
Primers for second-stage PCR (containing optionally an index)	
Outer_P1 (with Illumina P5)	AATGATACGGCGACCACCGAGATCTACAC XXXXXXXX TCGTCGGCAGCGTC
Outer_P2 (with Illumina P7)	CAAGCAGAAGACGGCATACGAGAT XXXXXXXX GTCTCGTGGGCTCGG

Y: nucleotides representing the priming sequence
X: nucleotides representing an optional index
N: nucleotides representing random unique molecular identifiers

5. Two sets of primers at 10 μM: one priming to the genome flanking the barcode locus and containing the unique-molecular identifiers, a DNA sequence handle and optionally a multiplexing index, and another set priming on the DNA sequence handles that also contain the Illumina adapter sequences that bind to the flowcell, and optionally a multiplexing index. The unique-molecular identifiers should be random nucleotides of equal base frequency and the total length should be about 16 bp (or 8 bp on each primers). We recommend multiplexing indexes of different lengths (6–11 bp), but many options are possible (*see* Table 1 and **Note 9**).
6. AMPure XP beads washing buffer: 85% Ethanol, 10 mM Tris–HCl, pH 8.

2.3 Growth Media

1. YPD liquid media: 20 g/L peptone, 10 g/L yeast extract, 20 g/L glucose. Autoclave first the peptone and yeast extract mix in 90% of the final desired volume and then add sterile glucose concentrated solution and sterile water up to the final volume.
2. Growth environment media: media should be prepared similarly to YPD and be free of particulate (*see* **Note 10**).

2.4 Softwares

1. Python version 3.6 or more recent.
2. Python module Regex [10].
3. gcc version 6 or more recent.
4. tsl robin_map hash table package [11].

3 Methods

3.1 Growth Assays

1. Barcoded yeast cells form a population that can be used to inoculate fresh YPD media to define a start culture. In general, the number of cells in YPD at saturation is approximately 10^8 cells/mL, and this number is used as a reference for seeding this start culture. Inoculate at least 1000 times more cells than the number of barcodes in the populations and grow for about 5–7 generations. For example, if the number of genotypes whose fitness is being measured is 10^4, then ten million cells (100 μL) can be inoculated into approximately 5 mL of YPD to be grown for 24–48 h on a rotating culture drum or on a platform shaker. Larger assays can be scaled appropriately (*see* **Notes 11** and **12**).
2. When yeasts from the start culture are fully saturated, passage at least 1000 times more cells than the number of barcodes to the growth media of interest (e.g., a stressful growth media) into two parallel cultures to produce technical replicates. The inoculation volume to final volume should be at a fraction of 1/32 to 1/1024, and grown for 24 or 48 h (*see* **Notes 13–15**).
3. Continue passaging the culture at least twice (and up to ten times) for no more than 50 total doublings and aliquot a portion (we usually do at least the inoculation volume, but 1.5 mL is typical) of the saturated culture for DNA extraction at each re-inoculation. Pellet the cells to remove the supernatant. The pellet can be processed immediately for DNA extraction (*see* Subheading 3.2) or stored at −20 °C (*see* **Notes 16** and **17**).

3.2 Genomic Extractions of Yeasts

1. To the cell pellet, add 100–200 μL of yeast lysis buffer. Generally, 100 μL of yeast lysis buffer is sufficient to lyse over 10^8 yeast cells. Vortex well to resuspend, and place at 37 °C with mild shaking for at least 30 min and up to 2 h.
2. Add four volumes of binding buffer to the lysed cells (if 100 μL of yeast lysis buffer was used, add 400 μL of binding buffer) and vortex extensively. The lysed cell solution should clarify significantly indicating DNA release. If some cells remain un-lysed, centrifuge at 16,000 × *g* for 1 min to pellet them.
3. Pipet the clarified solution to a silica mini-preparative column fitted in a collection tube and centrifuge at 16,000 × *g* for 1 min. Discard the flowthrough.
4. Pipet 400 μL of wash buffer 1 onto the silica mini-preparative column and centrifuge at 16,000 × *g* for 1 min. Discard the flowthrough.

5. Pipet 600 μL of wash buffer 2 onto the silica mini-preparative column and centrifuge at 16,000 × *g* for 1 min. Discard the flowthrough.
6. Centrifuge at 16,000 × *g* for 3 min to dry the column.
7. Place the mini-preparative column into a clean 1.5-mL microcentrifuge tube, add 50 μL elution buffer to the center of the column and centrifuge at 10,000 × *g* for 2 min. Discard the column and store the eluted DNA, which is typically at 20 ng/μL, at −20 °C until further use.

3.3 Two-Step PCR Amplification of Barcodes

1. Prepare a 20 μL PCR reaction using the Q5 polymerase: 10 μL of genomic DNA (approximately 200 ng of DNA, or the DNA belonging to about 10^7 yeast cells, *see* **Note 18**), 4 μL 5× Q5 polymerase buffer, 0.4 μL 10 mM dNTP mixture, 1 μL each of primers priming around the barcode and containing the UMIs, 0.2 μL Q5 polymerase and water to 20 μL. Cycle as follows:
   (a) 98 °C for 1 min.
   (b) 98 °C for 10 s.
   (c) 50 °C for 10 s (or 5 °C lower than the T_m of the primers).
   (d) 72 °C for 30 s.
   (e) Repeat **steps 2–4**, twice (*see* **Note 19**).
   (f) 72 °C for 1 min.
   (g) 4 °C indefinitely.
2. Use AMPure XP beads to purify the reaction at a 1.25× bead ratio (*see* **Notes 20** and **21**):
   (a) Add 25 μL AMPure XP beads and vortex well (25 μL = 1.25 × 20 μL).
   (b) Place PCR tube on a magnet to collect the beads.
   (c) Pipette out the supernatant.
   (d) Add 180 μL AMPure XP beads washing buffer. The beads can be washed by taking the tube off the magnet, rotating the tube, and replacing it back on the magnet.
   (e) Pipette out the washing buffer and let dry at room temperature for about 2 min. Ensure that there is no remaining washing buffer at the bottom of the PCR tube.
   (f) Add 33 μL water to the beads and vortex well.
   (g) Place the PCR tube on a magnet to collect the beads.
   (h) Pipette the supernatant to a fresh PCR tube and store at 4 °C if not proceeding to the next step immediately (*see* **Note 22**).
3. Prepare a 50 μL PCR reaction using the KAPA polymerase: 33 μL of the AMPure XP cleaned reaction made in **step 2** in

this section, 10 μL 5× KAPA HiFi polymerase buffer, 1 μL 10 mM dNTP mixture, 2.5 μL each of primers containing the Illumina adapters binding to the flowcell and 1 μL of the KAPA HiFi enzyme. Cycle as follows:

(a) 98 °C for 1 min.

(b) 98 °C for 10 s.

(c) 62 °C for 10 s (or the T_m of the primers).

(d) 72 °C for 30 s.

(e) Repeat **steps 2–4**, 34 times.

(f) 72 °C for 2 min.

(g) 4 °C indefinitely.

4. Use AMPure XP beads to purify the reaction at a 0.8× bead ratio (*see* **step 2** in this section).
5. The cleaned PCR should be run on an agarose gel and visualized for successful reaction (*see* Fig. 3 and **Note 23**). It is ready to be quantified, pooled, and sequenced with an Illumina sequencing platform (*see* **Notes 24** and **25**).

3.4 Parsing Sequencing Reads into Barcodes

1. Parse the demultiplexed sequencing reads using the python fuzzy regex package [10] (*see* **Note 26**). The following python functions may be used as example for the described protocol. Prepare a python script resembling the following (adapting the barcode and UMI lengths, indexes, priming sites, and downstream sequences for your needs). This script will parse the barcode, the UMI, and the optional inline index for the first read and second read.

```
import regex

inner_indexes =
["XXXXXX","XXXXXXX","XXXXXXXX","XXXXXXXXX","XXXXXXXXXX","XXX-
XXXXXXXX","XXXXXX","XXXXXXX","XXXXXXXX","XXXXXXXXX","XXXXXXX-
XXX","XXXXXXXXXXX"]

inner_indexes_2 =
["XXXXXX","XXXXXXX","XXXXXXXX","XXXXXXXXX","XXXXXXXXXX","XXX-
XXXXXXXX","XXXXXX","XXXXXXX","XXXXXXXX","XXXXXXXXX","XXXXXXX-
XXX","XXXXXXXXXXX"]

def OR(inner_indexes):
 return "("+"|".join(["("+index+")" for index in inner_in-
dexes])+")"
UMI = "(.{8})"
barcode = "(.{16})"
inner_P1_priming_site = "YYYYYYYYYYYYYYYYYYY"
```

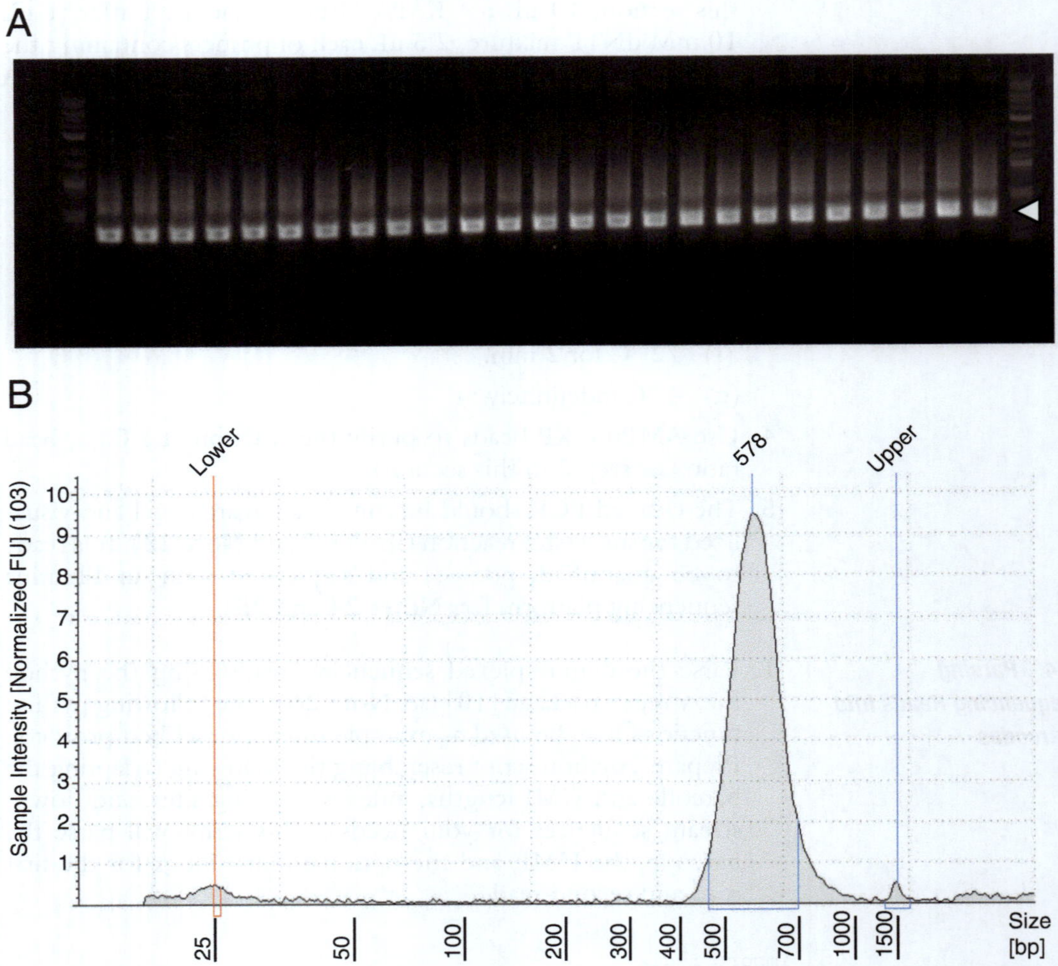

Fig. 3 Successful reactions. (**a**) Agarose electrophoresis of the final two-stage PCR reaction for 25 parallel reactions. The white triangle indicates the desired band, with expected fragment size of approximately 290 bp. 1 kb Plus DNA ladder from New-England Biolabs in first and last lane, with major bands representing 500 bp, 1000 bp, and 3000 bp. A faint smear or band at two times the fragment size can also be observed. (**b**) Agilent TapeStation analysis of the pooled two-stage PCR reaction from **a**. Measured fragment size by the TapeStation frequently shows a peak of two times the expected size

```
inner_P2_priming_site = "YYYYYYYYYYYYYYYYYYY"
sequence_downstream = "ZZZZZZZZZZZZZZZ"

barcode_locus = (
 UMI +
 OR(inner_indexes[0:12]) +
 inner_P1_priming_site + "{e<=3}" +
 barcode +
 sequence_downstream + "{e<=3}")
```

```
barcode_regex_object = regex.compile(barcode_locus)
inner_p2_regex = ( UMI + OR(inner_indexes_2[0:12]) + in-
ner_P2_priming_site + "{e<=3}")
inner_p2_regex_object = regex.compile(inner_p2_regex)
for line in sys.stdin:
 reads = line.split("\t")
 m = barcode_regex_object.match(reads[0])
 m2 = inner_p2_regex_object.match(reads[1])
 if m and m2:
  UMI_val = m.groups()[0] + m.groups2()[0]
  index_val = m.groups()[1] + m.groups2()[1]
  bc_val = m.groups()[2]
   print(str(UMI_val) + " " + str(index_val) + " " + str
(bc_val))
```

2. Run the above script using the following commands to parse one pair of demultiplexed read files:

```
paste <(gunzip -c reads_illumina_1.R1.gz) <(gunzip -c read-
s_illumina_1.R2.gz) | python above_script.py >init_parsed_1.
txt
```

3. If using inline indexes, it may be useful here to demultiplex the files further. This command will only output reads corresponding to the desired index (replacing the XXXXXX with the indexes of choice).

```
grep -P "\tXXXXXX\t" init_parsed_1.txt >parsed_1.txt
```

4. Once barcodes are parsed, they should be "error-corrected." When the list of barcodes is known (there is a dictionary), such as in the yeast deletion collection, correction of barcodes is done by using the lowest Levenshtein distance to the dictionary up to two mismatches. This can be done efficiently using a software package such as FLAMINGO [12] that uses N-grams or error-correcting tries such as used in short-read mapping strategies [13]. We provide a code in the Supplementary information that can efficiently perform this operation. Compile the software and prerequisites:

```
g++ error_correct_from_dict.cpp -o error_correct_from_dict -O3
-std=c++0x -I$HOME/local/include -lpthread -fopenmp -D_GLIB-
CXX_PARALLEL
```

This software has the prerequisite tsl robin_map freely available [11] which should be unpacked in your $HOME/local/include directory.

5. Run the software, which will error correct by default at two Levenshtein distance and discard reads that do not map to known barcodes (*see* **Notes 27** and **28**).

```
error_correct_from_dict -bc 3 -dict dictionary.txt parsed_1.
txt >corrected_1.txt
```

The -bc flag specifies the column in the parsed_1.txt file that contains the barcode of interest, and the dictionary.txt file contains a single column of known barcodes.

6. After error correction, barcode counts are obtained by counting their distinct unique molecular identifiers. This can be performed efficiently by keeping unique lines:

```
sort -u corrected_1.txt >unique_1.txt
```

7. The above process should be repeated for each time-point and each replicate assay. Counts can be obtained by simply counting each barcode for each time-point, and frequencies can be obtained by dividing the counts by the total counts for each time-point. A function such as this one is useful as it will provide counts based on the barcode field only (replacing the cut flag with the proper column):

```
cut -f3 unique_1.txt | sort | uniq -c | awk '{printf("%s\t%s
\n",$2,$1)}' >counts_1.txt
```

3.5 Fitness Inference

1. Following this parsing step, the data useful for fitness inference is the number of reads for each barcode, at each time-point, and for each replicate assays. We define the relative selection coefficients, *s*, based on the deterministic continuous model for logistic growth of alleles against each other with no frequency dependence. The general formula for the change in frequency of an allele over time:

$$\frac{\mathrm{d}f_i}{\mathrm{d}t} = f_i(s_i - \overline{s}).$$

Where we define:

$$\overline{s} = \sum_i f_i(t) s_i$$

as the population mean fitness at time t.

If we define $f_i(0)$ as the initial frequency of lineage i, the solution to the above equations is:

$$f_i(t) = \frac{f_i(0)e^{s_i t}}{\sum_j f_j(0)e^{s_j t}}$$

The fitness for every lineage is obtained by maximizing the likelihood of observed count data:

$$likelihood = C * \Pi\, f_i(t)^{n_i(t)}$$

where C is a constant factor, and $n_i(t)$ is the number of reads for lineage i at time t.

For two time-points, the relative fitness of a strain (s_i) compared to a reference strain ($s_j = 0$) is the change in logarithmic ratio of frequencies over time and is solved as:

$$s_i = \frac{\log\left(f_i(t_2)/f_j(t_2) \right) - \log\left(f_i(t_1)/f_j(t_1) \right)}{t_2 - t_1}.$$

For more than two time-points, the solution for multiple alleles is a joint-inference problem and numerical methods to maximize the likelihood of the observed count data can be used. We provide here a program that will do the full numerical solve, (*see* **Notes 29–31**) for a discussion on faster approximations that are roughly equivalent to each other and are useful when the number of reads is sufficiently large as in described in this protocol. A correlation over 0.99 between these approximations and a full numerical solve is typical when barcodes have at least 20 reads per time-point (*see* **Note 32**).

First, assemble counts into a tab-delimited text file with the following format:

```
BC 10 20 30 40 50
NNNNNNNNNNNNNNNNN 1534 1079 602 346 101
NNNNNNNNNNNNNNNNN 1061 1573 2200 3044 3509
```

which consists of a first header row which describes the following fields. There should be more than two time-points, and the header row should have integer values labeling the generations. The following rows should have the barcode (BC) as the first field, and integer values for the counts at each time-point. Ensure that all rows have the same number of fields. Ensure that barcodes are unique and do not appear multiple times throughout the file. Remove barcodes that have only a single time-point with non-zero counts: their fitness cannot be estimated.

2. Compile the program as follows, which has the same prerequisite as the error_correct script:

```
g++ -std=c++11 -O3 -I$HOME/local/include BFA_solve.cpp -o
BFA_solve -fopenmp
```

3. Run the software, including both the replicates of the same assay (replicates are not required and fitness values can be estimated independently, but estimation is better with replicates for lower frequency strains):

```
BFA_solve arranged_counts_1.txt arranged_counts_1_rep.txt
>fitness_1.txt
```

4. The software describes the following row header:

```
Lineage s stderr(s) f0_array
```

Following rows have the barcode, the fitness, the estimated error on this fitness, and the estimated starting frequencies of the barcode (when running the software with replicates). A reference strain barcode is chosen based on the highest read counts. To adjust all fitness values based on a wanted reference barcode, subtract the fitness values of all strains by the fitness of the reference barcode.

Some statistics are also given by the software, which describe the fit of the data to the model.

The software may take a long time to converge depending on the count values and the number of strains to estimate. We suggest using the noted approximations in cases where the software takes an unacceptably long time to converge.

4 Notes

1. The yeast lysis buffer can be made using 1 mg/mL of Zymolyase 100 T instead, but this reagent has been found to be more expensive.
2. The detergent (SB3-14) can be made as a 10% stock solution and dissolved with the help of moderate heat. Avoid mixing too vigorously as bubbles easily form.
3. The enzymatic solution is cloudy when thawing and turns into a brown, transparent liquid above approximately 10 °C. Using the lysis buffer when cloudy does not harm the reaction. The lysis buffer is stable upon multiple freeze-thaws.
4. The binding buffer may be more easily made from a concentrated solution of 6 M guanidine thiocyanate. Stir on a warm plate when dissolving as dissolving guanidine thiocyanate is endothermic.
5. It is important that the final pH of the solution be about 5 as DNA binding to silica preparative column is more efficient in acidic conditions.

6. The concentration of ethanol should remain above 75%, so it is important to seal the bottle containing the washing buffers accordingly.
7. Experience has shown that mini-preparative silica columns of different brands or of different types have different binding properties, such as binding capacities and fragment length retention, but that this protocol will yield sufficient high-quality DNA in all instances. Use columns that have specifications for large DNA fragments when possible. Alternatively, magnetic silica beads can also be used.
8. The two different polymerases used here have been optimized for sensitivity and library yield.
9. We use 12 inline inner indexes, which in combination with the Illumina outer indexes facilitates multiplexing large number of assays. The inline indexes should be of different length to prevent homogenous base compositions for each Illumina sequencing cycle.
10. We usually add ampicillin at 100 μg/mL to reduce the occurrence of bacterial contamination.
11. Barcodes should ideally uniquely represent one strain or genotype.
12. The barcoded yeast population can be maintained indefinitely in 5–20% glycerol at −80 °C, thawed at room temperature and vortexed before inoculation.
13. Any media that can sustain yeast growth can be used here, but for convenience it is recommended that the doubling time of the average yeast population in this media be no more than 8 h. In all cases, it is best to calibrate the media such that yeast do not spend more than 12 h at saturation as cell lysis is greatly reduced on saturated cultures.
14. The number of cells passaged per barcode (N) dictates the effect of genetic drift on fitness measurements [14]. To measure relative fitness with average errors in the order of 1% requires at least 100 cells passaged per barcode (and preferably much more). Increasing this number of cells passaged does not however decrease this error bound indefinitely as errors such as media batch variation, new mutations occurring in the assay, and inaccuracies in sampling time tend to dominate. In practice, we hardly obtain measurements more precise than 0.1%.
15. The inoculation ratio determines the number of doublings (or generations) until saturation, which can be determined by the number of doublings to carrying capacity. For example, an inoculation fraction of $1/2^5$ leads to 5 generations per day. The desired ratio is dependent on the fitness variance of the

population, with a higher relative volume of inoculum to fresh media being useful for large fitness variances, and a lower relative volume being convenient for small fitness variances, which can be determined empirically. Large fitness variance leads to higher rate of adaptation (the mean population fitness increases faster), which greatly reduces the frequency of several low-fitness lineages, and, as explained in **Note 12**, the error in fitness measurement is directly related to the number of cells passaged. Low fitness variance, on the other hand, only leads to deterministic changes in lineage frequencies over long periods of time. Experience has shown us that 5–10 generations per day usually lead to adequate fitness determinations. Because higher relative inoculation volume dilutes the stressful media, cells can be washed with fresh media prior to inoculation.

16. We typically passage 5–7 times total. The higher sampling and sequencing frequency leads to more accurate fitness measurements because errors in frequency determination by sequencing are reduced by the regression on multiple time-points.
17. Passaging for fewer than 50 generations limits the effect of new beneficial mutations within the growth assay.
18. The reaction will usually work well if the number of templates in the reaction is at least 10^6 (approximately 10 ng for yeast) and up to 1 μg of total template DNA.
19. The limited cycling ensures that true DNA barcode frequency in the genomic DNA is not distorted by exponential amplification, as duplicate UMI–barcode pairs will be removed bioinformatically. We have at times increased the number of cycles here and have yet to see large discrepancies.
20. AMPure XP bead ratios to use can be determined empirically or with useful tables that show retention of desired fragments to the magnetic beads [15]. In practice, we use ratios that maximize the retention of desirable fragments.
21. In many cases, the addition of carrier molecules such as highly purified rRNA from another species such as bacteria (at a final 50 ng/μL carrier RNA) can dramatically improve yield of this purification. However, we found that this sometimes leads to excess of primer dimers.
22. The PCR product is not visible at this stage if visualized on an agarose gel due to the limited cycling.
23. When the library is visualized by electrophoresis using the Agilent TapeStation and occasionally in agarose gels, we frequently observe one peak corresponding to the fragment of desired size, and a second peak at twice the size of the desired fragment. Sometimes the second peak is the only visible peak

on the TapeStation. We interpret the second peak to be due to incoherent migration of non-complementary DNA fragments (so-called "bubble" products) frequently observed in overamplified libraries. These do not interfere with downstream measurements.

24. We quantify each reaction with a Qubit dsDNA HS kit or equivalent, pool at equimolar concentration and quantify the final pool again.
25. The accuracy of the fitness measurement is strongly dependent on the number of reads or the sequencing coverage per barcode per time-point; however, the assay is less sensitive to the sequencing depth than the number of cells passaged per barcode given that the number of sampling time-points is relatively high. We usually aim for at least 100 reads per barcode per time-point, and it is usually not useful to go beyond the number of cells passaged per barcode, or beyond the number of genomic DNA fragments per reaction (which is typically around 10^7 total per time-point).
26. Sequencing reads containing barcodes can be parsed and demultiplexed with a variety of bioinformatics solutions including in-house scripts (e.g., *see* [16] or [17] for pipelines that includes error correction, chimera detection, duplicate read removal and frequency parsing) or using published methods designed for this task [18]. The example here is meant to show the procedure using the simplest pipeline possible and a variety of modifications depending on the primer and experimental designs are anticipated.
27. When barcodes are unknown, a clustering step can be used to first build this dictionary. In practice, we assume that well-represented barcodes that are all two Levenshtein distance from each other belong to the dictionary.
28. The error rate of barcode reads is dependent on the Illumina platform used and the quality of the data obtained. At optimal clustering on the NextSeq platform, about 0.1% of barcode reads of 16 bp will have one mismatch to a known barcode, and 0.01% will have two mismatches.
29. For the approximations: obtain the average s_i from all possible pairs of time-points using the above formula. This is approximately the same as performing simple linear regression on the logarithm of observed ratio of frequencies against the number of generations.
30. The two-point approximation cannot be used when a barcode is not observed for a time-point (no reads map to the barcode). Either the point should be ignored, or the second

approximation that simply performs linear regression on the logarithm of ratios should be used. Finally, the full joint-inference algorithm provided here accounts for this.

31. The approximations given here ignore the increased noise at lower sequencing read counts. Thus, data points at low read counts should have lower weights in the regression because they do not accurately estimate the actual lineage frequencies. The mentioned joint-inference algorithm accounts for this.
32. This fitness model can be made more accurate by incorporating genetic drift which becomes important when the number of cells at the passaging bottleneck is low [16, 19].

References

1. Petropavlovskiy AA, Tauro MG, Lajoie P, Duennwald ML (2020) A quantitative imaging-based protocol for yeast growth and survival on agar plates. STAR Protoc 1: 100182. https://doi.org/10.1016/j.xpro.2020.100182
2. Hall BG, Acar H, Nandipati A, Barlow M (2014) Growth rates made easy. Mol Biol Evol 31:232–238. https://doi.org/10.1093/molbev/mst187
3. Lenski RE, Rose MR, Simpson SC, Tadler SC (1991) Long-term experimental evolution in Escherichia coli. I. Adaptation and divergence during 2,000 generations. Am Nat 138:1315–1341
4. Thompson DA, Desai MM, Murray AW (2006) Ploidy controls the success of mutators and nature of mutations during budding yeast evolution. Curr Biol 16:1581–1590. https://doi.org/10.1016/j.cub.2006.06.070
5. Wiser MJ, Lenski RE (2015) A comparison of methods to measure fitness in Escherichia coli. PLoS One 10:e0126210. https://doi.org/10.1371/journal.pone.0126210
6. Shoemaker DD, Lashkari DA, Morris D et al (1996) Quantitative phenotypic analysis of yeast deletion mutants using a highly parallel molecular bar-coding strategy. Nat Genet 14: 450–456. https://doi.org/10.1038/ng1296-450
7. Venkataram S, Dunn B, Li Y et al (2016) Development of a comprehensive genotype-to-fitness map of adaptation-driving mutations in yeast. Cell 166:1585–1596.e22. https://doi.org/10.1016/j.cell.2016.08.002
8. Giaever G, Nislow C (2014) The yeast deletion collection: a decade of functional genomics. Genetics 197:451–465. https://doi.org/10.1534/genetics.114.161620
9. Lundberg DS, Yourstone S, Mieczkowski P et al (2013) Practical innovations for high-throughput amplicon sequencing. Nat Methods 10:999–1002. https://doi.org/10.1038/nmeth.2634
10. Barnett M regex: Alternative regular expression module, to replace re. https://pypi.org/project/regex/. Accessed 8 Apr 2021
11. Goetghebuer-Planchon T (2021) Tessil/robin-map. https://github.com/Tessil/robin-map/. Accessed 8 Apr 2021
12. FLAMINGO Package (Approximate String Matching) Release 4.1. http://flamingo.ics.uci.edu/releases/4.1/. Accessed 8 Apr 2021
13. Li H, Durbin R (2009) Fast and accurate short read alignment with burrows-wheeler transform. Bioinformatics 25:1754–1760. https://doi.org/10.1093/bioinformatics/btp324
14. Desai MM, Fisher DS (2007) Beneficial mutation selection balance and the effect of linkage on positive selection. Genetics 176:1759–1798. https://doi.org/10.1534/genetics.106.067678
15. Lundin S, Stranneheim H, Pettersson E et al (2010) Increased throughput by parallelization of library preparation for massive sequencing. PLoS One 5:e10029. https://doi.org/10.1371/journal.pone.0010029
16. Nguyen Ba AN, Cvijović I, Rojas Echenique JI et al (2019) High-resolution lineage tracking reveals travelling wave of adaptation in laboratory yeast. Nature 575:494–499. https://doi.org/10.1038/s41586-019-1749-3

17. Johnson MS, Martsul A, Kryazhimskiy S, Desai MM (2019) Higher-fitness yeast genotypes are less robust to deleterious mutations. Science 366:490–493. https://doi.org/10.1126/science.aay4199
18. Zhao L, Liu Z, Levy SF, Wu S (2018) Bartender: a fast and accurate clustering algorithm to count barcode reads. Bioinformatics 34:739–747. https://doi.org/10.1093/bioinformatics/btx655
19. Levy SF, Blundell JR, Venkataram S et al (2015) Quantitative evolutionary dynamics using high-resolution lineage tracking. Nature 519:181–186. https://doi.org/10.1038/nature14279

Part V

In Silico Integration of Functional Genomics Data

Chapter 23

Prediction of Gene and Genomic Regulation in *Candida* Species, Using the PathoYeastract Database: A Comparative Genomics Approach

Pedro Pais, Jorge Oliveira, Romeu Viana, Inês V. Costa, Isabel Sá-Correia, Pedro T. Monteiro, and Miguel C. Teixeira

Abstract

The ability of living organisms to survive changing environmental conditions is dependent on the implementation of gene expression programs underlying adaptation and fitness. Transcriptional networks can be exceptionally complex: a single transcription factor (TF) may regulate hundreds of genes, and multiple TFs may regulate a single gene—depending on the environmental conditions. Moreover, the same TF may act as an activator or repressor in distinct conditions. In turn, the activity of regulators themselves may be dependent on other TFs, as well as posttranscriptional and posttranslational regulation. These traits greatly contribute to the intricate networks governing gene expression programs.

In this chapter, a step-by-step guide of how to use PathoYeastract, one of several interconnecting databases within the YEASTRACT+ portal, to predict gene and genomic regulation in *Candida* spp. is provided. PathoYeastract contains a set of analysis tools to study regulatory associations in human pathogenic yeasts, enabling: (1) the prediction and ranking of TFs that contribute to the regulation of individual genes; (2) the prediction of the genes regulated by a given TF; and (3) the prediction and ranking of TFs that regulate a genome-wide transcriptional response. These capabilities are illustrated, respectively, with the analysis of: (1) the TF network controlling the *C. glabrata QDR2* gene; (2) the regulon controlled by the *C. glabrata* TF Rpn4; and (3) the regulatory network controlling the *C. glabrata* transcriptome-wide changes induced upon exposure to the antifungal drug fluconazole. The newest potentialities of this information system are explored, including cross-species network comparison. The results are discussed considering the performed queries and integrated with the current knowledge on the biological data for each case-study.

Key words YEASTRACT+, PathoYeastract, Transcriptional regulation prediction, Transcription regulatory networks, *Candida* species, Cross-species comparison

1 Introduction

Transcription regulatory networks are key for organisms to respond to environmental queues. The advent of transcriptomics granted us the power to look at transcriptome-wide changes and facilitated our

Frédéric Devaux (ed.), *Yeast Functional Genomics: Methods and Protocols*,
Methods in Molecular Biology, vol. 2477, https://doi.org/10.1007/978-1-0716-2257-5_23,

understanding on how transcriptional regulation is crucial to every Kingdom of life. Not only that, but the study of regulatory networks also has important implications in most biological sciences fields, from research to industry.

The YEASTRACT+ portal (http://yeastract-plus.org/) is an information system dedicated to the analysis and prediction of transcription regulatory associations in yeasts of human health or industrial pertinence [1]. The system is comprised of four distinct yet interconnected databases: YEASTRACT (http://yeastract-plus.org/yeastract/scerevisiae/), which focuses on the model yeast *Saccharomyces cerevisiae* [2, 3]; PathoYeastract (http://yeastract-plus.org/pathoyeastract/), which focuses on the *Candida* pathogenic yeasts [4]; N.C.YEASTRACT (http://yeastract-plus.org/ncyeastract/), focusing on non-conventional yeasts with biotechnological relevance [1]; and CommunityYeastract (http://yeastract-plus.org/community/sboulardii/index.php), a repository of automatically generated YEASTRACT-like databases, for yeast species or strains, according to the request of community members, currently including the oleaginous yeast *Rhodosporidium toruloides* and the probiotic yeast *S. cerevisiae var. boulardii* [5].

The exploitation of the PathoYeastract database for analysis and prediction of regulatory associations at gene and genomic levels in the pathogenic yeasts from *Candida* spp. is here presented. It makes use of a curated repository of known regulatory associations between transcription factors (TFs) and target genes based on hundreds of bibliographic references, currently providing information on the four most prevalent species: *C. albicans*, *C. glabrata*, *C. parapsilosis*, and *C. tropicalis*. The database currently includes 45,000+ documented regulatory associations, 120+ DNA binding sites and 150+ transcription factors. For cross species network comparison, data on predicted protein orthology is retrieved from Candida Genome Database (CGD) [6], Candida Gene Order Browser (CGOB) [7], and associated Gene Ontology (GO) terms [8].

In this chapter, the use of the PathoYeastract database is described, step by step, and illustrated with three case studies in *C. glabrata*. This pathogen is known for its ability to rapidly acquire azole antifungal resistance [9–11], making it a good model to showcase the tools available at PathoYeastract applied to study the regulation of recently identified drug resistance determinants. First, the analysis of the transcriptional control of the *QDR2* gene encoding a Multidrug Resistant Transporter (MDR) from the Major Facilitator Superfamily (MFS) [12, 13], identified as a determinant of azole resistance by reducing antifungal accumulation within yeast cells [14]. The second case study comprises the analysis of the regulatory network underlined by the TF Rpn4. This TF was recently proposed to control ergosterol levels in the yeast plasma membrane in response to the antifungal drug fluconazole, exerted by direct transcriptional control over the drug target [15]. The third example is based on the analysis of a recent dataset on the transcriptome-wide response to fluconazole [15], used here to highlight the potential of PathoYeastract to analyze and predict regulation of transcriptome-wide remodeling.

2 Methods

2.1 Predicting the Regulation of Individual Genes Based on Data for the Species under Study

1. Open the "Search for TFs" query in the *C. glabrata* portal from the PathoYeastract website (http://pathoyeastract.org/cglabrata/index.php).
2. Insert the gene name in the "Regulated Genes" box, in this case "*QDR2*" being used as an example.
3. Not changing any of the default parameters, press the "Search" button. The resulting table includes all the transcription factors (TFs) that are demonstrated to control *QDR2*, either directly (TF binding to target promoter sequence) or indirectly (TF affecting target gene expression). The references that provide evidence to support each regulatory association are also provided.
4. To visualize the network composed of all the *QDR2* regulators, hover over the "Menu" box and select "Interactive Network Image," in the bottom of the menu options. The resulting image is illustrated in Fig. 1.

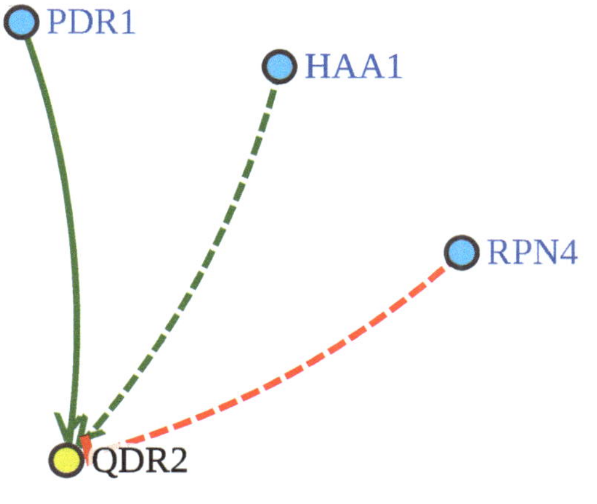

Fig. 1 Regulation of the *QDR2* gene from *C. glabrata*, considering documented regulatory associations. Full lines: DNA-binding evidence; dashed lines: expression evidence; green lines: positive interaction; red lines: negative interaction

5. The prediction of the actual *QDR2* regulators can be filtered by changing some of the default parameters in **step 3**. This can be done at multiple levels: evidence underlying the regulatory association, experimental condition, or mode of action of the transcription factor. In this example, given the low number of regulators, these parameters were not changed. For more information on these filtering conditions, check [16].

6. Once the network image is generated (**step 4**), it is possible to filter by the type of supporting evidence (DNA binding, expression, or both), and by the type of regulation (positive, negative). This can be done in the "Supporting evidence:" and "Type:" menus.

   Select only "Positive" regulations and "DNA-binding and expression" evidence. Now, the network only includes one TF, Pdr1 (*see* **Note 1**).

7. In a complementary approach, the user may search for the regulators of a given gene, based on the occurrence of TF binding sites in its promoter region. To do that, open the "Search for TFs" query, insert "*QDR2*" in the "Regulated Genes" box, select "Potential" in the "Regulations filter" box, and press the "Search" button. The result is provided with a figure illustrating the distribution of the TF binding sites in the promoter region of the gene under analysis (*see* **Note 2**), plus a Table including the exact TF binding sequences considered and the exact position and strand within the promoter region (*see* **Note 3**). The resulting image is illustrated in Fig. 2.

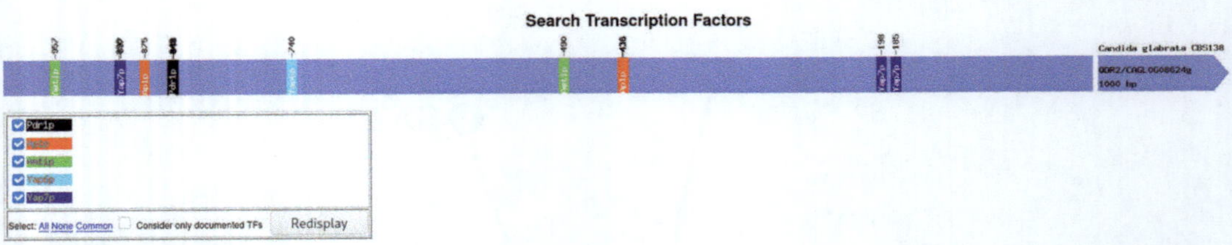

User inserted	Consensus	Position	Strand
Target Sequence: qdr2 (size 1000) Back to top			
Pdr1p	TCCRYGGA	-848	F
Pdr1p	TCCRYGGA	-848	R
Pdr1p	TCCACGGA	-848	R
Pdr1p	HYCCGTGGR	-849	F
Ap1p	TTACAAA	-435	F
Ap1p	VTTACWAAB	-436	F
Ap1p	VDTASTAA	-875	R
Amt1p	WTWHNGCTGW	-490	F
Amt1p	WTWHNGCTGW	-957	R
Yap6p	AATKACV	-740	R
Yap7p	MTKASTMA	-898	F
Yap7p	MTKASTMA	-185	R
Yap7p	MTKASTMA	-897	R
Yap7p	AGTCATM	-198	R

Fig. 2 Output of a Search for TFs query focused on the potential regulators of the *QDR2* gene. Graphic display of the transcription factor-binding sites found to occur in the promoter regions of the *QDR2* gene. The table below contains the list of all transcription factors identified as potential regulators of *PDR18*, followed by an indication of the exact TF binding site nucleotide sequence and position found in the *QDR2* promoter

2.2 Predicting the Regulation of Individual Genes Based on Data for Other Species in YEASTRACT+

1. YEASTRACT+ also allows the user to search for common and unique transcription regulators of orthologous genes in other species deposited in the database. For that, open the "Homologous network" query in the *C. glabrata* portal, within the PathoYeastract website (http://pathoyeastract.org/cglabrata/index.php).
2. Insert "*QDR2*" in the "Target ORF/Genes" box and tick the "Check for all TFs" box (*see* **Note 4**). Not changing any of the default parameters, press the "Search" button (*see* **Note 5**).
3. With this tool, a graphic representation for the regulatory network predicted for *QDR2*, based on the regulation of *QDR2* orthologs in other species, is generated. In the menu "Orthologous documented regulations in species:" the user can change the orthologous regulatory network for the different species in the database. Select *C. albicans* and note that the predicted *QDR2* network is considerably more complex than that of *C. glabrata*. If selecting *S. cerevisiae*, an even more extensive network can be predicted. Finally, if selecting *C. tropicalis* or *C. parapsilosis*, the predicted *QDR2* regulatory network is restricted to one or two TFs (*see* **Note 6**). The resulting images are illustrated in Fig. 3.

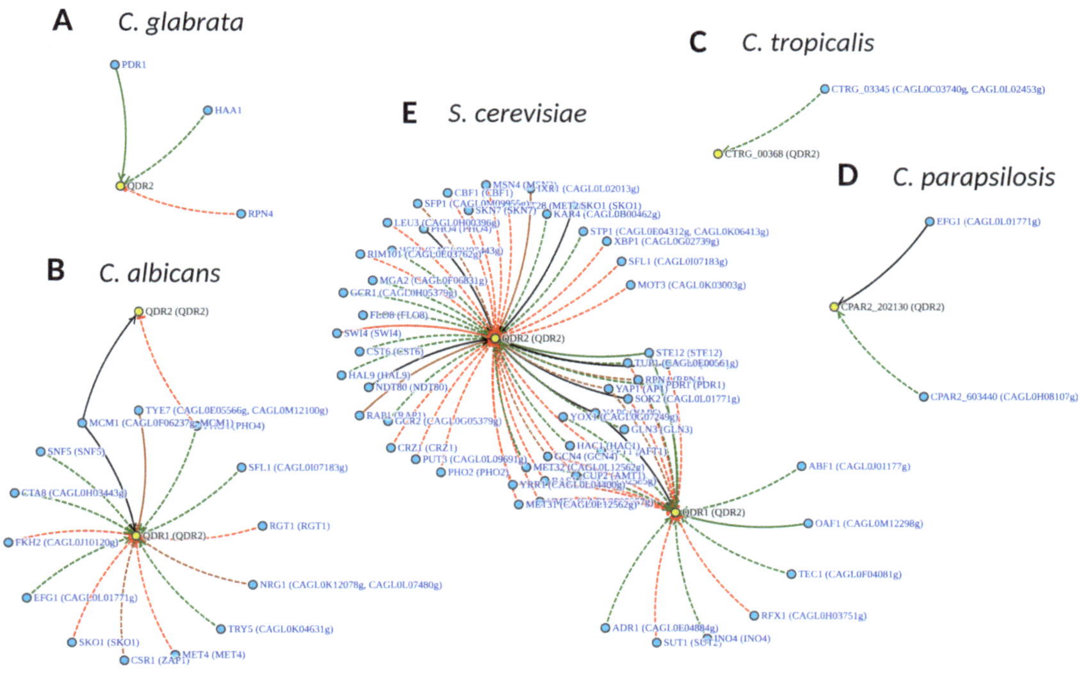

Fig. 3 Regulation of the *QDR2* gene from *Candida glabrata*, considering their known regulatory associations in *C. glabrata* (**a**) or the regulation of their orthologs in *C. albicans* (**b**), *C. tropicalis* (**c**), *C. parapsilosis* (**d**), and *S. cerevisiae* (**e**). Full lines: DNA-binding evidence; dashed lines: expression evidence; green lines: positive interaction; red lines: negative interaction; black lines: undefined interaction

4. TF binding sites can also be used in the prediction of TF target genes in closely related species [1]. The user can also use this approach on the YEASTRACT+ portal. Go to "Promoter analysis" query, insert "*QDR2*" in the "Target ORF/Genes" box, select "*Candida glabrata CBS138*" in the "Consider TF binding sites from strain:" menu and tick the box for the species: *Candida glabrata CBS138*, *Candida albicans SC5314*, *Candida parapsilosis CDC317*, *Candida tropicalis MYA-3404*, and *Saccharomyces cerevisiae S288c*. The result is provided as a table including all the orthologs for *QDR2* in each species and the unique TF consensus sequences that are present in the respective promoters (*see* **Note 7**). The results are illustrated in Fig. 4a.
5. In the last column of the generated table, click on "Promoter." A more detailed graphical representation is presented, containing all binding sites in each ortholog, also including a table with the exact TF binding sequences and the exact position and strand within the promoter region. The results are illustrated in Fig. 4b.

2.3 Predicting the Targets of Individual TFs Based on Data for the Species Under Study

1. Open the "Search for Genes" query in the *C. glabrata* portal from the PathoYeastract website (http://pathoyeastract.org/cglabrata/index.php).
2. Insert the transcription factor name in the "Transcription Factors" box, in this case, "Rpn4" being used as an example.
3. Without changing any default parameters, press the "Search" button. The resulting table includes all genes that are demonstrated to be regulated by Rpn4, either directly (DNA binding to promoter) or indirectly (expression evidence). The references that provide evidence for each regulatory association are also present.
4. To visualize the network composed of the Rpn4 regulated genes, hover over the "Menu" box and select "Interactive Network Image", the last option of the menu (*see* **Note 8**).
5. Since it is unlikely that a given transcription factor regulates every gene under every condition, the prediction of the genes regulated by Rpn4 can be filtered by changing some of the default parameters in **step 3** (*see* **Note 9**). This can be done at multiple levels: evidence of the regulatory association, environmental condition, or mode of action of the TF.
6. To filter the results by the experimental evidence underlying the observed regulatory associations select the box "Only Expression evidence" to find the genes regulated by Rpn4. Press the "Search" button. The resulting table shows all genes demonstrated to be regulated by Rpn4 based on expression studies. Select "Interactive Network Image" to visualize the network as illustrated in Fig. 5a.

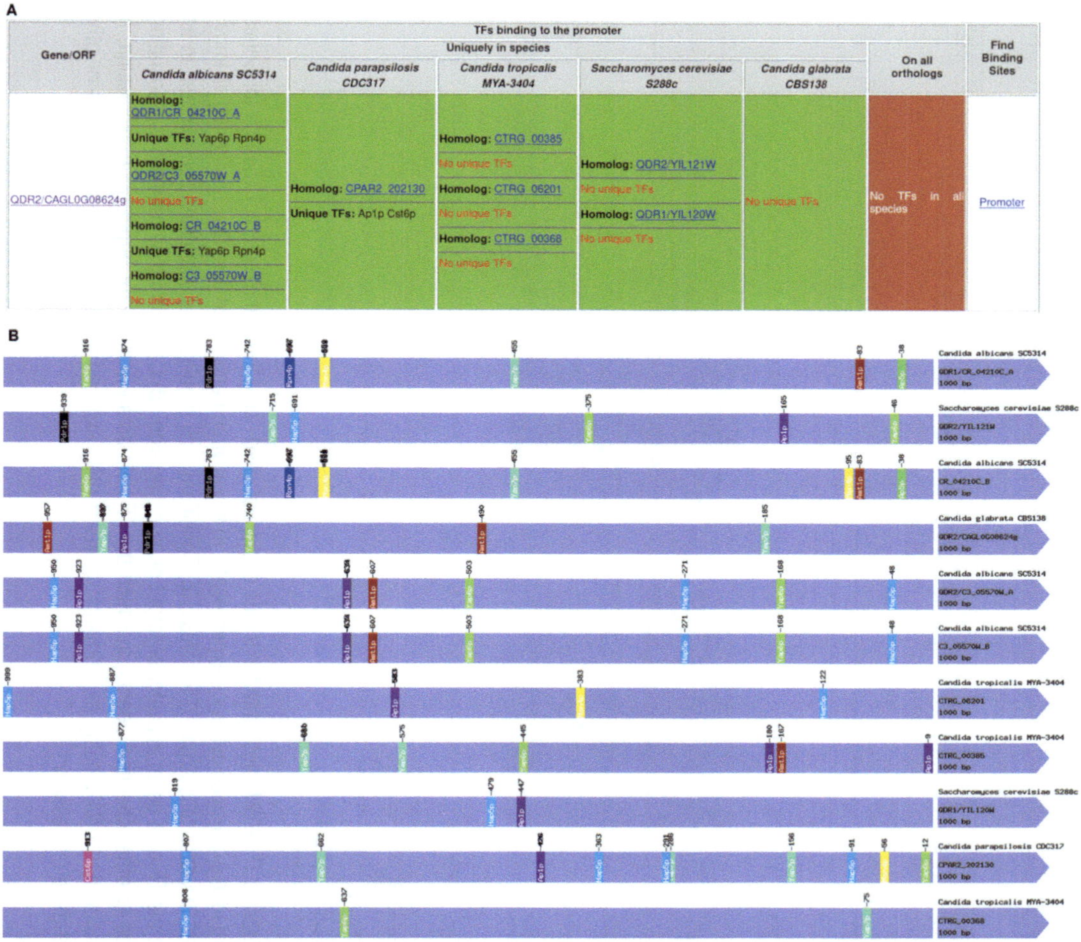

Fig. 4 (**a**) Cross-species comparison of the promoters of the *CDR1* orthologous genes in *Candida* species and *Saccharomyces cerevisiae*, as obtained in the "Promoter analysis" tool. Green in the species columns indicates that there is at least one TF that uniquely regulates the corresponding gene or one of its orthologs. In the last column, it means that there is at least one TF that regulates all the genes and orthologs, and red indicates the opposite. Yellow indicates that there is no ortholog for that gene in that species. (**b**) Detailed graphical representation containing all binding sites for each ortholog in each species

7. To filter the results by the environmental condition in which the regulatory associations were observed, select, for example, "Stress" from the "Group" box to find the genes regulated by Rpn4 in stress conditions. Press the "Search" button. The resulting table shows all genes demonstrated to be regulated, directly or indirectly, by Rpn4, in stress exposed cells. Select "Interactive Network Image" to visualize the network as illustrated in Fig. 5b. Additional filtering can also be performed by selecting a particular stress condition from the "Subgroup" box.

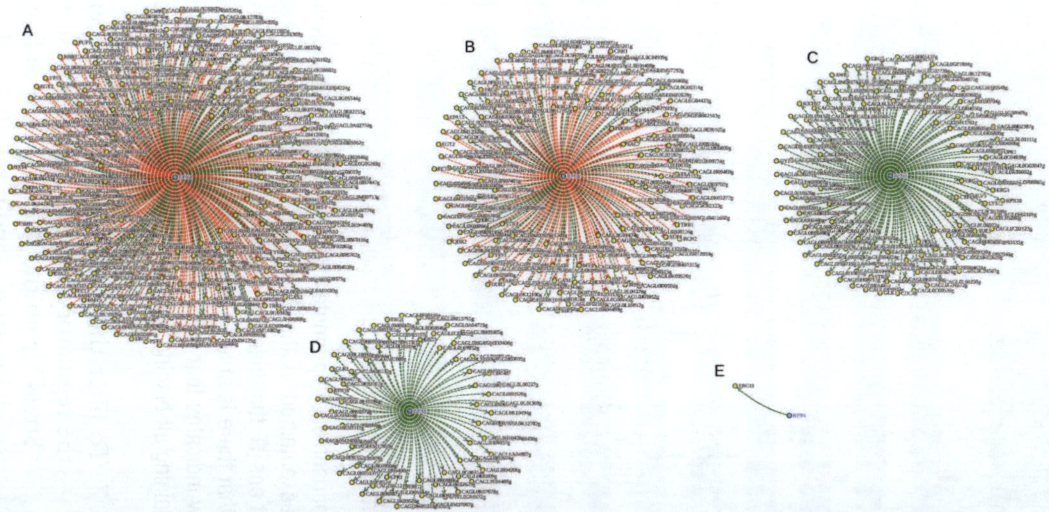

Fig. 5 Five different regulatory networks controlled by the *C. glabrata* TF *RPN4*, highlighted in a blue circle, obtained using the PathoYeastract "Search for Genes" tool. (**a**) Considering all known regulatory associations documented in expression evidence experiments, in any environmental condition. (**b**) Considering all known regulatory associations documented, in any experimental setup, in yeast cells exposed to stress. (**c**) Considering all regulatory associations documented in any experimental setup and any environmental condition, in which the TF acts as an activator of target gene expression. (**d**) Considering all regulatory associations documented in yeast cells exposed to stress induced by drugs or other chemical agents, in which the TF acts as an activator of target gene expression. (**e**) Considering all regulatory associations documented in expression and DNA-binding experiments, in yeast cells exposed to stress induced by drugs or other chemical agents, in which the TF acts as an activator of target gene expression

8. To filter the results by the mode of action of the transcription factor, deselect "TF acting as repressor" in the "Regulations Filter" box to filter out the genes repressed by the transcription factor. Press the "Search" button. The resulting table shows the genes demonstrated to be activated, directly or indirectly, by Rpn4. Select "Interactive Network Image" to visualize the network as illustrated in Fig. 5c.
9. To perform a stricter prediction, it is possible to combine more than one filtering approach within the "Regulations Filter" and the "Filter Documented Regulations by environmental condition" boxes. For example, it is possible to select only "TF acting as activator" induced by "Stress", particularly "Drug/chemical stress exposure." Press the "Search" button. The resulting table shows the genes demonstrated to be, directly or indirectly, activated by Rpn4 under the intended stress condition. Select "Interactive Network Image" to visualize the network as illustrated in Fig. 5d. Further, the user can investigate which of the Rpn4 activated genes under drug stress are directly regulated by the TF. To do so, select "DNA binding and expression evidence" in the "Regulations Filter" box and select only "TF acting as activator" induced by "Stress," particularly "Drug/chemical stress exposure." Pressing the "Visualize Network" button, at the bottom of the results table, the obtained results are illustrated in Fig. 5e (*see* **Note 10**).

2.4 Predicting the Targets of Individual TFs Based on Data for Other Species in YEASTRACT+

1. Follow the instructions in **steps 1** and **2** of the previous section, accessing to the *C. glabrata* portal of the PathoYeastract database and the "Search for Genes" tool.
2. The user may also use this tool to predict regulatory associations in one species based on known associations of orthologous TFs in other species. For example, consider that no Rpn4 target genes were known in *C. glabrata*. The user may make a prediction on what the *C. glabrata* Rpn4 regulon would be based on documented regulations of the *C. albicans* Rpn4 ortholog, here used as an example. To do so, insert the TF name in the "Transcription Factors" box and select "*Candida albicans* SC5314" from the "Search for Orthologous Regulations in" box. Press the "Search" button. The resulting table will feature the predicted Rpn4 regulon in *C. glabrata* based on the documented *C. albicans* Rpn4 regulon (*see* **Note 11**). Select "Visualize Network" to visualize the predicted *C. glabrata* Rpn4 regulon, with the correspondent *C. glabrata* orthologs in parentheses, as illustrated in Fig. 6a (*see* **Note 12**).
3. It is also possible to use the same tool to assess orthologous regulations across two distinct species. For example, it is

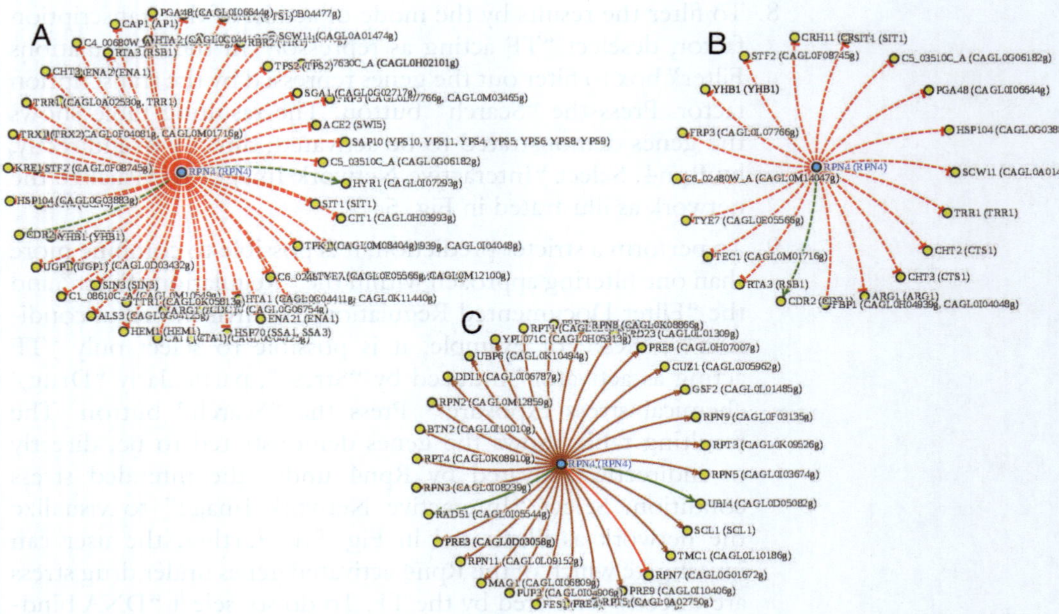

Fig. 6 Cross-species regulatory network analysis using the PathoYeastract "Search for Genes" tool. (**a**) Predicted regulatory network controlled by the *C. glabrata* TF *RPN4*, highlighted in a blue circle, based on the documented regulon of its ortholog in *C. albicans*. (**b**) Orthologous Rpn4 regulon based on documented associations between the *C. glabrata* and *C. albicans* networks. (**c**) Orthologous Rpn4 regulon based on documented associations, supported by DNA binding and expression evidence, between the *C. glabrata* and *S. cerevisiae* networks

possible to use the documented *C. glabrata* Rpn4 regulon and evaluate orthologous regulations documented in *C. albicans*. To do so, insert the TF name in the "Transcription Factors" box and press the "Search" button. The resulting table will feature all documented genes to be regulated by Rpn4 in *C. glabrata*. Select "Visualize Network" followed by "*Candida albicans* SC5314" from the "Orthologous documented regulations in:" to visualize the orthologous Rpn4 regulon in *C. albicans*, with the corresponding *C. glabrata* orthologs in parentheses, as illustrated in Fig. 6b (*see* **Note 13**). To assess orthologous regulation of the *C. glabrata RPN4* regulon with the model *S. cerevisiae*, follow the same procedure but selecting "*Saccharomyces cerevisiae* S288c" from the "Orthologous documented regulations in:" box upon network visualization. To narrow down the regulations to be analyzed, select only "DNA Binding + Expression" from the "Supporting evidence" box to attain the orthologous Rpn4 regulon in *S. cerevisiae*, with the correspondent *C. glabrata* orthologs in parentheses, as illustrated in Fig. 6c (*see* **Note 14**).

Transcription Factor	% in user set	% in cglabrata	p-value ▲	Target ORF/Genes
☑ Rpn4p	56.82%	6.60%	0.000000000000000	CAGL0A01089g CAGL0A03740g HMX1 CAGL0B01875g CAGL0F08965g CAGL0G02739g CAGL0G05632g CAGL0H05005g CAGL0H09504g CAGL0J03916g MEP2 CAGL0K07337g CAGL0K12100g CAGL0L03982g CAGL0L06776g CAGL0L10692g CAGL0M04499g CAGL0M11660g CDR1 ERG1 ERG11 ERG2 ERG3 RTA1 SRO7 ≡Menu
☑ Upc2ap	15.91%	70.00%	0.000000000000000	ERG1 ERG11 ERG2 ERG3 ERG4 ERG5 ERG6 ≡Menu
☑ Pdr1p	43.18%	2.19%	0.000002052828025	ATF2 CAGL0A01089g CAGL0A03740g HMX1 CAGL0G05632g CAGL0J03916g CAGL0K12100g CAGL0L03828g CAGL0L06776g CAGL0M04499g CAGL0M05995g CDR1 ERG11 ERG3 ERG4 ERG6 MUP1 RTA1 SSA3 ≡Menu
☑ Gln3p	2.27%	33.33%	0.000179163011234	MEP2 ≡Menu
☑ Ap1p	18.18%	2.97%	0.000187813090511	CAGL0A01089g CAGL0G05632g CAGL0K03927g ERG3 ERG6 HBN1 MUP1 SSA3 ≡Menu
☑ Ada2p	25.00%	2.28%	0.000234004977451	CAGL0A03740g CAGL0G02739g CAGL0G05632g CAGL0H09504g CAGL0K07337g CAGL0M05995g CAGL0M11660g ERG6 HBN1 RTA1 SSA3 ≡Menu
☑ Aft1p	4.55%	9.52%	0.000541174316941	HMX1 ERG11 ≡Menu
☑ Stb5p	4.55%	5.88%	0.002267934954677	CDR1 RTA1 ≡Menu
☑ Hap5p	9.09%	3.01%	0.003578077382040	CAGL0A02277g CAGL0B01875g CAGL0K12100g ERG11 ≡Menu
☑ Hal9p	13.64%	1.97%	0.008799733771152	HMX1 CAGL0H05005g CAGL0L08448g CAGL0L10692g ERG4 SSA3 ≡Menu
☑ Haa1p	18.18%	1.60%	0.013948536161738	CAGL0B01875g CAGL0G05632g MEP2 CAGL0K07337g CAGL0K12100g ERG3 RTA1 SSA3 ≡Menu
☑ Pho4p	4.55%	2.53%	0.023480807796437	CAGL0M11660g RTA1 ≡Menu
☑ Yap6p	2.27%	1.49%	0.096392660330062	CAGL0M11660g ≡Menu
☑ CAGL0G08844g	6.82%	0.88%	0.275280082518440	CAGL0A03740g CAGL0M11660g ERG2 ≡Menu
☑ Yap7p	2.27%	0.58%	0.392351913818258	ERG11 ≡Menu

Fig. 7 Table obtained by ranking the genes by TF. The first column lists all the TFs that regulate genes in the dataset; the second column shows the percentage (%) of TF target genes relative to the user set; the third column shows the enrichment of the TF targets in the dataset provided by the user, when compared with the genome, as measured by a *p*-value; and the fourth column shows the target ORF/genes of each TF

2.5 Transcriptomics Data Analysis Based on Documented Data of Regulatory Interactions for the Species Under Study

1. Open the "Rank by TF" query in the *C. glabrata* portal from the PathoYeastract website (http://pathoyeastract.org/cglabrata/index.php).
2. Insert the user's list of genes in the "Target ORF/Genes" box. As a case study, we will use the list of 44 genes whose expression is altered in *C. glabrata* cells exposed to fluconazole [15].
3. Not changing any of the default parameters, tick the "Check for all TFs" checkbox and press the "Search" button. The resulting table (Fig. 7) shows the *C. glabrata* transcription factors that are documented to regulate the ORFs/genes in the list (*see* **Notes 15–19**).
4. To visualize the network composed by TF and regulated ORFs/genes, illustrated in Fig. 8a, select the "Visualize Network" button, at the bottom of the table (*see* **Note 20**).
5. To filter the results, the user can step back and change some of the default parameters in **step 3**, or alternatively change the parameters directly in the interactive image. Filtering can be done at multiple levels: evidence of the regulatory association, experimental condition, mode of action of the transcription factor. To filter the results by the experimental evidence underlying the observed regulatory associations, select "Only DNA binding evidence" or "Only Expression evidence" in the "Filter Documented Regulations by" box. This will filter the network

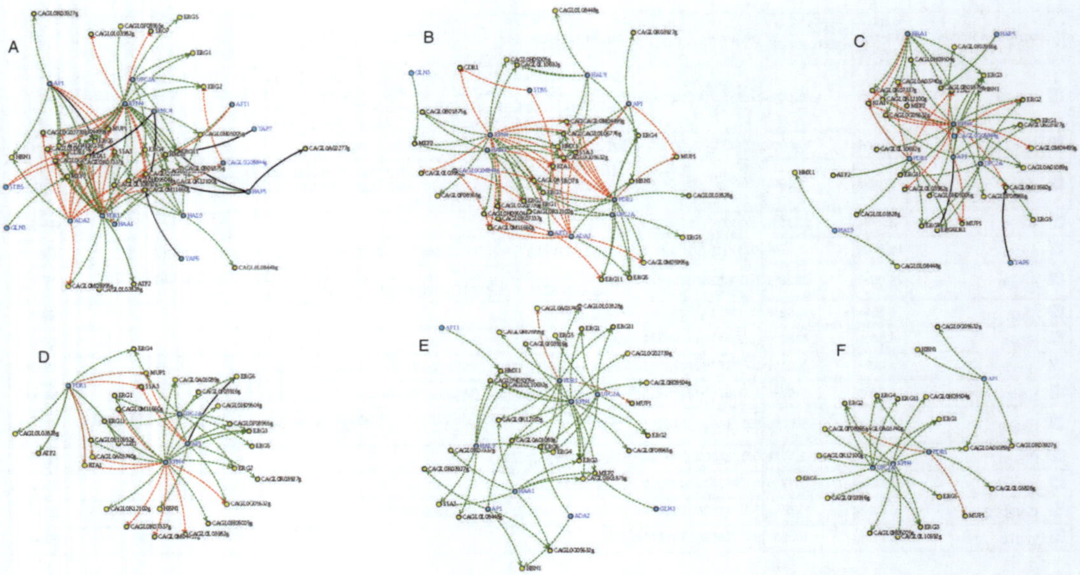

Fig. 8 Predicted network of interactions between TFs and target genes in *C. glabrata*. (**a**) The unfiltered network. (**b**) Interaction network only between ORF/genes and respective TF, based on expression studies. (**c**) Interaction network filtered by the environmental condition group "Stress." (**d**) Interaction network filtered by the environmental condition group "Stress" and subgroup "Drug/Chemical stress exposure." (**e**) Interaction network with only TFs acting as activators. (**f**) Combines multiple filtering options in the regulation network, namely TF acting only as activators induced by "Stress," particularly "Drug/chemical stress exposure"

of interactions between the TFs and respective regulated genes from the list of ORFs/genes, to show the interactions based on a given experimental evidence. Alternatively, in the interactive image, the user can select or deselect the boxes corresponding to a given experimental evidence, in the "Supporting evidence:" box. As an example, select "Only Expression evidence" and then press the "Search" button. The resulting table now only shows the ORFs/genes from the list, who have been demonstrated to be regulated by their respective TF, based on expression studies. Select "Interactive Network Image" to visualize the network as illustrated in Fig. 8b.

6. To filter the results by the environmental condition, for example, select "Stress" from the "Group" box to find the TFs and respective regulated genes, interacting under stress conditions. Press the "Search" button. The resulting table shows all TFs and genes demonstrated to be interacting, directly or indirectly, in stress conditions. Select "Visualize Network" to visualize the network as illustrated in Fig. 8c. Additional filtering can also be performed by selecting a particular stress condition from the "Subgroup" box, for example, the "Drug/chemical stress exposure" condition, as illustrated in Fig. 8d, which is interesting in this case due to the nature of the study from which the dataset was obtained.

7. To filter the results by the mode of action of the TF, as an example, deselect "TF acting as inhibitor" and press the "Search" button. The resulting table shows the genes in the list demonstrated to be activated, directly or indirectly, by their respective TF. Select "Visualize Network" to visualize the network as illustrated in Fig. 8e.
8. To perform a stricter prediction, it is possible to combine more than one filtering approach within the "Regulations Filter" and the "Filter Documented Regulations by environmental condition" boxes. For example, it is possible to select only "TF acting as activator" induced by "Stress," particularly "Drug/chemical stress exposure." Press the "Search" button. The resulting table shows the genes in the list demonstrated to be, directly or indirectly, activated by their respective TF under the intended stress condition. Select "Visualize Network" to visualize the network as illustrated in Fig. 8f.

2.6 Transcriptomics Data Analysis Based on Data for Other Species in PathoYeastract

1. Follow the instructions present in **steps 1** and **2** from the previous section, accessing to the *C. glabrata* portal of the PathoYeastract database and the "Rank by TF" tool.
2. The user may also use this tool to predict regulatory associations of orthologous genes, in documented regulations in other species (*see* **Note 21**). For example, using this dataset, it is possible to predict an orthologous regulatory association of the 44 genes under analysis in the following species: *C. albicans* (Fig. 9a), *C. parapsilosis* (Fig. 9b), *Saccharomyces cerevisiae* (Fig. 9c), *Zygosaccharomyces bailii* (Fig. 9d), *Kluyveromyces lactis* (Fig. 9e), *Yarrowia lipolytica* (Fig. 9f), *and Komagotaella phaffii* (Fig. 9g). To do so, select the species of interest using the "Search for Orthologous Regulations in:" box, and then press the "Search button." By comparing each species orthologous association network, some patterns of conservation arise between species. Namely TFs that are documented to regulate genes associated with drug/xenobiotic and stress resistance, which are conserved in pathogenic and non-pathogenic species, or even TFs that are unique to a particular species (*see* **Note 22**).

3 Notes

1. Pdr1 is an activator of drug resistance genes via pleiotropic drug response elements, including *QDR2*, an MFS transporter involved in multidrug efflux [14]. Therefore, as expected, this transcription factor is a positive regulator of *QDR2* with evidence at both levels.

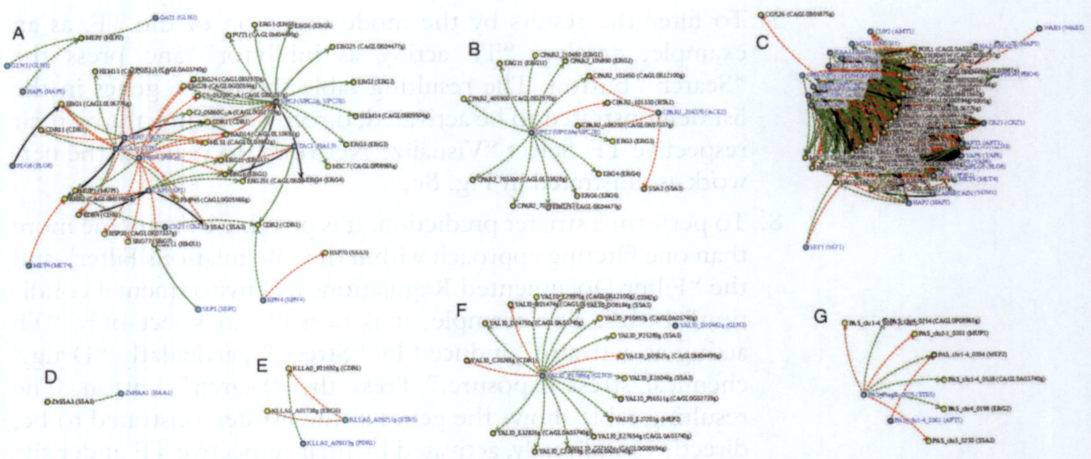

Fig. 9 Predicted orthologous regulations in several species present in the YEASTRACT+ portal. (**a**) Predicted orthologous regulation in *C. albicans*. (**b**) Predicted orthologous regulations in *C. parapsilosis*. (**c**) Predicted orthologous regulation in *S. cerevisiae*. (**d**) Predicted orthologous regulation in *Z. bailii*. (**e**) Predicted orthologous regulation in *K. lactis*. (**f**) Predicted orthologous regulation in *Y. lipolytica*. (**g**) Predicted orthologous regulation in *K. phaffii*

2. In all YEASTRACT+ databases, the promoter region is always considered to be the 1000 bp immediately preceding the START codon.
3. Rpn4 is not present in this new table; however, it was identified in the table from **step 4** as a repressor of *QDR2*. This difference is because the negative regulation identified was based only at the expression level and not by DNA binding. Possibly, Rpn4 acts as an indirect repressor of *QDR2*.
4. Alternatively, it is possible to define a group of specific transcription factors. To do this, unselect the box "Check for all TFs" and insert the list of TFs in the "Transcription factors" box.
5. In this step, it is possible to change the default parameters, the user can specify the type of evidence, selecting "Only DNA binding evidence," "Only Expression evidence," "DNA binding plus expression evidence," and "DNA binding and expression evidence." Another parameter that can be changed is the synteny level, changing the "Synteny Level" box, to select distinct levels of stringency in the definition of homologous/orthologous genes.
6. The observable differences between the orthologous regulatory networks are due to the amount of documentation related to each species. It is clear that *S. cerevisiae* and *C. albicans* are more extensively studied than the remaining species; therefore, a larger and more complete regulatory network is known, enabling a larger predictive power.

7. Notice that there are no TF consensus sequences present in all the promoters, this may indicate that they are not under a core conserved regulation. There are specific TFs that are potential regulators of only one of the orthologous genes. This occurs for *C. albicans* and *C. parapsilosis* suggesting that there might be some degree of specialization of this gene in different species.
8. Searching for genes using this tool with default options will attain every gene to have any documented regulatory association with the transcription factor. This corresponds to the totality of data in the database regarding a regulon for a given transcription factor. Depending on the amount of information available, the visualization of the network in these circumstances can be bulky (e.g., hundreds of regulated genes).
9. The authors do not recommend any of the filtering approaches in detriment of the others. However, the user should consider that their application may, in some cases, spare the user of unnecessary experiments, which may result from testing all possible target genes.
10. Interestingly, only one gene (*ERG11*) is currently documented to be directly activated by Rpn4 during drug stress. By investigating the reference associated with this regulatory associations, it is possible to note that this regulation was documented upon stress by the widely used antifungal fluconazole [15]. Notably, Erg11 is the target of azole drugs [17], thus positioning Rpn4 as a promising regulator of azole resistance.
11. Note that upon predicting regulations using this method, all regulation filters are ignored. The table provides orthology information: it shows the predicted *C. glabrata* network by identifying the orthologs of the *C. albicans* regulon, presenting the *C. glabrata* gene names in parentheses. The results from this analysis are therefore dependent on the regulatory associations available for the species used for the prediction, in this case, *C. albicans*. It is also possible to select distinct levels of stringency to define orthologous genes in the "Synteny Level" box. If the user wishes to filter the results by, for example, environmental condition, it can be done in the "Search for Genes" tool. The generated table shows all the predicted regulon independently of environmental condition, however, that filter will be applied upon network visualization.
12. Upon visualizing the predicted Rpn4 regulatory network based on the *C. albicans* regulon, only the genes that possess a *C. glabrata* ortholog are shown, with their respective designations in parentheses. Note that the positive or negative regulation over target genes is based on the species used for the prediction, in this case, *C. albicans*. Notably, the prediction of

the Rpn4 network is much smaller than the actual regulon presented in Fig. 5. This may be due to several reasons: the documented *C. albicans* regulon is smaller than the *C. glabrata* regulon; or there is limited orthology between the two species thus hindering prediction of the corresponding targets in *C. glabrata*. Regardless, this example aims to illustrate how regulatory data from one species can be used as a starting point to investigate regulatory networks in another species within the database. In this example, only two targets are found to be activated. Interestingly, one of the activated genes is the MDR transporter *CDR2*, a known determinant of fluconazole resistance [18]. This raises the hypothesis of Rpn4 also being a regulator of antifungal drug resistance in *C. albicans*.

13. Upon visualizing the orthologous Rpn4 regulatory network in *C. albicans*, note that only genes that possess a *C. glabrata* ortholog are present, with their respective designations in parentheses. For this reason, if the user aims to analyze the complete *C. albicans* regulon, it is recommended to use the "Search for Genes" tool in the *C. albicans* page of PathoYeastract. Note that the positive or negative regulation over target genes is based on the species being probed, in this case, *C. albicans*.
14. Upon visualizing the orthologous Rpn4 regulatory network in *S. cerevisiae*, supported by DNA binding and expression evidence, a more diverse network can be observed. In this example, we have not only positive (green) or negative (red) associations, but also cases of genes with both positive and negative associations (brown arrows). This is the result of evidence for both activation and repression of those target genes, which can be the result of more than one study finding contradictory regulation evidence, or differential regulation according to distinct environmental conditions. In this example, the user can see an enrichment of proteasome genes that comprise the orthologous regulon of Rpn4 in both species, which is the hallmark of the documented Rpn4 network in *S. cerevisiae* [19, 20].
15. Additional information shown is the percentage (%) of user inserted ORF/genes that are not regulated by any TF. In the example dataset, this corresponds to the ORFs CAGL0K04477g, CAGL0I02970g, CAGL0G05566g, CAGL0G04499g, CAGL0L08547g, CAGL0J00297g, and CAGL0G00594g, or 16% of the ORFs/genes in the example dataset.
16. Since our dataset is small, there is not much of a need to highlight or filter the results. However, in the case of bigger datasets for example, the user can rank the results using three methods: ranking based on the percentage (%) of TF target genes to the user set; ranking based in the percentage of TF

target genes relative to the whole genome; and ranking based on the enrichment of the TF targets in the dataset provided by the user, when compared with the genome, as measured by a p-value. This can be useful for example, if the user chooses to select only the associations associated to a low *p*-value.

17. The authors do not recommend any of the filtering approaches in detriment of the others. However, the user should consider their application may, in some cases, spare the user of unnecessary experiments, which may result from testing all possible target genes.
18. This table can be downloaded as a .csv file, however the information about "% in user set," "% in cglabrata," and "*p*-value" is not exported.
19. The "Synteny level:" option of the tool is not explored in this example, but it could be another way to explore the data and filter the results when searching for orthologous regulations.
20. Another option that is not mentioned in the examples is the "show additional regulatory associations between TFs" present in the interactive images. The user can select this option, and in the image the self-regulatory interactions that TFs have with themselves are shown.
21. When predicting the orthologous associations, note that the results obtained with this query are merely indicative, given that: (1) the orthologs from different species are predicted, at CGD, based on sequence homology; (2) the conservation of regulatory associations among different species is not complete, and that all regulation filters are deliberately ignored, if previously selected.
22. For example, Upc2, documented to be a regulator of ergosterol biosynthetic genes and sterol uptake, has predicted homologies in both *C. albicans* and *C. parapsilosis*, as well as in *S. cerevisiae*. Another interesting association concerns the TF Hal9, which in *C. glabrata* and *S. cerevisiae* is not associated with azole resistance. Its homologue *C. albicans*, however, is Tac1, the master regulator of multidrug resistance genes. Rpn4, known to have a role in azole resistance, is also conserved in *C. albicans* and *S. cerevisiae*, playing a wider role in stress resistance as a proteasome gene regulator. Stb5, a known negative regulator of azole resistance, is also conserved in *S. cerevisiae*, *Kluyveromyces lactis*, and *Komagataella phaffii*; and TF Pdr1, a known activator of drug resistance genes, is also conserved in *S. cerevisiae* and *K. lactis*. This suggests that drug/xenobiotic resistance mechanisms are widely represented, and somewhat conserved, in all yeast species with important implications for their pathogenicity, but also for their robustness as microbial cell factories in industrial applications.

Acknowledgments

The YEASTRACT+ database is a Node Service of the Portuguese distributed *infrastructure* for biological data BioData.pt. funded by Programa Operacional Regional de Lisboa 2020 (LISBOA-01-0145-FEDER-022231). The work on pathogenic yeasts at iBB-IST is currently supported by national funds through Fundação para a Ciência e a Tecnologia (FCT) (PTDC/BII-BIO/28216/2017). Funding (multiannual) received from FCT by iBB (the Institute for Bioengineering and Biosciences) (UIDB/04565/2020) and INESC-ID (UIDB/50021/2020) is also acknowledged.

References

1. Monteiro PT, Oliveira J, Pais P et al (2020) YEASTRACT+: a portal for cross-species comparative genomics of transcription regulation in yeasts. Nucleic Acids Res 48:D642–D649
2. Teixeira MC, Monteiro P, Jain P et al (2006) The YEASTRACT database: a tool for the analysis of transcription regulatory associations in Saccharomyces cerevisiae. Nucleic Acids Res 34:D446–D451
3. Teixeira MC, Monteiro PT, Palma M et al (2018) YEASTRACT: an upgraded database for the analysis of transcription regulatory networks in Saccharomyces cerevisiae. Nucleic Acids Res 46:D348–D353
4. Monteiro PT, Pais P, Costa C et al (2017) The PathoYeastract database: an information system for the analysis of gene and genomic transcription regulation in pathogenic yeasts. Nucleic Acids Res 45:D597–D603
5. Pais P, Oliveira J, Almeida V et al (2021) Transcriptome-wide differences between Saccharomyces cerevisiae and Saccharomyces cerevisiae var. boulardii: clues on host survival and probiotic activity based on promoter sequence variability. Genomics 113:530–539
6. Skrzypek MS, Binkley J, Binkley G et al (2017) The Candida genome database (CGD): incorporation of assembly 22, systematic identifiers and visualization of high throughput sequencing data. Nucleic Acids Res 45:D592–D596
7. Maguire SL, ÓhÉigeartaigh SS, Byrne KP et al (2013) Comparative genome analysis and gene finding in Candida species using CGOB. Mol Biol Evol 30:1281–1291
8. Carbon S, Douglass E, Dunn N et al (2019) The gene ontology resource: 20 years and still GOing strong. Nucleic Acids Res 47:D330–D338
9. Azie N, Neofytos D, Pfaller M et al (2012) The PATH (prospective antifungal therapy) Alliance® registry and invasive fungal infections: update 2012. Diagn Microbiol Infect Dis 73: 293–300
10. Whaley SG, Rogers PD (2016) Azole Resistance in Candida glabrata. Curr Infect Dis Rep 18:41
11. Castanheira M, Messer SA, Rhomberg PR et al (2016) Antifungal susceptibility patterns of a global collection of fungal isolates: results of the SENTRY antifungal surveillance program (2013). Diagn Microbiol Infect Dis 85:200–204
12. Cavalheiro M, Pais P, Galocha M et al (2018) Host-pathogen interactions mediated by MDR transporters in fungi: as pleiotropic as it gets! Genes (Basel) 9:332
13. Costa C, Dias PJ, SÃ¡-Correia I et al (2014) MFS multidrug transporters in pathogenic fungi: do they have real clinical impact? Front Physiol 5:197
14. Costa C, Pires C, Cabrito TR et al (2013) Candida glabrata drug: H+ antiporter CgQdr2 confers imidazole drug resistance, being activated by transcription factor CgPdr1. Antimicrob Agents Chemother 57: 3159–3167
15. Pais P, Califórnia R, Galocha M et al (2020) Candida glabrata transcription factor Rpn4 mediates fluconazole resistance through regulation of ergosterol biosynthesis and plasma membrane permeability. Antimicrob Agents Chemother 64:e00554–e00520
16. Teixeira MC, Monteiro PT, Sá-Correia I (2016) Predicting gene and genomic regulation in Saccharomyces cerevisiae, using the YEASTRACT database: a step-by-step guided

analysis. In: Methods in molecular biology. Humana Press Inc., Totowa, New Jersey, pp 391–404

17. Marichal P, Vanden Bossche H, Odds FC et al (1997) Molecular biological characterization of an azole-resistant Candida glabrata isolate. Antimicrob Agents Chemother 41:2229–2237
18. Sanglard D, Ischer F, Monod M et al (1997) Cloning of Candida albicans genes conferring resistance to azole antifungal agents: characterization of CDR2, a new multidrug ABC transporter gene. Microbiology 143:405–416
19. Mannhaupt G, Schnall R, Karpov V et al (1999) Rpn4p acts as a transcription factor by binding to PACE, a nonamer box found upstream of 26S proteasomal and other genes in yeast. FEBS Lett 450:27–34
20. Owsianik G, Lisabetta Balzi I, Ghislain M, Balzi l L, Ghislain M. (2002) Control of 26S proteasome expression by transcription factors regulating multidrug resistance in Saccharomyces cerevisiae. Mol Microbiol 43:1295–1308

Chapter 24

Enabling Studies of Genome-Scale Regulatory Network Evolution in Large Phylogenies with MRTLE

Shilu Zhang, Sara Knaack, and Sushmita Roy

Abstract

Transcriptional regulatory networks specify context-specific patterns of genes and play a central role in how species evolve and adapt. Inferring genome-scale regulatory networks in non-model species is the first step for examining patterns of conservation and divergence of regulatory networks. Transcriptomic data obtained under varying environmental stimuli in multiple species are becoming increasingly available, which can be used to infer regulatory networks. However, inference and analysis of multiple gene regulatory networks in a phylogenetic setting remains challenging. We developed an algorithm, Multi-species Regulatory neTwork LEarning (MRTLE), to facilitate such studies of regulatory network evolution. MRTLE is a probabilistic graphical model-based algorithm that uses phylogenetic structure, transcriptomic data for multiple species, and sequence-specific motifs in each species to simultaneously infer genome-scale regulatory networks across multiple species. We applied MRTLE to study regulatory network evolution across six ascomycete yeasts using transcriptomic measurements collected across different stress conditions. MRTLE networks recapitulated experimentally derived interactions in the model organism *S. cerevisiae* as well as non-model species, and it was more beneficial for network inference than methods that do not use phylogenetic information. We examined the regulatory networks across species and found that regulators associated with significant expression and network changes are involved in stress-related processes. MTRLE and its associated downstream analysis provide a scalable and principled framework to examine evolutionary dynamics of transcriptional regulatory networks across multiple species in a large phylogeny.

Key words Gene regulation, Regulatory networks, Evolutionary dynamics, Comparative analysis, Probabilistic models, Multi-task learning, Ascomycete yeasts, Phylogeny

1 Introduction

Transcriptional regulatory networks define interactions between regulators, such as transcription factors (TFs) and signaling proteins and target genes [1], and control context-specific gene expression patterns important in various developmental, environmental, evolutionary, and disease processes. Recently, several studies have measured genome-wide expression levels in multiple species for comparative functional genomic analyses [2–4], facilitating a comprehensive study of how regulatory networks vary across species in

Frédéric Devaux (ed.), *Yeast Functional Genomics: Methods and Protocols*,
Methods in Molecular Biology, vol. 2477, https://doi.org/10.1007/978-1-0716-2257-5_24,

response to environmental stimulus. However, computational tools to define and compare regulatory networks across multiple species in a complex phylogeny are in their early stages. To address this need, we developed Multi-species Regulatory neTwork LEarning (MRTLE), a computational approach, based on probabilistic graphical models [5] and multi-task learning to infer regulatory networks for multiple species in a phylogeny. The multi-task learning framework treats each species network inference as a separate task and uses the phylogenetic relationships as a probabilistic prior to constrain the inferred network topologies. A multi-task learning framework enables us to share information between tasks, which is beneficial for species with limited samples. MRTLE's probabilistic framework also handles complex orthology relations arising due to loss and gene duplications that occur in large phylogenies spanning millions of years [6].

The inputs to MRTLE are a phylogenetic tree, branch-specific probabilities for gain or loss of a regulatory edge, gene expression data for each species, and finally gene orthology relationships. One optional input is sequence-specific transcription factor binding motifs for each species, which can be incorporated as prior knowledge to improve the quality of the inferred networks. The outputs of MRTLE are the regulatory networks of each species. The regulatory network of each species is modeled as a probabilistic graphical model [7]. The phylogenetic information is incorporated as a model structure prior to capture the evolutionary dynamics of gaining or losing edges from the ancestral species to the extant species. In this chapter, we describe how to apply MRTLE and several downstream analyses to examine network evolutionary dynamics across large phylogenies in detail. We demonstrate MRTLE's application with a case study of six ascomycete yeast species and how these networks can be validated and analyzed to identify patterns of change that can be associated to different phylogenetic points as well as species-specific traits.

2 Materials

MRLTE is developed in C++, and the source code is publicly available at https://github.com/Roy-lab/mrtle. To compile MRLTE, a C++ compiler (e.g. GCC) and the GNU scientific library (https://www.gnu.org/software/gsl/) are needed to be installed with the option setting to C++14 with GNU extensions (-std=gnu++14). To run MRTLE on expression data in three species, with 1000 genes and 100 regulators, with 30 measurements per species, a laptop or desktop with adequate memory (<1 GB) and disk space (<1 GB) is sufficient. For large datasets, high-throughput computing resources are recommended. The memory requirements increase with the number of species, number of genes and regulators. For each species, MRTLE needs the expression data, the set of target genes, regulators, an orthology mapping (OGIDs file, below), and an optional collection of regulator-gene relationships based on sequence-specific motifs.

The following section provides the steps for installing and applying MTRLE using real expression data from six ascomycete yeasts species: *Saccharomyces cerevisiae (S. cerevisiae), Candida glabrata (C. glabrata), Saccharomyces castellii (S. castellii), Candida albicans (C. albicans), Kluyveromyces lactis (K. lactis), and Schizosaccharomyces (S. pombe)*. The expression data measure the transcriptional response of genes to four stress conditions over time: glucose depletion, heat shock, osmotic stress, and oxidative stress and were originally presented in the MRLTE paper [6]. This dataset has 30 measurements for *S. cerevisiae*, *C. glabrata*, *S. castellii*, *K. lactis*, and *C. albicans* and 21 measurements for the sixth species, *S. pombe*. Species-specific sequence motifs identified by the Cladeoscope algorithm by Habib et al. [8] were used as prior information.

3 Methods

We first introduce some theory of the MRTLE model to help better understand the outputs of MRTLE (Subheading 3.1) and then describe the main steps of using MRTLE (Subheadings 3.2–3.5). To use MRTLE, there are four main steps: downloading and installing MRTLE (Subheading 3.2), preparing input files for MRTLE (Subheading 3.3), application of MRTLE to multi-species datasets (Subheading 3.4), evaluation and validation of results (Subheading 3.5).

3.1 The Theory of MRTLE Model

MRTLE is based on a probabilistic framework for multi-task learning [9], where each task is to learn the network for each species. Each network is modeled as a dependency network [10], which is a class of probabilistic graphical models for representing predictive statistical relationships among random variables. Each gene is modeled as a random variable X_j which encodes the expression level of gene j. A conditional probability distribution $P(X_j \mid X_i)$ models the strength of the interaction between a regulator i and a gene j. In MRTLE, these distributions are estimated using conditional Gaussian distributions. MRTLE incorporates two priors, one is the per-species prior that controls the sparsity of the network and favors a regulatory edge if there exists a species-specific motif instance of the regulator in the promoter region, and another one is the phylogenetic prior. The per-species prior probability is modeled using a logistic function:

$$P(R_{ij} = 1) = \frac{1}{1 + e^{-(\beta_0 + \beta_1 * m_{ij})}}$$

where R_{ij} is a binary variable defining whether there is an edge between regulator i and target j (1 means present, 0 means absent). The β_0 parameter is a sparsity prior that controls the penalty of adding of a new edge to the network, which takes a negative value ($\beta_0 < 0$). A smaller value of β_0 will result in a higher penalty on adding new edges. The β_1 parameter controls how strongly motifs are incorporated as prior ($\beta_1 \geq 0$). A higher value of β_1 will result in motif presence being valued more strongly to select an edge. β_1 is set to 0 when there is no species-specific motif information available. m_{ij} represents the motif instance score if gene j is bound by a motif of regulator i in its promoter region.

The phylogenetic prior is modeled using a continuous-time Markov process, parameterized by the gain and loss probabilities for each branch on the tree defined as:

$$P\left(R_{ij}^{Z} \middle| R_{ij}^{A}\right) = e^{-Qt_z}$$

R_{ij}^{Z} denotes the state of the edge between regulator i and target gene j in species Z. A is the ancestor of species Z, and R_{ij}^{A} denotes the state of the edge between regulator i and target gene j in ancestor A. t_z denotes the branch length for species Z. The gain and loss probabilities for each branch on the tree depend upon the branch length (t_z) and the rate matrix Q which specifies the rates at which we expect regulators to gain or lose targets per unit time. To estimate the rate matrix Q, expectation-maximization-based approaches can be applied, such as SiPhy developed by Garber et al. [11, 12].

3.2 Downloading and Installing MRTLE

Below are the Unix commands to download and install MRTLE in a standard command line terminal. There are three steps: (1) Download the source code from GitHub. (2) Enter the MRTLE code directory using cd. (3) Compile the code.

```
1) git clone https://github.com/Roy-lab/mrtle.git
2) cd mrtle/code/
3) make
```

3.3 Preparing Inputs for MRTLE

MRTLE needs several input files. Below we list these and describe its contents and how it can be created. All the files mentioned are available at https://github.com/Roy-lab/mrtle/tree/master/data.

3.3.1 Configuration File

This file specifies the location of the species-specific input data files and the location of the outputs. Each line corresponds to a species and should have the following tab-delimited species-specific entries: 1) species name, 2) filename of the expression data for each species, 3) the output directory, 4) filename of a list of regulators to be considered for a given target gene, 5) filename of a list of target genes to be used, 6) filename of prior network for each species. Here is an example of three lines in the configuration file (horizontal and vertical lines are just for clarity):

Scer	data/expression/Scer_geneexp.txt	out/Scer/	data/TFs/Scer_tf_mrna_binding.txt	data/genelists/Scer_allGenes.txt	data/motifs/Scer_motifs.txt
Cgla	data/expression/Cgla_geneexp.txt	out/Cgla/	data/TFs/Cgla_tf_mrna_binding.txt	data/genelists/Cgla_allGenes.txt	data/motifs/Cgla_motifs.txt
Scas	data/expression/Scas_geneexp.txt	out/Scas/	data/TFs/Scas_tf_mrna_binding.txt	data/genelists/Scas_allGenes.txt	data/motifs/Scas_motifs.txt

3.3.2 Expression Data File

In the file of the expression data for each species, each row represents one gene, and each column represents one condition. The first row is the header. Pre-processing of the expression data is described in **Note 1**. Here is an example of the expression data file for the *S. cerevisiae* species, `Scer_geneexp.txt` mentioned above.

Name	LAG	LL	DS	PS	PLAT	HS5
YAL016W	-2.57316	0.18674	0.351552	0.884193	0.461141	0.305003
YAL042W	-0.827777	-0.068987	-0.223257	-1.07268	-1.00647	0.0169562
YAL053W	0.981654	0.421282	0.0732861	-0.274487	0.344917	0.0213995
YBL030C	-0.956715	0.738636	1.99264	2.5738	2.57844	-0.077775
YKL112W	0.91353	-0.441429	-0.512388	-2.79075	-1.76492	0.264341

3.3.3 Species-Specific Regulators File

This file contains the list of the genes to be considered as potential regulators for each species. A regulator per line is specified. Example lines from the species-specific regulators file `Scer_tf_mrna_binding.txt`:

```
YKL112W
YAL021C
YBR130C
YAL011W
```

3.3.4 Species-Specific Targets File

This file contains the list of the genes to be considered as potential targets for each species. Each line lists the name of the target gene.

Example lines from the species-specific target genes file `Scer_allGenes.txt`:

```
YAL016W
YAL042W
YAL053W
YBL030C
```

3.3.5 Motif-Based Prior Network File (Optional)

The motif-based prior network file has three tab-separated columns, listing the regulator, target gene, and motif score. Example lines from the prior network file `Scer_motifs.txt`:

YKL112W	YAL016W	0.868431
YKL112W	YAL042W	0.902258
YKL112W	YAL053W	0.836423
YKL112W	YBL030C	0.807526

3.3.6 Tree File

This file specifies the phylogenetic tree structure and can be obtained from public resources like the Ensembl Compara database (http://useast.ensembl.org/info/genome/compara/index.html). For the six yeasts species example, the species tree was obtained from Wapinski et al. [13]. To generate the species tree, multiple sequence alignments from orthogroups were concatenated [13], and the tree topology was learned using the Phylip package [14]. The PAML [15] package is another tool that can be used for the species-tree-inference task from gene orthology data. The tree file should have five columns: (1) the child species, (2) the relation of the child species to the parent species, whether it is a left or right child, (3) the parent species, (4) branch-specific gain probability p_g, ranging from 0 to 1, (5) branch-specific loss probability p_l, ranging from 0 to 1. In **Note 2**, we discuss how to determine branch-specific gain probability p_g and branch-specific loss probability p_l. Example lines from the file `yeast_tree_rates.txt` are as below:

Anc11	left	Anc14	0.0529786	0.882977
Spom	right	Anc14	0.0538444	0.897406
Anc9	left	Anc11	0.0465452	0.775753
Calb	right	Anc11	0.0543971	0.906619
Anc5	left	Anc9	0.022713	0.378551
Klac	right	Anc9	0.0515299	0.858832
Anc4	left	Anc5	0.0139051	0.231751

(continued)

Scas	right	Anc5	0.0423569	0.705948
Scer	left	Anc4	0.0454312	0.757187
Cgla	right	Anc4	0.0471082	0.785137

3.3.7 Gene Orthology Mappings File (OGID File)

This file has the orthogroup information for genes. The gene trees can be obtained from public resources like the Ensembl Compara database (http://useast.ensembl.org/info/genome/compara/index.html), and the Inparanoid eukaryotic ortholog database (http://inparanoid.sbc.su.se/cgi-bin/index.cgi) [16]. For the six yeasts species example, the gene orthology mappings were created using the SYNERGY algorithm [17]. Each row represents an orthogroup of genes. The first column of this file is of the format OG{ID}_{DUP}. Each orthogroup is denoted by an ID, which takes a numeric value. For orthogroups with duplications, DUP denotes the ith duplication. If there are no duplications in the dataset, DUP is set to 1. The second column of this file is the gene name for each orthogroup from the respective species and separated by a comma across all species. The gene names should be in the same order as in the species list file described next (Subheading 3.3.8). If an orthogroup does not exist in one species, NONE is used as the gene name. Gene names must be unique across species. Example lines from the file `OGid_members.txt` are as follows:

Gene_OGID	NAME
OG16_1	YLL043W,CAGL0C03267g,Scas400.1, KLLA0E00550g,NONE,SPAC977.17
OG16_2	YLL043W,CAGL0C03267g,Scas400.1, KLLA0E00550g,NONE,SPAC977.17
OG23_1	NONE,NONE,NONE,NONE,NONE,SPAC1F8.07c

3.3.8 Species List File

This file contains the list of all the species with orthology relationships in the OGID file. The order should be the same as the species listed in the second column of the OGID file. Example lines from the file `specorder_allclade.txt` are as below:

```
Scer
Cgla
Scas
Klac
Calb
Spom
```

3.3.9 Potential Regulator Orthogroups ID File

This file contains the list of IDs of the orthogroups to be considered as potential regulators. The ID is the same number as in the OGID file (Subheading 3.3.7). A regulator must also be present in the species-specific list of regulators given in the species-specific configuration file (Subheading 3.3.1). Example lines from the file `TFs_OGs.txt` are as below:

```
10261
1033
104
105
```

3.3.10 Potential Target Orthogroups ID File

This file specifies the list of IDs of the orthogroups to be considered as potential targets. This option can be used as a way to parallelize MRTLE on separate gene sets (*see* **Note 4**). The ID is same number as in the OGID file describing the orthology relationships (Subheading 3.3.7). A target must also be present in the species-specific list of targets given in the species-specific configuration file (Subheading 3.3.1). Example lines from the file `AllGenes.txt` are as below:

```
100
1000
1001
10013
```

3.4 Applying MRTLE to Multi-Species Data

MRTLE takes the following inputs: (1) expression data for each species, (2) a phylogenetic tree relating the species, (3) branch-specific probabilities for gain or loss of a regulatory edge, (4) gene orthology relationships specified in an OrthoGroup ID (OGID) file, (5) list of regulators in each species, (6) optional species-specific regulatory information, for example, from motif instances in gene promoters (Fig. 1a). Example inputs are available from the website https://github.com/Roy-lab/mrtle/tree/master/data. These are specified through various MRTLE arguments and file contents described in Subheading 3.3.

The arguments of MRTLE are: (1) a configuration file, (2) maximum number of regulators to be considered for a given target gene, (3) the number of cross-validation, (4) the gene orthology relationship file, (5) a list of the orthogroups to be considered as potential regulators, (6) a list of orthogroups to be considered as target genes, (7) the phylogenetic tree, (8) a list of the species, (9) the sparsity parameter, and (10) the parameter that controls how strongly the motif networks are incorporated as prior.

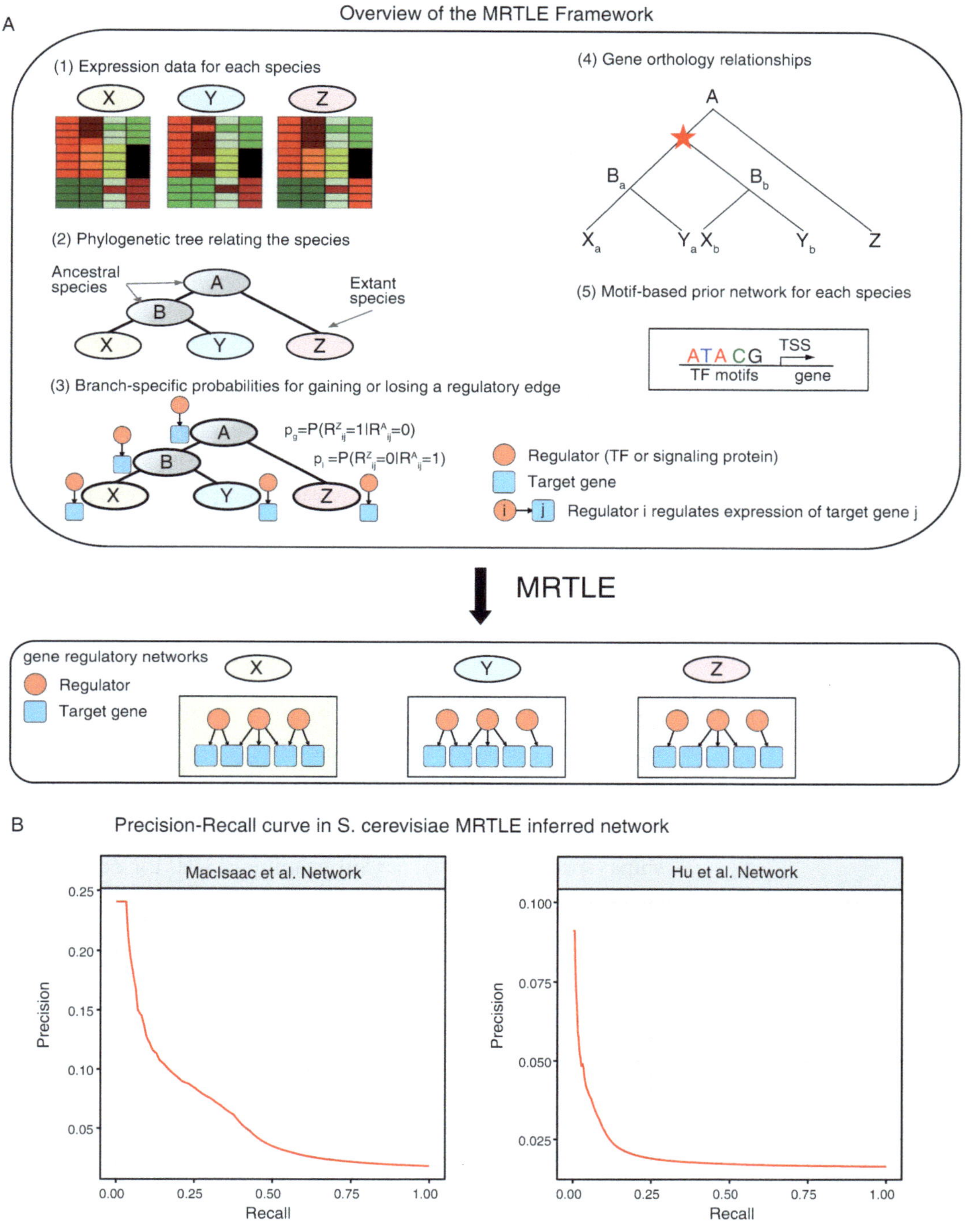

Fig. 1 (**a**) Overview of the MRTLE framework. MRTLE takes the following inputs: (1) expression data for each species, (2) a phylogenetic tree relating the species, (3) branch-specific probabilities for gain or loss of a regulatory edge, (4) gene orthology relationships, and, (5) optional motif-based prior network for each species. The outputs of MRTLE are regulatory networks for each species. (**b**) The precision and recall curves of comparing the MRTLE-inferred network for *S. cerevisiae* to two gold standard networks from ChIP-chip [19] and regulators knockout [20] assays

An example usage of running MRTLE using the example data is as below:

```
./mrtle -f data/speciesconf.txt -x300 -v1 -m data/
  OGid_members.txt -l data/TFs/TFs_OGs.txt -n data/
  genelists/AllGenes.txt -d data/
  yeast_tree_rates.txt -s data/
  specorder_allclade.txt -b -0.9 -q 4.0
```

Below we describe each of these arguments to the MRTLE program in detail.

1. **-f** takes the configuration file in which each row represents information of one species. *See* Subheading 3.3.1.
2. **-x** specifies the maximum number of regulators to be considered for a given target gene. This can be a large number, e.g., 200, to enable prediction of as many regulators as possible.
3. **-v** specifies the number of cross-validation, which can be set to 1 as well. If it is set to 1, the model is trained on the entire data.
4. **-m** specifies the name of the OGID file, describing the gene orthology relationships. *See* Subheading 3.3.7 for the description of how to make this file.
5. **-l** specifies the list of IDs of the orthogroups to be considered as potential regulators. The ID number is the ID in the OGID file (specified by **-m**). *See* Subheading 3.3.9 for details.
6. **-n** specifies the list of IDs of the orthogroups to be considered as potential targets. This option can be used as a way to run MRTLE in parallel on separate gene sets (*see* **Note 4**). The ID number is the ID in the file describing the orthology relationships (specified by **-m**). *See* Subheading 3.3.10 for details.
7. **-d** takes the file of the phylogenetic tree. This file provides the phylogenetic relationship of the species, and branch-specific gain and loss probabilities. *See* Subheading 3.3.6 for how to prepare the tree file.
8. **-s** specifies the list of all the species present in the file describing the orthology relationships (specified by **-m**). *See* Subheading 3.3.8 for how to prepare the species list file.
9. **-b** specifies the β_0 parameter which controls the sparsity of the network, a penalty for adding new edges. In **Note 3**, we discuss how to choose the β_0 parameter.
10. **-q** specifies the β_1 parameter which controls how strongly motif instances are incorporated as prior. A higher value will result in motif network being valued more strongly. *See* **Note 3** about how to choose the β_1 parameter.

3.5 MRTLE Outputs and their Interpretation

Assuming MRTLE is run according to the usage provided in Subheading 3.4, the MRTLE outputs will be available in the `out/` directory where the command is executed. These results include (1) inferred regulatory network for each species within its own species-specific folder: `var_mb_pw_k299.txt`, (2) model parameters: `modelparams.txt`, (3) the data likelihood of the MRTLE model: `scoreFile.txt`. Most of the interpretation is done on the network file.

In the regulatory network file, the first column denotes the regulator, the second column denotes the target gene, and the third column denotes the regression edge weight capturing the strength of the interaction between the regulator and the target gene (*see* Subheading 3.1). Example lines from the regulatory network file `var_mb_pw_k299.txt` are given below:

YDR443C	Q0045	-0.0392211
YPL128C	Q0045	0.16192

In order to get reliable results, we typically run MRTLE algorithm in a stability selection framework. We randomly subsample the expression data for each species 25 times keeping 2/3 of the samples each time, and subsequently apply MRTLE algorithm on each subsample to infer a regulatory network. More subsamples could be beneficial if there are more measurements in the expression data. A confidence score is then calculated for each edge as the fraction of how frequently that edge is present in the inferred networks across all subsamples.

MRTLE outputs can be interpreted using several downstream analyses:

1. Compare MRTLE inferred regulatory networks to gold standard networks, such as transcription factor knockout network and ChIP-chip-based regulatory network (Fig. 1b).
2. Identify target genes of some TFs of interest, visualize the regulatory subnetworks in each species, and compare the dynamics of targets (Fig. 2).
3. Compare MRTLE inferred regulatory networks across species to get patterns of conservation and divergence across a phylogenetic tree (Fig. 3a).
4. Visualize the inferred regulatory networks in each species to identify important hubs (Fig. 3b).
5. Identify transcriptional modules using algorithms such as Arboretum [18] and combine MRTLE inferred regulatory networks with gene modules to get context-specific networks associated with different gene modules.

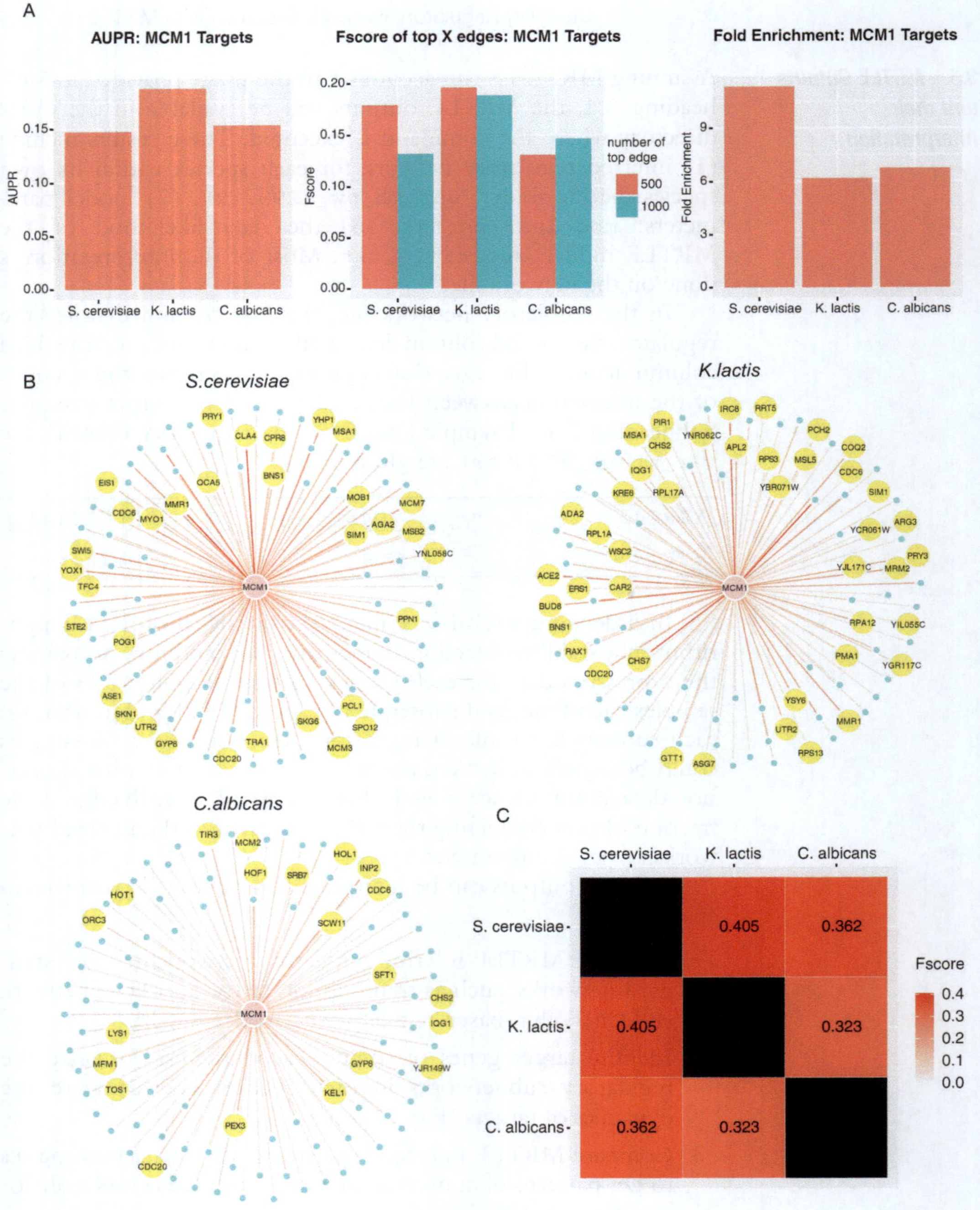

Fig. 2 (**a**) The Area Under the Precision Recall Curve (AUPR) comparing the MRTLE-inferred targets of MCM1 to the ChIP-chip MCM1 targets (left). The F-score comparing the top 500, 1000 predicted MCM1 targets to the ChIP-chip MCM1 targets (middle). The fold enrichment of MCM1 ChIP-chip gene targets in the MRTLE-inferred MCM1 targets among the top 30k edges (right). (**b**) The predicted targets of MCM1 in *S. cerevisiae*, *K. lactis*, and *C. albicans*. The top 100 target genes of MCM1 are shown and nodes that are also ChIP-chip hits are labeled with common names. If a common name does not exist, it is replaced with the gene name of *S. cerevisiae*. For genes with more than one orthologs, the one with the maximum confidence score among all orthologs is shown. (**c**) The patterns of how MRTLE-inferred MCM1 targets change across species. Shown are the F-score similarities between the top predicted MCM1 targets for every two species

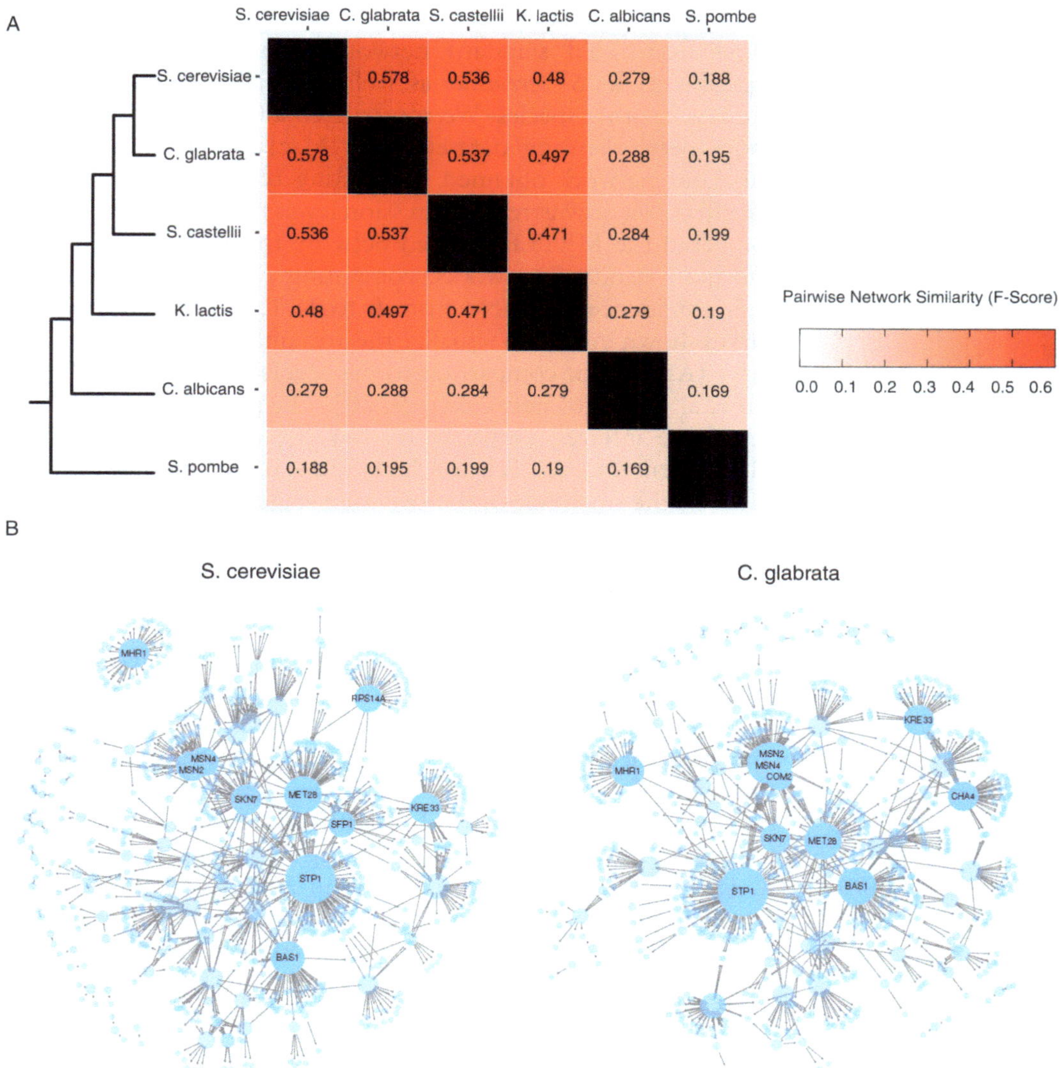

Fig. 3 (**a**) The patterns of conservation and divergence of MRTLE inferred regulatory networks across a phylogenetic tree. Shown are the F-score similarities for all pairs of species' networks. (**b**) The MRTLE inferred networks for *S. cerevisiae* and *C. glabrata.* Shown are the top 1k edges in each species. The size of the node is proportional to the number of targets. Gene names of top 10 hubs are shown in the network

6. Assess the role of a regulator as an activating or repressive factor by computing the significance of overlap between the targets of this regulator in MRTLE inferred networks and the targets in activating module versus repressive module based on hypergeometric distribution. We demonstrate the first four analyses as the examples in this chapter.

First, we evaluate how well MRTLE inferred networks recover edges from gold standard networks, derived from large-scale experiments in *S. cerevisiae* as well as smaller regulator specific datasets in other species. We constructed a ChIP-chip-based regulatory network using datasets from MacIsaac et al. [19] and a knockout network obtained by deleting regulators and assessing the effects on downstream expression levels from Hu et al. [20]. We compared the MRTLE-inferred network for *S. cerevisiae* to these two gold standard networks, computing precision and recall for each confidence threshold from 0 to 1. The precision and recall curves shown in Fig. 1b visualize the area under the precision–recall curve (AUPR) for comparison to the MacIsaac et al. (0.0607) and Hu et al. (0.0206) networks [19, 20]. The greater the AUPR, the better the inferred networks recover the experimental validated edges. Here the AUPR is not high and is understood to be dominated by the small size of the two gold standard networks available for comparison. In addition, we compared the target genes inferred for specific regulators with ChIP-chip data in species other than *S. cerevisiae.* We obtained binding gene targets for a set of TFs measured with ChIP-chip experiments by Tuch et al. in *S. cerevisiae, K. lactis, and C. albicans* [21], and tested how well the MRTLE-inferred networks recapitulate these experimentally derived target sets. We compared how many targets are recovered in each species using AUPR, F-score, and fold enrichment metrics. Among 227 targets of MCM1 in *S. cerevisiae*, 99 target genes are recovered in *S. cerevisiae* MRTLE inferred network considering all edges (AUPR: 0.148, Fig. 2a). Among 562 targets of MCM1 in *K. lactis*, 183 target genes are found in *K. lactis* MRTLE inferred network (AUPR: 0.187, Fig. 2a). 225 target genes out of 680 targets of MCM1 in *C. albicans* are recovered in MRTLE inferred network (AUPR: 0.153, Fig. 2a). Among all predicted MCM1 targets, we picked top 500, 1000 MCM1 targets with highest confidence and obtained an F-score ranging from 0.130 to 0.196 compared to ChIP-chip MCM1 targets (Fig. 2a). Moreover, we selected the top 30k edges inferred in each species ranking by the confidence score and calculated the fold enrichment of the gene targets of MCM1 from ChIP-chip data in the predicted MCM1 targets (Fig. 2a). We found that the predicted targets were enriched for MCM1 targets from ChIP-chip data in all three species.

MRTLE inferred networks can be compared to assess the global pattern of conservation and divergence based on the similarity of the networks of each pair of species (Fig. 3a). We picked the top 30k edges based on confidence in each species, and computed the F-score of similarity for each pair of networks to quantify the variation in the networks across species. We found that the inferred networks exhibit patterns of conservation and divergence consistent with the phylogenetic tree. For example, *S. cerevisiae* is most similar to *C. glabrata* (F-score 0.578), its similarity to *S. castellii*

(F-score 0.536), *K. lactis* (F-score 0.480), and *C. albicans* (F-score 0.279) decreases and is least similar to *S. pombe* (F-score 0.188). Similarly, the network similarity of *C. glabrata* with *S. castellii*, *K. lactis*, *C. albicans*, and *S. pombe* decreases along the phylogenetic tree. This indicates that the networks inferred by MRTLE obey a phylogenetic trend.

MRTLE networks can also be compared by visualizing the subnetworks of known and important regulators across species (e.g., in our study, regulators involved in stress response are useful to compare across species). As an example, we visualize the MRTLE subnetworks for targets genes of the transcription factor MCM1 in *S. cerevisiae, K. lactis, and C. albicans* (Fig. 2b). The top 100 target genes of MCM1 are shown and the nodes supported by the MCM1 targets from ChIP-chip experiments are named. To compare how the targets change across species, we picked top 1000 target genes from each inferred network and computed the F-score similarity. We see that target genes in *S. cerevisiae* and *K. lactis* are more conserved (F-score 0.405), compared to *S. cerevisiae* and *C. albicans* (F-score 0.362), indicating the phylogenetic trend is also recapitulated at a local factor-specific level (Fig. 2c). Additionally, we visualize the MRTLE networks of top 1k edges in *S. cerevisiae* and *C. glabrata* (Fig. 3b) and identify top important hubs such as MSN2, MSN4, STP1, KRE33, and BAS1 which are conserved in both the species. MSN2 and MSN4 have been found to be general stress response regulators and are key transcriptional regulators of yeast glycolysis by regulating the expression of glycolytic enzymes [22]. STP1 plays a key role in pre-tRNA processing in yeast, regulating the expression of genes encoding amino-acid permeases and controlling the uptake of several amino acids [23].

4 Notes

1. Pre-processing of Gene Expression Data.

   The input expression data for MRTLE can be Microarray data or RNA-sequencing (RNA-seq) data. MRTLE model assumes that the expression values are from a multivariate Gaussian distribution, so we typically perform some normalization and transformation on the expression data. For microarray data, the values are log ratio of gene expression level from an experimental sample and gene expression level in the reference sample, which can be used as input expression data without further transformation. The transcriptomic data in the six yeast species example are microarray data. For RNA-seq data, we recommend normalizing the data across all samples for each species by applying a log transformation and doing quantile normalization.

2. Estimating the Parameters: Branch-Specific Gain and Loss Probabilities.

   Setting the appropriate prior probability of gaining or losing an edge for each branch along the phylogenetic tree may need some tunings and knowledge about the phylogeny. For our example data with six yeasts species, the rate matrix Q was estimated using the average rate of gaining or losing a motif binding site as in the Habib et al. paper [8]. The branch-specific gain probability $p_g = P\left(R_{ij}^{Z} = 1 \middle| R_{ij}^{A} = 0\right)$ and branch-specific loss probability $p_l = P\left(R_{ij}^{Z} = 0 \middle| R_{ij}^{A} = 1\right)$ were then calculated based on this rate matrix and the branch length for each branch in the phylogenetic tree.

3. Setting the Parameters: $\boldsymbol{\beta_0}$ and $\boldsymbol{\beta_1}$.

   The β_0 parameter is a sparsity prior that controls the penalty of adding of a new edge to the network, which takes a negative value ($\beta_0 < 0$). We suggest trying β_0 for values between -5 and -0.1. A smaller (more negative) value will add a greater cost to edge addition and end up with a sparser network. The β_1 parameter controls how strongly motifs are incorporated as prior ($\beta_1 \geq 0$). We suggest setting β_1 to a value between 0 and 5. A higher value will increase the weight of the motif prior network. MRTLE can be run with different configurations of β_0 and β_1, and one can choose the configuration that gives the highest data likelihood score or recapitulates the known phylogenetic trends.

4. Running MRTLE in Parallel for Multiple Gene Subsets.

   To make MRTLE run faster, we can split the target gene set into subsets, e.g., of 50 genes and run MRTLE in parallel on each set. This can be implemented by splitting the potential target orthogroups into non-overlapping sets of 50 orthogroups per set and storing the corresponding OGID numbers into `AllGenes0.txt`, `AllGenes1.txt`, etc. (Subheading 3.3.10). Next, MRTLE can be run in parallel using each of the potential target orthogroup ID file (`AllGenes` `${i}.txt`) while keeping the rest of the arguments the same. After all of the parallel runs are finished, we merge the inferred networks of each 50 target genes together as the final inferred network.

Acknowledgments

This work is supported by the National Institutes of Health (R01-HG010045-01), James McDonnell Foundation and the National Science Foundation (DBI-1350677).

References

1. Kim HD, Shay T, O'Shea EK et al (2009) Transcriptional regulatory circuits: predicting numbers from alphabets. Science 325:429–432
2. Brawand D, Soumillon M, Necsulea A et al (2011) The evolution of gene expression levels in mammalian organs. Nature 478:343–348
3. Brawand D, Wagner CE, Li YI et al (2014) The genomic substrate for adaptive radiation in African cichlid fish. Nature 513:375–381
4. Thompson DA, Roy S, Chan M et al (2013) Evolutionary principles of modular gene regulation in yeasts. elife 2:e00603
5. Markowetz F, Spang R (2007) Inferring cellular networks—a review. BMC Bioinformatics 8: S5
6. Koch C, Konieczka J, Delorey T et al (2017) Inference and evolutionary analysis of genome-scale regulatory networks in large phylogenies. Cell Syst 4:543–558.e8
7. Friedman N (2004) Inferring cellular networks using probabilistic graphical models. Science 303:799–805
8. Habib N, Wapinski I, Margalit H et al (2012) A functional selection model explains evolutionary robustness despite plasticity in regulatory networks. Mol Syst Biol 8:619
9. Caruana R (1997) Multitask Learning. Mach Learn 28:41–75
10. Heckerman D, Chickering DM, Meek C et al (2001) Dependency networks for inference, collaborative filtering, and data visualization. J Mach Learn Res 1:49–75
11. Hobolth A, Jensen JL (2005) Statistical inference in evolutionary models of DNA sequences via the EM algorithm. Stat Appl Genet Mol Biol 4:18
12. Garber M, Guttman M, Clamp M et al (2009) Identifying novel constrained elements by exploiting biased substitution patterns. Bioinformatics 25:i54–i62
13. Wapinski I, Pfeffer A, Friedman N et al (2007) Natural history and evolutionary principles of gene duplication in fungi. Nature 449:54–61
14. Felsenstein J (1989) PHYLIP-phylogeny inference package (version 3.2). Cladistics 5:164–166
15. Yang Z (2007) PAML 4: phylogenetic analysis by maximum likelihood. Mol Biol Evol 24: 1586–1591
16. O'Brien KP, Remm M, Sonnhammer ELL (2005) Inparanoid: a comprehensive database of eukaryotic orthologs. Nucleic Acids Res 33: D476–D480
17. Wapinski I, Pfeffer A, Friedman N et al (2007) Automatic genome-wide reconstruction of phylogenetic gene trees. Bioinformatics 23: i549–i558
18. Roy S, Wapinski I, Pfiffner J et al (2013) Arboretum: reconstruction and analysis of the evolutionary history of condition-specific transcriptional modules. Genome Res 23: 1039–1050
19. MacIsaac KD, Wang T, Gordon DB et al (2006) An improved map of conserved regulatory sites for Saccharomyces cerevisiae. BMC Bioinformatics 7:113
20. Hu Z, Killion PJ, Iyer VR (2007) Genetic reconstruction of a functional transcriptional regulatory network. Nat Genet 39:683–687
21. Tuch BB, Galgoczy DJ, Hernday AD et al (2008) The evolution of combinatorial gene regulation in fungi. PLoS Biol 6:e38
22. Kuang Z, Pinglay S, Ji H et al (2017) Msn2/4 regulate expression of glycolytic enzymes and control transition from quiescence to growth. elife 6:e29938
23. Jorgensen MU, Gjermansen C, Andersen HA et al (1997) STP1, a gene involved in pre-tRNA processing in yeast, is important for amino-acid uptake and transcription of the permease gene BAP2. Curr Genet 31:241–247

Chapter 25

Omics Analyses: How to Navigate Through a Constant Data Deluge

Thomas Denecker and Gaëlle Lelandais

Abstract

Omics data are very valuable for researchers in biology, but the work required to develop a solid expertise in their analysis contrasts with the rapidity with which the omics technologies evolve. Data accumulate in public databases, and despite significant advances in bioinformatics softwares to integrate them, data analysis remains a burden for those who perform experiments. Beyond the issue of dealing with a very large number of results, we believe that working with omics data requires a change in the way scientific problems are solved. In this chapter, we explain pitfalls and tips we found during our functional genomics projects in yeasts. Our main lesson is that, if applying a protocol does not guarantee a successful project, following simple rules can help to become strategic and intentional, thus avoiding an endless drift into an ocean of possibilities.

Key words Omics, Data, Information, Knowledge, Visualization, Analysis cycle

1 Introduction

In 2021, there should be no need to explain that scientists' daily routines are challenged by a constant data deluge. Email messages, conferences, social media, web and smartphone alerts from around the world all bring information about new articles to read, discoveries to understand, datasets to analyze, or hypotheses to evaluate, all the time. The rapidity with which data acquisition technologies evolve contrasts with the amount of work required to develop solid expertise in a domain. Researchers are struggling to find a balance between the ever-increasing flow of data they receive and the limited time they have, to process and understand the data.

The terminology "data deluge" first appeared in the literature over 10 years ago [1]. In biology and life sciences, it has been tightly coupled to the emergence of multi-omics experimental methods, i.e., genomics, transcriptomics, proteomics, metabolomics, and more recently epigenomics, glycomics, lipidomics, or microbiomics [2]. The common element of these disciplines is the

Frédéric Devaux (ed.), *Yeast Functional Genomics: Methods and Protocols*,
Methods in Molecular Biology, vol. 2477, https://doi.org/10.1007/978-1-0716-2257-5_25, © The Author(s) 2022

simultaneous acquisition of experimental measurements for several thousands of cellular components. Year after year, the number of technical solutions grows considerably, and the associated biological data accumulate in public databases. As illustrations, the website “Enseqlopdia” references more than 350 different Next-Generation Sequencing (NGS) methods [3], and the scientific journal *Nucleic Acids Research* (NAR) publishes descriptions of more than 1600 molecular biology databases [4]. And this is very likely just the tip of the iceberg. How much data has been left behind on hard drives of scientists' computers? It is not surprising that journals like *BMC Research Notes* have created a new category called “data notes” to encourage data sharing [5].

Omics data are very valuable for research in biology, providing incredible opportunities to better understand cell function, and the underlying genomic regulatory systems, but like any innovative technology, omics data follows the “hype cycle” [6]. After a peak of inflated expectations in the 2000s (human genome project), many researchers experienced the trough of disillusionment [7]. It became clear that the accumulated omics datasets cannot speak by themselves and thus, the discipline of “data science” emerged as the new hype [8]. Ten years later, despite great perspectives raised by machine learning and artificial intelligence in medical fields [9], multi-omics data analysis remains a major bottleneck in most research projects [10].

We define data analysis as the process of inspecting, cleaning, transforming, and modeling a dataset, with the aim of revealing new information. In an omics project, full interpretation of the data that were generated constitutes the major expense [11]. It requires a multidisciplinary team of bioinformaticians, statisticians, and biologists to transform the raw data into useful biological knowledge. With this in mind, many integrated computational tools were developed to help researchers with no particular informatic skills to work on their data. The most emblematic is the Galaxy web platform [12], which is based on a tool shed gathering several thousand software packages. Clearly, technical resources exist today, as well as training programs to learn how to use them [13]. Exhaustive courses in Data Science, Bioinformatics, Statistics, and Computational Biology are freely available on MOOC platforms like Coursera [14], for both beginners and advanced researchers. In this context, we may wonder why data analysis remains such a burden for those who produce the data. We believe that beyond the issue of dealing with a very large number of experimental results, working with omics data requires a change in the way we solve scientific problems.

Most biologists were taught to follow the scientific method, which consists of (1) making an observation; (2) formulating hypotheses to explain this observation; (3) designing experiments to test the hypotheses; (4) performing the experiments; and

(5) drawing conclusions from the experiments which support or reject the hypotheses. This is a difficult process, which can last several weeks, months, or even years. Therefore, how can we imagine generalizing such a method in the context of omics data? In the literature, the answer to this question instigates disagreements between researchers over how science should be properly conducted [15]. More pragmatically, if a researcher does not want to spend his/her entire scientific career testing the hypotheses which can be raised after only a single omics experiment, deductive and inductive logics are both indispensable. And this is a new way of thinking especially difficult to acquire, because it is not explicitly described or even understood. In a word, there is no standard procedure.

In this chapter, our aim is to share pitfalls and tips from our experience of functional genomics in yeasts, working with multi-omics datasets [16–19]. Even though our colleagues refer to us as "bioinformaticians," we have very different scientific backgrounds and histories. Fifteen years separate our first contributions to functional genomics projects [20, 21], 15 years during which bioinformatics has changed considerably [22]. From a field essentially devoted to the study of sequences (genes and proteins), bioinformatics has become a discipline centered around the data (Fig. 1). We each have a different conception of what "we should be doing" to improve what "we already do." Above all, we have learned by working together that following a protocol does not guarantee a successful project. More than ever, there is no right or wrong answer; hence, data analysts must be creative scientists and feel free to think outside the box. The following text is organized into three main sections. In the first one, we present three basics for data analysis. We believe they can provide a foundation for assessment of the concepts that underlie the challenges of any data analysis. In the second section, we remind the reader that "cherry picking" and "p-hacking" are two major issues they must be aware of, especially in omics data analyses. Finally, in the third section, we recommend "10 simple rules," which we have experienced to be helpful in our daily work.

2 Basics for Data Analysis

2.1 Basics #1: Differentiate "Data," "Information," and "Knowledge"

Data analysis is the procedure by which new information and knowledge are discovered from the data, and hence, it helps make decisions and highlight conclusions. In this context, the words "data," "information," and "knowledge" are often connected, but it is worth considering that they refer to very different steps and aims in a data analysis [23].

Data, by definition, represents the starting point of the whole story. The term is plural, equivalent to "facts." Therefore, data are

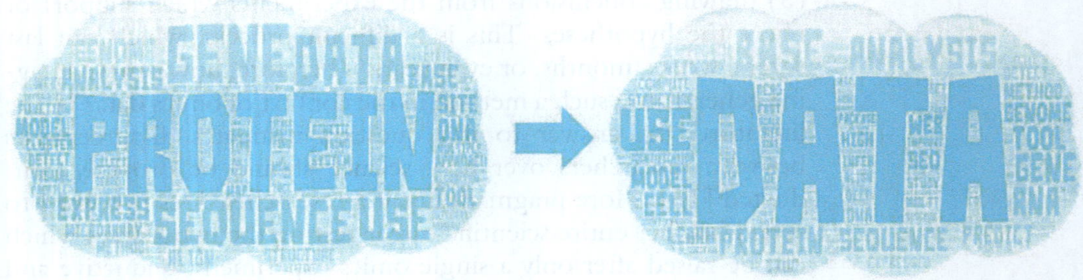

Fig. 1 Word clouds from the titles of articles published in the field of "bioinformatics" between 2000–2005 (left) and 2015–2020 (right). The list of journals from which the articles were taken is: *Bioinformatics*, *BMC Bioinformatics*, *Briefings in Bioinformatics*, and *Journal of Bioinformatics and Computational Biology*. Note that the larger the word, the more frequently it is used. The period 2000–2005 corresponds to the time G. Lelandais prepared her PhD and 2015–2020 corresponds to the time T. Denecker prepared his PhD

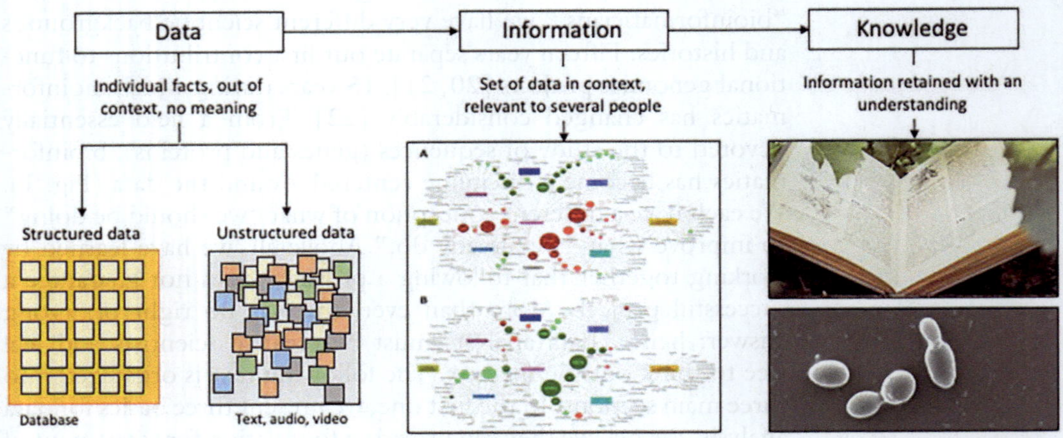

Fig. 2 Illustrations of basics #1 for data analysis. The progression from data to information and knowledge is illustrated here. Data are individual facts, out of context, and with no meaning. They can be either structured data or unstructured data (see the main text). Information arises from a set of data, which have been placed in a context. Information is therefore relevant to several people. In this picture, two lists of genes for which expression is activated (red color) or repressed (green color) are represented. They are classified around functional categories. More details for this analysis can be found in [16]. Knowledge is the final stage of any data analysis. It is the most challenging to achieve

the individual facts, which are still out of any context. In biology, they rely on the raw experimental results or observations. Data can be of different types, i.e., numerical or textual, but they can also be images, audio recordings, videos, etc. The diversity of data sources is certainly one of the biggest challenges for their analysis. At this stage, it is important to distinguish structured data from unstructured data (Fig. 2). Structured data are easily organized in databases and can be efficiently retrieved and manipulated. Unstructured data are more inconsistent and complicated to use. Preliminary work is required to find and/or define their underlying organization, and for that, manual verification or curation is often necessary. This makes the process uncertain and time consuming. Whether structured or unstructured, the key idea behind the concept of data is that they have no meaning.

This is the exact difference with the term "information." Considered to be singular, information can be seen as a set of data placed in a particular context. Hence, information has a relevance to one or more people and, for them, means something. Information can, for instance, be that a gene is more expressed when cells are in a particular condition than it is in another condition or that the overall organization of a network of genes has a particular topology (Fig. 2). Whatever the signification is (simple or more complex), an information arises from inspection of the data, applying procedures decided by the scientist doing the data analysis. Importantly, this implies that multiple information can arise from a single dataset, depending on the processes used to inspect the data. Information therefore represents a great advance in a data analysis, but it should not be confused with the word "knowledge."

This last idea is a key concept to understand what data analysis really is. Indeed, by definition, knowledge represents an information which is understood. It means that we know why and how the information is obtained, and we can often make predictions based on this understanding. Considering previous examples, the knowledge can be, for instance, the existence of a regulatory protein which impacts the transcriptional activity of a set of genes in the studied condition and drives the overall topology of the associated gene network.

Data analysis is a very lexically challenged discipline. The terms "data," "information," and "knowledge" are often misused or used interchangeably. To limit misunderstanding among collaborators in a project, these three terms must be formally defined and consistently used. Indeed, if one person is expecting new knowledge from an analysis, whereas the other is working on collecting new information, it could be difficult to reconcile both their expectations. Also, it is our experience that trying to obtain new knowledge without first generating new information is a very risky strategy, a bit like looking for a needle in a haystack.

2.2 Basics #2: Recognize the Importance of Data Visualization

At all stages of a data analysis, visualization plays a critical role. This is well illustrated by considering Anscombe's quartet [24]. Whereas the four datasets have almost identical descriptive statistics (in terms of individual mean, variance, and pairwise correlation), they appear to have very different distributions when they are plotted on a graph (Fig. 3a). Graphics are therefore very important for increasing perspectives on a set of observations.

We have often observed that data visualization is underestimated by our collaborators. The main reason is probably that we

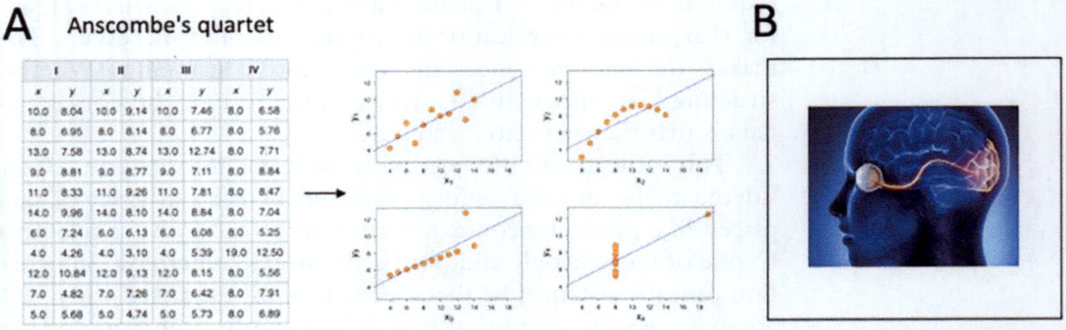

I		II		III		IV	
x	y	x	y	x	y	x	y
10.0	8.04	10.0	9.14	10.0	7.46	8.0	6.58
8.0	6.95	8.0	8.14	8.0	6.77	8.0	5.76
13.0	7.58	13.0	8.74	13.0	12.74	8.0	7.71
9.0	8.81	9.0	8.77	9.0	7.11	8.0	8.84
11.0	8.33	11.0	9.26	11.0	7.81	8.0	8.47
14.0	9.96	14.0	8.10	14.0	8.84	8.0	7.04
6.0	7.24	6.0	6.13	6.0	6.08	8.0	5.25
4.0	4.26	4.0	3.10	4.0	5.39	19.0	12.50
12.0	10.84	12.0	9.13	12.0	8.15	8.0	5.56
7.0	4.82	7.0	7.26	7.0	6.42	8.0	7.91
5.0	5.68	5.0	4.74	5.0	5.73	8.0	6.89

Fig. 3 Illustrations of basics #2 for data analysis. (**a**) Presentation of Anscombe's quartet, which is an emblematic example of the necessity of converting numbers into graphics. Whereas all sets of data (*x*, *y*) have identical means, standard deviation, and pairwise correlations, they are very different in terms of graphical shapes. (**b**) Illustration of the direct connection between data visualization and cognition [42]

are so immersed in a world of images that we no longer notice them. But more than a simple process to "show" or to "communicate" ideas, visualization triggers a cognitive process (Fig. 3b). Applied to data, visualization (often referred to as "dataviz") helps to detect potential bias in the collected datasets, to define underlying structures or to anticipate the significance of the arising information. Applied to information (often referred to as "infographics"), visualization helps to integrate heterogeneous results, compare, and confront them with current knowledge in a particular field of interest. Finally, applied to knowledge, visualization helps to explain and communicate new understanding of processes. While some visualizations are quick procedures, possibly automated with the use of graphics software, others are very slow. They require deep thought to realistically synthesize and convey a complex message.

Again, visualization is above all a cerebral process. It allows us to form mental images of concepts, supports thinking, and amplifies cognition. Visualization thus gives new perspectives and drives further analyses. This is worth considering for the benefit of a data analysis.

2.3 Basics #3: Think of Data Analysis as a Cycle

Our research projects have taught us a lesson: data analysis never happens as originally planned. More complicated or time consuming than anticipated, something is always missing. What is the effect of a method parameter on the retrieved list of genes? The genes are not those expected, what shall we do? Should we change raw data normalization? These are examples of questions raised in our analyses of transcriptomics data. Writing this chapter was an opportunity for us to take a step back from our practices and explain the different steps of a more reasoned and efficient strategy.

In that respect, our main insight is that data analysis is not a linear process, but rather a cycle. The challenge for researchers is to

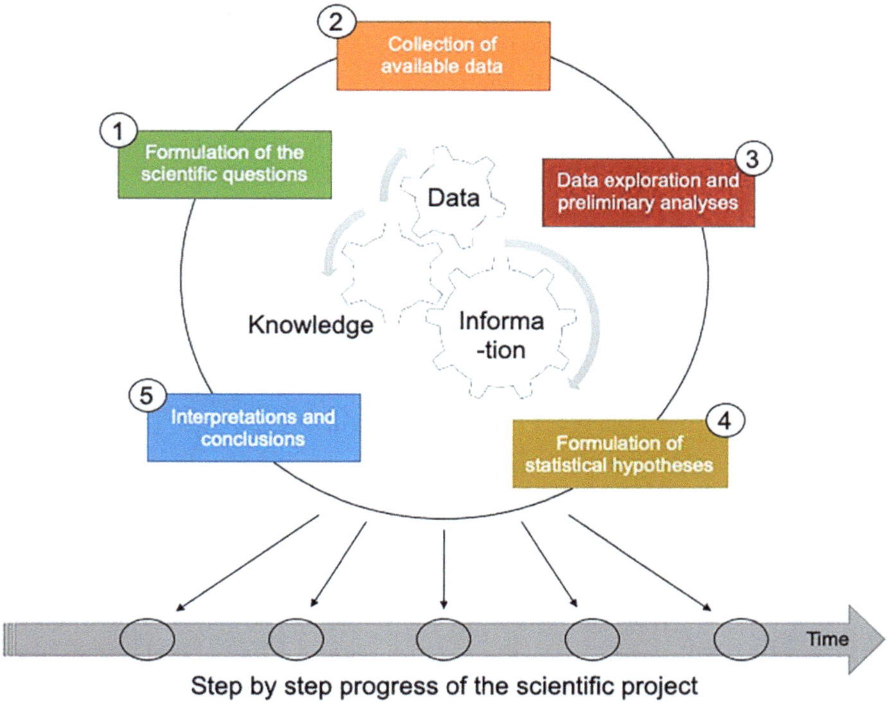

Fig. 4 Representation of the data analysis cycle that underlies the step-by-step progression of a scientific project. Five main steps are required: (1) Formulation of the scientific question; (2) Collection of available data; (3) Data exploration and preliminary analyses; (4) Formulation of statistical hypotheses; and (5) Interpretations and conclusions. This last step often opens new perspectives in the project, thus initiating a new cycle of data analysis. The succession of coherent cycles allows a step-by-step progression of the project

avoid getting stuck in a perpetual cycle. We identified five different steps, which are represented in Fig. 4 and detailed below.

2.3.1 Formulation of the Scientific Question

This is the most important step. A clear formulation of the scientific question helps to delineate the purpose of the data analysis and to anticipate the information that will be obtained and the associated potential issues. A typical mistake is to directly target the creation of new knowledge, forgetting the information intermediate. Even if they are very interdependent notions (see previous section), keeping in mind their specificities is at this step critical. Information is associated with "data meaning," whereas knowledge is associated with "data understanding." These are very different aims. In case of a transcriptomics data analysis, for instance, it can be illustrated as follows. Imagine you are comparing gene expression between two conditions; it is very different to say "I want to know how cells adapt in response to environmental changes?" (vague question, multiple answers can be formulated) than to say "I want to know the set of genes for which normalized gene expression measurements are significantly different, with an associated type I error rate of 1%?" (precise question associated with classical statistical procedures). Of course, the first question is more attractive from a scientific point of view. The underlying idea is clearly to create new knowledge, the goal for all researchers. But we can anticipate the first question will be very hard to fully answer, especially if several collaborators share different points of view concerning "cell adaptation." There is no such risk with the second goal. Even more, we can anticipate, based on the number of replicates available, the power of the statistical test, and hence, our ability to detect genes that are truly differentially expressed [25]. The results may be disappointing, but they will still be results, i.e., new information collected that can be associated with other information, as part of a more ambitious scientific project.

2.3.2 Collection (and Cleaning) of Available Data

Being a data analyst in 2021 is a strong advantage for this step. Many sources of biological data exist and are freely available in public databases [4]. It is thus possible to answer a lot of scientific questions without carrying out new experiments. Of course, the experiments of others can still be considered imperfect and unsatisfactory, but this is not so important if some new information can be obtained from the data. A typical mistake at this step is to collect more data than we actually need. It is important to remain focused on the scientific question only. Indeed, with too much data, the risk is to spend more time than necessary on purely technical (informatics) issues. Data cleaning is also a necessary step, whose cost increases with the amount of data collected. On this subject, interesting advice is presented here [26]. To ensure good data quality, it is necessary to check validity, accuracy, completeness, and consistency. This takes time but deserves to be taken into account because the quality of cleaning and data collection has a direct impact on the quality of the information, i.e., the adage "garbage in, garbage out" [27]. In this context, being specific about what we are trying to accomplish is a great help in deciding what actions to take.

2.3.3 Data Exploration and Preliminary Analyses

This step is our favorite. For the first time, the collected dataset is handled without any restraint regarding what is right or wrong to do. It consists of exploring the data, plotting multiple graphs, calculating many parameters, etc. We like the idea that this step is about "getting to know" the collected data. The visualization methods described above are of great interest at this stage, while ensuring the reproducibility of the process is not yet necessary (see next section). This greatly reduces the work of the data analyst and provides very appreciable freedom. A typical mistake is to spend too much time at this step and over-interpret the observations. Indeed, no solid conclusions will be drawn from these explorations and preliminary analyses, the objective of this step is only to develop intuitions, which will help for the following steps.

2.3.4 Formulation of Statistical Hypotheses

This is undoubtedly the most challenging part of the cycle. The problem is that we are going to subject the collected data to more complex calculations, applying procedures that involve mathematical details, often beyond our general understanding. At this step, the hard work is purely mental. It consists of converting the scientific question into relevant statistical hypotheses, prerequisite to using the right methodology. Once done, all the rest is mere computation, which simply requires statistical software. Note that at this step of the data analysis cycle, the issues of repeatability, replicability, and reproducibility must be carefully considered [28]. While the previous exploration step can remain "quick and dirty" [29], this part of the analysis must be rigorous and well documented [30]. A typical mistake is to confuse "scientific hypotheses" with "statistical hypotheses." Again, data analysis is a very lexically challenged discipline. As an illustration, a scientific hypothesis could be "I think that the function of this gene is required for the adaptation of cells to this environmental change," whereas a derived statistical hypothesis could be "the mean of logFC replicates in RNAseq data for this gene is different from 0." These are two very different statements. We have often encountered colleagues who were already convinced that their scientific hypothesis was true (for very good reasons which we do not question), and thus strongly rejected the results of a statistical analysis that did not go in the expected direction. Such misunderstandings lead to difficult discussions, which could be avoided through a better understanding of statistics [31]. Rejecting a statistical hypothesis, based on a *p*-value calculation and/or a threshold for type I error rate, does not imply the scientific hypothesis is wrong [32]. But the information cannot simply be ignored, on the contrary, it can help decide the course of further analyses.

2.3.5 Interpretation and Conclusions

Finally, the last step. It is an essential time to describe the results and to interpret them in the general context of the scientific project. Expertise in the scientific field for which the data analysis is performed is very important. Links with already established knowledge are identified, discussed, and may lead to new scientific questions. A new cycle of analysis can be performed and, in this way, the scientific project advances one step (or cycle) after another.

3 Specific Statements in Omics Data Analyses

Below, we alert the reader about two important issues, which are specific to the analysis of omics data. Although detailed explanation is beyond the scope of this chapter, we provide references for those who wish to pursue these topics.

3.1 Be Aware of Cherry Picking

Over the past 10 years, Big Data has revolutionized many areas of the life sciences [33]. But is it an infinite source of new information? The answer is a priori yes, if appropriate analytic procedures are applied. Otherwise, Big Data can become an important source of false discoveries. Indeed, when the mind is motivated to search through a mass of data, it always ends up discovering a path, usually a straight line, to the narrative it initially wished to find. In other words, if we are already convinced that our scientific hypothesis is true (see previous section), there is a risk of retaining from the multiple explorations of a large data set, only the information that supports the hypothesis, while rejecting (consciously or unconsciously) the information that contradicts the initial conviction. This is called "cherry picking" [34], i.e., the action of pointing to individual cases that confirm a particular idea, while ignoring a significant portion of related and similar cases that may contradict that idea. In this context, statistics have an important role to play, helping to discriminate observations that have a high probability of being random from those which, on the contrary, have a low probability.

3.2 Be Aware of P-Hacking

P-hacking relies on an inappropriate manipulation of data which favors the result to be presented as statistically significant [35]. Extensively discussed in the literature (for instance [36]), there are several P-hacking scenarios. In the first one, several statistical tests are performed on the same dataset (e.g., a parametric test and a nonparametric test), and only the results of the most significant test (with the lowest P-value) are shown. In the second one, observations are added (or removed) from a sample used as statistical test input, until the calculated *P*-value is below a targeted threshold (generally 0.05). In the last scenario, a single dataset composed of multiple variables is used to test multiple statistical hypotheses until one of them is significant. It is very easy to fall into the P-hacking trap, even for experienced data analysts and especially with P-hacking scenario 3, which is very close to the exploration step of the data analysis cycle described above. This is the reason why this exploration step must be followed by a rigorous validation step. To delve deeper into this subject, we suggest the very comprehensive article of [37].

4 Our Ten Recommended Rules

In the last section, we list 10 rules that help us move forward, despite the constant deluge of data. We are not always able to follow them consistently, but they represent a common objective.

1. **Take the time to set your long-term goal and break it into smaller, very specific parts.**

While it is important to dream big, it is also important to define the different stages that will allow you to move forward in the desired direction. Each of these steps must be specific enough to be carried out in a simple and effective way, without too much effort.

2. **Do not bypass elementary analyses.**

   Even if there is a temptation to directly apply "hype" methodologies (in 2021, this is AI), simple analyses are still very valuable. Because they are easy to understand, you can stay focused on the data rather than the underlying assumptions that guide the methodology. You also limit the risk of misunderstanding what the method can tell you. Elementary analyses are also interesting for guiding subsequent choices of more complex methodologies. Taking a little time with this can save a lot of time later. If your data is not worth considering for instance, we might not put as much energy into using it. Note here that the choice of methods (simple or complex) must always be guided to achieve the stated objective, no more, no less, hence the importance of the previous rule.

3. **Use unlimited data visualization.**

   We explained in previous sections how data visualization can drive our thinking in an analysis. Be careful not to restrict your data explorations to graphics that are automatically and systematically generated by a single tool. To be fully useful, data visualization should be as flexible and broad as possible, allowing data to be viewed from very different angles. Online tools [38, 39] exist to support your creativity. Keep in mind the importance of distinguishing "dataviz" (rapid process used for exploration) and "infographics" (lengthy process to support communication and help other people understand your message).

4. **Use statistics because you need them, not because others tell you to.**

   Do not ask a question if you do not want to listen to the answer. This is the best advice to remember when the issue of carrying out a statistical test is discussed. Do we need statistics to convince ourselves or to convince others? We deeply believe that statistics are a powerful and indispensable tool for performing rigorous data analysis. However, poorly mastered or applied for the wrong reasons, their use can lead to borderline situations. "P-hacking" and "Cherry picking" are perfect illustrations (see previous section). Do not hesitate to consult a statistician to verify the accuracy of the analyses and keep in mind that, at the end, it is you and you alone who make the decision, statistics are only one part of your overall thinking.

5. **Choose the right time to discuss your results, neither too early nor too late.**

   This is an important advice. Presenting your results (whether they are preliminary or advanced) requires clear thinking. Explaining your motivations, justifying your choices, and expressing your doubts cannot be improvised. We control how a message is sent, but we do not control how it is received. Communication therefore deserves a little preparation. Also, to be able to listen to feedback and comments, and thus identify weak points in the analysis, you cannot be in a hurry because of an impending deadline. Timing is therefore important. Finally, it is essential to take into account your audience. The less people know about your project, the more you will have to explain. This is worth considering if you are expecting the discussion to focus on a specific element of your project.

6. **Listen to the opinions of others but take full responsibility for all decisions.**

   Collaboration in a multidisciplinary context can sometimes be difficult. While having different perspectives on the same issue can be very rich, it can also be crippling. At some point, therefore, it is important to be able to decide and move forward. At the end of the study, a critical review process can be applied in order to assess the relevance of the decisions that have been taken and therefore to learn from any errors.

7. **Guarantee the reproducibility of analyses but pay attention to not unnecessarily burden the process.**

   The reproducibility of analyses is one of the major challenges of current bioinformatics. However, it can be arduous to test and very constraining in a situation where creativity can make all the difference. This is why we clearly separate two stages of the data analysis cycle. When exploring the data, it is important to work as freely as possible and then, only if the preliminary results are worth it, to implement good reproducibility practices. Keep in mind that the more these practices are mastered, the less the effort required to use them.

8. **Remain constantly intentional regarding where you are in a cycle.**

   Working in analysis cycles has been a decisive change of habit for us. Our experience is that performing short cycles (a few days) is most effective.

9. **Take your time, but not too much, perfection is stagnation.**

   You understood that to navigate through a constant deluge of data, you must be able to slow down. Not run faster and faster just to stay in the same place, like the Red Queen [40]. Be careful, however, not to stop completely. Doing nothing but reading and thinking, looking for the perfect solution to the problem being addressed is ultimately useless if it does not turn into concrete action.

10. **Remember that being a data analyst is an amazing job!**
    Data are nowadays everywhere and allow endless creativity for those who appreciate it. Combining computer science, statistics, and biology, omics data analysis is fascinating interdisciplinary work with many attractive perspectives. Isn't data the new (s)oil?

5 Conclusion

In this chapter, we have shared key ideas and convictions that help us work on a daily basis. They have been developed both during the experiences of our own omics data analyses and in the context of collaborative projects with biologists, informaticians, and statisticians. Overwhelmed by the constant flow of data and challenged with the methodological evolution in bioinformatics to follow the FAIR Guiding Principles [41], we started to feel "stuck," as if our work was never enough. The desire to automate our analyses was very strong at that point and we did make significant progress in that direction. While this has been helpful, it was not enough. Today, we accept that we cannot compete with the pace at which the omics technologies evolve, and we recognize that the burden represented by omics data is much too heavy to be carried alone. Again, data analysis is a complex process for which there is no universal protocol. It requires strong expertise, a lot of curiosity, and above all, patience. We believe that developing one's own "know-how" is one key to success. It helps to become strategic and intentional, thus avoiding an endless drift into the ocean of possibilities. Taking full responsibility for the different choices required at each analysis step is another key to success. It is not easy in a multidisciplinary context, where we often think that others are more legitimate to say that we are proceeding in the right direction. But in practice, we learned that if we become the person in the room who asks the right questions, we greatly help the experts to give us accurate answers. Finally, we like the idea that data analysis is as much an art as a science [29], there are no right or wrong answers, and that it is ok to have differences in opinion. So, listening, reading, brainstorming, and going one step at a time are our main recommendations, and we suggest constantly keeping in mind that if we often overestimate what we can do in a day, we also underestimate what we can do in a year.

Acknowledgments

We thank Linda Sperling for her careful proofreading of the text and her very wise advice, Bertrand Cosson and Pierre Poulain for stimulating discussions during the DUO training program (www. https://omics-school.net/), and Fabienne Malagnac, Pierre Grognet and other members of the research team "Fungal Epigenomics and Development" (I2BC, Orsay) for their daily trust. From a more personal point of view, we also thank our respective spouses, Céline Chasport (for Thomas) and Benjamin Bussière (for Gaëlle). In recent years, our concerns and questions have often occupied discussions at our family tables. Their listening and their patience were a great support. Finally, we would like to thank Frédéric Devaux, editor-in-chief of this book, for having accepted our proposal to write on this subject and thus for giving us the unique opportunity to publish this text.

References

1. Bell G, Hey T, Szalay A (2009) Computer science. Beyond the data deluge. Science 323(5919):1297–1298
2. Veenstra TD (2021) Omics in systems biology: current Progress and future outlook. Proteomics 21(3–4):e2000235
3. http://enseqlopedia.com/enseqlopedia/
4. Rigden DJ, Fernández XM (2021) The 2021 nucleic acids research database issue and the online molecular biology database collection. Nucleic Acids Res 49(D1):D1–D9
5. https://bmcresnotes.biomedcentral.com/data-notes
6. Johnson AT (2015) The technology hype cycle. IEEE Pulse 6(2):50
7. Chakravarti A (2011) Genomics is not enough. Science 334(6052):15
8. Davenport TH, Patil DJ (2012) Data scientist: the sexiest job of the 21st century. Harv Bus Rev 90(10):70–76, 128
9. MacEachern SJ, Forkert ND (2020) Machine learning for precision medicine. Genome:1–10
10. Krassowski M, Das V, Sahu SK, Misra BB (2020) State of the field in multi-omics research: from computational needs to data mining and sharing. Front Genet 11:610798
11. Sboner A, Mu XJ, Greenbaum D, Auerbach RK, Gerstein MB (2011) The real cost of sequencing: higher than you think! Genome Biol 12(8):125
12. Ostrovsky A, Hillman-Jackson J, Bouvier D, Clements D, Afgan E, Blankenberg D et al (2021) Using galaxy to perform large-scale interactive data analyses-an update. Curr Protoc 1(2):e31
13. https://usegalaxy.org/training-material/
14. https://www.coursera.org/
15. Caroll S (2009) Defining the scientific method. Nat Methods 6(4):237
16. Denecker T, Zhou Li Y, Fairhead C, Budin K, Camadro J-M, Bolotin-Fukuhara M et al (2020) Functional networks of co-expressed genes to explore iron homeostasis processes in the pathogenic yeast Candida glabrata. NAR Genom Bioinform 2(2):lqaa027
17. Denecker T, Durand W, Maupetit J, Hébert C, Camadro J-M, Poulain P et al (2019) Pixel: a content management platform for quantitative omics data. PeerJ 7:e6623
18. Denecker T, Lelandais G (2018) Empowering the detection of ChIP-seq "basic peaks" (bPeaks) in small eukaryotic genomes with a web user-interactive interface. BMC Res Notes 11(1):698
19. Lelandais G, Denecker T, Garcia C, Danila N, Léger T, Camadro J-M (2019) Label-free quantitative proteomics in Candida yeast species: technical and biological replicates to assess data reproducibility. BMC Res Notes 12(1):470
20. Lucau-Danila A, Delaveau T, Lelandais G, Devaux F, Jacq C (2003) Competitive promoter occupancy by two yeast paralogous transcription factors controlling the multidrug resistance phenomenon. J Biol Chem 278(52):52641–52650

21. Hadjadj D, Denecker T, Maric C, Fauchereau F, Baldacci G, Cadoret J-C (2016) Characterization of the replication timing program of 6 human model cell lines. Genomics Data 9:113–117
22. Bourne PE (2021) Is "bioinformatics" dead? PLoS Biol 19(3):e3001165
23. Kempe S. (2013) The data – information – knowledge cycle [internet]. DATAVERSITY. [cited 28th june 2021]. Available at: https://www.dataversity.net/the-data-information-knowledge-cycle/
24. Anscombe FJ (1973) Graphs in statistical analysis. Am Stat 27(1):17–21
25. Krzywinski M, Altman N (2013) Power and sample size. Nat Methods 10(12):1139–1140
26. The Ultimate Guide to Data Cleaning|by Omar Elgabry|Towards Data Science [Internet]. https://towardsdatascience.com/the-ultimate-guide-to-data-cleaning-3969843991d4
27. Kilkenny MF, Robinson KM (2018) Data quality: "garbage in - garbage out". Health Inf Manag 47(3):103–105
28. Lewis J, Breeze CE, Charlesworth J, Maclaren OJ, Cooper J (2016) Where next for the reproducibility agenda in computational biology? BMC Syst Biol 10(1):52
29. Peng RD. Exploratory Data Analysis with R [Internet]. https://bookdown.org/rdpeng/exdata/
30. Kim Y-M, Poline J-B, Dumas G (2018) Experimenting with reproducibility: a case study of robustness in bioinformatics. GigaScience 7(7):giy077
31. Makin TR, Orban de Xivry J-J (2019) Ten common statistical mistakes to watch out for when writing or reviewing a manuscript. elife 8:e48175
32. Amrhein V, Greenland S, McShane B (2019) Scientists rise up against statistical significance. Nature 567(7748):305–307
33. Fillinger S, de la Garza L, Peltzer A, Kohlbacher O, Nahnsen S (2019) Challenges of big data integration in the life sciences. Anal Bioanal Chem 411(26):6791–6800
34. Morse JM (2010) « cherry picking »: writing from thin data. Qual Health Res 20(1):3
35. Teixeira PM (2018) P-hacking - a call for ethics. Pulmonology 24(3):207–208
36. Dodson TB (2019) The problem with P-hacking. J Oral Maxillofac Surg 77(3):459–460
37. Head ML, Holman L, Lanfear R, Kahn AT, Jennions MD (2015) The extent and consequences of p-hacking in science. PLoS Biol 13(3):e1002106
38. https://www.r-graph-gallery.com/index.html
39. https://www.python-graph-gallery.com/
40. Carroll L (1971) Alice's adventures in wonderland and through the looking glass. Oxford University Press, London
41. Boeckhout M, Zielhuis GA, Bredenoord AL (2018) The FAIR guiding principles for data stewardship: fair enough? Eur J Hum Genet 26(7):931–936
42. https://www.gastonsanchez.com/

INDEX

A

Ac/Ds....350, 364
Affinity purification (AP)....43–47, 59, 60, 180, 181, 183–185, 192, 195–221, 225–227, 264
Analysis cycle....463, 465, 466, 468

B

Base editing....331–347
Biotin-phenol....261, 264
Bulk fitness assays....399–414

C

Cap analysis of gene expression (CAGE)....22, 31
Cell lysis....4, 153, 157–159, 167, 196, 197, 204, 205, 228, 262, 271, 273, 296, 299, 304, 411
Cell viability....375
Chromatin immunoprecipitation (ChIP)....6, 35, 58, 130–132, 134, 135, 137, 139–143, 145, 149, 150, 155, 157, 164, 166, 167, 170, 171, 182, 186, 190
Chromatin profiling....129–146
Colony....93, 111, 135, 156, 241, 243, 245, 246, 252, 255, 256, 265, 267, 318–320, 323–328, 332, 340, 346, 351, 354–359, 370–375, 381–397
Comparative transcriptomics....v, 91–102
Convolutional neural network (CNN)....119, 122
Cotranslational interactions....179–181
CRISPR-Cas9....238, 243, 255, 256, 317, 331
Crosslinking analysis of cDNAs (CRAC)....35–54
Cross-species comparison....425
Cryo-milling....196, 198, 204, 205

D

Data....v, 4, 7, 9, 10, 13, 14, 27–31, 35, 37, 39, 42, 48, 51, 52, 85, 91–102, 116, 117, 119–123, 126, 127, 130, 132, 141, 143, 146, 151, 152, 156, 162, 165, 169, 170, 172, 191, 198, 211–215, 226, 227, 233, 238, 257, 270, 273, 279–284, 286–291, 303–304, 328, 334, 347, 350, 381–384, 388–390, 392–396, 406, 409, 410, 413, 414, 420–425, 427–431, 433–435, 440–444, 446–449, 452–454, 457–470
Deep sequencing....57–69, 130, 179, 181, 184–191, 246, 337, 349, 350
De novo assembly....92, 95, 97, 101
Direct RNA sequencing....71, 73, 74, 86
DNA barcodes....412
DNA replication....107–128

E

Evolutionary dynamics....440

F

Fitness....257, 322, 333, 345, 347, 349, 381–384, 389, 395, 396, 399, 400, 403, 406, 409–414
Full-length cDNA....80–83, 86
Functional genomics....v, 275, 439, 457–469

G

Gene-regulatory networks....149–173
Growth curves....383, 389, 390, 399

H

High-throughput....v, 3, 107, 318, 320, 332, 345, 351, 440

I

Information....16, 51, 52, 76, 80, 85, 86, 107, 122, 125, 165, 169–173, 196, 205, 213, 214, 235, 276, 279, 289, 333, 376, 384, 387, 392, 393, 420, 422, 433–435, 440–442, 445, 446, 448, 457–466

K

KNIME....279, 283, 284, 288, 290
Knowledge....v, 276, 440, 454, 458–465

L

Long-read sequencing....79–86

Frederic Devaux (ed.), *Yeast Functional Genomics: Methods and Protocols*,
Methods in Molecular Biology, vol. 2477, https://doi.org/10.1007/978-1-0716-2257-5,

M

Mass spectrometry 150, 162, 169, 195–221, 225–235, 262, 264, 265, 267, 269, 283, 287, 295, 304
Metabolism 275, 276, 278, 351
Minora 280, 283, 288, 289
Multi-task learning 440, 441
Mutagenesis 238, 241, 246, 254–256, 331–347, 349, 351

N

Nanopore sequencing 79–87, 108, 113–118
Nascent-chain 179
Nascent RNA (nRNA) 35, 36, 39, 58, 59, 61, 69
Next-generation sequencing (NGS) 94, 134–136, 151, 155, 162–165, 167, 172, 191, 192, 243, 247, 253, 257, 332–334, 346, 347, 458
N-terminal acetylation 212
N-terminal acetyltransferases (NATs) 293, 294, 296
N-terminomics 293–307
Nucleosomes 63, 129, 130, 132–134, 143, 145, 146, 172

O

Omics 229, 419–435, 457–469

P

PathoYeastract 173, 419–436
Phenomics v
Phylogeny 99, 439–454
Probabilistic models 440, 441
Protein complexes .... v, 37, 150, 151, 160, 166, 225–235, 314
Protein-fragment complementation assay (PCA) 238, 240, 252–255, 257
Protein–protein interactions (PPIs) .... v, 195, 237–257
Proteome discoverer 156, 162, 169, 170, 201, 212, 275–291
Proteomics .... v, 170, 192, 195, 212, 262, 267, 269, 270, 272, 273, 275–291, 294, 457
Proximity labeling 261–274
Python software 81, 85, 119, 191, 334, 388, 391, 402, 406

Q

Quantification 98, 113, 125, 126, 136, 170, 190, 191, 226, 233, 235, 248, 270, 271, 275–291, 338, 343, 347, 360, 362, 375, 376, 387, 396

R

Replication origins 107
RNA interference (RNAi) .... v, 313–328
RNA modifications 71–75
RNA polymerase II .... v, 36
RNA polymerase III 254
RNA-seq 7, 22, 28, 62, 66, 80, 191, 347, 453

S

Screen .... v, 93, 95, 101, 115, 136, 213, 214, 220, 331–347, 349–351, 370, 381–397
Simple light isotope metabolic labeling (bSLIM) 275–291
Single-cell RNA-sequencing 3, 6
Stable isotope labeling by amino acids in cell culture (SILAC) 262–264, 267, 269, 270, 273, 276
STRIPE-seq 21–32
Systems biology 276

T

Target-AID 331–333, 335, 336, 340, 341
Transcription 4, 6, 16, 21, 22, 24–26, 28, 31, 35–54, 57–69, 74, 75, 188, 315, 420, 422, 423
Transcription factors 129, 130, 145, 146, 150–152, 163, 166, 167, 172, 420–422, 424, 425, 427–430, 432, 433, 439, 440, 449, 453
Transcription initiation 21–32
Transcription regulatory networks 419
Transcription start site (TSS) 21–32, 63
Transcriptome reconstruction 79–86
Transposon .... v, 315, 316, 349–377

U

Unique molecular identifiers (UMIs) 9, 30, 400, 402, 404, 408

V

Visualization 7, 13, 14, 32, 209, 226, 227, 233, 235, 262, 428, 433, 460, 462, 464, 467

Y

Yeast .... v, 3–19, 38, 39, 42, 57–69, 79–86, 91–102, 108–112, 123, 129–146, 149–173, 195–221, 226, 228, 230, 238–240, 243, 244, 251–253, 255, 256, 261–291, 299, 304, 313–328, 331–347, 349–377, 399–414, 420, 426, 435, 440, 441, 444, 445, 453, 454, 459
YEASTRACT+ 420, 423–425, 428, 432

MIX
Papier aus verantwortungsvollen Quellen
Paper from responsible sources
FSC® C105338

If you have any concerns about our products,
you can contact us on
ProductSafety@springernature.com

In case Publisher is established outside the EU,
the EU authorized representative is:
Springer Nature Customer Service Center GmbH
Europaplatz 3, 69115 Heidelberg, Germany

Printed by Libri Plureos GmbH
in Hamburg, Germany